Hayner Public Library District-Alton
T5-CQZ-539

# COLD WAR SUBMARINES

*The Project 941/Typhoon was the largest undersea craft ever built. These two—of six completed—are at their remote base of Nerpichya, about six miles from the entrance to Guba Zapadnaya Litsa on the Kola Peninsula, close to the borders with Finland and Norway.* (Rubin CDB ME)

**Other Submarine-Related Books by Norman Polmar**

*The American Submarine*

*Atomic Submarines*

*Death of the* Thresher

*Guide to the Soviet Navy* (4 editions)

*Rickover: Controversy and Genius* with Thomas B. Allen

*Ships and Aircraft of the U.S. Fleet* (7 editions)

*Submarines of the Imperial Japanese Navy* with Dorr Carpenter

*Submarines of the Russian and Soviet Navies, 1718-1990* with Jurrien Noot

# COLD WAR SUBMARINES

## The Design and Construction of U.S. and Soviet Submarines

**Norman Polmar**
**and**
**Kenneth J. Moore**

BRASSEY'S, INC.
WASHINGTON, D.C.

**Library of Congress Cataloging-in-Publication Data**
Polmar, Norman.
Cold War submarines: U.S. and Soviet design and construction / Norman Polmar and Kenneth J. Moore.—1st ed.
p. cm.
Includes bibliographical references and index.
ISBN 1-57488-594-4 (alk. paper)
1. Submarines (Ships)—United States—History. 2. Submarines (Ships)—Soviet Union. 3. Cold War. I. Moore, Kenneth J., 1942– II. Title.

V858.P63 2003
359.9′383′094709045—dc21
2003013123

Printed in the United States of America on acid-free paper that meets the American National Standards Institute Z39-48 Standard.

Brassey's, Inc.
22841 Quicksilver Drive
Dulles, Virginia 20166

First Edition

10 9 8 7 6 5 4 3 2 1

# Epigraph

It is the nature of human memory to rid itself of the superfluous, to retain only what has proved to be most important in the light of later events. Yet that is also its weak side. Being biased it cannot help adjusting past reality to fit present needs and future hopes.

—Milovan Djilas
*Conversations with Stalin*

# Dedication

This book is dedicated to the officers, warrant officers, and enlisted men who served in Soviet and U.S. submarines during the Cold War. They did their duty courageously and steadfastly. Although no shots were fired, many of these submariners lost their lives to fires, flooding, and other accidents. It was the price paid for 45 years of Cold War confrontation and for the rapid development and application of technology.

# Contents

**Appendixes**

# Perspective

Submarines had a vital role in the 45 years of the Cold War—1946–1991. Throughout that conflict, U.S. and Soviet submarines carried out intelligence collection operations and sought out and prepared to destroy opposing surface ships and submarines. In addition, from the early 1960s missile-carrying submarines threatened nuclear attacks on their opponent's homeland.

The possibility of Soviet submarines near American coasts firing ballistic missiles, with a much shorter time of flight than missiles launched from the Soviet Union, presented a direct threat to U.S. command centers, bomber bases, and even land-based strategic missiles, thereby forcing major changes in the U.S. strategic posture. Similarly, U.S. missile submarines forced major changes in Soviet naval development and strategic forces deployments.

Soviet and U.S. submarines of the Cold War era had the same origins: Their antecedents were German U-boat developments of 1943–1945, especially the Type XXI, the most advanced submarine to go to sea during World War II. The U.S. version of the Type XXI was the *Tang* (SS 563) class, with similar features being incorporated in the *K1*-class "killer" submarines and 52 conversions of war-built submarines in the so-called GUPPY program.[1] The Soviets adopted Type XXI features in the Whiskey and Zulu designs and their successors.[2] The U.S. and Soviet (as well as British) submarine communities also had major interest in German closed-cycle, or "single-drive," submarine propulsion systems, with these submarines being evaluated in the postwar period by Britain and the Soviet Union.

But the undersea craft produced by the respective navies rapidly diverged in their designs from the Type XXI model. In 1985 the U.S. Assistant Secretary of the Navy for Research, Engineering, and Systems Melvin R. Paisley observed:

> The Soviet submarine technological advantages for quieting, strengthened double hulls, higher speed, higher reserve buoyancy, and deeper operations are advances which are by and large not stolen or bought from the United States. Some technologies [classified deletion] are Soviet design decisions which are different from our decisions. Other technologies . . . are the result of Soviet engineered high-power-density material and hull strength material. The Soviets are ahead of us in these technologies.[3]

By the end of the Cold War, U.S. and Soviet submarines were radically different in design and capabilities. In view of the importance of undersea craft—both torpedo/cruise missile attack and strategic missile submarines—it is useful to examine how and why this divergence occurred. The causes of this divergence in submarine design can be found in (1) differing naval missions, (2) differing technical/operational priorities, (3) differing levels of industrial competence, and (4) differing approaches to submarine design organizations and management.

Significantly, for much of the Cold War the U.S. Navy had a highly centralized, authoritarian organization. The head of Naval Nuclear Propulsion held de facto control of submarine development, with virtually unqualified power to veto—if not enforce—key design decisions.[4] Indeed, the incumbent of this position, Admiral H. G. Rickover, by the early 1960s was able to deter any infusion of design ideas or concepts from outside of the senior officers of the nuclear submarine community—the so-called submarine mafia—unless it corresponded with his views and goals. Those goals were based largely on a U.S. Navy plan of *1950* to develop a series of submarine propulsion plants progressing from the 13,400-shaft-horsepower plant of the pioneer *Nautilus* (SSN 571) to 30,000, 45,000, and, ultimately, 60,000 horsepower.[5]

In contrast, the Soviet Union had several design bureaus engaged in submarine development during the Cold War (see Appendix D). Those bureaus were, to a large degree, in competition in submarine design, although ostensibly each specialized in different types of submarines. Further, the Soviet regime pursued to various stages of fruition innovative proposals from qualified (and at times unqualified) submarine designers and naval officers. This, in turn, led to the examination of innumerable submarine designs and concepts, which contributed to the highly innovative submarines produced by Soviet design bureaus and shipyards.

Innovation in itself does not necessarily result in quality or capability. This volume seeks to examine the process of submarine design and the results of those efforts to produce submarines by the two super powers during the Cold War, and to assess their success in translating innovation into capability.

During the 45 years of the Cold War (1945–1991) the United States and the Soviet Union put to sea a combined total of 936 submarines, of which 401 were nuclear propelled (see table 1). The Soviet regime allocated considerably more resources to the design and construction of submarines during the Cold War than did the United States. This Soviet emphasis on undersea craft continued after the 1970s, when major efforts were undertaken to construct large surface warships, including nuclear-propelled missile cruisers and aircraft carriers.

Until the 1970s and the initiation of large warship programs, the Soviet Union committed far fewer resources to surface combatants and aircraft carriers than did the United States. The subsequent Soviet carrier construction and the nuclear "battle cruisers" of the *Kirov* class required a commitment of resources to surface warships of the same order-of-magnitude as that of the U.S. Navy, while continuing to emphasize submarine construction.

It is convenient to address nuclear-propelled submarines of the Cold War in terms of "generations" (see table 2). While this categorization is not precise, it is a useful tool.

Many individuals and organizations assisted in the writing of this book. They are gratefully acknowledged in the following pages. But all opinions and conclusions—and such errors that may appear—are the responsibility of the authors.

—Norman Polmar
Kenneth J. Moore

TABLE 1
**Cold War Submarine Construction***

| | United States | Soviet Union |
|---|---|---|
| World War II programs completed Aug. 1945–1951 | | |
| diesel-electric | 23 | 61 |
| closed-cycle | — | 1** |
| *total* | *23* | *62* |
| Cold War programs completed 1945–1991 | | |
| diesel-electric | 21 | 399 |
| closed-cycle | — | 31 |
| nuclear-propelled | 170 | 231 |
| *total* | *191* | *661* |
| Cold War programs completed 1992–2001 | | |
| diesel-electric | — | 3*** |
| nuclear-propelled | 22# | 17## |
| *total* | *22* | *20* |

Notes: * Does not include two Soviet midget submarines (Project 865) and the single U.S. Navy's midget submarine *X-1* (SSX 1). Submarines built by both nations specifically for foreign transfer are not included.
** Project 95 closed-cycle propulsion submarine.
*** All SS Project 636 (Kilo).
# 15 SSN 688 (Improved) *Los Angeles.*
5 SSBN 726 *Ohio.*
2 SSN 21 *Seawolf.*
## 1 SSBN Project 667BDRM (Delta IV).
1 SSN Project 671RTM (Victor III).
1 SSN Project 945A (Sierra II).
5 SSGN Project 949A (Oscar II).
5 SSN Project 971 (Akula I/II).
2 SSAN Project 1083 (Paltus).
2 SSAN Project 1910 (Uniform).

TABLE 2
**Nuclear-Propelled Submarine Generations (Major Combat Designs)**

| Generation | United States | | Soviet Union | (NATO name) |
|---|---|---|---|---|
| I | SSN 571 | *Nautilus* | SSN Proj. 627 | (November) |
| | SSN 575 | *Seawolf* | SSN Proj. 645 | (mod. November) |
| | SSN 578 | *Skate* | SSBN Proj. 658 | (Hotel) |
| | SSRN 586 | *Triton* | SSGN Proj. 659 | (Echo I) |
| | SSGN 587 | *Halibut* | SSGN Proj. 675 | (Echo II) |
| II | SSN 585 | *Skipjack* | SSGN Proj. 661 | (Papa) |
| | SSN 593 | *Thresher* | SSBN Proj. 667A | (Yankee) |
| | SSN 597 | *Tullibee* | SSBN Proj. 667B | (Delta I) |
| | SSBN 598 | *Geo. Washington* | SSBN Proj. 667BD | (Delta II) |
| | SSBN 602 | *Ethan Allen* | SSBN Proj. 667BDR | (Delta III) |
| | SSBN 616 | *Lafayette* | SSGN Proj. 670 | (Charlie) |
| | SSN 637 | *Sturgeon* | SSN Proj. 671 | (Victor) |
| | SSN 671 | *Narwhal* | SSN Proj. 705 | (Alfa) |
| III | SSN 685 | *Lipscomb* | SSBN Proj. 667BDRM | (Delta IV) |
| | SSN 688 | *Los Angeles* | SSN Proj. 685 | *Komsomolets* (Mike) |
| | SSN 688I | Imprv. *Los Angeles* | SSBN Proj. 941 | (Typhoon) |
| | SSBN 726 | *Ohio* | SSN Proj. 945 | (Sierra) |
| | | | SSGN Proj. 949 | (Oscar) |
| | | | SSN Proj. 971 | (Akula) |
| IV | SSN 21 | *Seawolf* | SSN Proj. 885 | *Severodvinsk* |
| | SSN 774 | *Virginia* | SSBN Proj. 955 | *Yuri Dolgorukiy* |

# Acknowledgments

Many individuals contributed to the preparation of this book, which originated during discussions of Academician Igor Spassky and his friend and colleague Viktor Semyonov with Norman Polmar at the Rubin Central Design Bureau in St. Petersburg in 1992.

Mr. Polmar was most fortunate in having many hours of discussions about submarine design and development with Academician Spassky, chief designer and head of the Rubin design bureau, as well as several hours of discussions with Academician Anatoly V. Kuteinikov, chief designer and head of the Malachite design bureau. Both men facilitated interviews with the senior designers and engineers from their respective bureaus as well as from other Russian institutions and agencies.

Their subordinates—Semyonov at Rubin and Alexandr M. Antonov at Malachite—were invaluable to this project with respect to the time and effort they took to help educate the authors in the history of Soviet submarine design and construction.

Mr. K. J. Moore also has been a guest of the Malachite design bureau for technical discussions as well as several Russian and Ukrainian research institutes.

A special debt is owed to Larry Lynn, who encouraged the undertaking of this project.

In the following listing, asterisks indicate persons interviewed by Polmar for previous submarine projects, with some of the material provided by them having been used in this volume.

*The beginning of this book: Academician I. D. Spassky makes a point to Norman Polmar, coauthor of this book, during their first meeting, which took place at the Rubin Central Design Bureau in St. Petersburg in 1992. Michael Polmar is seen between them.* (Rubin CDB ME)

## Interviews and Correspondence

### Germany

Prof. Fritz-Folkmar Abels, senior designer, IKL design bureau
*Dr. Ernest A. Steinhoff, Peenemünde rocket research center
*Hellmuth Walter, engineer and developer of closed-cycle submarine propulsion plants

### Great Britain

Comdr. Richard Compton-Hall, RN, submarine commander, author, and Director of the Royal Navy Submarine Museum
Comdr. Michael Davis-Marks, RN, Staff Officer Submarines, British Defence Staff, Washington, and commanding officer of HMS *Turbulent*
Commo. Robin W. Garson, RN, Staff Officer Submarines, British Naval Staff, Washington, and submarine commander
Rear Adm. John Hervey, RN, naval attaché in Washington, submarine commander, and author
Comdr. Jonathan (Jonty) Powis, RN, Staff Officer Submarines, British Defence Staff, Washington, and commanding officer of HMS *Unseen, Resolution,* and *Victorious*
Capt. Gordan A. S. C. Wilson, RN, head of Defence Studies (Royal Navy)
Michael Wilson, commanding officer of HMS *Explorer* and author

### Netherlands

*Comdr. Jurrien Noot, RNethN, naval analyst and submarine historian

**Russia**

Capt. 1/Rank I. P. Bogachenko, Admiralty supervisor for Project 971 (Akula) and Project 945 (Sierra) submarines

Alexandr Churilin, senior counselor, Embassy of the Russian Federation in Washington, D.C., and senior Soviet strategic arms negotiator

Fleet Adm. V. N. Chernavin, CinC Soviet Navy and submarine commander

L. P. Chumak, director, Admiralty Shipyard museum

V. G. Davydov, technical director, Admiralty Shipyard

Capt. 2/Rank Vasiliy V. Doroshenko, commanding officer of a Project 675 (Echo II) submarine

Sergei N. Khrushchev, missile guidance engineer in the Chelomei bureau, historian, and biographer for his father, Soviet Premier Nikita Khrushchev

L. Y. Khudiakov, Chief Scientist, First Central Research Institute (Naval Shipbuilding)

Dr. Eugene Miasnikov, Russian political scientist

Capt. 1/Rank V. Nikitin, staff, Pushkin Higher Naval School

Capt. 1/Rank B. Rodionov, historian and submarine commanding officer

Dr. George I. Sviatov, submarine designer, political scientist, and historian

Yuri Shestakov, curator, Admiralty Shipyard museum

Capt. 1/Rank Nikolai Vorobjev, submarine designer

**Rubin Central Design Bureau for Marine Engineering**

(Saint Petersburg, née Leningrad)

E. A. Gorigledzhan, Chief Designer

O. A. Gladkov, chief designer

Anna Kipyatkova, director of the bureau's museum

Academician S. N. Kovalev, general designer and chief designer of strategic missile submarines

Guri Malafeyev, designer

Capt. 1/Rank S. A. Novoselov, history staff; senior designer

A. Pinégin, senior designer

D. A. Romanov, deputy chief designer of the submarine *Komsomolets*

**Malachite Marine Engineering Bureau**

(Saint Petersburg, née Leningrad)

Dr. V. I. Barantsev, chief designer

B. F. Dronov, head, Design Department

V. V. Kryov, chief designer

G. D. Morozkin, chief designer

Y. K. Mineev, chief designer

L. A. Samarkin, deputy chief designer

R. A. Shmakov, chief designer

**Lazurit Central Design Bureau**

(Nizhny Novgorod, née Gor'kiy)

N. I. Kvasha, general designer

A. A. Postnov, chief designer

Natalia Solbodyanyuk, assistant to the general designer

**United States**

*Capt. William R. Anderson, USN, who commanded the USS *Nautilus* on her historic voyage to the North Pole

*Rear Adm. Dean A. Axene, USN, first commanding officer of the USS *Thresher*

A. D. Baker III, naval analyst and author of the reference book *Combat Fleets of the World*

Anthony Battista, senior staff member, House Armed Services Committee; member of Aspin panel

Capt. Edward L. Beach, USN (Ret), first commanding officer of the USS *Triton,* historian, and novelist

Richard J. Boyle, U.S. submariner and historian; engineer and, subsequently, officer-in-charge midget submarine *X-1*

Capt. Richard Brooks, USCG (Ret), submarine technology analyst

Rear Adm. Thomas A. Brooks, USN (Ret), Director of Naval Intelligence

*Vice Adm. James F. Calvert, USN, who commanded the USS *Skate* on her first two Arctic voyages

Capt. Myron Eckhart Jr., USN (Ret), head of preliminary ship design, Naval Sea Systems Command

Alan Ellinthorpe, technical consultant

Rear Adm. Thomas W. Evans, USN (Ret), commanding officer of the USS *Batfish,* Deputy Commander for ASW and Undersea Warfare Programs and, subsequently, and Deputy Commander for Submarines, Naval Sea Systems Command

Dr. John Foster, Director Defense Research and Engineering.

Dr. H. H. Gaffney, analyst and team leader, Center for Naval Analyses; CNA liaison to the Russian Institute of USA Studies

*Adm. I. J. Galantin, USN, submarine commander and Director of the Special Projects Office
Dr. Paris Genalis, director, Naval Warfare, Office of the Secretary of Defense (Acquisition and Technology)
Peter Gorin, Russian missile historian
*Ross Gunn, Navy scientist and the first man to explore the potential of employing nuclear energy to propel a submarine
*Vice Adm. Frederick J. Harlfinger, USN, submarine commander and Director of Naval Intelligence
Dean A. Horn, senior U.S. submarine designer
Mark R. Henry, head of preliminary design, Submarine Branch, Naval Sea Systems Command
Capt. Harry Jackson, USN (Ret), senior U.S. submarine designer
Capt. Donald Kern, USN (Ret), head of the submarine design branch in the Bureau of Ships
*Dr. Donald Kerr, director, Los Alamos National Laboratory
Capt. Richard B. Laning, USN (Ret), first commanding officer of the USS *Seawolf* (SSN 575)
*Rear Adm. George H. Miller, USN, Director of Navy Strategic Systems
Michael Neufeld, curator at the National Air and Space Museum and V-2 missile historian
*Vice Adm. Jon H. Nicholson, USN, who took the USS *Sargo* on her remarkable Arctic cruise
Ambassador Paul H. Nitze, Secretary of the Navy, Under Secretary of Defense, and senior U.S. strategic arms negotiator
Ronald O'Rourke, senior naval analyst, Congressional Research Service, Library of Congress
*Melvyn R. Paisley, Assistant Secretary of the Navy (Research, Development, and Systems)
*Dr. D. A. Paolucci, submarine commanding officer and strategic analyst
Robin B. Pirie, commanding officer of the USS *Skipjack,* Under Secretary of the Navy, and Acting Secretary of the Navy
Capt. John P. Prisley, USN (Ret), submariner and intelligence specialist
Raymond Robinson, intelligence specialist
Dr. David A. Rosenberg, distinguished naval historian and biographer of Adm. Arleigh A. Burke
Jeffery Sands, naval analyst, Center for Naval Analyses
Thomas Schoene, senior analyst, Anteon Corp., and editor *par excellance*
Rear Adm. Edward Shafer, USN, Director of Naval Intelligence
Frank Uhlig, senior editor of the U.S. Naval Institute and editor, Naval War College *Review*
Dr. William Von Winkle, leading U.S. Navy sonar expert
Dr. Edward Whitman, Assistant Secretary of the Navy and senior editor, *Undersea Warfare*
John Whipple, managing editor, *Undersea Warfare*
Lowell Wood, physicist; member of the Aspin panel
Bruce Wooden, naval architect and engineer
Adm. Elmo R. Zumwalt, USN (Ret), Chief of Naval Operations

Ms. Irina Alexandrovna Vorbyova of the Rubin bureau provided outstanding service as an interpreter during several visits to St. Petersburg. George E. Federoff, Andrew H. Skinner, and Jonathan E. Acus as well as Dr. Sviatov have provided translations in the United States. Samuel Loring Morison and Gary Steinhaus undertook research for this project.

Many individuals at the Operational Archives of the U.S. Naval Historical Center (NHC) have provided assistance and encouragement for this project, especially Bernard F. (Cal) Cavalcante, Kathy Lloyd, Ella Nargele, and Mike Walker; and Richard Russell, long the Russian expert of the NHC staff, provided special help and, subsequently, as acquisitions editor of Brassey's USA, had a key role in bringing this book to fruition.

David Brown, late head of the Naval Historical Branch, Ministry of Defence, and Capt. Christopher Page, RN, his successor, also were helpful in this project as was Comdr. W. J. R. (Jock) Gardner, RN, of their staff. Page previously served as head of Defence Studies (Royal Navy), and Gardner is a distinguished naval author.

Providing photographs for this project, in addition to the Rubin and Malachite design bureaus, were Capt. Ian Hewitt, OBE, RN; Capt. A. S. L. Smith, RN; and Capt. Simon Martin, RN, at the Ministry of Defence. Also, the late Russell Egnor of the U.S. Navy Office of Information and his most able assistants—Lt. Chris Madden, USN, his successor; Journalist 2d Class Todd Stevens, USN; Henrietta Wright; and Diane Mason. Charles Haberlein, Jack A. Green, Ed Feeney, and Robert Hanshew of the U.S. Naval Historical Center searched out and provided several important historical photos, while Dawn Stitzel and Jennifer Till of the U.S. Naval Institute were always ready to assist in photo research.

Chapter 15, "Aircraft-Carrying Submarines," benefited from a review by aviation writer Peter B. Mersky.

Jeffrey T. Richelson, leading intelligence writer, has kindly shared with us the voluminous fruits of his efforts to have appropriate official documents declassified. Similarly, Mr. and Mrs. Armin K. Wetterhahn have made available their voluminous files on Soviet submarines and weapons.

*Anatoly V. Kuteinikov*
(Malachite SPMBM)

The U.S. Naval Institute (USNI), the professional association of the U.S. Navy, made its oral history collection available through the auspices of Paul Stillwell, director of history, USNI; that collection includes numerous interviews with senior U.S. submarine officers.

Materials used in the creation of the drawings by A. D. Baker were from various unclassified U.S. and Russian publications and from material provided by Richard J. Boyle, Jim Christley, Norman Friedman, and David Hill.

Lt. Jensin W. Sommer, USN, Lt. Rick Hupt, USN, and Lt. Brauna R. Carl, USN, of the U.S. Navy's Office of Information sought out numerous answers to inquiries, a pleasant contrast to the U.S. Navy's usual attitude toward inquiries about submarine issues.

Teresa Metcalfe did an excellent job of shepherding the book through the Brassey's publishing complex, and Margie Moore and Pepper Murray of the Cortana Corp. provided valuable editorial assistance. And, a special thanks to Michael Polmar for his research activities in support of this book.

Finally, the authors are also are in debt to those in the West who earlier wrote about Soviet submarines, generally with very limited information available to help them, especially Claude Hahn, Michael MccGwire, and Jurrien Noot.

# Glossary

| | |
|---|---|
| AEC | Atomic Energy Commission (U.S.; 1948-1974) |
| AGSS | auxiliary submarine (non-combat) |
| AIP | Air Independent Propulsion |
| *Alberich* | *anechoic* hull coating used for German U-boats |
| AN/ | Prefix for U.S. electronic equipment |
| anechoic coating | coating on submarines to reduce the effectiveness of an enemy's active sonar by absorbing acoustic energy to reduce the reflection; on Soviet submarines also used between *double hulls* to absorb internal noise. May be a rubber-like coating or tiles. |
| AOSS | submarine oiler (also SSO) |
| APS | transport submarine (also APSS, ASSP, LPSS, SSP) |
| APSS | transport submarine (also APS, ASSP LPSS, SSP) |
| ASDS | Advanced SEAL Delivery System |
| ASSA | cargo submarine (also SSA) |
| ASSP | transport submarine (also APS, APSS, LPSS, SSP) |
| ASTOR | Anti-Submarine Torpedo (U.S. Mk 45) |
| ASW | Anti-Submarine Warfare |
| beam | maximum width of hull; control and stern surfaces may be of greater width |
| BuShips | Bureau of Ships (U.S. Navy; 1940 to 1966) |
| cavitation | formation and collapse of bubbles produced in seawater as a result of dissolved gases being released in low-pressure regions such as those produced by the high velocity of propeller or propulsor blades |
| CEP | Circular Error Probable (measure of weapons accuracy) |
| CIA | Central Intelligence Agency (U.S.) |
| CNO | Chief of Naval Operations (U.S.) |
| conning tower | *see* sail |
| DARPA | Defense Advanced Research Projects Agency (U.S.) |
| depth | *Working* depth is the Russian term for the normal maximum operating depth, approximately 0.8 times limiting depth. *Test* depth is the deepest that a submarine is |

designed to operate during trials or in combat situations; the Russian term is *limiting depth.* Repeated submergence to that depth can take place only a limited number of times (estimated at about 300 times for the entire service life of a submarine). *Collapse* depth is the deepest that a submarine's pressure hull is expected to survive; in the U.S. Navy this is 1.5 times the submarine's test depth; it is approximately the same in the Russian Navy.

| | |
|---|---|
| dimensions | U.S. submarine dimensions are based on the English system (i.e., feet, inches); Soviet submarine dimensions are based on the metric system, with English measurements approximated. |
| displacement | U.S. submarine displacements in long tons (2,240 pounds); Soviet submarine displacements are based on metric tons (1,000 kg or 2,205 lb). In Russian terminology the term *normal* is used for surface displacement and *water* for submerged displacement. |
| DoD | Department of Defense |
| double hull | Submarine hull configuration in which a non-pressure hull surrounds all or portions of the inner pressure hull. The between-hull volume may be free-flooding or contain ballast tanks, fuel tanks, and possibly weapons and equipment (at ambient sea pressure). |
| draft | maximum draft while on surface |
| EB | Electric Boat (shipyard) |
| fairwater | *see* sail |
| FBM | Fleet Ballistic Missile (early U.S. term for Submarine-Launched Ballistic Missile [SLBM]) |
| fin | *see* sail |
| GAO | General Accounting Office (U.S.) |
| GIUK Gap | Greenland-Iceland-United Kingdom (passages between those land masses) |
| GUPPY | Greater Underwater Propulsive Power (the letter *y* added for pronunciation) |
| HEN | Hotel-Echo-November (NATO designation of first-generation of Soviet nuclear-propelled submarines and their propulsion plants) |
| HF | High Frequency |
| HMS | His/Her Majesty's Ship (British) |
| hp | horsepower |
| HTP | High-Test Peroxide |
| HTS | High-Tensile Steel |
| HY | High Yield (steel measured in terms of pounds per square inch in thousands; e.g., HY-80, HY-100) |
| kgst | kilograms static thrust |
| knots | one nautical mile per hour (1.15 statute miles per hour) |

| | |
|---|---|
| KT | Kiloton (equivalent of 1,000 tons of high explosive) |
| lbst | pounds static thrust |
| length | length overall |
| limber holes | holes near the waterline of a submarine for draining the superstructure when on the surface |
| LOA | Length Overall |
| LPSS | transport submarine (also APS, APSS, ASSP, SSP) |
| LST | tank landing ship |
| MIRV | Multiple Independently targeted Re-entry Vehicle (warhead) |
| Mk | Mark |
| MG | Machine Gun(s) |
| MRV | Multiple Re-entry Vehicle (warhead) |
| MT | Megaton (equivalent of 1,000,000 tons of high explosive) |
| NATO | North Atlantic Treaty Organization |
| NavFac | Naval Facility (U.S. Navy) |
| NavSea | Naval Sea Systems Command (U.S. Navy; from 1974) |
| NavShips | Naval Ship Systems Command (U.S. Navy; 1966–1974) |
| NII | scientific research institute (Soviet) |
| n.mile | nautical mile (1.15 statute miles) |
| OKB | Experimental Design Bureau (Soviet) |
| ONR | Office of Naval Research (U.S.) |
| polymer | fluid ejected from a submarine as a means of reducing skin friction drag |
| pressure hull | submarine hull that provides protection against pressure for crew, machinery, weapons, and equipment; it may be encased within an outer, non-pressure hull, that is, *double hull* configuration |
| PWR | Pressurized Water Reactor |
| RN | Royal Navy |
| rpm | revolutions per minute |
| sail | Upper appendage or *fairwater* of a submarine. The sail replaced the *conning tower* of earlier submarines, which was a watertight compartment placed on top of the hull, usually serving as the attack center. There is a small bridge atop the sail, with the structure serving as a streamlined housing for periscopes, masts, and *snorkel* induction tube; in some submarines the forward diving planes are mounted on the sail. Also called *bridge fairwater* (i.e., a streamlined structure to support the bridge); called *fin* in the Royal Navy. |

| | | |
|---|---|---|
| SA-N-( ) | | U.S./NATO designation for a Soviet/Russian naval surface-to-air missile |
| SAR | | Synthetic Aperture Radar |
| SKB | | Special Design Bureau (Soviet) |
| shp | | shaft horsepower |
| shutters | | closing over torpedo tube openings, hull-mounted forward diving planes, and (on Soviet submarines) over *limber holes* to enhance hydrodynamic and acoustic performance of submarine |
| SKB | | Special Design Bureau (Soviet) |
| SLBM | | Submarine-Launched Ballistic Missile |
| SM | | minelaying submarine (also SSM) |
| snorkel | | intake and exhaust tubes to permit the operation of a diesel engine from a submerged submarine (operating at periscope depth) |
| sonar | | Sound Navigation And Ranging |
| SOSUS | | Sound Surveillance System (U.S.) |
| SPO | | Special Projects Office (U.S.) |
| SS | SSN | attack (torpedo) submarine (*N* suffix indicates nuclear propulsion) |
| SSA | SSAN | (1) auxiliary (special-purpose) submarine<br>(2) cargo submarine (later ASSA) (*N* suffix indicates nuclear propulsion) |
| SSAG | | auxiliary submarine (retains combat capability) |
| SSB | SSBN | (1) bombardment submarine<br>(2) ballistic missile submarine (*N* suffix indicates nuclear propulsion) |
| SSC | | coastal submarine |
| SSE | | (1) ammunition submarine<br>(2) electronic reconnaissance submarine |
| SSG | SSGN | guided (cruise) missile submarine (*N* suffix indicates nuclear propulsion) |
| SSK | SSKN | hunter-killer submarine (*N* suffix indicates nuclear propulsion) |
| SSM | | minelaying submarine (also SM) |
| SS-N-( ) | | U.S./NATO designation for a Soviet/Russian naval surface- or subsurface-to-surface missile, either tactical or strategic; NX indicates a missile estimated to be in development and not in service |
| SSO | | submarine oiler (also AOSS) |
| SSP | | submarine transport (also APS, ASSP, LPSS) |
| SSPO | | Strategic Systems Project Office (U.S.) |
| SSQ | SSQN | communications submarine (*N* suffix indicates nuclear propulsion) |

| | |
|---|---|
| SSR SSRN | radar picket submarine (*N* suffix indicates nuclear propulsion) |
| SST | target-training submarine |
| SSV | submarine aircraft carrier |
| SSX | submarine with undefined propulsion system |
| SubDevGru | Submarine Development Group (U.S.) |
| SUBROC | Submarine Rocket (U.S.) |
| SUBSAFE | Submarine Safety (U.S. modification program) |
| SUBTECH | Submarine Technology (U.S. program) |
| TEDS | Turbine Electric-Drive Submarine (USS *Glenard P. Lipscomb*/SSN 685) |
| TASM | Tomahawk Anti-Ship Missile |
| TLAM | Tomahawk Land-Attack Missile |
| TsKB | Central Design Bureau (Soviet) |
| TTE | Tactical-Technical Elements (requirements) |
| USN | U.S. Navy |
| USNR | U.S. Naval Reserves |
| VLS | Vertical-Launching System |

# 1

# Genesis

*The German Type XXI was the most advanced combat submarine to go to sea during World War II. A single Type XXI began an abortive war patrol shortly before the conflict in Europe ended. Type XXI features would have a major impact on Cold War submarine design.* (Imperial War Museum)

The Second World War was in large part a naval war. In every theater of the war, submarines played major roles, especially Soviet and German submarines in the Northern regions; German submarines in the Atlantic; British, German, and Italian submarines in the Mediterranean; and U.S. and Japanese submarines in the Pacific theaters.

In the Atlantic the German U-boats by themselves came very close to defeating Britain. Winston Churchill, Britain's wartime Prime Minister, said of the Battle of the Atlantic: "The only thing that ever really frightened me during the war was the U-boat peril. . . . I was even more anxious about this battle than I had been about the glorious air fight called the Battle of Britain."[1] And "The U-boat attack was our worst evil. It would have been wise for the Germans to stake all upon it."[2]

In the spring of 1943 the U-boats brought Britain perilously close to defeat. British cabinet historian S. W. Roskill, describing the situation in March 1943, related:

> 'It appeared possible' wrote the Naval Staff after the crisis had passed, 'that we should not be able to continue [to regard] convoy as an effective system of defence'. It had, during three-and-a-half years of war, slowly become the lynch pin of our maritime strategy. Where could the Admiralty turn if the convoy system had lost its effectiveness? They did not know; but they must have felt, though no one admitted it, that defeat then stared them in the face.[3]

But the Allies did triumph over the U-boats in May 1943, driving them from the major North Atlantic shipping lanes. In the period 10 April to 24 May 1943, when German naval commander-in-chief Karl Dönitz withdrew his U-boats from the Atlantic convoy routes, the German submarine command had lost 22 U-boats, while sinking only 28 Allied merchant ships and one destroyer, an unacceptable exchange ratio. (Many other U-boats were damaged in that period and forced to return to port.)

But the defeated submarines were little more than updated World War I–era undersea craft. The principal U-boat of World War II was the 770-ton Type VIIC, essentially an updated and slightly larger version of their Type UBIII that first went to sea in 1917.[4] German shipyards produced 661 of the Type VIIC from 1940 to 1944. The Type VIIC variant could travel 6,500 nautical miles (12,045 km) at 12 knots on the surface, being propelled by two diesel engines (maximum surface speed was 17 knots); underwater, on battery-supplied electric motors, the VIIC could travel only 80 nautical miles (148 km) at four knots; top underwater speed was 7.5 knots.[5]

Despite these and other U-boat limitations, it had taken 44 months of war at sea to clear the North Atlantic of Dönitz's U-boats.[6] The limitations of these U-boats were long understood by the German naval high command. As early as 1936, the year after Germany was "legally" allowed to possess submarines, German engineer Helmuth Walter proposed a revolutionary *underwater* warship.[7] "Submarines" of the era were in reality surface ships that could submerge for a few hours at a time, operating underwater on electric motors with energy provided by storage batteries. Those batteries could provide propulsion for only a few hours at relatively slow speeds.

After perhaps a half hour at maximum submerged speed, or a day at "creeping" speeds, the submarine had to surface to recharge the battery, using diesel engines, which also propelled the submarine while on the surface. Walter would achieve greater underwater endurance and higher speeds through the use of hydrogen-peroxide propulsion. Under his scheme submarines would have a streamlined hull and would be propelled by a closed-cycle turbine plant using the thermal energy produced by the decomposition of a high concentration of hydrogen peroxide (perhydrol). It was a complex system that enabled a turbine to be operated in a closed (submerged) atmosphere to provide sustained high underwater speeds. At the urging of Walter, an experimental submarine, the *V-80,*

*Helmuth Walter* (Courtesy Fritz-Folkmar Abels)

was built in 1940 with such a turbine plant. That submarine reached more than 26 knots submerged for short periods of time—by a significant margin the fastest speed achieved to that date by a manned underwater vehicle. In 1942 several Type XVIIA Walter experimental boats were ordered and plans were drawn up for building 24 operational Type XVIIB submarines. The bureaucracy of the German military establishment slowed the program, while skepticism among many naval engineers and others led to further delays. In November 1943 the first two Walter Type XVIIA submarines, the *U-792* and *U-794,* were commissioned for sea trials. They attained 25-knot underwater speeds for short periods; their longest fully submerged run was 5½ hours at 20 knots (twice the underwater speed of U.S. fleet submarines, which could maintain their maximum speed of nine knots for about one hour).

Even before the massive U-boat losses of the spring of 1943, Admiral Dönitz realized that it was only a matter of time until his existing U-boats would be driven from the sea, primarily, he believed, by radar fitted in surface escorts and aircraft. Radar deprived a submarine of its *surface* mobility and hence effectiveness, even at night.

In November 1942 Dönitz convened a conference at his command post near Paris to determine how soon Walter U-boats could become available for combat. In addition to Walter, the senior German U-boat constructors attended—Fritz Bröcking, propulsion expert, and Friedrich Schürer, hull designer.

In his memoirs, Dönitz wrote:

> At this conference I learnt to my regret that the Walter U-boat was nowhere ready for service.... To U-boat Command, who

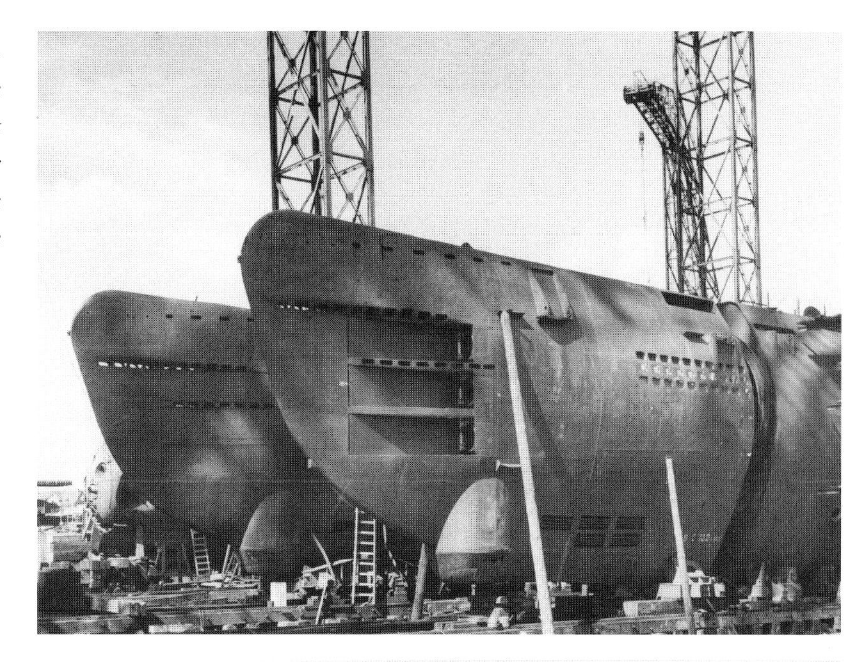

*Bow sections of Type XXI submarines on the assembly ways in the Deschimag shipyard in Bremen at the end of the war. The shutters for the bow torpedo tubes and the under-keel "balcony" housing for the GHG sonar are visible.* (U.S. Navy, courtesy Capt. H. A. Arnold, USN Ret)

> viewed with anxiety the clearly recognizable extent to which the enemy's defensive measures [radar] against 'the surface U-boat' were being further developed, this came as a great disappointment.[8]

As a near-term substitute for the Walter submarines, Bröcking and Schürer proposed adopting Walter's streamlined hull, which had been fully tested, and doubling the number of electric storage batteries. While such a submarine would fall short of the Walter U-boat performance, it would provide greater underwater speed and endurance than conventional submarines. Its huge electric battery gave rise to the term "electro" U-boat.

At the same time Walter proposed fitting the electro submarine with a *schnorchel* (snorkel) device that could draw in air from the surface for the diesel engines and expel the diesel exhaust gases just below the surface. This would permit the operation of diesel engines while submerged, to propel the boat and recharge the batteries, giving the U-boat the possibility of sustained underwater operations. Further, the head of the snorkel induction mast could be coated with radar absorbent material to prevent detection of the snorkel mast extending above the surface.[9]

## A Revolutionary Submarine

Design began on what would evolve as the Type XXI U-boat—the most advanced undersea craft of the war. According to the U-Boat Command's manual for employing the electro submarine:

*A Type XXI hull section awaiting assembly at the Deschimag shipyard. The hull sections arrived at the assembly yards in an advanced state of fitting out. The similarity to an inverted figure eight and three deck levels are evident.* (U.S. Navy, courtesy Capt. H. A. Arnold, USN Ret)

> Type XXI is a boat with strongly pronounced underwater fighting qualities which are capable of largely eliminating the superiority of the enemy's A/S [Anti-Submarine]

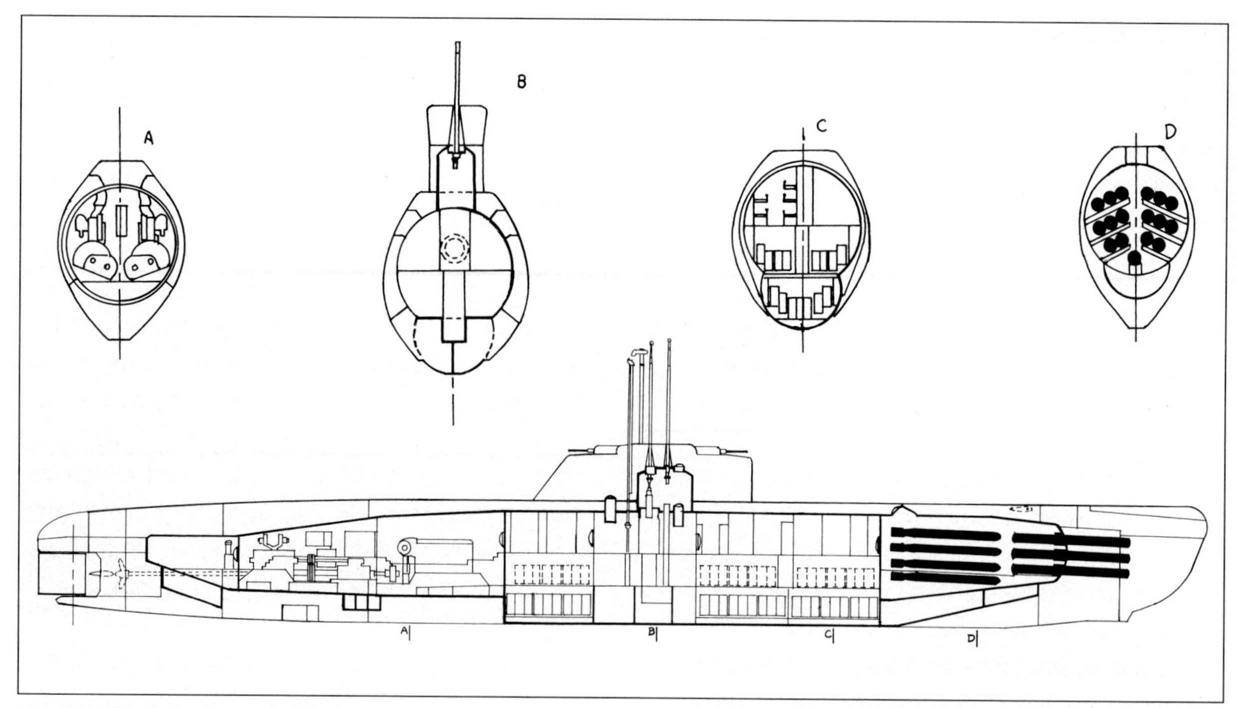


*Type XXI U-boat. LOA 251 ft 7 in (76.7 m)* (©A.D. Baker, III)

operations, resulting from his command of air and surface radar. With this boat and [Walter] types, it will be possible to start a new successful U-boat war.[10]

The Type XXI was a large submarine, displacing 1,621 tons on the surface, more than twice the displacement of a Type VII. The Type XXI has a streamlined hull, devoid of protuberances such as chocks, cleats, or gun mounts. Instead of a large conning tower with gun platforms and an internal pressure chamber that served as an attack center, the Type XXI had a streamlined sail, or fairwater, around the shears that supported the periscopes and other masts and antennae. These features reduced the drag above the waterline to about one-sixth that of earlier submarines. The openings in ballast tanks were reduced to further reduce drag. These modifications, coupled with improvements in batteries and increasing the voltage to the main motors, resulted in a near doubling of the Type XXI underwater speeds over previous U-boats:

16 knots for 25 n.miles (46 km)
12 knots for 60 n.miles (111 km)
6 knots for 280 n.miles (520 km)

Beyond two standard electric motors, the Type XXI submarine had two "crawling" *(schleich)* motors for quiet operation. These could propel the submarine at 5½ knots for 320 nautical miles (595 km), an unprecedented underwater endurance for a submarine. The creeping speed motors were mounted on bonded rubber mounts and had other features to reduce noise output. Even when using its standard electric motors, the Type XXI was significantly quieter than contemporary U.S. submarines.

The submarine had a designed operating depth of 440 feet (135 m), slightly greater than its foreign contemporaries. However, the Type XXI had a safety factor of 2.5—that is, a crush depth predicted at almost 1,110 feet (340 m), far greater than any other submarine.

The basic Type XXI hull design differed greatly from conventional submarines. For part of its length the pressure hull was like an inverted figure eight and for the remainder cylindrical.[11] It had a rounded bow for enhanced underwater performance, unlike the ship-like prows of conventional submarines, which were intended for high surface speeds. The Type XXI stern was very deep and very narrow, with no stern tubes fitted because of the fine lines. The extremity formed the vertical stabilizer/rudder, called the *schneiden,* or "cutting" stern, also referred to as a "knife" configuration.

The purpose of the Type XXI was to destroy Allied merchant shipping. In both World Wars submarines destroyed shipping with deck guns and torpedoes. Deck guns were useful against unarmed merchant ships and trawlers, with a couple of dozen shells being much cheaper than a torpedo. The Allied convoys of World War II, well protected by radar-equipped surface escorts and aircraft, made surface attacks deadly for the U-boat. During World War II submarines also carried light anti-aircraft guns—up to a caliber of 40 mm—to fight off hostile aircraft.

The Type XXI—intended for completely submerged war patrols—carried no deck guns except for light anti-aircraft guns, fitted in power-operated turrets at the forward and after ends of the sail structure. These guns were for use against attacking aircraft only as the U-boat was transiting to and from its protected, bomb-proof shelter to deep water.[12] (Although designed to mount twin 30-mm gun turrets, all Type XXIs were completed with 20-mm guns.)

The Type XXI's attack weapon was the torpedo: Six torpedo tubes were fitted in the bow; the torpedo room, the forwardmost of the six compartments, could accommodate 17 reload torpedoes. However, because of the need on long patrols to extract torpedoes from the tubes to check and maintain them, a loadout of only 20 torpedoes would be normal. While comparable to the number of torpedoes in a U.S. fleet submarine of the era (which had ten tubes), the Type XXI used hydraulic power to move the torpedoes from their stowage position into the tubes at a time when U.S. and other submarines used rollers and manpower to move torpedoes.

A semi-automatic, hydraulic reloading system enabled a Type XXI, after firing six torpedoes in one salvo, to fire a second salvo after only ten minutes, and a third salvo after a period of half an hour. This was a far faster firing rate than previous submarines; U.S. fleet boats of the war era required 35 to 40 minutes to reload the six bow torpedo tubes. Modified Type XXI designs—never completed—had differing arrangements: The Type XXIC had 6 bow torpedo tubes plus 12 amidships tubes firing aft; a total of 18 torpedoes would be carried, that is, no reloads. Such a "convoy killer" would only have to penetrate a convoy escort screen once to deliver a devastating attack with 18 torpedoes.

The Type XXI's torpedoes consisted of the *Lüt*, a pattern-running torpedo, and the T11, a passive acoustic homing weapon. The latter was believed to be immune to the "Foxer" and other acoustic decoys used by the Allies. Under development for future U-boat use were active acoustic homing and wire-guided torpedoes. To help the Type XXI detect hostile ships, the submarine was fitted with radar and the so-called GHG sonar, the most advanced acoustic detection system in service with any navy.[13] The sonar was mounted in an under-keel "balcony," and hence was referred to as *Balkon*.

The GHG was key to an advanced fire control system fitted in the Type XXI. The submarine's echo-ranging gear and plotting table, specifically designed for such attacks, were linked to a special device for so-called "programmed firing" in attacking convoys. As soon as a U-boat had succeeded in getting beneath a convoy, data collected by sonar was converted and automatically set in the *Lüt* torpedoes, which were then fired in spreads of six. After launching, the torpedoes fanned out until their spread covered the extent of the convoy, when they began running loops across its mean course. In this manner the torpedoes covered the entire convoy. In theory these torpedoes were certain of hitting six ships of from 197 to 328 feet (60 to 100 m) in length with the theoretical success rate of 95 to 99 percent. In firing trials such high scores were in fact achieved.

The crew of 57 officers and enlisted men lived in accommodations that were—by German naval standards—"virtually palatial."[14] Intended for long-range operations, the Type XXI represented an attempt to improve living conditions, and while many of the objectionable features of the previous designs were retained, the net results are in certain aspects superior to those on contemporary U.S. submarines. Some privacy was provided for crewmen by eliminating large sleeping compartments and by dividing the off-watch personnel into fairly small groups, each within its own quarters that were segregated from passageways. Against these features there were bunks for only 47 men, meaning that ten men had to have hammocks or some crew members had to share the same berth. The galley represented an improvement over earlier submarines. The cooking range permitted a greater variety of foods to be served, and twin sinks made it possible to get the mess and galley gear cleaner after use.

Still, the sanitary arrangements were inadequate by U.S. standards. Further, because of the interconnection of the washing and drinking water systems, the latter was considered unsafe by U.S. submarine experts.[15]

*Otto Merker*

(Courtesy Fritz-Folkmar Abels)

An important aspect of the revolutionary nature of the Type XXI was its production. Initial Type XXI production plans provided for two prototypes to be completed at the end of 1944, with mass production to begin the following year. The first Type XXIs were to be ready for combat at the end of 1946. This plan was based on the assumptions that German dictator Adolf Hitler would give U-boat construction priority over all other military programs, that all materials would be available, and that Allied air raids would not interfere with construction. All of these assumptions were flawed.

Such a schedule was unacceptable to Admiral Dönitz. He took the Type XXI production issue to Albert Speer, Hitler's personal architect, who, since January 1942, had astutely performed the duties of Minister of Armaments and Munitions. Speer, in turn, put Otto Merker, the managing director of the Magirus Works—who had absolutely no knowledge of the shipbuilding industry—in charge of the program. Merker had been highly successful in applying mass-production methods to the manufacture of automobiles and fire trucks. Dönitz would later write:

> Merker suggested that the U-boats should not be built on the slipways of the shipbuilding yards as had hitherto been done, but should be built in sections at other suitable industrial plants and then sent to the yards for final assembly. (This method was then being successfully applied by the American, [Henry] Kaiser, to the mass-production of merchantmen in the United States.) The method had the advantage of saving a great deal of time. Later, we found that boats of the size of Type XXI could be completed by mass-production methods in from 260,000 to 300,000 man-hours, whereas under the old method a boat of similar size required 460,000 man-hours. Under the Merker plan the first Type XXI boat was to be ready by the spring of 1944. Merker was also prepared to accept responsibility for putting these boats into mass-production at once. This meant that large numbers of them would be ready in the autumn of 1944.[16]

The ambitious Type XXI construction program—to provide a monthly output of 40 submarines—was personally approved by Hitler on 29 July 1943. Construction of a Type XXI submarine was to have taken six months from the start of rolling steel for the pressure hull to completion. The first stage in fabrication of the Type XXI hull was done at steel works, which rolled and cut the necessary steel plates and manufactured the pressure hull sections. The sections were assembled at 32 factories that specialized in steel and boiler production. These section assembly facilities were selected not because of their prior experience in shipbuilding or from considerations of their dispersed locations, but because of their having access to the inland waterways, as the large sections had to be transported by barge.

The 11 fitting-out yards completed the sections—eight per submarine—-and installed all appropriate equipment and machinery, except for those items that were either too heavy (diesel engines) or would extend over two or more sections (main propeller shafting). None of the eight sections could weigh more than 165 tons after completion and fitting because of the capacity of the available cranes. These 11 shipyards all had experience in submarine construction, which was necessary because of the installation of wiring and the fitting of main electric motors, gearing, switchboards, and other specialized submarine equipment. The three final assembly yards were Blohm &

Voss in Hamburg; Deschimag AG-Weser in Bremen; and Schichauwerft in Danzig. The first Type XXI, the *U-3501,* was launched on 19 April 1944 at the Danzig yard. The craft was incomplete, however, as the openings in the hull had been temporarily covered by wooden blocks and she had to be towed into a floating drydock immediately. The *U-3501* was not commissioned until 29 July 1944.

Although mass production of the Type XXI was initiated, through early 1944 it was believed that the available supplies of lead and rubber in Germany would support increased battery production for only some 250 Type XXI submarines, after which production would have to shift over to Walter boats.

In addition to the Type XXI, production began of a smaller, coastal version of the electro submarine—the Type XXIII—while work was to continue on the ultimate underwater warships to be propelled by Walter closed-cycle turbines. Finally, with increased Allied operations in European coastal waters, Germany also commenced the production of large numbers of midget submarines.

Significantly, there was little hesitancy on behalf of Admiral Dönitz and his colleagues to make changes in the U-boat program to meet changing defensive and offensive strategic conditions as they arose, with little regard for the continuity of existing construction. According to a U.S. Navy evaluation, "Superseded programs were ruthlessly cast aside to make way for succeeding designs. Several older type hulls were left unfinished on the ways when the yard in question was given an assignment in a new program which did not require these ways. In other instances, fabricated sections in sufficient number to complete several boats were discarded in order to make way for a new program."[17]

A second, exceptionally noteworthy characteristic of the U-boat command was that whenever a new concept was presented, designs were usually solicited from several agencies. Ideas submitted at any time by recognized submarine engineers were given detailed consideration.

Actual deliveries of the Type XXI and Type XXIII electro submarines were far short of the numbers planned. Allied bombing raids on German industry increased steadily from the fall of 1943, hence critical materials were not available in the required amounts. Even with a virtual halt in the production of earlier U-boat designs, Type XXI construction lagged because of material problems and Allied air attacks; the numbers of units completed were:

| | |
|---|---|
| June 1944 | 1 |
| July 1944 | 6 |
| Aug 1944 | 6 |
| Sep 1944 | 12 |
| Oct 1944 | 18 |
| Nov 1944 | 9 |
| Dec 1944 | 28 |
| Jan 1945 | 16 |
| Feb 1945 | 14 |
| Mar 1945 | 8 |
| Apr 1945 | 1 |

Admiral Dönitz demanded extensive training in the Baltic Sea before the submarines could go on patrol. Meanwhile, U.S. and British aircraft were attacking submarines on trials and training, and destroyed several Type XXIs on the assembly ways.

The first operational Type XXI was the *U-2511,* commanded by *Korvettenkapitän* Adalbert Schnee.[18] She sailed from Kiel on 18 March 1945 en route to Norway for final preparations for her first combat patrol, to the tanker shipping lanes of the Caribbean Sea. But upon arrival in Norway, Schnee had to correct problems with a periscope and diesel engines as well as repairing slight damage suffered during the *U-2511*'s deep-diving trials. The U-boat finally went to sea on patrol on 30 April. At the time another 12 Type XXI submarines

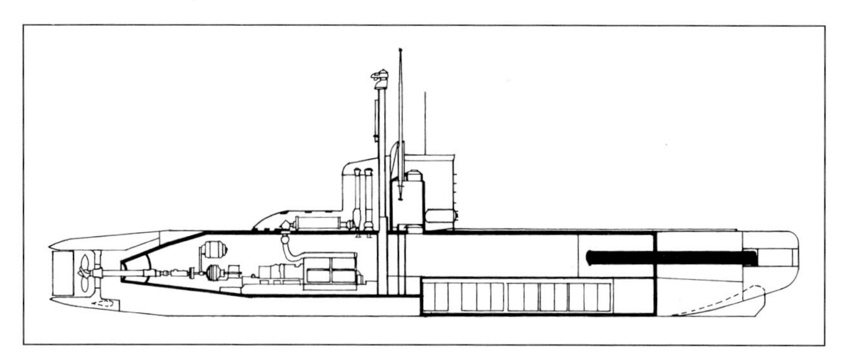

*Type XXIII U-boat. LOA 113 ft 10 in (34.7 m)* (©A.D. Baker, III)

were in Norwegian ports being readied for war patrols, while some 50 more were being readied in German ports to sail to Norway for final combat preparations.

At sea on 4 May 1945, the *U-2511* received Admiral Dönitz's order for all U-boats to cease operations, dispose of their weapons, and return to port. According to Schnee:

> First contact with the enemy was in the North Sea with a hunter killer-group. It was obvious that with its high under-water speed, the boat could not come to harm at the hands of these killer-groups. With a minor course alteration of 30°, proceeding submerged I evaded the group with the greatest ease. On receipt of order to cease fire on May 4, I turned back for Bergen; a few hours later I made contact with a British cruiser and several destroyers. I delivered a dummy under-water attack and in complete safety came within 500 yards [457 m] of the cruiser. As I found out later during a conversation when I was being interrogated by the British in Bergen, my action had passed completely unnoticed. From my own experience, the boat was first class in attack and defence; it was something completely new to any submariner.[19]

The first Type XXIII, the *U-2324,* put to sea in January 1945, and the design proved to be a success. Five of the 300-ton, 12-knot submarines went to sea, carrying out eight short-duration patrols against Allied merchant shipping. These electro boats sank six Allied ships without a loss, a harbinger of what might have been had more of the advanced submarines gone to sea. At the end of the war, 63 Type XXIII submarines had been completed, with scores more still on the building ways.

**The Type XXI and its diminutive cousin, the Type XXIII, were truly revolutionary *undersea* craft. A squadron, flotilla, or even a fleet of these submarines could not have won World War II. However, had they been available in late 1944 or even early 1945 they would have significantly slowed the advance of the Western Allies, perhaps delaying the end of the war in the West by several months or even a year. This, in turn, could have given the Soviet Union a more advantageous position in Europe when the fighting ended.**

**By 1944–1945 the Allied anti-submarine effort was too large and had too much momentum to have lost a new Battle of the Atlantic against the Type XXIs, while U.S. shipyards could replace merchant losses at a prodigious rate. Still, such a campaign by advanced U-boats would have seriously hurt the flow of weapons, matériel, and fuels to the Allied forces fighting in Western Europe, possibly opening the way for Red Army advances farther westward.**

**Other factors influencing such a scenario were that British and Soviet troops were rapidly overrunning the U-boat manufacture and assembly facilities in northern Germany, Allied tactical aircraft were denying the Baltic to the U-boats as a training and work-up area, and the overall chaos in Germany was denying supplies and provisions to the new U-boats.**

**Rather than affect the course of the war, the Type XXI's place in history was to serve as the progenitor to the Cold War–era submarines designed and produced by the United States and the Soviet Union. U.S. Navy Department historian Gary Weir observed of the Type XXI:**

> **For the first time since John Holland's act of invention [in the late 1800s], a submarine had spent more time operating below the waves than on the surface. The paradigm shift completed with the nuclear-propelled USS *Nautilus* (SSN 571) in 1955 began with the type 21.[20]**

TABLE 1-1
**1945 Submarine Designs**

| | German Type XXI | German Type XXIII | U.S. *Tench* SS 417 | Soviet K Series XIV |
|---|---|---|---|---|
| Operational | 1945 | 1945 | 1944 | 1940 |
| Displacement | | | | |
| surface | 1,621 tons | 234 tons | 1,570 tons | 1,480 tons |
| submerged | 1,819 tons | 258 tons | 2,415 tons | 2,095 tons |
| Length | 251 ft 7 in (76.7 m) | 113 ft 10 in (34.7 m) | 311 ft 8 in (95.0 m) | 320 ft 4 in (97.65 m) |
| Beam | 21 ft 8 in (6.6 m) | 9 ft 10 in (3.0 m) | 27 ft 3 in (8.3 m) | 24 ft 3 in (7.4 m) |
| Draft | 20 ft 8 in (6.3 m) | 12 ft 2 in (3.7 m) | 17 ft (5.18 m) | 14 ft 10 in (4.51 m) |
| Diesel engines | 2 | 1 | 4 | 2 |
| horsepower | 4,000 | 580 | 5,400 | 8,400 |
| Electric motors | 2* | 1* | 2 | 2 |
| horsepower | 5,000 | 580 | 4,600 | 4,400 |
| Shafts | 2 | 1 | 2 | 2 |
| Speed | | | | |
| surface | 15.6 kts | 9.75 kts | 20.25 kts | 20 kts |
| submerged | 17.2 kts | 12.5 kts | 8.75 kts | 10 kts |
| Range (n.miles/knots) | | | | |
| surface | 11,150/12 | 2,600/8 | 11,000/10 | |
| submerged | 285/6 | 175/4 | 96/2 | |
| Test depth | 440 ft (135 m) | 330 ft (100 m) | 400 ft (120 m) | 330 ft (100 m) |
| Guns | 4 x 20 mm | nil | 2 127-mm**<br>2 40-mm<br>4 MG | 2 100-mm<br>2 45-mm |
| Torpedo tubes*** | 6 533-mm B | 2 533-mm B | 6 533-mm B<br>4 533-mm S | 6 533-mm B<br>2 533-mm S<br>2 533-mm D |
| Torpedoes | 20 | 2 | 24 | 24# |
| Complement | 57 | 14 | 81 | 62 |

Notes: * Plus "creeping" motor.
** Guns varied; the "ultimate" gun armament approved in 1945 is listed.
*** Bow + Stern + Deck torpedo tubes.
# In addition, the K class carried 20 chute-laid mines.

2

# Advanced Diesel Submarines

*The* K1 *at New London, Connecticut. She was the lead submarine for a planned massive program to produce small hunter-killer submarines. Her large bow housed the BQR-4 sonar array; the small, BQS-3 "single-ping" ranging sonar was fitted atop the BQR-4 dome.* (U.S. Navy)

The U.S. Navy constructed a large number of submarines during World War II: 201 submarines were delivered between 7 December 1941 and 15 August 1945; another 23 war-program submarines were completed after the war.[1] These all were "fleet boats," large, long-range submarines originally designed to operate across the broad Pacific, scouting out the Japanese Fleet and then attriting Japanese capital ships before the major clash of U.S. and Japanese dreadnoughts.

Even taking into account wartime losses and immediate postwar disposals of worn out or heavily damaged fleet boats, the U.S. Navy emerged from World War II with about 150 relatively modern, long-range submarines. These fleet boats were of similar design; the principal difference was that the 38 surviving boats of the *Gato* (SS 212) class had an operating depth of 300 feet (90 m), while the 114 submarines of the similar *Balao* (SS 285) and *Tench* (SS 417) classes were known as "thick-hull" submarines and were rated at 400 feet (120 m).[2] (See table 2-1.)

The depth increase to 400 feet was achieved by shifting from mild steel to High-Tensile Steel (HTS) and increasing the thickness of the pressure hull. HTS provided a yield strength of about 50,000 pounds per square inch. The increase in pressure hull weight was compensated for by meticulous attention to detail in every part of the submarine.[3] On an operational basis the value of the depth increase was to evade an enemy depth charge attack, as depth charges were preset to detonate at a specific depth. The stronger hull could help reduce the effects of other ASW weapons—hedgehogs and acoustic homing torpedoes. And, all weapons took more time to reach greater depths. The fleet submarine *Chopper* (SS 342), in 1969, made an uncontrolled dive off Cuba. The submarine's bow reached a depth of 1,050 feet (320 m); she was able to make it back to the surface. She suffered some damage,

*The U.S.* Gato *and* Balao *"fleet boats" were the most successful Allied submarines of World War II. The* Ronquil *of the latter class is shown with two 5-inch guns and two 40-mm guns; the masts aft of the two periscope shears mount the SV air search and (larger) SJ surface search radar antennas.* (U.S. Navy)

but the integrity of her pressure hull was intact.

The U.S. fleet boat could be considered the most capable long-range submarine of any navy except for the German Type XXI. However, the Type XXI was far superior in underwater performance to the American fleet boat. Beyond the technical obsolescence of the fleet boat was the vital question of the role of the U.S. Navy and, especially, submarines in the post–World War II environment. U.S. submarines had a major role against Japanese warships and merchant shipping in the Pacific.[4] With Germany and Japan vanquished, only the Soviet Union appeared on the horizon as a potential antagonist of the United States. Soviet Russia had virtually no naval fleet or merchant fleet that would be the targets for the U.S. submarine force. Indeed, seeking a rationale for modernizing the U.S. Navy, in 1947 the Director of Naval Intelligence told a classified meeting of senior officers:

> It's quite conceivable that long before 2000 A.D. Russia may either through military measures or political measures overrun Western Europe and obtain the advantage of the brains and technical know-how [of] the industry of Western Europe, probably including the United Kingdom. If that were to come to pass, *thinking 50 years hence,* it would change this whole strategic picture. It would bring Russia then into contact with United States seapower, so I don't feel that we can, without any reservations whatsoever, accept a situation which you now have where Russia is impotent on the sea and the United States has potentially complete control of the sea. . . .[5] [Emphasis added]

The U.S. Navy's search for roles for submarines could be seen in an earlier memorandum from the Office of the Chief of Naval Operations:

> In World War II the primary task of our submarine force was the destruction of enemy shipping. The success achieved in this direction should not, however, be allowed to influence our planning for any future war. It is conceivable for example that we might be up against a land power whose economy does not depend on extensive seaborne commerce. This was more or less the position in which Great Britain found herself in the last war against Germany. Under these circumstances full use may still be made of the submarine as an instrument of stealth and surprise without regard to its properties as an anti-shipping weapon.[6]

The memorandum went on to propose the development of five specialized types of submarines: (1) torpedo attack, (2) guided missile, (3) cargo-troop carrier, (4) reconnaissance, and (5) midget. In addition, preparations already were under way for the conversion of several fleet boats to a radar picket configuration, intended to provide task forces with early warning of air attacks.

Primary attention, however, centered on the torpedo-attack submarine. With the end of the Pacific War, the Navy's General Board—the princi-

Table 2-1
**U.S. Submarine Concepts, 1945–1946**

| | *Tench* SS 417 | General Board | Submarine Officers Conference |
|---|---|---|---|
| Displacement | | | |
| surface | 1,570 tons | 1,960 tons | 800 to 1,000 tons |
| submerged | 2,415 tons | | |
| Length | 311 ft 8 in (95.0 m) | 337 ft (102.7 m) | |
| Propulsion | | | |
| surface | 4 diesel engines | 4 diesel engines | diesel-electric or turbine |
| submerged | 2 electric motors | 2 electric motors | |
| Shafts | 2 | 2 | |
| Speed | | | |
| surface | 20.25 kts | 22.5 kts | 14 kts |
| submerged | 8.75 kts | 9 kts | 26 kts |
| Test depth | 400 ft (120 m) | 500 ft (150 m) | 800 to 1,000 ft (240 to 300 m) |
| Guns | 2 127-mm | 2 127-mm | none |
| Torpedo tubes* | 6 533-mm B<br>4 533-mm S | 6 533-mm B**<br>6 533-mm S** | 4 to 6 533-mm B<br>plus 533-mm amidships tubes |
| Torpedoes | 24 | approx. 30 | |

Notes: * Bow + Stern + Amidships torpedo tubes.
** At the suggestion of the Chief of Naval Operations, 24-inch (610-mm) torpedo tubes to be considered as an alternative to the standard 21-inch (533-mm) tubes.

pal advisory body to the Navy's leadership—proposed the construction of an enlarged fleet-type submarine. Employing the basic fleet boat configuration, which dated from the 1920s, the General Board's submarine was to be a simply larger and hence more capable fleet boat. But the submarine community—manifested in the Submarine Officers Conference—opposed the concept of an enlarged fleet boat.[7] (See table 2-1.)

During World War II several senior U.S. and Royal Navy officials knew of the German efforts to develop advanced, high-speed submarines. The source for this information was primarily Vice Admiral Katsuo Abe, the head of the Japanese military mission to Germany from 1943 to 1945. *Grossadmiral* Karl Dönitz, the head of the German Navy, personally met with Abe to brief him on new U-boat programs and permitted Abe to visit submarine building yards. Abe then sent details of the German programs to Tokyo by radio. Those enciphered radio messages were promptly intercepted and deciphered by the British and Americans, revealing details of the Type XXI and other U-boat designs and programs.[8]

Allied knowledge of the Type XXI design was soon exceeded by the actual acquisition of the advanced submarines. As Allied armies overran German shipyards, submarine blueprints, components, and other material were scooped up by the victors, as were some German submarine engineers and technicians. Immediately after the war, in accord with the Potsdam Agreement of July 1945, Great Britain, the Soviet Union, and the United States each took possession of ten completed U-boats; among those 30 submarines were 11 of the electro boats:

| | | |
|---|---|---|
| Type XXI | Great Britain | *U-2518*,[9] *U-3017* |
| | Soviet Union | *U-2529, U-3035, U-3041, U-3515* |
| | United States | *U-2513, U-3008* |
| Type XXIII | Great Britain | *U-2326, U-2348* |
| | Soviet Union | *U-2353* |

The U.S. Navy conducted extensive trials with the *U-2513*. Among her passengers were senior naval officers, including Chief of Naval Operations Chester W. Nimitz, a submariner, as well as President Harry S. Truman. She was operated until 1949. The *U-3008* was similarly employed on trials until 1948.

Based on this cornucopia of German submarine technology, the U.S. submariners undertook a two-track program: First, existing, relatively new fleet boats would be given enhanced underwater performance under the Greater Underwater Propulsive Power (GUPPY) program and, second, a new, high-performance submarine would be developed.[10]

This immediate postwar emphasis on submarines was justified for the Anti-Submarine Warfare (ASW) role to counter an anticipated Soviet buildup of submarines—employing German technology—that could again threaten merchant shipping connecting the resources of America with the battlefields of Europe in a future conflict. U.S. Navy intelligence predicted that by the 1960s it was possible for the Soviets to have 1,200 or even 2,000 submarines of all types at sea. One U.S. admiral in discussing these numbers made

> an assumption that the Russians will maintain their numbers of submarines in approximately the same amount that they have now but improve their types and replace older types with new ones, and the second assumption, that by 1960 or within ten years, 1958, that the Russians could have two thousand up-to-date submarines. I have chosen that [latter] figure because I believe it is within their industrial capability of producing that number and I believe if they really intend to employ the submarines as a means of preventing the United States or her Allies from operating overseas that two thousand would be the number they would require for their forces.[11]

A Soviet admiral, also in 1948, reportedly alluded to the possibility of a Red undersea force of 1,200 submarines. However, references to such numbers cannot be found in official Soviet documents.[12]

The major limitations on the Soviet force envisioned by U.S. naval intelligence were dock space, fuel, and distilled battery water. These factors would, it was believed, probably limit the force to 400 submarines.[13] To reach 2,000 submarines it was estimated that Soviet shipyards would have to produce more than 16 submarines per month; U-boat production in Germany during World War II had reached a maximum average of some 25 submarines per month.

*The* Tench *following her GUPPY IA conversion. She has a stepped fairwater enclosing her masts and periscopes, rounded prow, deck protuberances removed, increased battery power, and other Type XXI features. A longer, non-stepped sail was fitted to most GUPPYs.* (U.S. Navy)

The U.S. and British Navies envisioned employing submarines as an ASW weapon on the basis of the 58 Axis submarines reported sunk by Allied undersea craft in World War II. Significantly, in all but one encounter, the target submarine was on the surface. Only HMS *Venturer* sank the *U-864* off Norway in early 1945 when both of the submarines were fully submerged.[14]

To enhance its submarine ASW capability, the U.S. Navy proposed to modernize up to 90 fleet boats to the GUPPY configuration. The GUPPY conversions included many Type XXI features—a more rounded bow; the conning tower and bridge encased in a streamlined fairwater, or "sail," housing the

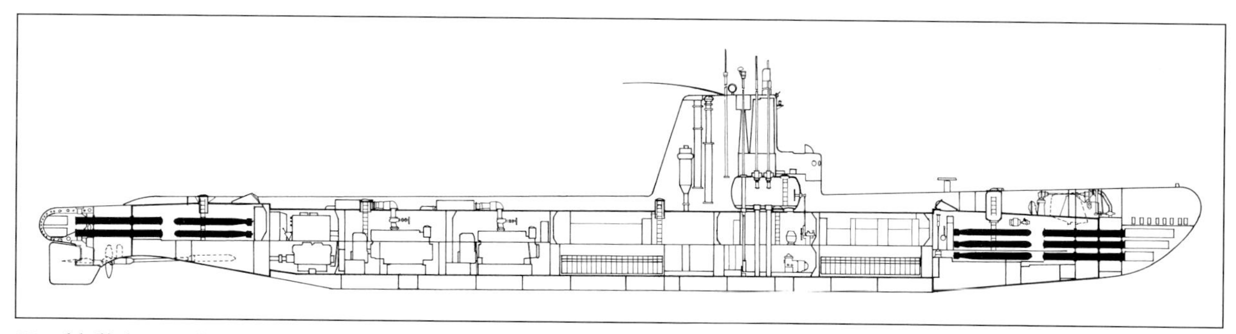

Hawkbill *(SS 366) GUPPY IB conversion. LOA 308 ft (93.9 m)* (©A.D. Baker, III)

periscopes and snorkel intake; deck guns and other projections were removed; and, in some variants, one diesel engine was removed to provide more space for a new sonar room, air-conditioning, and the Prairie Masker quieting equipment.[15] And, especially important, the electric battery capacity was increased.

The GHG sonar taken from the *U-3008* was installed in the USS *Cochino* (SS 345), a converted GUPPY II; the other GUPPYs initially retained their inadequate, war-era sonars.[16] Eventually, all surviving U.S. GUPPYs would be fitted with a variant of the BQR-2 sonar, a duplication of the GHG.

GUPPY submarines could reach 14 to 16 knots submerged, compared to ten knots for non-converted fleet boats. The GUPPYs suffered major vibrations at higher speeds; one submariner recalls "almost loosing my teeth to the vibration." Some efforts were undertaken to quiet the machinery noises in the GUPPYs, because the Type XXI was found at 12 knots to be quieter than the best U.S. fleet submarines at six knots. The GUPPY configurations retained the full armament of ten torpedo tubes, although several reload torpedoes were removed to provide more berthing space. There were, however, several problems with the modernized submarines—initially major difficulties were encountered in snorkel operations. Also, in comparison with the Type XXI, the GUPPY boats still were noisy.

In the event, 52 GUPPY conversions were completed to several configurations from 1946 to 1963, with some early conversions subsequently being upgraded to later variants. GUPPY submarines served in the U.S. fleet until 1975, with several transferred to other navies after U.S. service, four of which remained in service into the 21st Century.[17]

(The Royal Navy preempted the U.S. Navy in the conversion of high-speed submarines. Based on intelligence reports of German development of the Type XXI and closed-cycle undersea craft, in 1944 the British began the conversion of eight S-class submarines to high-speed underwater targets to train ASW forces. First was HMS *Seraph.* Her electric motors were upgraded, higher-capacity batteries were provided, and the hull and conning tower were streamlined, giving the boat an underwater speed of just over 12.5 knots compared to 8.8 knots in her original configuration. Unlike the later American GUPPYs, these target submarines had their torpedo tubes deactivated and thus were suitable only for the ASW target role.)

Separate from the GUPPY program, the extensive conversions of U.S. fleet boats were undertaken for a variety of specialized roles as the submarine community searched for missions in the postwar environment: cargo (designated SSA/ASSA), guided missile (SSG), hunter-killer (SSK), oiler (SSO/ASSO), radar picket (SSR), various research roles (AGSS), and troop transport (SSP/ASSP). A proposed minelayer conversion (SSM) was not undertaken, while preliminary designs for a new construction electronic reconnaissance (SSE) and aircraft carrier (SSV) were quickly abandoned. The older fleet boats were especially valuable for service in various research configurations. Most of the remaining fleet boats were fitted with a snorkel (as were most of the specialized conversions) and had minimal streamlining—with guns removed—to provide a slight increase in their underwater performance.[18]

While the large GUPPY and specialized fleet boat conversion programs were under way, the U.S. Navy sought to produce a smaller, faster, more

*The* Tang *class introduced Type XXI features to new-construction U.S. submarines. The* Gudgeon *of the class shows her fully retracted bow diving planes; a dome covers her QHB-1 transducer on the forward deck. Her fairwater, or sail structure, was similar to those fitted in most GUPPY submarines.* (U.S. Navy)

maneuverable successor to the fleet boat. The smaller submarine would be more suitable for forward operations in northern European waters and would be several knots faster underwater than the fleet boat. The submarine community also wanted to employ a closed-cycle propulsion plant to attain speeds of some 25 knots and wanted consideration of eliminating the conning tower to enhance underwater performance (as planned for the German Type XXI derivative designated Type XXX, which was to have had no fairwater structure). The 25-knot underwater speed was sought to permit U.S. submarines to make submerged approaches to surface targets that had required surface approaches with fleet boats. Transits to forward areas would still be made on the surface.

However, U.S. efforts to evaluate closed-cycle submarine propulsion was limited to land tests. (See Chapter 3.) The U.S. Navy decided to pursue a submarine of Type XXI size with advanced diesel-electric propulsion as an "interim" submarine until a closed-cycle propulsion plant was proven and available. This was the *Tang* (SS 563) class, with the first two submarines being ordered in August 1947, which were the start of series production.[19] The Portsmouth Naval Shipyard (Kittery, Maine) was responsible for the contract design as well as construction of the lead ship of the class.

Like the Type XXI, the *Tang* would be designed for optimum underwater performance. The design provided a rounded bow with a small, streamlined fairwater without a conning tower; chocks, cleats, and other topside projections were recessed or retractable. Consideration was given to deleting the fairwater structure.[20] No deck guns were to be fitted. The *Tang* would have six full-length bow torpedo tubes and two short stern tubes, only two fewer than the fleet boats; 22 torpedoes would be carried. The BQR-2 and BQS-4 sonars were fitted.[21]

The smaller size of the *Tang* in comparison with the previous fleet boats required smaller propulsion machinery. Instead of the four large diesels (4,600 hp total) in the fleet boats, the *Tang* would have four so-called "pancake," or radial, 16-cylinder diesel engines, each generating 850 horsepower. These were for surface operation and, using a snorkel mast, to charge the electric batteries during submerged operation. Submerged, the *Tang* would be driven by two electric motors (4,610 hp total). Four high-capacity electric battery sets were provided, each with 126 cells (as in the later GUPPYs; the fleet boats and early GUPPYs had two sets with 126 cells each). Twin propeller shafts were supported outside the pressure hull by struts; only a lower rudder being fitted, aft of the propellers. This was the same arrangement as in the fleet and GUPPY boats.

This power plant was to drive the *Tang* at a submerged speed of 18.3 knots and 15.5 knots on the surface (less than the approximately 20 knots of the fleet boats and GUPPYs). The *Tang* was thus the first modern U.S. submarine designed with greater underwater than surface performance. An automatic hovering system was provided for a quiet and energy-saving operation when the submarine was in a sonar listening mode.

Problems with the *Tang*'s hydraulic torpedo ejection system required modifications that added some 24 tons to the forward torpedo room, disturbing the submarine's longitudinal stability. The solution was to add six feet (1.83 m) to the craft, which increased the length to 268 feet (81.7 m) and changed surface displacement from 1,575 tons to 1,617 tons (virtually identical to the Type XXI).[22]

Significantly, in these first U.S. postwar submarines, the test depth was increased to 700 feet (215 m), that is, 75 percent greater than the previous *Balao-Tench* designs. Given a 1.5 "safety margin," this meant a predicted collapse depth of 1,100 feet (335 m).

The *Tang* was placed in commission on 25 October 1951. Her radial diesel engines soon proved to be a disaster. Their unusual "pancake" design and light weight compared to the conventional engines made them highly attractive for the shorter length of the *Tang*. But they were a nightmare to service and were constantly breaking down, at times forcing submarines to be towed back to port. The diesels had electric generators suspended under the engines, where they were vulnerable to oil seal leaks, they were extremely loud in the engine room (over 140 decibels), and they vibrated considerably. After limited operational service, the first four *Tang*-class submarines had their radial diesels removed, and each boat was given three conventional diesel engines. The alteration required that the submarines be cut in half and a nine-foot (2.7-m) section added to their machinery spaces. The last two submarines of the class were built with conventional, in-line diesel engines.

The replacement diesels were not very accessible because of the small engine room, and they too were subject to periodic failures, albeit not as often as the "pancakes." (The one other submarine to have the pancake diesels was the research craft *Albacore* [AGSS 569].)

Re-engined, these six *Tang*-class submarines served effectively in the U.S. Navy for two to three decades, after which most were transferred to foreign fleets. One modified submarine of the *Tang* design was built—the *Darter* (SS 576)—before the revolutionary and more efficient *Albacore* design was adopted.

The high cost of the *Tang*-class submarines limited the production rate to two boats per year.[23] At the time predictions of the Soviet Union producing more than a thousand submarines of Type XXI performance called for a new approach to ASW. War-built destroyers and frigates and even ASW aircraft would have great difficulty countering such high-performance submarines. A complementary anti-submarine strategy would be to destroy Soviet submarines en route to Allied shipping lanes and when the survivors returned to port for more torpedoes and provisions. Previous attempts at such ASW tactics included the Anglo-American North Sea minefield of World War I that sought to bottle up German U-boats, and the Anglo-American air patrols over the Bay of Biscay in World War II that sought to attack German submarines departing and returning to bases in France.

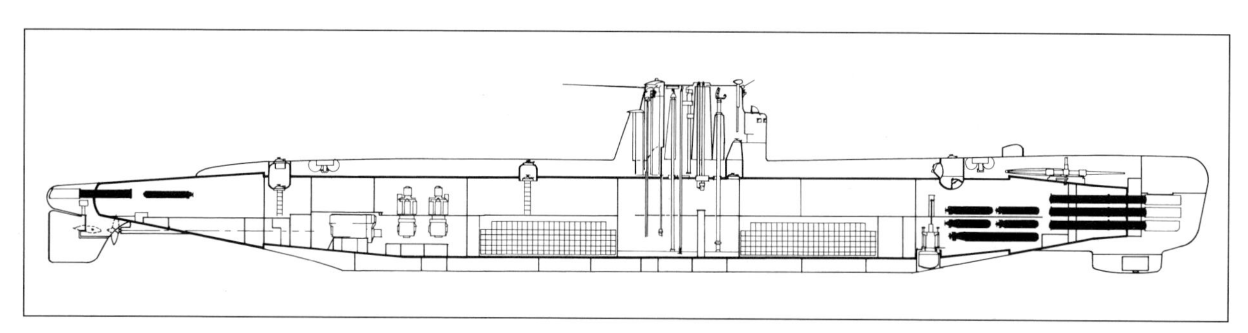

Tang *(SS 563) in 1952 configuration. LOA 269 ft 2 in (82.06 m)* (©A.D. Baker, III)

## Hunter-Killer Submarines

As early as 1946 the U.S. Navy's Operational Evaluation Group had proposed the use of submarines in ASW, and that September the chairman of the planning group for the Submarine Officers Conference noted that "with the further development and construction in effective numbers of new submarines by any foreign power the employment of our submarines in anti-submarine work may well become imperative."[24] Also in 1946 the Navy's ASW Conference proposed equal priority for a specialized, small ASW submarine as well as the new attack submarine (i.e., *Tang*).

The specialized "hunter-killer" submarines (SSK) would lay in wait to ambush enemy submarines off Soviet ports and in channels and straits where Soviet submarines would transit—on the surface or snorkeling—en route to and from the Atlantic shipping routes. The concept of specialized ASW submarines date to the British "R" class of World War I, when ten hunter-killer submarines were built, all launched in 1918 with only one being completed in time to see active service.[25] In the U.S. Navy the use of an ASW submarine was proposed in a 1946 report of the Navy's Operational Evaluation Group. The proposal resulted from the erroneous belief that the Japanese had sunk several U.S. submarines in World War II by employing such craft.

A series of Navy ASW conferences and exercises that began in 1947 in both the U.S. Atlantic and Pacific Fleets led to proposals for a hunter-killer submarine (SSK) force to counter the Soviet undersea fleet. The central component of the American SSK design was long-range, passive sonar, which would be coupled with effective torpedoes that "would destroy any submarine which passed within detection range" with a very high degree of probability.[26] The SSK was envisioned as a relatively small, simply constructed submarine capable of mass production by shipyards not previously engaged in building submarines.

Several SSK preliminary designs were developed; the smallest would have had a surface displacement of only 250 tons, with a large sonar, minimal torpedo armament, and a crew of two officers and 12 enlisted men. The Chief of Naval Operations (CNO) initially accepted a proposal for a submarine of 450 tons with a pressure hull 14 feet (4.27 m) in diameter, but further study by the Submarine Officers Conference revealed that the submerged endurance of this submarine would be wholly inadequate.[27] To provide sufficient endurance the SSK characteristics ultimately approved by the CNO, on 27 May 1948, provided for a surface displacement of 740 tons—close to the German Type VII—with a pressure hull diameter of 15½ feet (4.65 m).[28]

The principal SSK sonar was the large BQR-4, the first array sonar developed by the U.S. Navy. Produced by the Edo Corporation, this was an enlarged version of the GHG/BQR-2 sonar. The BQR-4 had 58 hydrophones, each ten feet (3.0 m) high, mounted in a circular arrangement, similar to the BQR-2. These both had significant advantages over earlier, simple, horizontal-line hydrophones. It was more sensitive to the direction of a target, and, the electronic steering (by directing the sonar beams) rather than being mechanically trained was a quieter process.

Early SSK design sketches showed an array of the BQR-4 hydrophones ten-feet (3-m) long wrapped around the submarine's sail structure. The final SSK configuration placed the sonar in a dome at the extreme bow of the submarine, as far as possible from the noise-generating machinery and propellers of the submarine. The estimated passive (listening) range of the BQR-4 was up to 20 n.miles (37 km) against a surfaced or snorkeling submarine (i.e., using diesel engines). Under perfect conditions, ranges out to 35 n.miles (65 km) were expected.[29] The BQR-4 could track targets to within five degrees of accuracy. Of course, effective U.S. torpedo ranges at the time were a few thousand yards, far short of expected target detection ranges. (See below.) And, the *SSK*'s slow submerged speed—8.5 knots—would make it difficult to close with targets detected at greater ranges.

The massive BQR-4 in the SSKs would be supplemented by the high-frequency BQR-2—a copy of the German GHG—mounted in a keel dome, as in the Type XXI. The BQR-2 had 48 hydrophones forming a circle eight feet (2.44 m) in diameter. It was credited with ranges up to ten n.miles (18.5 km) with a bearing accuracy of ⅒th of a degree, making it useful for fire control in torpedo attacks. Also fitted in the SSK would be the small BQR-3, an improved version of the U.S. Navy's wartime JT passive sonar, intended as a backup for the newer sets. The small, active BQS-3 sonar would be fitted to transmit an acoustic "ping" toward a target submarine to obtain a precise measurement of range. Also, a hydrophone suspended by cable from the submarine to provide long-range, non-directional listening was planned, but not installed. With some 1,000 feet (305 m) of cable, the hydrophone could be lowered away from submarine-generated noises. A key factor in SSK effectiveness was to be self-quieting, with very quiet refrigeration and air-conditioning equipment being specially developed.

A Navy analysis indicated that a "minimum" of 25 to 70 surface ships would be required on station per 100 n.miles (185 km) of barrier to pose more than a negligible threat to snorkeling submarines. In comparison, three to five SSKs per 100 miles could be expected to detect practically all of the transiting submarines.[30] The Navy's SSK proposal of 1948 to meet the perceived threat of 2,000 modern Soviet submarines in the 1960s called for 964 hunter-killer boats! This number included SSKs in transit to and from patrol areas, undergoing overhaul, and being rearmed:[31]

| Operating Area | On Station | Total |
|---|---|---|
| Greenland-Iceland-Scotland | 124 | 372 |
| Southern England-Spain | 86 | 258 |
| Northeast Pacific-Kamchatka Peninsula | 10 | 30 |
| Petropavlovsk | 6 | 18 |
| Kurile Islands | 30 | 90 |
| Kyushu (Japan)-China | 42 | 126 |
| Training | — | 70 |
| *Totals* | 298 | 964 |

SSK armament would consist of four bow torpedo tubes with eight torpedoes being carried. The submarine would carry straight-running Mk 16 torpedoes and the new, acoustic-homing Mk 35. The latter, which entered service in 1949, was primarily an anti-surface ship weapon. The Mk 16 had a speed of 46 knots and a range of 11,000 yards (10,060 m); the smaller Mk 35 had a speed of only 27 knots for 15,000 yards (13,700 m).

The tactics envisioned the killer submarines operating in forward areas, virtually motionless and hence noiseless when on their patrol station, seeking to detect Soviet submarines transiting to ocean areas. One method considered for hovering on station was to employ an anchor for buoyancy control. With an operating depth of 400 feet (120 m), the K-boats would be able to anchor in water as deep as 3,400 feet (1,040 m). The *SSK*s also were intended for operation in Arctic waters in the marginal-ice area, with fathometers being fitted in the keel and atop the sail.

The SSK concept provided for a retractable buoy for radio communications with other SSKs. Two submarines in contact would be able to solve torpedo fire control solutions using only bearings (i.e., passive sonar).

Congress authorized construction of the first SSK—to be "named" *K1*—in fiscal year 1948 (which began on 30 June 1947) and two more were authorized the following year.[32] These three K-boats were authorized in place of one additional *Tang*-class submarine. To mature the K-boat design before it was turned over to non-submarine shipyards, the *K1* was ordered from the privately owned Electric Boat yard (Groton, Connecticut), while the *K2* and *K3* were ordered from the Mare Island Naval Shipyard (near San Francisco). Proposals to build some of this trio at the New York Shipbuilding yard in Camden, New Jersey, did not work out.

In 1948 the Navy planned a most ambitious construction program for both the *K1* and *Tang* classes; these submarines would be in addition to several special-purpose undersea craft and a large fleet boat conversion program. Construction rates of the *Tang*-class would increase in 1960 to begin replacing GUPPYs that would be retired.[33]

| Fiscal Year | *K1* Class | *Tang* Class |
|---|---|---|
| 1947 | — | 2 |
| 1948 | 1 | 2 |
| 1949 | 2 | 2 |
| 1950 | 5 | — |
| 1951 | 3 | 2 |
| 1952 | 2 | 2 |
| 1953 | 2 | 2 |
| 1954 | 3 | 2 |
| 1955–1959 | 2 annum | 2 annum |
| 1960 | 2 | 6 |

However, after the 1948–1949 programs, further construction of the *K1* class was delayed until the first units were evaluated in fleet operations. As an interim step, seven fleet boats of the *Gato* class were converted to an SSK configuration, their principal alteration being installation of the large BQR-4 sonar and special sound-isolation mountings provided for their auxiliary and propulsion machinery. These large SSKs proved to be highly effective hunter-killers for their time, being superior in performance and habitability in comparison to the small K-boats. (More fleet boat/SSK conversions were planned, but not undertaken, because of the use of nuclear-propelled submarines for the hunter-killer role.)

In January 1949 the Chief of Naval Operations directed both the Atlantic and Pacific Fleets to create a submarine division to develop techniques for submarines to detect and destroy enemy undersea craft. Named Project Kayo, this led to the establishment of Submarine Development Group (SubDevGru) 2 in the Atlantic and SubDevGru 11 in the Pacific, with the sole mission of solving submarine ASW problems. Initially each group was assigned two fleet submarines and two GUPPY conversions. Both individual and multiple-submarine tactics were investigated under Kayo, often in Arctic waters.

A multitude of problems were identified by Project Kayo and by other ASW exercises. Submarine communications were found to be completely unsatisfactory, preventing coordinated efforts with aircraft and surface ships. Also, in the SSK role submarines only could detect diesel submarines that were moving at high speeds (over eight knots). Although Project Kayo was soon reduced to only SubDevGru 2, the Korean War, which erupted in June 1950, increased interest in submarine ASW. The three submarines of the *K1* class were completed in 1951–1952. Their anti-submarine performance was most impressive for the time: In exercises off Bermuda in 1952, the prototype *K1* detected a snorkeling submarine at 30 n.miles (55.5 m) and was able to track the target for five hours. However, the small K-boats were cramped and uncomfortable, and their slow transit speed limited their being sent into forward areas during a crisis or when there were intelligence indications of a possible conflict. Criticism of their range and endurance was met by proposals to base the K-boats at friendly European and Asian ports within 1,000 n.miles (1,853 km) of their patrol areas, and to employ submarine tankers (SSO) to refuel them—while submerged—on station.[34]

But their ability to detect a snorkeling submarine at long range was not enough. If Soviet submarines could transit through critical areas submerged on battery/electric power or had a closed-cycle propulsion system, they would likely evade K-boat detection. And the SSKs would be severely limited by weaknesses in SSK-to-SSK communications and the short range of their torpedoes.

An epitaph to the K-boats was written by Captain Ned Kellogg, who had served aboard the *K3* as a young officer:[35]

> Some of the good features of the class were its simplicity . . . . It had a dry induction mast, no main induction valve . . . no conning tower and therefore no safety tank, no low pressure blower for the ballast tanks, instead a diesel exhaust blow system similar to what the German submarine force used during World War II, a simple remotely operated electrical control panel which kept the battery always available for propulsion, the

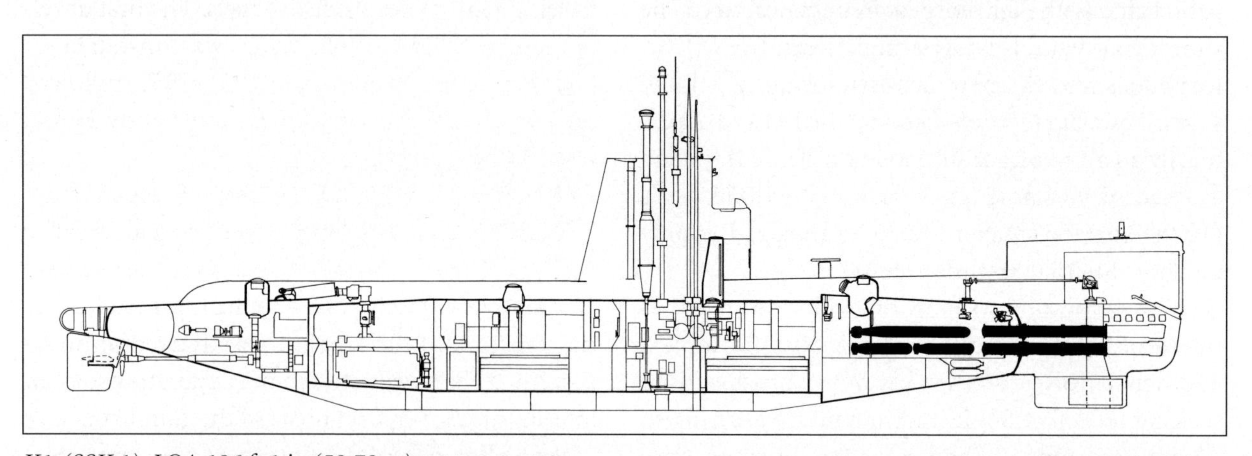

K1 *(SSK 1). LOA 196 ft 1 in (59.78 m)* (©A.D. Baker, III)

newest fire control system . . . all AC power rather than split between AC and DC.[36]

But the submarine suffered from having diesel engines that were difficult to maintain, an unreliable and insufficient fresh water plant, undependable electrical generators, and slow speed. Kellogg's conclusion: "You just can't build an inexpensive submarine that is worth much at all, unless you man her with a crew of courage and heart."[37]

## U.S. Weapons and Sensors

The Navy's Operational Evaluation Group reported to the Submarine Officers Conference in 1952 that if Soviet submarines transited SSK barriers submerged on battery (at slow speeds) rather than on the surface or while snorkeling, the number of SSKs required for effective barriers would increase by a factor of ten! Thus the SSK program, intended to provide 15 to 20 small SSKs by the mid-1950s and "several hundred" hunter-killer submarines in wartime, was halted with three purpose-built SSKs plus the seven older, fleet-boat SSK conversions. Subsequent exercises demonstrated that the greater underwater speed and endurance (due to shape and large batteries) of a GUPPY fitted with BQR-2 sonar gave that submarine the same overall effectiveness as the *K1* or a converted fleet boat/SSK.[38] Coupled with improvements in torpedoes and the plotting/tracking of submarine targets, by the mid-1950s the effectiveness of the GUPPYs and the re-engined *Tang*s caused the concept of a specialized SSK to be abandoned until the nuclear-propelled submarine appeared on the scene.[39]

Acoustic homing torpedoes that would seek submarine propeller noises were introduced by the German and U.S. Navies in World War II.[40] An SSK could detect a high-speed or snorkeling submarine at greater ranges than the sonar fitted in the torpedo. Accordingly, in 1950 the Bureau of Ordnance awarded a contract for development of a wire-guided torpedo for the SSK that initially would be guided by the launching submarine until the torpedo's sonar could acquire the target. However, the Bureau initially emphasized that it was only a research program and would not be expedited.[41]

Limited torpedo improvement was achieved with the belated acquisition of a wire-guided homing torpedo. In 1956 some 120 Mk 27 Mod 4 acoustic homing torpedoes of wartime design were refitted with wire guidance and redesignated Mk 39 Mod 1 for the purpose of development and fleet familiarization. The nuclear-propelled submarine *Nautilus* (SSN 571) had gone to sea in January 1955, before the Mk 27 had been upgraded. It quickly became obvious that even the new Mk 37 torpedo—in development at the time—would not be able to counter nuclear submarines.

The Mk 37 would have wire guidance to enable the launching submarine's sonar to guide the weapon toward an enemy submarine; the wire would then break, and the torpedo's sonar would seek out the target. The basic Mk 37 was rated at 26 knots; with wire guidance the Mk 37 had a speed of only 14.7 knots, too slow to counter the expected Soviet nuclear submarines. Accordingly, the Navy initiated development of the Mk 45 ASTOR (Anti-Submarine Torpedo) nuclear torpedo and the SUBROC (Submarine Rocket) nuclear depth bomb.[42]

While hunter-killer submarines were increasingly being considered as the "best" ASW platform by some U.S. Navy officials, major resources also were being allocated to advanced ASW surface ships, to ASW aircraft—both carrier-based and land-based, and to the seafloor Sound Surveillance System (SOSUS).[43] The last was a network of hydrophones emplaced on the ocean floor to detect low-frequency noise sources. During World War II the U.S., British, and Soviet Navies installed limited-capability acoustic arrays on the ocean floor in shallow waters, primarily at the entrances to harbors. After the war the U.S. Navy began the development of deep-ocean arrays. The first developmental SOSUS-type array was installed at Eleuthera in the Bahamas in 1951–1952, followed by a small, experimental array off Sandy Hook, south of Manhattan.[44]

The first operational test of SOSUS was conducted from 26 April to 7 June 1954 during an exercise labeled ASDevEx 1-54. Additional SOSUS arrays were placed along the Atlantic coast and, from 1958, along the Pacific coast of the United States and off of Hawaii. In 1960 arrays were emplaced in Hudson Bay to detect Soviet submarines operating in that area.[45] Overseas installations followed in areas that Soviet submarines were expected to transit, for example, the North Cape, the Greenland-Iceland-United Kingdom (GIUK) gaps, the deep channel running north to south in the Atlantic basin, and near the straits leading into the Sea of Okhotsk in the Far East.[46]

The SOSUS arrays are linked to shore stations by cable. At the stations—called "Naval Facilities," or NavFacs—technicians scrutinize the readouts of the cacophony of ocean noises and attempt to discern sounds, or "signatures," of submarine types and even specific boats. The first NavFacs were established in 1954 at the Ramey military facilities in Puerto Rico, Grand Turk, and San Salvador. Eventually, at the height of the Cold War, there were some 20 NavFacs located around the world.

The NavFacs would advise or cue regional ASW commanders of their detections to enable ASW air, surface, and submarine forces to be directed toward suspected Soviet submarine locations. SOSUS was vital for the effective use of aircraft and submarines in the anti-submarine role because of the limited search rates (area per unit time) of those platforms. According to a 1954 report on the performance of the SOSUS in the Bahamas, "Ranges out to 600 miles [1,110 km] were obtained. However, this was not the average. The average was about three to four hundred miles [555 to 740 km] reliable."[47] Later improvements in arrays and processing increased detection in SOSUS ranges. Published material indicates that in optimum SOSUS areas a submarine could be localized to within a radius of 50 n.miles (92.6 km).[48] Significantly, it was found that SOSUS also could detect and track overflying aircraft![49]

The SOSUS arrays were being planted by surface cable-laying ships, probably assisted from 1970 onward by the nuclear-propelled submersible *NR-1*. Soviet naval and research ships observed these operations with considerable interest. SOSUS was vulnerable in peacetime as well as in war. In 1978 alone the Navy asked Congress for $191 million for the repair of SOSUS cables believed to have been damaged by fishing craft, with the suspicion that they may have been cut intentionally by Soviet trawlers. As early as 26 February 1959 the Soviet fishing trawler *Novorossisk* was accused by the U.S. government of cutting American transatlantic communications cables off Newfoundland, Canada. The trawler was boarded by Navy personnel from the U.S. escort ship *Roy O. Hale*, but no evidence of cable cutting was found. There were press reports that the *Novorossisk* had actually cut SOSUS cables.[50]

By the mid-1950s the GUPPY and fleet boat conversion programs and the start-up of the *Tang* and *K1* programs were employing more skilled workers and using more industrial facilities than had the U.S. submarine construction effort of World War II. Seven shipyards were involved—the Electric Boat yard and the naval shipyards at Boston (Massachusetts), Charleston (South Carolina), Mare Island (California), Philadelphia (Pennsylvania), Portsmouth (Maine), and San Francisco (California).

The massive design effort to support submarine conversions and new construction in the postwar era was mostly undertaken by the design offices at the Portsmouth and Electric Boat (EB) yards, and to a lesser degree at Mare Island. Through World War II submarine design had been carried out by the Bureau of Ships in Washington, D.C. Known as "BuShips," the bureau had overall responsibility for the procurement of all Navy surface ships and submarines as well as their sensors.

A series of Navy Department reorganizations in the early postwar period led to changes in submarine design responsibility and instituted a competitive attitude between the Portsmouth and EB designers. This created a healthy competition for the most efficient designs and processes. By the mid-1950s, however, new designs were assigned alternatively to the two yards until the Portsmouth yard was phased out of new construction work in the late 1960s, at which time the yard was almost closed down completely.

## Soviet Projects

As with the U.S. Navy, after victory over the Axis was achieved in 1945, the Soviet Navy continued to take delivery of submarines designed before the war. However, Soviet submarine production during the war was slowed by the German occupation of much of the western portion of the country, including the Ukraine. The shipbuilding yards of the Black Sea were devastated. The shipbuilding center of Leningrad (St. Petersburg) was surrounded and besieged by the Germans for some 900 days. There were major construction yards that were away from the front at Molotovsk (Severodvinsk) in the Arctic, at Komsomol'sk on the Amur River in the Far East, and at Gor'kiy, far inland on the Volga River. But even their construction was impaired by the German siege of Leningrad and occupation of other western areas, which halted the flow of steel and components to these yards.

Those submarine designers of bureau TsKB-18 (later Rubin) in Leningrad who were not called to the front—with Leningrad being the front—continued both limited design efforts for shipyards that were able to carry on submarine work, and some new designs. That bureau, formally established in 1926, had designed all Soviet submarines through World War II except for a small effort under the auspices of the security police, the NKVD. (See Appendix D.)

During the period of conflict—June 1941 to August 1945—Soviet shipyards produced 54 submarines of several classes. Whereas the U.S. Navy built only fleet boats during the war, multiple operational requirements led the Soviets to build submarines of several designs, with the construction of four classes being continued after the war. (See table 2-2.) The war programs produced mainly coastal submarines of the *malyutka* (small) designs; mid-size and large submarines were produced in smaller numbers. Sixty-two submarines of wartime programs were completed after the war.

TABLE 2-2
**Soviet War Programs Completed, August 1945–1951**

| Units | Type | Series* | First Unit | Displacement** |
|---|---|---|---|---|
| 7 | S | IX-*bis* | 1939 | 856/1,090 tons |
| 1 | Shch | X-*bis* | 1939 | 593/705 tons |
| 53 | M*** | XV | 1943 | 283/350 tons |
| 1 | Project | 95 | 1946 | —/102 tons |

Notes: * *bis* indicates modification to basic design.
** Surface/submerged displacement.
*** Referred to as *Malyutka* ("small") submarines.

The rehabilitation of the Soviet shipbuilding industry had been given high priority by Soviet dictator Josef Stalin in the postwar period. He approved the building of a major fleet, including aircraft carriers, battle cruisers, lesser cruisers, destroyers, and submarines. At the end of the war, Soviet submarine designers and researchers concentrated on increasing submerged speed, which was considered to be a decisive factor in naval warfare. To achieve higher speeds for longer duration, the following approaches were taken: (1) increasing the propulsion (electric) motor power and the storage battery capacity; (2) providing diesel engine operation while submerged; and (3) employing turbine plants for submerged operations.[51]

The models for these efforts were mainly German submarines, especially a new Type VIIC sunk by Soviet forces in the Baltic in 1944 and subsequently salvaged;[52] the submarines found in German shipyards as Soviet troops occupied the eastern coast of the Baltic Sea; and the ten U-boats formally assigned to the Soviet Navy under the Potsdam Agreement. These ten were:

| | |
|---|---|
| Type VIIC | *U-1057, U-1058, U-1064, U-1305* |
| Type IXC | *U-1231* |
| Type XXI | *U-2529, U-3035, U-3041, U-3515* |
| Type XXIII | *U-2353* |

Several Western intelligence reports credit the Soviets with having actually acquired several more Type XXI submarines: an intelligence review by the U.S. Joint Intelligence Committee for the Joint Chiefs of Staff in January 1948 estimated that at the time the Soviets had 15 Type XXIs operational, could complete another 6 within two months, and could assemble 39 more within 18 months from prefabricated components.[53] Several German factories producing Type XXI components and the Schichau assembly yard at Danzig were occupied by the Soviets when the war ended. There were a number of unfinished Type XXI U-boats in the Danzig yard, plus numerous sections and components. Apparently the unfinished *U-3538* through *U-3542* were already on the assembly ways at Schichau when Soviet troops entered the yard. In addition, there were relatively complete sections for at least eight additional Type XXI submarines in the yard.

Western estimates of the fate of these undersea craft are conflicting. One German engineer contends that at least two near-complete Type XXI submarines at Schichau were launched after the war under Soviet direction and towed to Kronshtadt.[54]

One of his colleagues, however, wrote that "The unfinished U-Boats were left in that state, but everything appertaining to them and their equipment was submitted to exhaustive study and research. . . ."[55] This appears to be closer to the truth; Soviet records and statements by senior designers indicate that the Type XXIs were employed in trials and tests and then scuttled or expended in weapon tests, as were the unfinished submarine sections.[56]

In addition, the Soviets had the opportunity to examine three contemporary British submarines of the "U" class that had been transferred to the USSR in 1944.[57] Soviet sonar development also took advantage of technology available from British surface warships that had been loaned to the USSR during the war.

This bounty of submarine material was examined minutely at a number of research institutes as well as at submarine design bureau TsKB-18. The first postwar design undertaken at TsKB-18 was Project 614, essentially a copy of the Type XXI. The design was not completed, because it was considered to have insufficient "reliability," that is, in the Soviet view, one combat patrol into a high-threat area did not justify the cost of the submarine.[58] Many of the U-boat components and structures had been developed on the basis of a short service life, especially the high-pressure piping and batteries with thin lead plates.

But the Project 614 design process did cause intensive study of the construction and engineering solutions employed by German designers. This led to a number of submarine technologies being pursued in the USSR, especially new types of steel and electric welding, which would permit a doubling of wartime operating depths, new ship control concepts, low-noise propellers, machinery shock installations, sonar, anechoic coatings for submarine hulls to reduce active sonar detection, and radar-absorbing materials to reduce detection of a snorkel head.

Of special interest, during World War II, the German Navy had introduced the *Alberich* rubber-like laminated hull coating to reduce the effectiveness of the British active sonar (called ASDIC). It became operational in 1944.[59] Although investigated in the United States after the war, such coatings were not pursued because of the difficulty in keeping the covering attached to the hull. In the 1950s the Soviet Union began providing Series XII Malyutka-type coastal submarines with anti-sonar or anechoic coatings. These evolved into multi-purpose coatings, which also could absorb internal machinery noise. This was particularly feasible in double-hull submarines, where coatings could be placed on multiple surfaces.

Many Type XXI characteristics were incorporated in TsKB-18's Project 613 submarine—known in the West as "Whiskey."[60] This design had been initiated in 1942 as Project 608, but was rejected by the naval high command because it displaced 50 tons more than specified in requirements. The redesign of Project 608 into 613 was begun in 1946 under the supervision of Captain 1st Rank Vladimir N.

*An early Project 613/Whiskey with twin 25-mm guns mounted forward of the conning tower; the twin 57-mm mount aft of the conning tower has been deleted (note the widened deck). The bow diving planes are retracted as the submarine moves through the water at high speed.* (French Navy)

*A Whiskey with modified fairwater and guns removed. More of these submarines were constructed than any other post–World War II design. They formed the backbone of the Soviet submarine force and served at sea in six other navies. Pennant numbers disappeared from Soviet submarines in the 1960s.* (U.S. Navy)

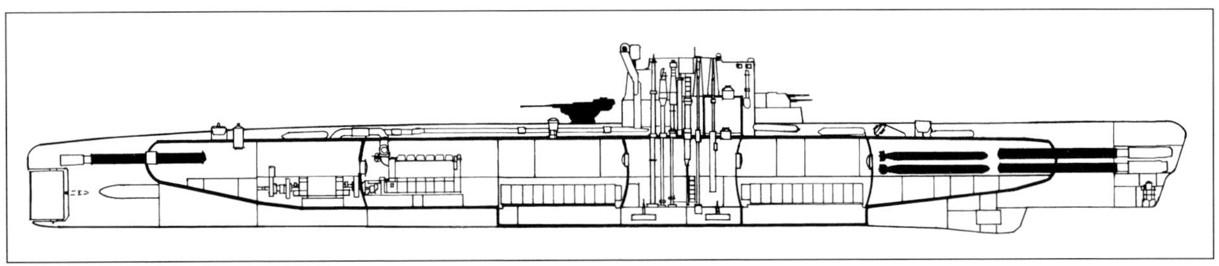

*Project 613/Whiskey SS as built with deck guns. LOA 249 ft 2 in (75.95 m)* (©A.D. Baker, III)

Peregudov, who incorporated several features derived after studies of Type VIIC and Type XXI U-boats.[61] One of the former, the *U-250,* had been sunk by the Soviets in the Gulf of Finland on 30 July 1944 and subsequently salvaged and carefully examined.[62]

The hull and fairwater of Project 613 were streamlined, and the stern was given a "knife" configuration, with the large rudder positioned aft of the twin propellers. The propeller shafts were supported outside of the hull by horizontal stabilizers rather than by struts (as used in most U.S. submarines). The stern diving (horizontal) planes were aft of the propellers. The "knife" arrangement provided the possibility of a more maneuverable submarine than the U.S. Fleet/GUPPY configurations.

A small attack center, or conning tower, was fitted in the Project 613 fairwater, a feature deleted from the Type XXI. When retracted, the various periscopes and masts were housed completely within the superstructure.

Propulsion on the surface was provided by two diesel engines with a total output of 4,000 horsepower; submerged propulsion normally was by two main electric motors producing 2,700 horsepower plus two smaller motors that provided 100 horsepower for silent or economical running. This feature—derived from the German "creeping" motors—was the first German feature to be incorporated into Soviet submarine designs.[63] Two large groups of batteries with 112 cells each were installed. Later a snorkel system would be installed for submerged operation of the diesel engines.[64] This propulsion system could drive the Whiskey at 18.25 knots on the surface and 13 knots submerged.

The principal combat capability of the Whiskey was the six torpedo tubes—four bow and two stern,

*A U.S. sailor stands guard on the unfinished "knife" stern section of a Type XXI submarine at the Deshimag shipyard in Bremen. The propeller shafts pass through the horizontal stabilizers. The stern diving (horizontal) planes and rudder were fitted aft of the propellers.* (Imperial War Museum)

with six reloads in the forward torpedo room—a total of 12 torpedoes. This torpedo loadout was small in comparison to U.S. submarines and the Type XXI, but was comparable to the five tubes and 15 torpedoes in the Type VIIC U-boat. The tubes were fitted with a pneumatic, wakeless firing system that could launch torpedoes from the surface down to almost 100 feet (30 m); in subsequent upgrades firing depth was increased to 230 feet (70 m). Previously the USSR, as other nations, had produced specialized minelaying submarines.[65] Beginning with the Whiskey, Soviet submarines could also lay mines through their torpedo tubes (as could U.S. submarines). In the minelaying role a Whiskey could have a loadout of two torpedoes for self-defense plus 20 tube-launched mines.

Early Project 608/613 designs had provided for a twin 76-mm gun mount for engaging surface ships. With the plan to conduct most or all of a combat patrol submerged, the gun armament was reduced to a twin 57-mm anti-aircraft mount aft of the conning tower and a twin 25-mm anti-aircraft mount on a forward step of the tower. (Guns were installed in Soviet submarines until 1956.)

With the use of a completely welded pressure hull using SKhL-4 alloy steel coupled with the design of its pressure hull, the Whiskey had a test depth of 655 feet (200 m) and a working depth of 560 feet (170 m).[66] This was considerably deeper than the Type XXI as well as the new U.S. *K1* class, and almost as deep as the *Tang* class. Unfortunately, in achieving the greatest feasible operating depth while restricting displacement, the designers excessively constrained the crew accommodations in the Whiskey (as in subsequent diesel-electric classes).

The Project 613/Whiskey introduced a new level of underwater performance to Soviet undersea craft, incorporating many German design features that would be found in future generations of Soviet submarines. The final TsKB-18 contract design was approved by the Navy in 1948, and construction began shortly afterward at the Krasnoye Sormovo shipyard in the inland city of Gor'kiy, some 200 miles (320 km) to the east of Moscow.[67] Submarines built at Gor'kiy would be taken down the Volga River by transporter dock for completion at Caspian and Black Sea yards.

The lead submarine of Project 613—the *S-80*—was laid down at Gor'kiy on 13 March 1950, followed by additional production at the Baltisky (Baltic) shipyard in Leningrad, the Chernomorskiy yard in Nikolayev on the Black Sea, and the Leninsky Komsomol yard at Komsomol'sk in the Far East. Automatic welding and prefabrication were widely used in Project 613 construction.

The *S-80* was put into the water—launched from a horizontal assembly facility—on 21 October 1950 when 70 percent complete. She was immediately transported by barge down the Volga River to the port of Baku on the Caspian Sea, arriving on 1 November. After completion and extensive trials, the *S-80* was commissioned on 2 December 1951, a very impressive peacetime accomplishment.

The massive Project 613/Whiskey program produced 215 submarines for the Soviet Navy through 1958 (i.e., an average of more than 2½ submarines per month of this design):

| Shipyard | | Proj. 613 Submarines | Completed |
|---|---|---|---|
| No. 112 | Krasnoye Sormovo | 113 | 1951–1956 |
| No. 444 | Chernomorskiy | 72 | 1952–1957 |
| No. 189 | Baltisky | 19 | 1953–1958 |
| No. 199 | Leninsky Komsomol | 11 | 1954–1957 |

This was the largest submarine program in Soviet history, exceeding in tonnage the combined programs of the Soviet era up to that time. Indeed, in number of hulls, Project 613 would be the world's largest submarine program of the Cold War era. (According to available records, a total of 340 submarines of this design were planned.)

In 1954 the documentation for Project 613 construction was given to China, and three additional submarines were fabricated in the USSR, dismantled, and shipped to China for assembly at Shanghai's Jiangnan shipyard. China then built 15 submarines at the inland shipyard at Wuhan on the Yangtze River, initially using Soviet-provided steel plates, sonar, armament, and other equipment. Soviet-built units also were transferred to Bulgaria (2), Egypt (8), Indonesia (14), North Korea (4), Poland (4), and Syria (1); Cuba and Syria each received one unit as a stationary battery charging platform to support other submarines. The Soviet Union transferred two submarines to Albania in 1960 and two additional units were seized in port by the Albanian government when relations with the USSR were broken for ideological reasons in 1961.[68]

The Project 613 submarines would form the basis for the first Soviet cruise missile submarines and would be configured for a number of specialized and research roles. Four submarines were converted to a radar picket (SSR) configuration at the Krasnoye Sormovo shipyard in Gor'kiy, with the first completed in 1957. These craft were fitted with the large Kasatka air-search radar (NATO Boat Sail) as well as additional radio equipment. Designated Project 640 (NATO Canvas Bag), these submarines initially were based at Baku on the Caspian Sea, apparently to provide air-defense radar coverage for that region.[69] One of the Project 640 submarines was provided with a satellite link at the Sevmorzavod shipyard in Sevastopol in 1966 (Project 640Ts).

In 1960 a submarine was converted to the Project 613S configuration to provide an advanced rescue system. That work also was undertaken at Gor'kiy. In 1962, at the same yard, another Project 613 submarine was modified to Project 666, a rescue submarine with a towed underwater chamber that had a depth of 655 feet (200 m). In 1969 that submarine was again modified to test prolonged exposure to pressure. One submarine was rebuilt to the Project 613Eh configuration to test a closed-cycle propulsion system. (See Chapter 13.)

And in the late 1950s one of these submarines, the *S-148,* was disarmed and converted to a civilian

*A Project 640/Whiskey radar picket submarine with her air-search radar extended. The lengthened conning tower has the Quad Loop RDF antenna moved to the forward end of the fairwater. When folded (not retracted), the radar was covered by a canvas bag, giving rise to the NATO code name for the SSR variant.* (U.S. Navy)

research ship. Renamed *Severyanka,* she was operated by the All-Union Institute for the Study of Fisheries and Oceanography with a civilian crew.[70]

Two Project 613 submarines were lost—the *S-80,* a radar picket craft, in the Barents Sea in 1961, and the *S-178* in the Pacific in 1981.

*An Mk 45 ASTOR torpedo—the West's only nuclear torpedo—being loaded in a U.S. submarine in 1960. The contra-rotating propellers have a circular ring (far right), which protects the guidance wire from the props. Almost 19 feet long and weighing more than a ton, the torpedo had a range of some 15,000 yards.* (U.S. Navy)

## Nuclear Torpedoes

The Whiskey-class submarine *S-144* made the first launch of a T-5 development torpedo, the first Soviet torpedo with a nuclear warhead.[71] Nuclear warheads in anti-ship torpedoes would improve their "kill" radius, meaning that a direct hit on an enemy ship would not be required. A nuclear warhead could thus compensate for poor acoustic homing by the torpedo or for last-minute maneuvering by the target, and later for overcoming countermeasures or decoys that could confuse a torpedo's guidance. The T-5 was initiated simultaneous with the massive T-15 *land-attack* nuclear torpedo. (See Chapter 5.)

The initial state test of the RDS-9 nuclear warhead for the T-5 torpedo took place at the Semipalatinsk experimental range in Kazakhstan in October 1954. It was a failure. The next test of the RDS-9 occurred on 21 September 1955 at Novaya Zemlya in the Arctic—the first underwater nuclear explosion in the USSR.[72]

Work on the T-5 continued, and on 10 October 1957 the Project 613 submarine *S-144* carried out the first test launch of the T-5. Again the test was at the Novaya Zemlya range (Chernaya Bay), with several discarded submarines used as targets. The nuclear explosion had a yield of ten kilotons at a distance of 6.2 miles (10 km) from the launching submarine.[73] The target submarines *S-20* and *S-34* were sunk and the *S-19* was heavily damaged.

The T-5 became the first nuclear weapon to enter service in Soviet submarines, becoming operational in 1958 as the Type 53-58. It was a 21-inch (533-mm) diameter weapon. Also in this period, a "universal" nuclear warhead, the ASB-30, was developed for 21-inch torpedoes that could be fitted to specific torpedoes in place of high-explosive warheads *while the submarine was at sea.* These were placed on board submarines beginning in late 1962. Initially two ASB-30 warheads were provided to each submarine pending the availability of torpedoes with nuclear warheads. Additional torpedoes were developed with nuclear warheads, including 25½-inch (650-mm) weapons. (See Chapter 18.)

In 1960—two years after the year the first Soviet nuclear torpedo became operational—the nuclear Mk 45 ASTOR (Anti-Submarine Torpedo) entered service in U.S. submarines. There was American interest in a nuclear torpedo as early as 1943, when Captain William S. Parsons, head of the ordnance division of the Manhattan (atomic bomb) project, proposed providing a "gun," or uranium-type nuclear warhead, in the Mk 13 aircraft-launched torpedo. In later variants the Mk 13 was a 2,250-pound (1,021-kg) weapon with a high-explosive warhead of 600 pounds (272 kg).[74]

The technical director of the Manhattan Project, J. Robert Oppenheimer, opposed Parsons's proposal: the Los Alamos laboratory was already strained with A-bomb projects and "we have no theoretical encouragement to believe that it will be an effective weapon, and we have what I regard as a reliable answer to the effect that it will produce inadequate water blast."[75] Surprisingly, Parsons did not propose the warhead for a submarine-launched torpedo, which could have been larger and have had a greater range.

Immediately after the war several U.S. submarine officers and engineers proposed nuclear torpedoes, and some preliminary research was undertaken on an "Atomic warhead which can be attached to a torpedo body and launched from a standard size submarine tube to provide a weapon for the neutralization of enemy harbors."[76]

Although there were later discussions of nuclear torpedoes within the U.S. Navy, none appeared until the Mk 45 ASTOR, the only nuclear torpedo produced in the West. A 19-inch (482.5-mm) diameter weapon, the Mk 45 carried a W34 warhead of 20 kilotons. It was launched from the standard 21-inch torpedo tubes and had a speed of 40 knots, with a maximum range of 15,000 yards (13.65 km). The wire-guided ASTOR had no homing capability and no contact or influence exploder; it was guided and detonated by signals sent from the submarine through its trailing wire. In a form of gallows humor, U.S. submariners often cited the ASTOR as having a probability of kill (pK) of *two*—the target submarine *and* the launching submarine!

Even as the Project 613/Whiskey design was being completed, in 1947–1948 TsKB-18 undertook the design of a larger submarine under chief designer S. A. Yegorov.[77] Project 611—known in the West as the Zulu—had a surface displacement of 1,830 tons and length of almost 297 feet (90.5 m). This was the largest submarine to be built in the USSR after the 18 K-class "cruisers" (Series XIV) that joined the fleet from 1940 to 1947. (Much larger submarine projects had been proposed; see Chapter 14.) The Zulu was generally similar in size and weapon capabilities to U.S. fleet submarines, developed more than a decade earlier, except that the Soviet

*A Project 611/Zulu IV showing the clean lines of this submarine. Production of this class as a torpedo-attack submarine was truncated, with the design providing the basis for the world's first ballistic missile submarines.*

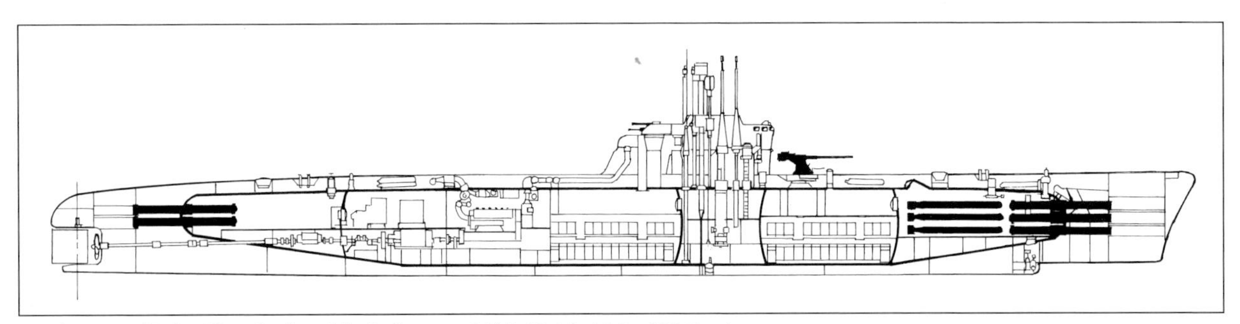

*Project 611/Zulu SS as built with deck guns. LOA 296 ft 10 in (90.5 m)* (©A.D. Baker, III)

craft was superior in underwater performance—speed, depth, and maneuverability.

In addition to ten torpedo tubes with 22 torpedoes (or an equivalent load of mines), the Project 611/Zulu was intended to carry a twin 76-mm gun mount; instead, the first units were armed with a twin 57-mm anti-aircraft mount and a twin 25-mm mount. These were fitted only to the first few ships and were soon deleted. (Removing the guns increased the underwater speed by almost one knot.)

Project 611 again reflected technologies gleaned from the Type XXI U-boat, refined by Soviet designers and adapted to their requirements and limitations. For example, the power limits of available diesel engines led to the Zulu being fitted with three diesel engines to drive three propeller shafts. Project 611 was the first three-shaft submarine to be built in Russia in more than four decades.[78] There was an effort to mount the machinery on sound-absorbing devices and other efforts were employed to reduce machinery noises.

The lead ship, the *B-61,* was laid down on 10 January 1951 at the Sudomekh shipyard in Leningrad. Her completion was delayed because of several defects being found in other Soviet submarines, some resulting in sinkings. Changes included the means of blowing main ballast tanks in an emergency, the hydraulic system, and strengthening the stern because of the increased vibration when all three shafts were rotating. The *B-61* was launched in 1953 and accepted by the Navy on 31 December 1953, still a remarkable building time in view of the size and complexity of Project 611.

Early planning called for 30 submarines of Project 611. However, only 13 submarines were laid down at the Sudomekh yard, of which eight were delivered at Leningrad to the Navy. The other five ships were transferred in an uncompleted state through the Belomor-Baltic Canal system to the Molotovsk yard for completion. The Molotovsk yard, above the Arctic Circle near the port of Arkhangel'sk, was created by Stalin in the 1930s to produce battleships. Another 13 submarines of Project 611 were constructed at Molotovsk, the first ships to be completely built at the yard.

Of these 26 Project 611/Zulu-class submarines, 21 went to sea as torpedo-attack craft from 1953 to 1958, and five were completed or refitted as ballistic missile submarines. (See Chapter 7.) The torpedo-armed units of Project 611 and the large number of the Project 613/Whiskey class became the mainstay of the Soviet submarine force during the 1950s and 1960s, well into the nuclear era. Also, this class brought to seven the number of Soviet shipyards engaged in submarine construction:

| Shipyard | | Location |
|---|---|---|
| No. 112 | Krasnoye Sormovo | Gor'kiy (Nizhny Novgorod) |
| No. 189 | Baltic | Leningrad (St. Petersburg) |
| No. 194 | Admiralty | Leningrad (St. Petersburg) |
| No. 196 | Sudomekh | Leningrad (St. Petersburg) |
| No. 199 | Komsomol'sk | Komsomol'sk-on-Amur |
| No. 402 | Molotovsk | Molotovsk (Severodvinsk) |
| No. 444 | Chernomorsky | Nikolayev |

With these new submarines would come new torpedoes, among them acoustic homing torpedoes for use against surface ships as well as submarines. Soviet interest in acoustic torpedoes began after the German *U-250* was salvaged and found to have on board three T5 acoustic homing torpedoes. These became important for Allied as well as Soviet torpedo development, with British technicians being given access to the recovered T5 weapons following a specific request from Prime Minister Winston Churchill to Stalin.

After the war Soviet engineers gained access to more German torpedo technology, and in 1950 the first Soviet acoustic homing torpedo was accepted for service, the SAET-50.[79] This weapon was roughly the equivalent of the German T5 acoustic torpedo of 1943. In 1958 the first fully Soviet-developed ASW torpedo was accepted, the SET-53. It had a speed of 23 knots, with a range of 6,560 yards (6 km); the range was soon increased with improved batteries. The SET-53M variant of 1963, which had a silver-zinc battery, had a speed of 29 knots with a range of 15,310 yards (14 km).

While advanced diesel-electric submarines and new torpedoes were being produced in the United States and the Soviet Union from the late 1940s, more advanced concepts in submarine propulsion were also being developed.

The first-generation Cold War submarines of the Soviet Union and United States—including the highly specialized U.S. K-boats and the GUPPY conversions—borrowed heavily from German designs and technologies. The large number of GUPPY and Project 613/Whiskey submarines, as well as the smaller numbers of *Tang*s, K-boats, and Project 611/Zulu-class submarines, indicated that undersea craft would have a major role in the strategies of both the United States and the USSR in the Cold War.

During this period the role of advanced Soviet diesel-electric submarines remained largely the same as in World War II, that is, the destruction of enemy shipping and coastal defense. However, the role of U.S. submarines shifted to ASW in anticipation of Soviet exploitation of German U-boat technology and construction techniques that could flood the ocean with advanced submarines. Further, geographic factors would force Soviet submarines to reach open ocean areas through narrow straits, which, it was believed, would facilitate the use of hunter-killer submarines (SSKs) to intercept Soviet submarines en route to their operating areas.

These initial postwar submarine programs of both navies led to the creation of large submarine construction and component production industries in each nation.

TABLE 2-3
**Advanced Diesel-Electric Submarines**

| | U.S. *Tang* SS 563 | U.S. *K1* SSK 1 | Soviet Project 613 NATO Whiskey | Soviet Project 611 NATO Zulu |
|---|---|---|---|---|
| Operational | 1951 | 1951 | 1951 | 1954 |
| Displacement | | | | |
| surface | 1,821 tons | 765 tons | 1,055 tons | 1,830 tons |
| submerged | 2,260 tons | 1,160 tons | 1,350 tons | 2,600 tons |
| Length | 269 ft 2 in (82.06 m) | 196 ft 1 in (59.78 m) | 249 ft 2 in (75.95 m) | 296 ft 10in (90.5 m) |
| Beam | 27 ft 2 in (8.28 m) | 24 ft 7 in (7.5 m) | 20 ft 8 in (6.3 m) | 24 ft 7 in (7.5 m) |
| Draft | 18 ft (5.5 m) | 14 ft 5 in (4.4 m) | 15 ft 1 in (4.6 m) | 16 ft 5 in (5.0 m) |
| Diesel engines | 4 | 3 | 2 | 3 |
| horsepower | 3,400 | 1,125 | 4,000 | 6,000 |
| Electric motors | 2 | 2 | 2* | 2** |
| horsepower | 4,700 | 1,050 | 2,700 | 2,700 |
| Shafts | 2 | 2 | 2 | 3 |
| Speed surface | 15.5 knots | 13 knots | 18.25 knots | 17 knots |
| submerged | 18 knots | 8.5 knots | 13 knots | 15 knots |
| Range (n.miles/kts) surface | 11,500/10 | 8,580/10 | 22,000/9 | |
| submerged | 129/3 | 358/2 | 443/2 | |
| Test depth | 700 ft (215 m) | 400 ft (120 m) | 655 ft (200 m) | 655 ft (200 m) |
| Torpedo tubes*** | 6 533-mm B<br>2 533-mm S | 4 533-mm B | 4 533-mm B<br>2 533-mm S | 6 533-mm B<br>4 533-mm S |
| Torpedoes | 26 | 8 | 12 | 22 |
| Guns | nil | nil | 2 57-mm<br>2 25-mm | 2 57-mm<br>2 25-mm |
| Complement | 83 | 37 | 52 | 72 |

Notes: * Plus two silent electric motors; 50 hp each.
** Plus one silent electric motor; 2,700 hp.
*** Bow; Stern.

3

# Closed-Cycle Submarines

*HMS* Meteorite *at sea, evaluating the German Type XVIIB closed-cycle propulsion plant. After the war several navies were interested in the Walter propulsion system, with the Royal Navy operating the ex-*U-1407*. The U.S. and Royal navies shared information of German closed-cycle submarines.* (Royal Navy Submarine Museum)

The Type XXI electro submarine and its smaller companion—the Type XXIII—were developed by the German Navy as interim submarines. The "ultimate" submarine envisioned by *Grossadmiral* Karl Dönitz and his colleagues was a closed-cycle, or single-drive, submarine that would employ an air-independent propulsion system to drive the submarine on both the surface and underwater, the latter at speeds of some 25 knots or more.

The principal closed-cycle propulsion system was developed in the 1930s by Helmuth Walter at the Germania shipyard. The Walter system was relatively complex, being based on the decomposition of highly concentrated hydrogen-peroxide *(perhydrol).*[1] *The perhydrol was carried in a tank below the main pressure hull; it was forced by water pressure up to a porcelain-lined chamber, where it was brought in contact with the catalyst necessary to cause decomposition. This produced steam and oxygen at high temperature—1,765°F (963°C)—which then passed into a combustion chamber, where they combined to ignite diesel oil, while water was sprayed on the gas to decrease its temperature and create additional steam. The combination steam-gas was then piped to the turbine, and from there into a condenser, where the water was extracted and the residual carbon dioxide generated in the combustion chamber was drawn off. These waste products were exhausted overboard. (See diagram.)*

As noted earlier, the highly successful 1940 trials of the first Walter-propulsion submarine, the *V-80*, was followed by an order for four Type XVIIA development submarines. (See Chapter 1.) Of these, the *U-792* and *U-794* were commissioned in October 1943 and were also successful, reaching 20.25 knots submerged. The other pair, the *U-793* and *U-795*, were commissioned in April 1944. The *U-793* reached a submerged speed of 22 knots in March 1944 with Admiral Dönitz on board. Dönitz commented, "with more courage and confidence at the Naval Command we should have had this [U-boat] a year or two ago."[2] In June 1944 the *U-792* traveled a measured mile at 25 knots. These Type XVIIA boats were found to be remarkably easy to handle at high speeds.

However, the Walter boats were plagued by mechanical and maintenance problems. Also, efficiency was low and significant power was lost because of the increase in back pressure on the exhaust system as the submarine went deeper.

*The* U-1406 *high and dry after being salvaged. Note the hull openings for hydrogen-peroxide storage. Larger submarines propelled by the Walter closed-cycle plant were planned by the German Navy as the "ultimate" U-boats.* (U.S. Navy)

Construction of operational Walter boats—the Type XVIIB—was begun at the Blohm & Voss shipyard in Hamburg. The initial order was for 12 submarines, the *U-1405* through *U-1416.* This choice of building yard was unfortunate, according to U-boat practitioner and historian Günter Hessler.[3] Blohm & Voss was already struggling to cope with the Type XXI program, and the Walter boats were neglected. The Navy cut the order to six boats.

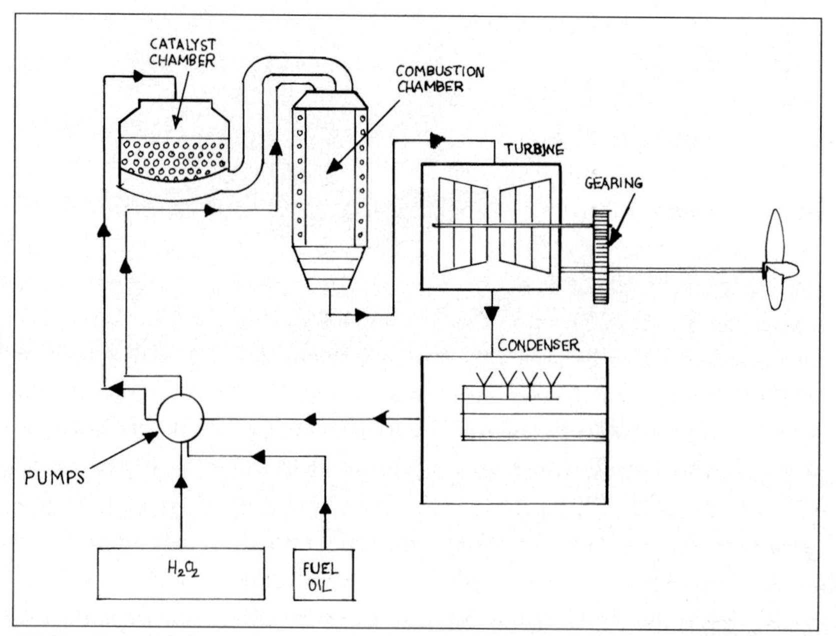


*Walter System Schematic*

Simultaneous with the Type XVIIB submarines of 312 tons surface displacement, the German Navy designed the larger, "Atlantic boats" of 1,485 tons that would have had two Walter turbines turning two shafts. Designated Type XVIII, two submarines were ordered, the *U-796* and *U-797;* but after the successful trials of the Type XVIIA U-boats, the larger submarines were discontinued in the spring of 1944. In their place, on 26 May 1944, contracts were placed for Walter boats of Type XXVIW to displace 842 tons. These single-shaft submarines were expected to achieve 24 knots submerged, with an endurance of almost 160 n.miles (300 km) at that speed.

The initial Type XXVIW program envisioned some 250 units to replace the Type XXI program. Anticipating a shortage of peroxide, the program was soon cut to 100 units, to begin with hull *U-4501.* All were to have been completed between March 1945 and October 1945, although as early as September 1944, it became apparent that the estimated supply of peroxide—whose production had to be shared with the Air Force—would be sufficient for only some 70 oceangoing Walter submarines.[4] The Type XXVIW submarines were started at the Blohm & Voss shipyard at Hamburg; few were begun and none was completed.

In addition to the Walter turbine submarines, several other closed-cycle concepts were examined and considered in wartime Germany. The Type XVIIK was to employ the Krieslauf closed-cycle diesel engine, which used stored oxygen for underwater operation; none of this design was built. The

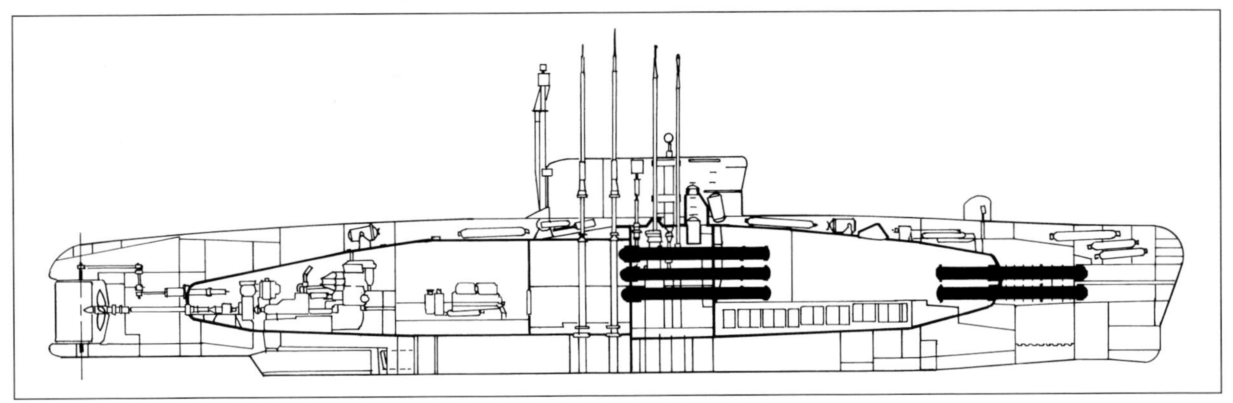

U-4501 *(Type XXVIW). LOA 184 ft 3 in (56.2 m)* (©A.D. Baker, III)

Type XXXVIW was to have had a two-shaft, closed-cycle system based on four fast-running diesel engines adapted from motor torpedo boats; again, stored oxygen would have provided the oxidizer for combustion. None of these U-boats was built.

Closed-cycle diesel systems had the advantage that oxygen was more readily available than was the peroxide needed for Walter propulsion. Still, oxygen systems had the disadvantage of the great weight of the oxygen containers when it was stored in a gaseous form. To solve this problem the possibility of producing oxygen on board was studied. Such a large number of closed-cycle propulsion schemes was considered because of the desire to produce the most efficient high-speed underwater warship with the limited resources available. In general, the Walter turbine offered high underwater speed, while the closed-cycle diesel systems offered great underwater endurance.

As the European war continued toward its desperate, violent climax of May 1945, fewer and fewer resources were available to German planners and commanders. Only three combat-configured Walter U-boats were completed: The Type XVIIB *U-1405* was completed in December 1944, the *U-1406* in February 1945, and the *U-1407* in March 1945. None became operational. But these U-boats marked the beginning of the next revolution in undersea warships. Although the Type XXI introduced the concept of a submarine designed entirely for underwater operation, it would be a closed-cycle (i.e., air-independent) propulsion plant that would bring about the "ultimate" undersea craft. In 1943–1945 the Walter turbine was the leading candidate for this revolution.

All three completed Type XVIIB submarines were scuttled in May 1945, the *U-1405* at Flensburg, and the *U-1406* and *U-1407* at Cuxhaven, all in territory occupied by British troops. At the Potsdam conference in July 1945, the *U-1406* was allocated to the United States and the *U-1407* to Britain, and both submarines were quickly raised from their shallow resting places.

British, Soviet, and U.S. naval officers and technicians picked through the ruins of the Third Reich, collecting blueprints and components of the Walter program. The unfinished, bomb-damaged *U-1408* and *U-1410* were found by the British at the Blohm & Voss shipyard in Hamburg. In Kiel, British troops took control of the Walterwerke, which was largely undamaged, and took into custody Professor Walter and his staff. U.S. and British technicians were soon interrogating them and making arrangements for their relocation to the West.

## Anglo-American Closed-Cycle Submarines

The U.S. Navy did not recondition and operate the *U-1406*, as it had the two Type XXI submarines. The Royal Navy did rehabilitate the *U-1407* and placed her in commission on 25 September 1945. Major changes were made in the submarine: a British escape system was provided, the ventilation system was changed, and all electric equipment was replaced (having been water damaged when she was scuttled). The Walter plant—referred to as High-Test Peroxide (HTP) by the British—was removed, rehabilitated, and reinstalled. The submarine was renamed HMS *Meteorite* in 1947 and carried out closed-cycle propulsion trials. Because the

*U-1407* was intended to be used only for trials and, possibly, as an anti-submarine target, her torpedo tubes were removed.

Based on intelligence obtained late in the war and the findings at German shipyards and research institutes, as early as 1945 the Royal Navy's new construction program included a submarine propelled by a hydrogen-peroxide/turbine plant. Although a second HTP submarine was included in the 1947–1948 shipbuilding program, all construction was deferred until the *Meteorite*'s trials were completed in 1949. This sizable British HTP program was of significance to the U.S. Navy because of the extensive exchange of data in this field between the two navies. U.S. naval officers visited the *Meteorite* and British officers visited the U.S. closed-cycle engineering projects.

The U.S. Navy was very interested in closed-cycle propulsion. As early as November 1945, the Submarine Officers Conference in Washington, D.C., addressed the advantages of closed-cycle submarines with tentative preliminary characteristics put forward for a 1,200-ton submarine fitted with a 7,500-horsepower Walter plant, which could provide a sustained underwater speed of 20 knots for 12 hours. Studies by the Bureau of Ships indicated a preference for closed-cycle propulsion employing stored oxygen because of costs—hydrogen-peroxide cost was 85 cents per horsepower hour as compared to only five cents for liquid oxygen.[5] (At the time the U.S. Navy was already discussing the potential of nuclear propulsion for submarines.)

Studies of closed-cycle systems were undertaken by various U.S. Navy agencies. A 1946 report listed six such candidates.[6] The 2,500-horsepower Walter plant from the *U-1406* and a 7,500-horsepower plant planned for the Type XXVI submarine were set up at the Naval Engineering Experiment Station in Annapolis, Maryland. Also set up at Annapolis was a 50-horsepower Krieslauf or recycle-diesel plant. This plant used part of the diesel exhaust products to dilute liquid oxygen or hydrogen-peroxide and lower their combustion temperatures for submerged operation with a diesel engine. From the outset the limitations of this plant were known to be significant: the resulting decrease in power and the high noise level of a diesel engine.

The closed-cycle plants tested at Annapolis suffered numerous problems. Still, several follow-on plants were proposed. The so-called Wolverine plant, considered for a *Tang* (SS 563) hull, was a closed-cycle gas turbine plant expected to provide 7,500 horsepower. As late as November 1948, the Navy's General Board heard testimony about Navy plans for closed-cycle propulsion:

> The *Tang* hull is suitable for 25 knots and its dimensions are being used as a design factor in developing the new closed cycle power plants. It is hoped that by 1952 we will have a closed cycle power plant which we can install in a *Tang* hull and which will give us a submerged speed of 25 knots for a period of 10 hours.[7]

It was also explained to the General Board that studies indicated that an increase in underwater speed to 25 knots (compared to the *Tang*'s 18 knots) will do "very little to increase the effectiveness of the submarine in making attacks." Rather, the speed increase would be a "tremendous" factor in escaping after an attack and in countering anti-submarine efforts.

By 1949–1950 comparative studies of nuclear and chemical closed-cycle propulsion plants were prepared by Navy offices. A committee of the Submarine Officers Conference addressing the alternative plants concluded:

> In general it appears that the Nuclear propelled submarine is tactically superior to the closed cycle submarine on almost every count. However, the true value of the Nuclear propelled submarine is not to be found in a discussion of tactical advantages. The effective employment of such a submarine will necessitate entirely new tactical concepts.[8]

The most derogatory aspect of the committee report addressed nuclear fuel: "It appears that the cost of fuel for the closed-cycle submarine will be less than that for the nuclear propelled submarine," and "Fissionable material is and probably will remain in critical supply, and in addition, the nuclear propelled submarine will have to compete with other applications for allocations. Supply for the closed cycle plants is far less critical."[9]

The comparative studies undertaken by the Navy all indicated that chemical closed-cycle plants were feasible, but fell far short of nuclear propulsion. A major comparison was developed by the Navy's Bureau of Ships. (See table 3-1.) At this time the U.S. Navy was already preparing to construct a landlocked prototype of a nuclear propulsion plant in the Idaho desert.

Such analyses demonstrated conclusively to the U.S. Navy's leadership that nuclear propulsion far exceeded the potential capabilities of chemical closed-cycle plants, if sufficient fissionable material could be made available. Accordingly, research and funding for non-nuclear efforts ceased and, in 1950, the U.S. Congress authorized the world's first nuclear-propelled submarine. Ironically, at almost the same time the Navy decided to construct a "submersible craft"—the midget submarine *X-1*—which would have a hydrogen-peroxide closed-cycle plant. (See Chapter 16.)

TABLE 3-1
**Propulsion Plant Comparisons**

| | ***Tang* class SS 563** | **Chemical ($H_2O_2$) Wolverine** | **Nuclear** |
|---|---|---|---|
| Displacement submerged | 2,170 tons | 2,960 tons | 2,940 tons |
| Length | 263 ft (80.2 m) | 286 ft (87.2 m) | 272 ft (82.9 m) |
| Beam | 27 ft (8.23 m) | 31 ft (9.45 m) | 27 ft (8.23 m) |
| Pressure hull diameter | 18 ft (5.5 m) | 20 ft (6.1 m) | 27 ft (8.23 m) |
| Test depth | 700 ft (215 m) | 700 ft (215 m) | 700 ft (215 m) |
| Propulsion plant | | | |
| weight | 402,000 lbs (182,347 kg) | 328,000 lbs (148,781 kg) | 1,471,680 lbs (667,554 kg) |
| battery | 533,000 lbs (241,769 kg) | 272,000 lbs (123,379 kg) | 403,000 lbs (182,801 kg) |
| volume (engineering spaces) | 9,160 $ft^3$ (256.5 $m^3$) | 12,500 $ft^3$ (350 $m^3$) | 23,400 $ft^3$ (655 $m^3$) |
| volume (with battery) | 17,300 $ft^3$ (484.4 $m^3$) | 17,300 $ft^3$ (484.4 $m^3$) | 28,700 $ft^3$ (803.6 $m^3$) |
| Horsepower | 3,200/4,700* | 15,000 | 15,000 |
| Endurance | | | |
| at slow speeds | ~ 40 hours | ~ 40 hours | — |
| at 10 knots | | 86 hours | 6 months** |
| at 25 knots | | | 600 hours |

*Source:* Chief, Bureau of Ships, to Director, Weapon Systems Evaluation Group, "Evaluation of NEPS Project; information for," 14 February 1950.
Notes: * Diesel engines/electric motors.
** Refuelings were estimated at approximately six-month intervals.

The Royal Navy continued to have a major interest in chemical closed-cycle propulsion long after it was abandoned by the U.S. Navy. Using data from the *Meteorite* trials as well as from German engineers and German files, and from the U.S. tests of the Walter plant at Annapolis, the keel was laid down in July 1951 for the HTP submarine *Explorer,* followed in February 1952 by the similar *Excalibur.*[10]

The British submarines were slightly larger boats than the German Type XXVIW, although they were intended specifically for trials and training. (See table 3-2.) Only one periscope was provided, and neither radar nor torpedo tubes were fitted. Their configuration included a small fairwater, conventional boat-like hull, and a diesel engine in the forward (normally torpedo) compartment for battery charging and for emergency surface transit; there were two electric motors. On-board accommodations were limited, hence for lengthy trials a support ship would accompany the submarines.

The *Explorer* was placed in commission on 28 November 1956, and the *Excalibur* on 22 March 1958. By that time the USS *Nautilus* (SSN 571), the world's first nuclear submarine, was already operational. On her initial sea trials, the *Explorer* exceeded 26 knots submerged. That speed had been exceeded by a significant margin several years earlier by the USS *Albacore* (AGSS 569).

The British submarines carried out trials into the mid-1960s. Although generally successful, frequent mishaps and problems led to these submarines commonly being referred to as the "Exploder" and "Excruciator." The British relearned—with difficulties—the many problems of handling HTP. On board the submarines the HTP was stored in plastic

*HMS* Explorer, *one of two British submarines built specifically to evaluate the feasibility of the Walter propulsion system. These submarines were not successful, and Anglo-American interest in closed-cycle propulsion ended with the success of nuclear propulsion.* (Royal Navy)

bags, surrounded by seawater in tanks external to the pressure hull. A full-power run lasting about an hour would use up all 111 tons of HTP carried in the submarine. A refueling ship, the *Sparbeck*, carrying sufficient HTP for one refueling, was also provided to support the submarines. According to Michael Wilson, a former *Explorer* commanding officer, "Any small leak in any of the plastic fuel bags needed a docking to change the whole lot. I do remember that we spent an awful lot of time in Scotts shipyard on the Clyde either changing bags or having various bits and pieces mended. It was VERY frustrating."[11]

There were special provisions to prevent the buildup of explosive gas. The smallest fragment of dust, rust, or other impurity in the propulsion system could initiate HTP decomposition and the release of oxygen, while the smallest spill could corrode metal, or burn through cloth or flesh.

While employed primarily for HTP trials, on rare occasions the submarines were employed in exercises. In one such exercise the *Explorer* "approached the 'Convoy,' fired our dummy torpedoes and then dashed away up-weather on the turbines at about 20 knots for about an hour. The escorts were not very happy when we surfaced miles from them to say 'here we are!' " recalled Wilson.[12]

Still, in view of the problems with HTP, the Royal Navy cancelled plans to convert the relativly new T-class submarines to a partial hydrogen-peroxide plant. These combined-plant submarines, with a section for the turbine added aft of the control room, would have had a 250 percent power increase to provide an increase in underwater speed from 9 knots to 14 or 15 knots. By 1948 the Royal Navy had planned up to 12 such "T" conversions plus construction of combined-plant submarines. The costs of this program, coupled with the difficulties experienced with the "Exploder" and "Excruciator," undoubtedly led to the cancellation of the "T" conversions.

British naval historian Antony Preston wrote of the hydrogen-peroxide turbine effort and the end of *Explorer* and *Excalibur* trials, "the achievement of the [Royal Navy] was considerable, but there can have been few tears at the boats' paying off."[13] The experiments ended in the 1960s and the Royal Navy belatedly wrote "closed" to almost two decades of research and development. Britain was already embarked on development of a nuclear-propelled submarine program, albeit with important American assistance.

Retirement of the *Explorer* and *Excalibur* marked the end of chemical closed-cycle submarine development in Western navies for some two decades. Not until the advent of Air-Independent Propulsion (AIP) schemes in the 1980s was there credible promise of alternative methods of high underwater performance other than nuclear propulsion.

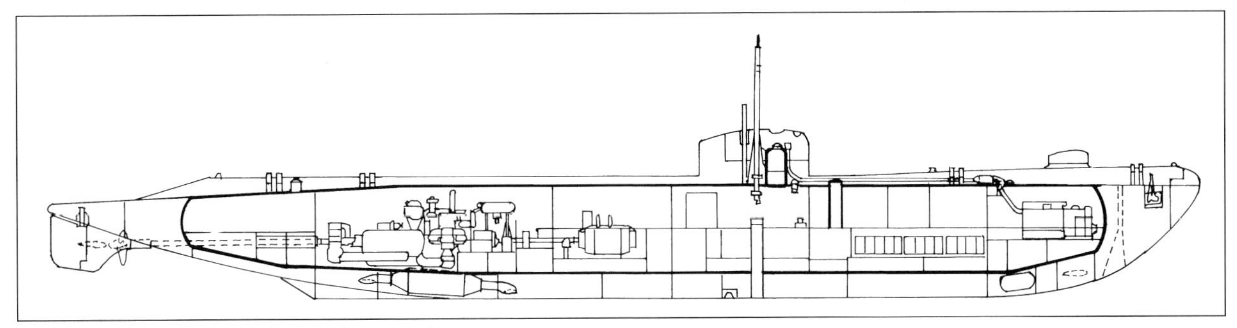

Explorer *SSX. LOA 225 ft 6 in (68.75 m)* (©A.D. Baker, III)

## Soviet Single-Drive Submarines

In the Soviet Union the concept of closed-cycle propulsion—called "single-drive" or "unified"—was initiated earlier than in the West and would survive longer. Investigations into submarine power plants capable of providing high underwater speed had been carried out in tsarist Russia and after that in the USSR over a lengthy period. These efforts were expanded in the 1930s, focusing primarily on the use of liquid oxygen to permit diesel engine operation underwater.

At the end of 1944 experiments were carried out in the use of hydrogen-peroxide, among other oxygen-carrying compositions, to oxidize fuel in the fuel chamber of a steam generator. The results of these experiments were not met with great enthusiasm by submarine designers because of low hydrogen-peroxide concentration and problems with the suggested technical approach.

In 1938 development began on Project 95, a small, 102-ton (submerged), 122⅓-foot (37.3-m) submarine employing solid lime as a chemical absorber for carbon dioxide exhaust. This absorber permitted the exhaust to be reused for the oxidation of the fuel (i.e., combustion) while the craft operated submerged with a diesel engine. This system used two lightweight diesel engines that could be employed for both surface and submerged propulsion.

Under chief designer Abram S. Kassatsier, Project 95 was begun in a design bureau operated by the NKVD—the Soviet secret police and intelligence agency.[14] Construction was undertaken at the Sudomekh shipyard in Leningrad. After launching, the unfinished submarine was transported via the extensive inland waterways to the Krasnoye Sormovo shipyard in the inland city of Gor'kiy; in November 1941 she was transported to Baku on the Caspian Sea. The submarine was completed in October 1944. During subsequent trials in the Caspian Sea through June 1945, the submarine, assigned the tactical number *M-401*, suffered several fires. The chief designer of her propulsion plant, V. S. Dmitrievskiy, was killed in one of the accidents. After the war, in 1946, responsibility for Project 95 was transferred to the TsKB-18 design bureau in Leningrad.

With the end of the war in Europe in May 1945, several groups of Soviet engineers—among them submarine designers and builders—were sent into Germany to master the German experience in military-related industries. All of the German shipyards involved in closed-cycle submarine construction were located in areas occupied by British or American troops.

In Dresden, at the Bruener-Kanis-Reder firm, which built the larger Walter turbines, Soviet engineers saw a turbine designed for submarine use. It had a power rating of 7,500 horsepower, with steam-gas as the working medium. The Soviet engineers were directed to the town of Blankenburg to obtain more detailed information. There they discovered the *Glück auf* bureau, which had played an important role in design of the Type XXVI submarine with Walter propulsion. Documents on submarine design as well as their Walter turbines were found.

With the help of the Soviet military commandant of the town, about 15 former employees of the *Glück auf* bureau were located, all of whom had important roles in development of the Walter submarines. The Germans were directed to make a report on the work of their bureau and on related submarine designs.

The Soviet reaction to this "find" was to establish a Soviet design bureau in Germany and to

*Akeksei A. Antipin* (Rubin CDB ME)

invite German specialists to participate. This plan also envisioned the placing of orders with German firms for a set propulsion equipment for a turbine-driven submarine. The new submarine design bureau was headed by the chief of the TsKB-18 design bureau in Leningrad, Engineer-Captain 1st Rank Aleksei A. Antipin. The chief engineer of the new bureau was B. D. Zlatopolsky; previously he was head of the department for special power plants at the Central Research Shipbuilding Institute in Leningrad, where the majority of work on submarine power plants had concentrated on achieving high underwater speeds.

The new design agency—referred to as the "Antipin Bureau"—was staffed with employees from TsKB-18 and the Central Research Shipbuilding Institute, and with German specialists headed by Dr. F. Stateshny, formerly of the *Glück auf.*[15] First the bureau began to collect plans, drawings, and documents related to the Type XXVI U-boat, and then it compiled lists of the required equipment. Firms that had produced equipment for the submarine were identified. Visits were made to all firms in East Germany that had manufactured equipment. One component not available for examination was the Lisholm screw compressor, as that firm was in Sweden. The work progressed rapidly, with all of the documentation developed by the Antipin Bureau being forwarded to Leningrad—drawings, technical descriptions, instruction manuals—as well as equipment available for the Walter plant.

In 1946 TsKB-18 in Leningrad began the actual design of a Soviet copy of the Type XXVI, which was designated Project 616. Some of the technical solutions adopted by the Germans for the Type XXVI would not satisfy Soviet designers and naval officers, such as the small reserve buoyancy margin, amidships torpedo tubes, and the large volume of the pressure hull compartments. Immediately after critical consideration of the design, TsKB-18 began to develop an original design of a submarine with a steam-gas turbine plant. The new design—Project 617—was considered to be of utmost importance within the Navy because the anticipated high underwater speed could expand the tactical use of submarines.

This submarine would have indigenous equipment with the exception of the turbine plant. The preliminary Project 617 design was completed at the end of 1947. This effort was developed under the leadership of the most experienced mechanical engineer, P. S. Savinov, who had participated in the development of all previous Soviet-era submarines, and young engineers named Sergei N. Kovalev and Georgi N. Chernyshov, who later would have major roles in the development of nuclear-propelled submarines. The submarine design was supervised by Boris M. Malinin, the chief designer of the first

*Project 617 was the first Soviet post–World War II submarine with closed-cycle or single-drive propulsion. Based on German technology, the* S-99 *continued earlier Soviet interest in this field. Western intelligence—which labeled the* S-99 *the "Whale"—speculated that she may have had nuclear propulsion.* (Rubin CDB ME)

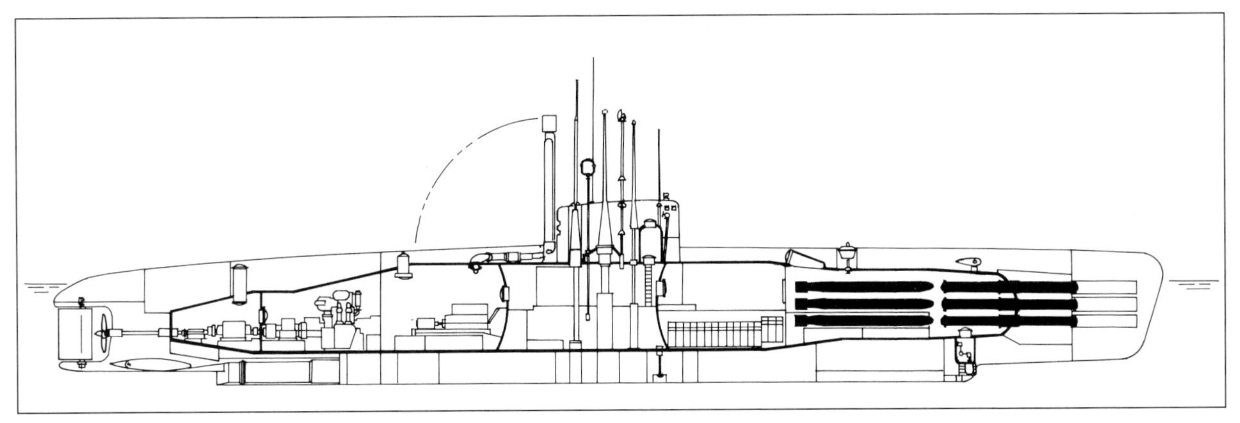

*Project 617/Whale SSX with folding snorkel intake. LOA 204 ft (62.2 m)* (©A.D. Baker, III)

Soviet submarines beginning in 1926, for whom this project became his last, as he died in 1949.

A third submarine design bureau was established in the USSR on 30 March 1948, for the further development of Project 617. This was SKB-143, a bureau set up specifically for the design of submarines with new types of power plants to provide high underwater speed. (See Appendix D.) The new bureau was staffed by specialists from TsKB-18 and the Antipin group in Germany (including ten German engineers), as well as by members of the special power plant department of the Central Research Shipbuilding Institute. Antipin was appointed to head SKB-143 and was appointed chief designer of Project 617; Kovalev became his deputy. The new bureau had two Leningrad locations, one in the Shuvalovo suburb and the other at the Sudomekh Shipyard, where the research departments were located that would develop new power plants and the related test facilities.

After the development of the design for Project 617, the bureau transferred plans to Sudomekh for construction of the submarine. The decision was made to first build a prototype, or experimental, submarine before combat units because of the innovative features of the design. Thus, series construction would be delayed until trials of the experimental boat.

During the construction phase of the Project 617 submarine, the design bureau assumed procurement functions that were not usually included in the functions of a design bureau. In one of the Sudomekh shops, a land prototype was erected with hydrogen-peroxide storage space and the hull section of the turbine compartment. This test-stand turbine plant was mounted in the hull section under conditions close to those of the submarine. The propulsion plant was assembled with equipment received from Germany; missing parts were manufactured in the workshop of the design bureau. German specialists took part in the work as consultants on technical problems; however, they worked in a separate area. Their role became less important as Soviet specialists gained experience. The last of the Germans returned home at the end of 1951.

The steam-gas turbine tests were completed at the beginning of 1951. The plant was dismantled in May of 1951, and all components were carefully examined and checked for defects. After modifications and the replacement of necessary components, the power plant and its control panel were prepared for installation in the Project 617 submarine.

The submarine was relatively short, with a small sail structure, there being no conning tower compartment; the hull was based on the Type XXI and XXVI designs, with a rounded bow and "knife" stern. There were free-flooding recesses for 32 plastic storage containers for hydrogen peroxide located in the space between the pressure hull and the outer hull. The craft would carry 103 tons of hydrogen-peroxide that would provide a submerged endurance of 6 to 23 hours at speeds of 20 to 10 knots, respectively. Unlike the British HTP submarines, the Soviet craft was a combat unit, fitted with sonar and six bow torpedo tubes with six reload torpedoes. The six-compartment submarine had a standard double-hull configuration, with the Soviet standard of "surface unsinkability," that is, able to remain afloat with any one compartment flooded. This was facilitated by a reserve buoyancy of 28 percent, more than 2½ times that of a German Type XXVI or the aborted Project 616. Project 617 was slightly larger than the Type XXVIW, with a submerged displacement of 950 tons.

The submarine—given the tactical number *S-99*—was laid down on 5 February 1951 and was launched precisely one year later. Trials began on 16 June 1952. The submarine was first sighted by Western intelligence in 1956 and was given the Allied code name "Whale."[16] Western intelligence was watching the progress of the *S-99* trials with avid attention. The streamlined hull shape, small sail structure, and vertical rudder of the submarine, coupled with reported high-speed runs, led some Western analysts to initially estimate that the submarine was nuclear propelled.

The power plant was the main distinguishing feature of the *S-99:* the 7,500-horsepower turbine could drive the *S-99* at a submerged speed in excess of 20 knots. The six-hour cruising range at this speed considerably increased the tactical possibilities of the submarines of this type.[17] In addition, the submarine demonstrated good maneuvering qualities. Submarine propulsion on the surface and at slow submerged speeds was provided by a diesel-electric plant. A snorkel was fitted to permit operation of the diesel for propulsion or for battery charging while at periscope depth.

In spite of the considerable time spent testing the turbine plant at the test-bed site ashore, major problems were experienced during *S-99* trials. These included leaking of the hydrogen-peroxide containers, resulting in fires and small explosions—called "claps"—caused by the rapid decomposition of the hydrogen-peroxide when it came in contact with dirt or oil as well as instability.

While the *S-99* trials were being conducted, in March 1953 the entire team that was working on Project 617 was returned to TsKB-18 together with its portfolio; from that moment SKB-143 received the task of designing the first Soviet nuclear-propelled submarine.

Only on 26 March 1956, after successfully completing Navy trials, was the *S-99* placed in commission for experimental operation. The development program had taken almost 12 years—only a few months less than the British program to develop the *Explorer* and *Excalibur.* While the British had developed a (Vickers) copy of the German turbine plant, the Soviets had used an actual German plant; still, Project 617 had a combat capability and was closer to being a warship prototype than were the British boats.

From 1956 to 1959 the *S-99* was in a special brigade of the Baltic Fleet for the training and overhaul of submarines. She went to sea 98 times and during those cruises traveled more than 6,000 n.miles (11,118 km) on the surface and about 800 n.miles (1,482 km) submerged, the latter including 315 n.miles (584 km) on turbine propulsion.

While the *S-99* was engaged in trials, the Project 617M design was developed. This project provided for enhanced weapons and sensors as well as increased endurance (i.e., greater reserves of hydrogen-peroxide and fuel). Other variants considered included Project 635, a twin-shaft, twin-turbine ship of some 1,660 tons surface displacement (speed 22 knots), and Project 643, an improved and enlarged twin-shaft ship of some 1,865 tons; both were propelled by paired, closed-cycle diesel engines. These variants were undertaken at TsKB-18 under the direction of Kovalev.

On 17 May 1959 a serious accident occurred in the *S-99.* During a routine starting of the plant at a depth of 260 feet (80 m), there was an explosion in the turbine compartment. The submarine was brought to the surface, down by the stern. Two reports were received from the diesel compartment: "A fire and explosion in compartment five [turbine]" and "Sprinkling [water spraying] has been initiated in compartment five." The alarm was sounded. Through glass ports in compartment bulkheads, it was observed that the aftermost compartment, No. 6—housing the propeller shaft and electric motor, as well as auxiliary equipment—also was flooded. When the damage was evaluated, the commanding officer, Captain 3d Rank V. P. Ryabov, decided to return to the base under the ship's own power.

After several hours the submarine reached the naval base at Leipaja (Libau) on the Baltic Sea. When the water was pumped out of the fifth (turbine) compartment, it was found that the hull valve of the hydrogen-peroxide supply pipeline had failed. The resulting explosion blew a hole just over three inches (80 mm) in the upper portion of the pressure hull, through which the turbine compartment had been partially flooded. The explosion had been caused by hydrogen-peroxide decomposition, initiated by mud that had gotten into the valve. The *S-99* "was saved by an astute commanding officer and a well-trained crew," said Viktor P. Semyonov, submarine designer and historian.[18]

After the accident the *S-99* was not repaired because a majority of the turbine components required replacement, which would have been very expensive. By that time the first nuclear-propelled submarine had joined the Soviet Navy. A complicated and interesting investigation of new types of power plants had been completed. The submarine *S-99* was cut up for scrap. Still, Project 617 provided valuable experience to Soviet submarine designers in the development of fast and maneuverable submarines, and the May 1959 accident confirmed the necessity for providing "unsinkability" characteristics in submarines.

At the same time Project 617 was under way, other attempts were being made to introduce steam turbines in submarines. In 1950 Project 611*bis*/Zulu was developed to place a 6,500-horsepower steam turbine on the center shaft of the submarine.[19] This scheme was not pursued because of delays in the turbine development, although when the original Project 611/Zulu submarine went to sea in 1954, many Western intelligence analysts believed that the three-shaft submarine had a Walter closed-cycle plant on the center shaft.[20]

Next was an attempt in 1954 to provide a steam turbine for an enlarged Project 611 submarine known as Project 631. This craft was to have a surface displacement of approximately 2,550 tons, with an underwater speed of 20 knots. This effort was also dropped after the problems with Project 617/Whale. Still another closed-cycle concept undertaken by SKB-143 in this period was Project 618, a small submarine with a closed-cycle diesel that was to provide an underwater speed of 17 knots. This design also was not pursued.

Concurrent with the Soviet efforts to put submarines to sea propelled by Walter-type steam turbines, work was under way to continue the 1930s efforts for submarines to employ their diesel engines while submerged. At the end of 1946, design work began at TsKB-18 on the small, 390-ton Project 615 submarine with Kassatsier as the chief designer. This submarine—later given the NATO code name Quebec—would have three diesel engines fitted to three propeller shafts plus an electric motor on the center shaft. The outer shafts each had an M-50 diesel engine generating 700 horsepower at 1,450 revolutions per minute. The center shaft was turned by a 32D diesel with a rating of 900 hp at 675 rpm for sustained propulsion on the surface or underwater. Liquid oxygen was carried to enable submerged operation of the 32D diesel engine for 100 hours at a speed of 3.5 knots. The maximum underwater speed of Project 615/Quebec was 15 knots, and that speed could be sustained for almost four hours. This was an impressive submerged performance for the time. (The M-50 engines had been developed for small, high-speed surface craft and had a limited service life, hence their employment in submarines was only for high submerged speeds.)

The first Project 615 submarine was laid down in 1950 at the Sudomekh yard, already known for the Project 617/Whale and other advanced technology propulsion plants, and was completed in 1953. After initial sea trials of the prototype—the *M-254*—series production was ordered.[21] The first production Project A615 submarine, the *M-255*,

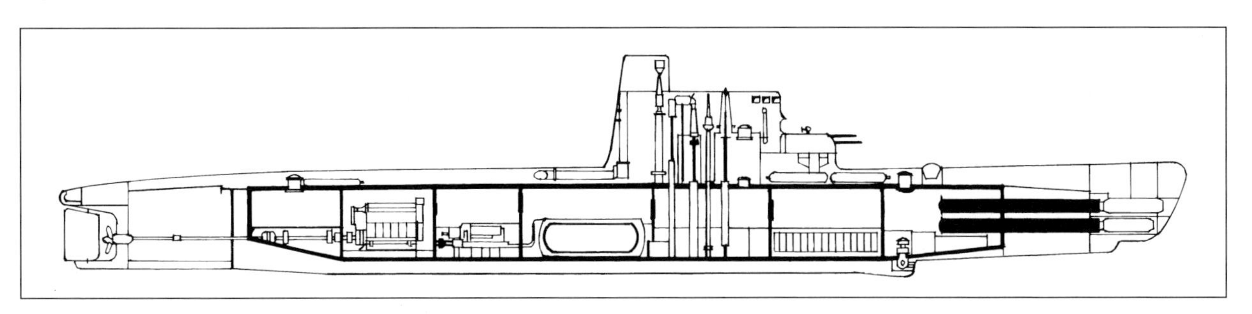

*Project A615/Quebec SSC as built with deck guns. LOA 186 ft 3 in (56.8 m)* (©A.D. Baker, III)

*A Project A615/Quebec single-drive submarine. These craft were not very successful, being plagued with propulsion plant problems, and having a very limited combat capability. These were the world's last submarines to be built with deck guns.* (U.S. Navy)

was completed in 1955.[22] Sudomekh produced 23 units through 1958, while the nearby Admiralty shipyard built its first modern submarines by producing six units through 1957. (Admiralty had constructed *Shch* and K-class submarines during the war.)

The torpedo armament of these submarines consisted of four bow torpedo tubes; no reloads were provided. Early units were completed with a twin 45-mm anti-aircraft gun mount on the forward step of the fairwater, making the Quebecs the world's last submarines to be completed with deck guns. (The guns were discarded by the end of the 1950s.) Thus combat capabilities were limited. They had seven compartments, and with a crew of 24 men, the craft were cramped and difficult to operate.

Project 615/Quebec submarines were plagued with engineering problems caused by the liquid oxygen. Oxygen evaporation limited endurance to about two weeks, and the continual leaking of liquid oxygen led to explosive "pops" throughout the engineering spaces. Like hydrogen-peroxide, the liquid oxygen was difficult to handle, and the propulsion plant suffered several serious accidents. Two submarines sank in 1957, the *M-256* in the Baltic and the *M-351* in the Black Sea. The former suffered a fire off Tallinn; the crew fought the blaze for 3 hours, 48 minutes before the submarine sank with the loss of 35 men; there were seven survivors. There were no fatalities in the *M-351* sinking. Other accidents often saw oxygen-fed flames spewing forth from the submarines, which led to their being referred to by their crews as *zazhigalka* (lighters) or "Zippos," the latter term derived from the popular American cigarette lighter of World War II.

In an effort to solve the Project 615/Quebec problems, from 1958 to 1960 one of the submarines, the *M-361,* was refitted to employ superoxidized sodium. The conversion was not completed. A modified design, Project 637 of 425 tons submerged displacement, was 95 percent complete when work halted in May 1960. The Project 637 submarine was cut into two sections and transported by rail to nearby Pushkin for use as a land training submarine at the Leningrad Higher Naval Engineering School to help submarine crews cope with the closed-cycle system. (With the oxygen system removed, that submarine continues in use today at Pushkin as an ashore training platform.) By the early 1970s the last Quebec submarines were taken out of service.

One Project 615/Quebec submarine served as a test bed for anechoic coating research. During World War II the Germans had experimented with rubber-like coatings on U-boat hulls to attenuate active sonar pulses (as well as coating snorkel intake heads to absorb radar pulses).[23] In 1947–1948, two Soviet research institutes—the Andreev and Krylov—initiated the development of hull coatings based on the earlier German efforts. Under the code name Medusa, this work included applying experimental coatings to small M or Malyuka ("baby") submarines and, subsequently, to one Project 615 submarine.

Initially these coatings were applied only to the exterior hull and sail structure to absorb active sonar pulses ("pings"). Later such coatings would be applied to the two internal surfaces of the double-hull configuration of Soviet submarines, reducing transmission of internal machinery noises that were vulnerable to detection by passive sonars. Beginning with the first Soviet nuclear submarines of Project 627/November, most Soviet combat submarines would have anechoic coatings.

During the late 1940s and 1950s, several other single-engine submarine projects were initiated in the USSR, mostly based on providing oxygen to diesel engines for submerged operation. One concept investigated from the early 1950s was based on burning "metallic fuel"—powdered aluminum and other materials. Most of these efforts were halted by about 1960 with the successful development of nuclear propulsion. Project 613Eh reflected interest in using electrochemical generators for submarine propulsion—in essence creating a giant storage battery to provide energy for electric propulsion. Interest in this field was "warmed up" in the 1960s with the use of large fuel cells in the American spacecraft of the Gemini and Apollo programs.[24] In the 1970s the Lazurit bureau was working on the design for an Air-Independent Propulsion (AIP) submarine, Project 947, with interest in an electrochemical generator that could produce electricity using hydrogen-peroxide as the reactant. Such fuel cells would have the advantages of high fuel efficiency, virually no pollution, low noise, high reliability, and relatively low cost. The reaction, in addition to providing electricity, produced heat and water, the latter being used to cool the electrodes.

*The Project 613Eh demonstration submarine for a fuel-cell submarine propulsion plant. Note the massive cryogenic storage tanks for hydrogen and oxygen, broad amidships section, and opening for retracted forward diving planes. The trials were successful, but more practicable AIP systems have become available.* (Malachite SPMBM)

In 1974 the Soviet government approved the development of shore and afloat facilities to develop electrochemical generators. One of the former, at the Krasnoye Sormovo yard in Gor'kiy, consisted of modeling a plant within a submarine hull section of a Project 613/Whiskey submarine. Subsequently, a complete electrochemical plant of 280 kilowatts was developed and fitted in an extensively converted submarine, the *S-273*, given the designation Project 613Eh and referred to as *Katran*.

The submarine was fitted with four large, cryogenic storage tanks for carrying four tons of hydrogen at –252°C and 32 tons of oxygen at –165°C. The loading of these tanks required more than 160 hours. One diesel engine was removed, and the electrochemical plant was installed in its place with an EhKhG-280 generator to provide direct drive for one of the submarine's two propeller shafts.

Extensive trials began in October 1988 and were carried out for six months. The trials demonstrated the efficiency and safety of the plant. Project 613 submarines had a rated submerged endurance of more than seven days traveling at two knots before they had to recharge their batteries, although in reality dives rarely exceeded three or four days. The *S-273* in the 613Eh configuration was capable of traveling submerged at 2.5 knots for 28 days. However, the concept was not pursued because of more competitive AIP systems. (See Chapter 13.)[25]

**The late 1940s and 1950s were a period of intensive submarine development in many navies, with great**

*The Project 613Eh submarine S-273 at sea. She retains the Project 613/Whiskey conning tower as well as bow and stern sections. The fuel-cell plant was a direct drive system.* (*Sudostroenie* magazine)

interest in the potential high-speed performance of closed-cycle submarines. The U.S. Navy quickly abandoned this path because of the obvious advantages of nuclear propulsion. The Royal Navy pursued it longer, investing considerable effort in rehabilitating a German Type XVIIB submarine and constructing two very troublesome test submarines.

Soviet submarine development and construction was remarkable in this period in view of the devastation to the shipbuilding industry in World War II. The USSR also examined closed-cycle propulsion, but with more intensity—and suffered more problems. What promise closed-cycle propulsion held in the 1950s was soon overshadowed by the development of nuclear propulsion.

Also during this period the rate of development and intensity of construction in the USSR was profound: From the end of World War II through 1959, some 350 submarines were completed in Soviet shipyards—with 74 in the peak year of 1955—as compared to a total of 50 submarines produced in that same period by American shipyards.

Three of the Soviet submarines had nuclear propulsion; eight U.S. submarines completed through 1959 were nuclear. But the nuclear submarine momentum was shifting.

This momentum was facilitated after World War II when the venerable TsKB-18 was joined in submarine design and development by the short-lived Antipin Bureau. Beginning in 1948, SKB-143 undertook the design of high-speed submarines. In 1953, following the death of Josef Stalin, the design and construction of large surface warships was largely halted; the TsKB-18 design bureau was accordingly reorganized to undertake submarine designs to incorporate missile systems. That same year the design office of the Krasnoye Sormovo shipyard in Gor'kiy became an independent submarine design bureau, SKB (later TsKB)-112. The new bureau's first assignment was to design a large, oceangoing submarine.

The Soviet Union thus entered the nuclear submarine era with four separate submarine design bureaus: TsKB-16 (shifted from surface ship design to submarine work in 1953), TsKB-18, SKB-112, and SKB-143. At the same time, seven Soviet shipyards were engaged in submarine construction.

TABLE 3-2
**Closed-Cycle Submarines**

| | German Type XVIIB | German Type XXVIW | British *Explorer* | Soviet Project 617 Whale | Soviet Project 615 Quebec |
|---|---|---|---|---|---|
| Operational | (1945) | — | 1956 | 1956 | 1956 |
| Displacement | | | | | |
| surface | 312 tons | 842 tons | 1,086 tons | | 406 tons |
| submerged | 337 tons | 926 tons | 1,203 tons | 950 tons | 504 tons |
| Length | 136 ft 2 in (41.5 m) | 184 ft 3 in (56.2 m) | 225 ft 6 in (68.75 m) | 204 ft (62.2 m) | 186 ft 3 in (56.76 m) |
| Beam | 10 ft 10 in (3.3 m) | 17 ft 9 in (5.4 m) | 15 ft 8 in (4.8 m) | 20 ft (6.08 m) | 14 ft 9 in (4.5 m) |
| Draft | 14 ft 1 in (4.3 m) | 19 ft 4 in (5.9 m) | 14 ft 5 in (4.4 m) | 16 ft 8 in (5.08 m) | 11 ft 10 in (3.6 m) |
| Turbines | 1 Walter | 1 Walter | 2 Vickers | 1 Walter | — |
| horsepower | 2,500 | 7,500 | 15,000 | 7,500 | |
| Diesel engines | 1 | 1* | 1 | 1 | 3 |
| horsepower | 210 | 2,580 | 400 | 1 600 | 2 1,400 |
| horsepower | | | | 2 450 | 1 900 |
| Electric motors | 1 | 1** | 1 | 1 | 1 |
| horsepower | 77 | 1,536 | | 540 | 100 |
| Shafts | 1 | 1 | 2 | 1 | 3 |
| Speed | | | | | |
| surface | 8.5 knots | 11 knots | 26 knots | 11 knots | 16 knots |
| submerged | 21.5 knots | 24 knots | 26 knots | 20 knots | 15 knots |
| Range (n.miles/kts) | | | | | |
| surface | 3,000/8 | 7,300/10 | 2,800/6.5 | 8,500/8.5 | 3,150/8.3 |
| submerged | 150/20 | 158/24 | 26/26<br>180/12 | 198/14.2*** | 410/3.5 |
| Test depth | 395 ft (120 m) | 435 ft (133 m) | 500 ft (150 m) | 655 ft (200 m) | 395 ft (120 m) |
| Torpedo tubes*** | 2 533-mm B | 4 533-mm B<br>6 533-mm side | nil | 6 533-mm B | 4 533-mm B |
| Torpedoes | 4 | 10 | nil | 12 | 4 |
| Guns | nil | nil | nil | nil | 2 x 25-mm |
| Complement | 19 | 35 | approx. 45 | 51 | 33 |

Notes: * A diesel generator of 100 hp also was fitted.
** An electric creeping motor of 75 hp was also fitted.
*** 130 n.miles at 20 knots.
# Bow; *side* indicates amidships tubes angled outboard.

4

# U.S. Nuclear-Propelled Submarines

*The USS* Nautilus *was the world's first nuclear-propelled vehicle in New York harbor. Beyond her remarkable nuclear propulsion plant, she had a conventional submarine design, based in large part on the Type XXI. Her success led to the U.S. Navy rapidly ceasing the construction of non-nuclear submarines.* (U.S. Navy)

While Germany excelled in submarine development and in several other advanced weapon areas, the German atomic bomb effort was inefficient and aborted. Still, the nuclear efforts of German scientists sparked the American atomic bomb effort and initiated the first interest in an atom-powered submarine.

In 1938 Otto Hahn and Fritz Strassman at the prestigious Kaiser Wilhelm Institute for Chemistry discovered nuclear fission—the potential release of vast amounts of energy by splitting atomic nuclei. Reports of the Hahn-Strassman discovery excited scientists around the world, among them Dr. George Pegram of Columbia University, one of the leading physicists in the United States.

In early 1939 Pegram proposed to Rear Admiral Harold G. Bowen, chief of the Navy's Bureau of Steam Engineering, which controlled the Naval Research Laboratory, that he meet with Navy researchers to discuss the practical uses of uranium fission. Pegram and Bowen met at the Navy Department buildings on Constitution Avenue in Washington, D.C., on 17 March 1939. Also present at that historic meeting were Dr. Enrico Fermi, the world's leading authority on the properties of neutrons, who had come close to discovering nuclear fission in 1934; Captain Hollis Cooley, head of the Naval Research Laboratory; and Dr. Ross Gunn, a physicist and head of the laboratory's mechanical and electrical division.[1]

The use of uranium fission to make a super-explosive bomb was discussed at the meeting. Fermi expressed the view that if certain problems relative to chemical purity could be solved, the chances were good that a nuclear chain reaction could be initiated.

"Hearing these outstanding scientists support the theory of a nuclear chain reaction gave us the guts necessary to present our plans for nuclear propulsion to the Navy," Gunn would recall. Three days later, on 20 March, Cooley and Gunn called on Admiral Bowen to outline a plan for a "fission chamber" that would generate steam to operate a turbine for a *submarine power plant.*[2] Gunn told the admiral that he never expected to seriously propose such a fantastic idea to a responsible Navy official. He asked for $1,500 to initiate research into the phenomenon of nuclear fission. Bowen, himself an innovator who had fought many senior officers to get the Navy to adopt improved steam turbines for surface ships, approved the funds.

*Ross Gunn*

*Philip A. Abelson*

The $1,500 was the first money spent by the U.S. government for the study of nuclear fission. That summer, after preliminary studies, Gunn submitted his first report on nuclear propulsion for submarines. His report came four months before delivery of the famous letter signed by Albert Einstein urging President Franklin D. Roosevelt to undertake a nuclear weapons program.

In his report, Gunn noted that an atomic power plant would not require oxygen, "a tremendous military advantage [that] would enormously increase the range and military effectiveness of a submarine."[3] Gunn's paper also pointed out the many problems and unknowns for such an effort, especially the need to develop the means for separating the lighter $U_{235}$ atoms, which would undergo fission when bombarded with neutrons from the heavier uranium atoms. Gunn then turned his efforts to solving the uranium separation problem.

Several civilian academic and research institutions were approached to work with the Naval Research Laboratory in the quest for a practical method of separating the elusive $U_{235}$ isotope. Beginning in January 1941, Philip A. Abelson of the Carnegie Institution worked on the separation problem. (The 28-year-old Abelson already was co-discoverer of Element 93, neptunium.) In July he joined the staff of the Naval Research Laboratory and, together with Gunn, developed a relatively simple and efficient method of $U_{235}$ isotope separation.

While the U.S. Navy and other institutions were taking the first steps toward the atomic age, the public was beginning to learn about the potential of atomic energy. In September 1940 science writer William Laurence wrote a highly perceptive article in the *Saturday Evening Post* about "the discovery of a new source of power, millions of times greater than anything known on earth." After describing atomic developments up to that time, Laurence wrote that "such a substance would not likely be wasted on explosives." In a somewhat naïve proposal, he wrote: "A five pound lump of only 10 to 50 percent purity would be sufficient to drive ocean liners and submarines back and forth across the seven seas without refueling for months."

With the war in Europe seeing Hitler's legions overrunning the continent, and with the belief that German scientists could be working on an atomic bomb, the United States and Britain began a joint A-bomb development program under the code name Manhattan Engineering District. The Navy's research efforts into nuclear fission were essentially halted. The uranium separation process that Gunn and Abelson had developed and put in place on a small scale in Philadelphia was adopted by the Manhattan Project and was developed on a massive scale at Oak Ridge, Tennessee.[4]

But the idea of propelling ships and submarines with atomic energy was not forgotten during the war. In August 1944 Brigadier General Leslie Groves, head of the Manhattan Project, appointed a five-man committee to look into potential nondestructive uses of atomic energy. Dr. Richard C. Tolman, the longtime dean of the California Institute of Technology's graduate school and chief scientific adviser to Groves, would head the committee; the other members would be two naval officers and two civilians. The influence of the naval officers would be considerable: Rear Admiral Earle W. Mills, the assistant chief of the Bureau of Ships (BuShips), and Captain Thorwald A. Solberg, also of BuShips.[5]

According to Groves, "One of the primary reasons why I appointed this committee was to have on the record a formal recommendation that a vigorous program looking towards an atomic powered submarine should be initiated when available personnel permitted. I wanted Mills on the committee to make certain that he, Tolman, and Solberg did just that."[6]

That fall the committee visited the Naval Research Laboratory in Washington and listened to Gunn and Abelson urge that a nuclear-powered submarine be given high priority in the committee's report. Tolman submitted the formal report in December 1944—seven months before the first atomic bomb was detonated—proposing that "the government should initiate and push, as an urgent project, research and development studies to provide power from nuclear sources for the propulsion of naval vessels."

A year later, with the war over, the subject of atomic energy for ship propulsion received public attention when the Senate established the Special Committee on Atomic Energy. In reporting the committee's hearings, *The New York Times* of 14 December 1945 quoted Gunn as declaring that "the main job of nuclear energy is to turn the world's wheels and run its ships." The *Times* also mentioned the possibility of "cargo submarines driven by atomic power."

By the end of 1945, many American scientists and engineers were discussing the possibility of nuclear submarines. A Navy report—apparently based on Gunn-Abelson data—dated 19 November 1945, listed the advantages of nuclear "transformations," which estimated a yield of three *billion* BTUs per pound of nuclear fuel compared to 18,000 BTUs per pound of diesel oil.[7] It listed these "outstanding characteristics" of nuclear propulsion:

1. Unlimited range.
2. Continuous submerged operation for weeks is possible with good living conditions.
3. High submerged and surface speeds.
4. No refueling at sea—an annual job.
5. No recharging of batteries.
6. More power available for same weight and volume.
7. A dry ship [i.e., no need to surface or use snorkel to charge batteries].
8. Control and handling about like present submarines.
9. Submarine clean and habitable [i.e., no diesel oil]
10. Space distribution not unlike present arrangements except forward section of submarine [forward of reactor] will be detachable for repair or replacement and forward [torpedo] tubes may have to be sacrificed.
11. Operation easier than with diesels.
12. Operation of power plant expected to be reliable.

Two "disadvantages" were listed:

1. Poisoned [*sic*] areas forward [reactor] cannot be entered by *any personnel* and all controls must be remote.
2. No repairs in the forward [reactor] compartment can be made away from a specially tooled yard.

Many naval officers were talking about nuclear propulsion. At the Bikini atomic bomb tests in the summer of 1946, Lieutenant Commander Richard B. Laning, commanding the fleet submarine *Pilotfish* (SS 386), talked about the feasibility of an atomic-powered submarine with Dr. George Gamow. Gamow calculated that such a craft could achieve underwater speeds of 30 knots and would have to be twice the size of existing submarines. Such craft could be developed, he said, "in ten years if we really put our heart into it." Laning immediately wrote a letter, through the chain of command, recommending a nuclear-propulsion program and volunteering to serve in it. (Laning did not recall having received a response.[8])

After the war Gunn and Abelson returned their attention to the possibility of an atomic submarine.

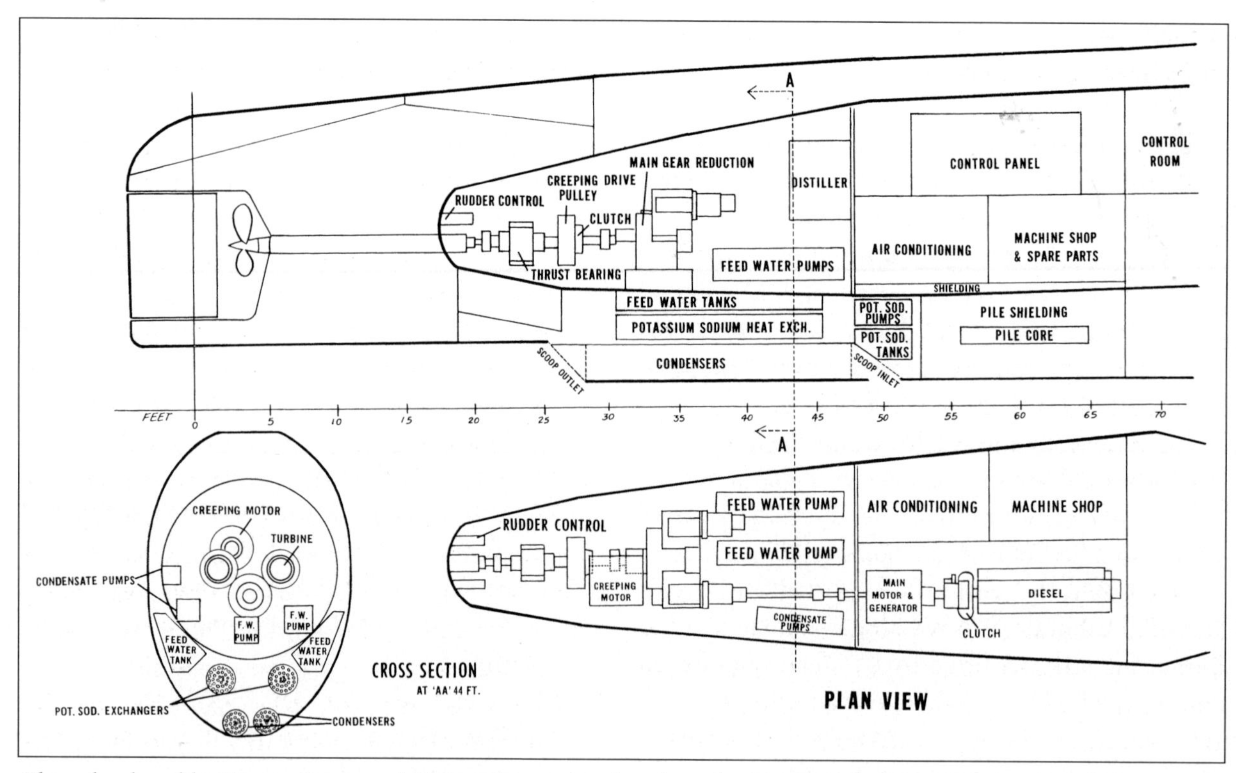


*Plans developed by Messrs. Gunn and Abelson for an atomic submarine based on the Type XXVI U-boat hull. Note the duplicate components of the power plant designed to keep the submarine operating even with part of the plant out of action. This is an artist's copy of the original plans.* (*Atomic Submarines*)

Abelson took a set of plans of a German Type XXVI (Walter) submarine and developed a scheme for a nuclear "pile" that could fit into the existing spaces with only minor changes in the basic submarine design. This approach, Abelson felt, would accelerate the development of a nuclear submarine. Much of the Abelson report was vague or—from a technical viewpoint—questionable. For example, there was little information about the design of the nuclear pile, or reactor, except that it would use a sodium-potassium alloy as the means to transfer heat from the reactor to the steam turbine.

Abelson's report was submitted in March 1946. Although in retrospect it had severe shortcomings, the report—"Atomic Energy Submarine"—would capture the imagination of many Navy men. His conclusions were:

> A technical survey conducted at the Naval Research Laboratory indicates that, with a proper program, only about two years would be required to put into operation an atomic-powered submarine mechanically capable of operating at 26 knots to 30 knots submerged for many years without surfacing or refueling. In five to ten years a submarine with probably twice that submerged speed could be developed.[9]

The 27-page report of the young physicist placed two "ifs" on his two-year timetable: Sufficient priority had to be given to the project by the Navy and the Manhattan Project, and "cooperation between the Manhattan District and the Navy is expanded somewhat to permit greater emphasis on Naval participation in design and construction of a Uranium pile reactor of proper characteristics for this application."

There was prophecy in the Abelson report's comment that a high-speed, atomic-powered submarine would operate at depths of about 1,000 feet (305 m) and "to function offensively, this fast submerged submarine will serve as an ideal carrier and launcher of rocketed atomic bombs."

Abelson and Gunn began briefing the Navy's submarine community on the proposal. About 30 senior submariners were told about the atomic submarine idea in March 1946. Vice Admiral Charles Lockwood, who had commanded U.S. submarines in the Pacific during World War II, recalled one of those briefings by Abelson:

> If I live to be a hundred, I shall never forget that meeting on March 28, 1946, in a large Bureau of Ships conference room, its walls lined with blackboards which, in turn, were covered by diagrams, blueprints, figures, and equations which Phil [Abelson] used to illustrate various points as he read from his document, the first ever submitted anywhere on nuclear-powered subs. It sounded like something out of Jules Verne's *Twenty Thousand Leagues under the Sea.*[10]

Simultaneous with the Gunn-Abelson work, Lieutenant Commander Charles N. G. Hendrix in the Navy's Office of Research and Inventions (later the Office of Naval Research), produced a 14-page memorandum on "Submarine Future Developments" that provided a succinct analysis of submarine operations in World War II to justify three types of future submarines:[11]

- small training submarine (600 to 800 tons displacement)
- medium fleet-type attack boat (1,200 to 1,500 tons)
- large special "rocket-launching type" (2,400 to 3,000 tons) to carry surface-launched rockets with atomic warheads

Hendrix's attack boat would have a submerged speed of 20 to 25 knots, an operating depth of 2,000 feet (610 m), and the "finest maneuverability." Significantly, both the attack submarine and the rocket-launching type would have nuclear propulsion. The Hendrix memorandum included a sketch of his attack submarine with nuclear propulsion (reactor forward), accompanied by weight estimates and a description of the power plant, albeit based mainly on the official but unclassified Smyth Report *Atomic Energy for Military Purposes.*[12]

Much of what Gunn, Abelson, and Hendrix were advocating, and what some officers in the submarine community were already supporting, was based almost entirely on theory and educated guesses. No engineering work had yet been undertaken. General Groves still would not provide nuclear information to military personnel or civilians who were not under his direct command unless specifically told to do so by the Chief of Naval Operations.[13] Groves believed that sharing information would violate President Franklin D. Roosevelt's directive on the security of atomic information. He also was aware of the very limited amounts of enriched uranium available and felt they could not—at the time—support both the nuclear weapons program and a Navy propulsion program. (This concern, even within the Navy, would constitute the principal argument against nuclear-propelled submarines.) Also, knowing that there would soon be some form of civilian agency to control U.S. atomic matters, Groves was reluctant to commit the Manhattan Project to a long-term policy concerning nuclear propulsion.

Meanwhile, Admiral Bowen, appointed after the war to ensure the continuation of close cooperation between the Navy and scientific community, and Commodore William (Deak) Parsons, who had been an executive in the Manhattan Project and the weapons officer on the Hiroshima A-bomb mission, prepared a letter from Secretary of the Navy James Forrestal to the Secretary of War, who was Groves's superior. The letter, dated 14 March 1946, stated that the Navy wished to undertake the engineering development of atomic power for ship propulsion.

In response, Groves and the Secretary of War responded that the Navy would best be served by assigning personnel to the atomic power pile program being set up at Oak Ridge. In discussing Navy participation in the Oak Ridge effort, the chief of the Bureau of Ships advised the General Board, the principal advisory body to the Chief of Naval Operations and the Secretary of the Navy, "It is the Bureau's opinion that the action being taken by the Manhattan District to develop an experimental power pile is the soundest possible approach to the problem and will produce the fastest results." Addressing the time factor, he noted that "at least 4–5 years will elapse before it will be possible to install atomic energy in a naval ship for propulsion purposes."

## Enter Captain Rickover

The Navy promptly assembled a team of five naval officers and three civilians to go to Oak Ridge. After

*Hyman G. Rickover*
(U.S. Navy)

some shuffling of assignments, Captain H. G. Rickover, an engineering specialist, was assigned as the senior officer to go to Oak Ridge. Rickover had served in destroyers, battleships, and submarines, and had briefly commanded a minesweeper before becoming an engineering specialist in 1938.[14] During most of World War II, he had been head of the electrical branch of BuShips, responsible for the development, procurement, and installation of electrical equipment in ships and submarines.

At the same time, other engineering officers were assigned to various firms working in the nuclear field. Rickover originally was named to go to the General Electric facility at Schenectady, New York. But Rear Admiral Mills, Rickover's wartime boss (and from November 1946 the chief of BuShips), decided to send him to Oak Ridge.

The Navy group arrived at Oak Ridge in June 1946. During the next four months, they were exposed to all aspects of nuclear technology. In this period Rear Admiral Bowen, who had become the Chief of Naval Research, and Captain Albert G. Mumma were the major spokesmen for nuclear power. Mumma was named coordinator for nuclear matters in BuShips, preempting the position Rickover expected to have.[15]

Rickover fought to gain the top position in naval nuclear matters, producing papers on the feasibility of nuclear propulsion. In this period he attempted to keep his Oak Ridge group intact, using them to help produce his reports. Also, Rickover obtained the support of Dr. Edward Teller, a leading nuclear physicist, for his submarine proposals. Finally, in September 1947 Mills appointed Rickover as his special assistant for nuclear matters.

Abelson, Bowen, Gunn, Mills, Rickover, and many others in the Navy Department believed that nuclear propulsion was feasible for surface ships and submarines. Approval to pursue a realistic nuclear-propulsion program, one that would result in the construction and operation of nuclear-propelled ships, had to come from the highest levels of the Navy, from the Secretary of the Navy and Chief of Naval Operations (CNO), and be supported by the administration and funded by Congress.

The first postwar CNO, Fleet Admiral Chester W. Nimitz, was immediately impressed by the promise of nuclear propulsion for the Navy.[16] However, Parsons, now a rear admiral, and others believed that the Navy's primary efforts in atomic energy should go into weapons development. They believed that the resources and knowledgable people available to the Navy were too limited to develop both nuclear weapons and nuclear propulsion. The former, they felt, were far more necessary if the Navy was to compete with the Army Air Forces for major military missions or, indeed, even to survive as a major military service in the postwar period. Few other voices were raised in opposition to nuclear propulsion. One was Vice Admiral Robert B. Carney, at the time Deputy CNO for logistics. He hoped for a worldwide ban on nuclear propulsion for warships, fearing that if the United States had them at some future time so would its enemies.

In the fall of 1946, Admiral Nimitz asked the Submarine Officers Conference to address the subject of nuclear propulsion. The request led to a major report, completed on 9 January 1947, that stated:

> Present anti-submarine techniques and new developments in submarine design have rendered our present fleet submarines obsolete, offensively and defensively, to a greater degree than any other type [of warship]. The development of a true submarine capable of operating submerged for unlimited periods, appears to be probable within the next ten years, provided nuclear power is made available for submarine propulsion.

The report fully supported nuclear propulsion. It recommended a multi-phase program, including the construction of advanced, conventionally propelled submarines pending the design and development of nuclear power plants. In addition, it recommended that future diesel-electric submarines be configured for subsequent conversion to nuclear propulsion. The report estimated that the first nuclear submarines could be ready for sea by the mid-1950s. Nimitz approved the report the day after it was submitted to him.

Now, with the help of many within the Navy Department, Rickover was becoming the chief spokesman for nuclear propulsion, and in August 1948, Admiral Mills established a Nuclear Power Branch within BuShips, with Rickover as its head. Mills's major problem was the newly established Atomic Energy Commission (AEC), placed in charge of all nuclear matters, which balked at near-term support of nuclear propulsion. In the view of the AEC commissioners, it was too early to make such a major commitment. Frustrated, Mills attacked the AEC in speeches and in discussions with members of Congress. In response, in February 1949, the Division of Reactor Development was established within the AEC with Rickover as its head and "double-hatted," that is, he was assigned to both the Navy and AEC (with a single staff).

Rickover and the Navy were thus in a position to begin development of a nuclear-propelled submarine. American industry was anxious to enter the nuclear era, and the Navy soon established contracts with Westinghouse, General Electric, and Combustion Engineering for the development of suitable reactor plants. The prototypes of submarine plants would be constructed on land, to be used to model the shipboard installations, identify and correct problems, and train crews. Further, these prototype plants would be assembled within actual submarine hull sections, to save time and ensure the "fit" in actual submarines.

The early selection of a shipyard was necessary for the fabrication of hull sections for those reactor prototypes. The design department of the Portsmouth Naval Shipyard—the Navy's lead submarine yard—was fully engaged in the GUPPY and *Tang* (SS 563) programs, as well as a variety of submarine conversions. Rickover turned to the Electric Boat yard, in Groton, Connecticut, which took on the detailed design and construction of the first U.S. nuclear submarines.

The Congress, impressed with testimony by Rickover and others from the Navy, in 1951 authorized the first nuclear submarine, to be designated SSN 571 and, subsequently, named *Nautilus*. This submarine would have a reactor employing pressurized water as the heat exchange medium between the reactor and the turbine. At the time the Bureau of Ships addressed the feasibility of four different sizes of nuclear plants, considered in increments of 15,000 horsepower—the goal for the *Nautilus* plant.

Captain Rickover selected a Pressurized-Water Reactor (PWR), initially designated STR, for Submarine Thermal Reactor, for the first nuclear submarine.[17] The STR Mk I would be the land

TABLE 4-1
**Theoretical Nuclear-propelled Submarines**

| Shaft horsepower | 15,000 | 30,000 | 45,000 | 60,000 |
|---|---|---|---|---|
| Submerged displacement | 2,935 tons | 4,105 tons | 5,377 tons | 6,658 tons |
| Length overall | 272 ft (82.9 m) | 290 ft (88.4 m) | 315 ft (96.0 m) | 330 ft (100.6 m) |
| Beam | 27 ft (8.23 m) | 31 ft (9.45 m) | 34 ft (10.37 m) | 37 ft (11.28 m) |
| Pressure hull diameter | 27 ft (8.23 m) | 31 ft (9.45 m) | 34 ft (10.37 m) | 37 ft (11.28 m) |
| Length machinery space | 55 ft (16.77 m) | 72 ft (21.95 m) | 88 ft (26.83 m) | 101 ft 6 in (30.94 m) |
| Maximum speed | 25.2 knots | 29.6 knots | 32.1 knots | 33.8 knots |
| Range at maximum speed | 15,100 n.mi (27,980 km) | 17,800 n.mi (32,983 km) | 19,300 n.mi (35,763 km) | 20,300 n.mi (37,616 km) |
| Range at 10 knots | 46,800 n.mi (86,720 km) | 61,600 n.mi (114,145 km) | 76,200 n.mi (141,200 km) | 82,000 n.mi (151,946 km) |

*Source:* Memorandum from Chief, Bureau of Ships, to Director, Weapon Systems Evaluation Group, "Evaluation of NEPS Project; information for," 14 February 1950, p. 16.

prototype and the STR Mk II would be installed in the *Nautilus*. The plants would be virtually identical.

Significantly, until about 1950 the first nuclear submarine was envisioned as exclusively a test submarine for a nuclear plant. Preliminary characteristics for the ship noted: "initially there will be no armament in the original [nuclear] ship, however, the design and construction should be such that armament . . . could be installed with a minimum of structural changes and at a minimum cost as a conversion project."[18]

*The first U.S. atomic "engine" to produce substantial quantities of power was the land-locked prototype of the* Nautilus *power plant, which was constructed inside a portion of a submarine hull at Arco, Idaho. The "sea tank" held 385,000 gallons of water. Note the circular cover over the reactor.* (U.S. Navy)

The Electric Boat design for the *Nautilus* provided for a larger submarine than had been envisioned in the 1950 BuShips study. The EB submarine would have a surface displacement of 3,180 tons and 3,500 tons submerged, with a length of almost 324 feet (98.7m). This was almost half again the displacement of the postwar *Tang* and more than 50 feet (15.2m) longer, the additional size being required for the nuclear plant.

The Mk I prototype of the *Nautilus* reactor plant was assembled at the National Reactor Test Station near Arco in the Idaho desert. The hull section housing the reactor itself was constructed in a tank some 50 feet (15.2 m) in diameter and almost 40 feet (12.2 m) high, holding some 385,000 gallons (1.46 million liters) of water. With this arrangement, tests with the reactor compartment at sea could be simulated. Provided in adjacent hull sections were the heat exchanger, pumps, and a turbine. The reactor was built with the same massive lead shielding that would be fitted in the submarine. Significantly, the 27⅔-foot (8.43-m) height of the reactor compartment (i.e., the pressure hull diameter at that point) was dictated in part by the problem of control rods. The control rods—fabricated of hafnium, a neutron-absorbing element—were raised to permit nuclear fission to occur, and lowered to halt fission. The rods were screw threaded, and would be moved up and down by an electromagnetic field, operating a rotor on the top of each rod. (This screw concept would provide the crew with maximum safety from radiation in comparison with the pulley concept subsequently adopted for Soviet submarines.)

The Arco reactor plant achieved criticality—self-sustained operation—on 30 March 1953. Extensive tests followed, with problems uncovered and corrected. Then in a highly impressive test, in June 1953, the plant successfully undertook a 96-

hour "voyage," in theory driving a submarine some 2,500 n.miles (4,633 km) across the Atlantic at an average speed of 26 knots.

## Building the *Nautilus*

The *Nautilus* herself was now under construction, the keel laying being officiated by President Truman on 14 June 1952, at the Electric Boat yard.[19] She was launched on 21 January 1954, with Mrs. Mamie Eisenhower breaking the champagne bottle against the ship. Already under construction on an adjacent building way at Electric Boat was the second atomic submarine, the *Seawolf* (SSN 575), also begun in 1952. These historic events were undertaken with a blaze of publicity and were well covered in the American and overseas press.

The *Nautilus* looked for the most part like an enlarged Type XXI with a rounded bow and streamlined hull and fairwater. Aft she had upper-and-lower rudders and twin screws. Her large 28-foot (8.5-m) diameter provided a large interior volume, with three levels in most of the submarine. She had a partial double hull—the reactor compartment extended to her outer dimensions—and six compartments: bow, main living quarters and galley, central operating, reactor, engine room, and stern compartments. The design reduced reserve buoyancy to about 16 percent.

Forward were six torpedo tubes with 26 torpedoes being carried (there were no stern tubes and, of course, no deck guns). A large BQR-4A passive sonar was fitted in the submarine's "chin" position, fully faired into the hull, with an SQS-4 active scanning sonar also fitted in the bow.[20] There briefly was thought of providing the *Nautilus* with a Regulus-type missile (see Chapter 6); however, that idea was quickly rejected to avoid complications in producing the first nuclear submarine. (The Revell model company produced a *Nautilus* fitted with a Regulus hangar and fixed launching ramp aft of the sail.)

There were accommodations for 12 officers and just over 90 enlisted men; the officers shared staterooms, except for the captain, who had a private cubicle, and there was a separate wardroom where the officers could eat and relax. Each sailor had his own bunk, and the crew's mess could accommodate 36 men at one sitting for meals, or up to 50 for movies and lectures. The *Nautilus* had a built-in ice-cream machine, Coca-Cola® dispenser, and nickel-a-play juke box connected to the ship's hi-fi system. Life on board would be luxurious by previous submarine standards, the nuclear plant being able to provide unlimited fresh water and air-conditioning.

Aft of the attack center and the control room (located on two levels beneath the sail structure), the after portion of the *Nautilus* was devoted to the propulsion plant and auxiliary machinery. The single STR reactor (later designated S2W) provided heat for the steam generators, which, in turn, provided steam to the two turbines. The plant produced 13,400 horsepower—enough to drive the large submarine at a maximum submerged speed of just over 23 knots.

At one point, according to Rickover, a twin reactor plant was considered to reduce the possibility that the submarine would be disabled or lost at sea because of a reactor failure. However, size was a constraint, and the *Nautilus* was built with only one reactor. An auxiliary diesel generator, complete with snorkel installation for submerged operation, and a half-GUPPY battery installation were provided to "bring home" the submarine in an emergency.

While deep-diving submarines demanded quality in materials and construction, the nuclear plant, with its radioactive and high-pressure steam components, demanded a new level of quality control. Rickover's mania for quality control in the *Nautilus* would establish a policy that would greatly enhance the safety record of U.S. nuclear submarines.[21]

Rickover's second major contribution to the nuclear program would be his effectiveness in mustering congressional support for the Navy's nuclear-propulsion program. In this latter role he would stand without equal in the Navy, invariably wearing civilian clothes, eschewing charts and briefing slides, and stressing that he was telling the solons the "truth," not what they wanted to hear.[22] AEC historians would write, for example, of Rickover's relationship with Senator Henry M. Jackson from the state of Washington, who earlier had been a member of the Joint Committee on Atomic Energy when in the House of Representatives, and would go on to be one of the most influential men in Washington:

> He had met Rickover briefly at committee hearings, but he had not come to know him

> well until the two men found themselves on the same airplane headed for the nuclear weapon tests in the Pacific in the fall of 1952. While waiting through the interminable hours for the test to occur, Rickover and Jackson had struck up some lively conversations. Jackson was intrigued by Rickover's candor and intensity and listened with rapt attention to Rickover's accounts of his incessant battles with the Navy's bureaucracy over nuclear power.[23]

The *Nautilus* was formally placed in commission on 30 September 1954, but she remained fast to the pier at Electric Boat. The ceremony was for public relations purposes—to demonstrate that the Rickover organization had met its schedule. The submarine's reactor plant was started up on 30 December, and on 17 January 1955 the *Nautilus* moved away from the pier. Despite a sudden engineering problem that was quickly solved as the submarine moved down the Thames River, her commanding officer, Commander Eugene P. Wilkinson, had a signal lamp flash the historic message: UNDERWAY ON NUCLEAR POWER.

The submarine's trials were highly successful, including a record submerged run of 1,381 n.miles (2,559 km) from New London, Connecticut, to San Juan, Puerto Rico, in 90 hours—an average of 15.3 knots. This was the fastest submerged transit yet undertaken by a submarine. Subsequently faster submerged passages were made, averaging close to her maximum speed. The *Nautilus* was a remarkable engineering achievement.

In exercises she demonstrated the value of nuclear propulsion for submarines. The *Nautilus* could close with an enemy or escape at will, being maneuverable in three dimensions, regardless of surface weather conditions. She could even outrun available U.S. anti-submarine homing torpedoes. Unlike the captains of "high-speed" submarines of the Type XXI, GUPPY, or *Tang* designs, the captain of the *Nautilus* did not have to concern himself with remaining battery power; he could steam at high speeds for days or even weeks rather than minutes or perhaps hours.

The *Nautilus,* however, was a noisy submarine. Extraordinary vibrations of uncertain origin afflicted the submarine. Vortex shedding—the eddy or whirlpool motion of water at the base of and behind the sail—caused the sail structure to vibrate. When at 180 cycles per minute the sail vibration frequency came dangerously close to the natural tendency of the hull to flex or vibrate as it passed through the water. If the two frequencies came into harmony, the *Nautilus* could have suf-

*The* Nautilus, *with her forward diving planes in the folded, or retracted, position. There are windows at two levels within her fairwater, a feature deleted in later generations of U.S. nuclear submarines. The* Nautilus *had a "conventional" hull design, unlike the first Soviet nuclear submarine, Project 627/November.* (U.S. Navy)

fered serious structural damage.[24] In the process of exploring the sail vibration, Navy engineers uncovered "excessive" vibrations in the hull at speeds above 16 knots.

Beyond the structural problems, these vibrations affected the submarine's performance. Commander Wilkinson wrote,

> Noise generated by hull and superstructure vibration is so great that *Nautilus* sonar capability is practically nil at any speed over 8 knots. This intolerable situation reduces its military effectiveness sufficiently to materially restrict the tactical advantages inherent in nuclear power. Furthermore, it endangers the safety of the ship.[25]

After riding the *Nautilus*, the Commander Submarine Force, U.S. Atlantic Fleet wrote:

> Vibration and superstructure noise prohibit normal conversation in the torpedo room at speeds in excess of 8 knots. It is necessary to shout to be heard in the torpedo room when the ship is in the 15–17 knot speed range. The noise renders worthless all of the installed sonar, active and passive. With the present bow configuration the high performance BQR-4 passive sonar is spare gear. The crude superstructure form is believed partially responsible for the unacceptable hydrodynamic noise generated at maximum speed. It is a serious problem because *Nautilus* realizes its greatest tactical advantages at flank speed where hydrodynamic noise is [at] the maximum.[26]

Modifications were made to remedy the *Nautilus*'s problems. But as a Navy historian wrote: "Even with its destructive forward hull vibration eliminated, *Nautilus* remained notoriously loud and easily detected. Thus it became a floating operational laboratory for a wide variety of self-noise investigations.[27]

During the two years that the *Nautilus* was the world's only nuclear-propelled submarine she proved to be a most valuable technical and tactical laboratory. After several operations under the edge of

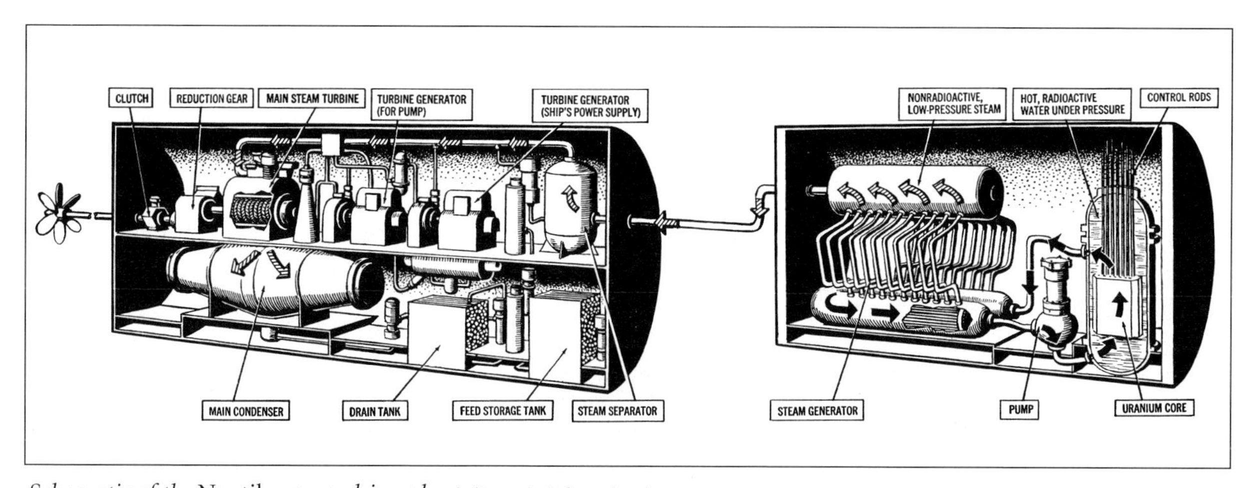


*Schematic of the* Nautilus *propulsion plant.* (Atomic Submarines)

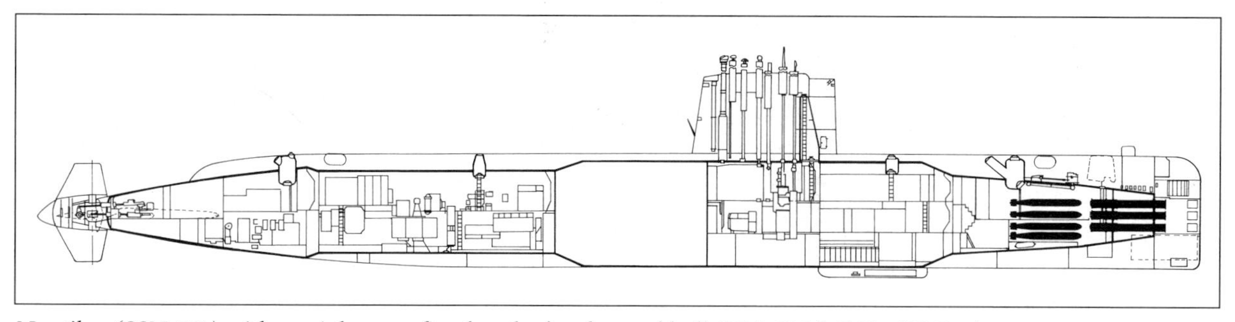

Nautilus *(SSN 571) with special sensor fitted under her forward hull. LOA 323 ft 8½ in (98.7 m)* (©A.D. Baker, III)

the Arctic ice pack, and one aborted effort to sail under the ice to the North Pole, the *Nautilus* transited from Pearl Harbor to England, passing under the North Pole on 3 August 1958—the first ship ever to reach the "top" of the world. Under her second commanding officer, Commander William R. Anderson, the ship operated under the ice for four days, steaming 1,830 n.miles (3,390 km). The polar transit was undertaken at the direct request of President Dwight D. Eisenhower, who was hoping to recapture the image of American technological leadership after the Soviet space triumphs of 1957–1958. Indeed, the mission was top secret until the submarine, having emerged from the ice, came to the surface near Iceland. There a helicopter picked up Anderson, flew him to the U.S. air base in Iceland, and a transport whisked him to Washington, D.C., where he was decorated in a White House ceremony at which the president revealed the polar operation. Anderson was then flown back to his submarine for the triumphant entry of the *Nautilus* into Portland, England.[28] Upon her return to the United States the *Nautilus* entered New York Harbor (at this writing the only nuclear-propelled ship to have visited the city).

The *Nautilus* continued in active Navy service as both a laboratory and an operational combat submarine until 1979.[29]

## The Liquid-Sodium Plant

Even while the *Nautilus* was on the building ways, the Navy ordered a second nuclear submarine, the *Seawolf*. She would be generally similar to the *Nautilus*, except that she would have a "sodium-cooled" Submarine Intermediate Reactor (SIR) plant. Sodium was a more efficient heat exchanger than pressurized water and held the promise of higher speeds. In place of pressurized water, liquid sodium would be employed as the heat exchange medium between the reactor and the steam generators. Another fluid system, using sodium-potassium, would be used to separate the sodium loop from the water-steam plant, adding further complexity to the plant. The Navy built a land-based SIR Mk I prototype at West Milton, New York. Later designated S1G, it was used for development and crew training. The SIR program would use the nation's entire annual production of liquid sodium (NAK), which was highly toxic and highly corrosive to steel.[30]

The almost identical SIR Mk II plant (S2G) would be installed in the *Seawolf*. The *Seawolf* was ordered from Electric Boat in July 1952, a little less than a year after the *Nautilus* was ordered. The second nuclear submarine would be slightly larger because of the additional shielding required for the sodium-cooled plant, displacing 3,420 tons on the surface and 4,287 tons submerged, with a length of 337½ feet (102.9 m). Also, her bow sonar configuration was different, and her sail structure had a stepped conning position. The *Seawolf*'s bow was modified because while the *Nautilus* was being built model tests showed that she would not reach

*The* Seawolf *in the Thames River, near her birthplace of Groton, Connecticut. She differed from the* Nautilus*—in addition to her propulsion plant—in having a different bow configuration and a step conning tower. The* Seawolf*'s liquid-sodium plant suffered problems, but many participants in the program felt that they could be solved.* (U.S. Navy)

her designed surface speed; the improved *Seawolf* bow provided a gain of three knots on the surface at a cost of 1/10th of a knot submerged, while providing a more effective bow sonar installation.

The *Seawolf* was launched on 21 July 1955. Completion of both the land-prototype and the *Seawolf* were delayed because of the complexity of the sodium plant. According to Captain Robert Lee Moore Jr., who became the Navy supervisor of shipbuilding at Electric Boat in 1952, the use of sodium-potassium was made "without paying any attention to the NRL [Naval Research Laboratory] reports on catastrophic failures of stainless steel in superheated steam with as little as ten parts per million of potassium present."[31]

The *Seawolf*'s reactor plant attained self-sustaining chain reaction (became critical) for the first time on 25 June 1956. During dockside tests on 20 August, a failure occurred in the starboard steam superheater, and two steam leaks appeared in the piping connected to the superheater of the steam-generating plant. The plant was shut down, and it was determined that the leaks were caused by the action of the sodium-potassium alloy, which had leaked into the superheated steam piping. During the shutdown other leaks were found in the starboard steam generator.

More than three months were needed to modify the *Seawolf*'s steam-generating system to correct the problems and resume dockside tests. The steam side of both superheaters was bypassed. On 1 December, however, another sodium-potassium leak was detected in the starboard steam generator, requiring further repairs.

The *Seawolf* belatedly began sea trials on 21 January 1957. She could operate at only 80 percent of her designed power. Finally, on 30 March 1957, the world's second nuclear-propelled submarine was placed in commission.

Commander "Dick" Laning took the *Seawolf* to sea. She operated primarily in anti-submarine exercises. On 26 September 1957, President Eisenhower became the first U.S. chief executive to go to sea in a nuclear submarine as the *Seawolf* cruised off New England.

There were periodic problems with the plant—and some unusual phenomena. Sometimes, for example, the *Seawolf*'s hull would glow in the dark. The glow was Cherenkov radiation—a bluish glow emitted by high-speed charged particles as they pass from one medium to another.[32] The radiation, more commonly observed in the water around a nuclear reactor, was not dangerous. But it was novel.

The *Seawolf* demonstrated the underwater endurance of nuclear submarines in the fall of 1958 when Laning took her on a record-breaking submerged run of 13,761 n.miles (25,500 km), remaining submerged for 60 days.[33] The *Seawolf* operated successfully for almost 21 months on her liquid-sodium reactor plant. According to Laning, the *Seawolf* plant was successful. He explained,

> The problem is that sodium is such a superb heat exchange fluid that it can cause very rapid temperature changes, and therefore strong gradients in the containment material.
>
> The 347 Stainless Steel used [in the plant] has a low heat transfer coefficient and a high coefficient of thermal expansion. This means that a sudden temperature swing can cause cracking, especially at points of varying thickness.
>
> Our very successful operation of *Seawolf* resulted from very close matching of sodium flow with power level and thus minimal temperature swings, and those as slow as possible.[34]

Then, "overriding technical and safety considerations indicated the abandonment of the sodium-cooled reactor as a means for propelling naval ships," according to Rickover. A second reactor core was available for the *Seawolf*, and the Bureau of Ships as well as Laning recommended that the submarine be refueled (a three-month process vice almost two years for replacing the sodium plant with a water plant) because of the demand for the early nuclear submarines for fleet ASW training. But Rickover had, on his own authority, already directed that the spare *Seawolf* fuel core be cut up to reclaim the uranium.

In late 1958 the *Seawolf* returned to the Electric Boat yard, where she was torn open, her sodium plant was removed, and a duplicate of the *Nautilus* PWR plant was installed (designated S2Wa).[35] She

*The* Skate-*class submarines—the* Sargo *shown here—were a production version of the basic* Nautilus *design. The* Sargo's *radar mast and snorkel induction mast are raised. She was the first nuclear submarine to be built on the West Coast (Mare Island Naval Shipyard).* (U.S. Navy)

returned to service on 30 September 1960. With her new plant the *Seawolf* continued in service until early 1987, most of the later stages of her career being spent in classified research, intelligence, and seafloor recovery work. For these special operations, the *Seawolf* was again taken into the Portsmouth Naval Shipyard where, from May 1965 to August 1967, she was refueled and was cut in half for an additional compartment to be added to provide access to the open sea for saturation divers, that is, divers using mixed gases for sustained periods to work at depths of 600 feet (185 m) and greater.[36]

Although Rickover had halted the further development of sodium-cooled reactor development within his organization, the Office of Naval Research (ONR) continued looking into alternate reactor plants because of the promise of lighter or more capable power plants with the use of more efficient heat transfer methods. Rickover became furious after the subject was raised at a conference sponsored by ONR in 1974. The Chief of Naval Research, Rear Admiral Merton D. Van Orden, personally delivered a transcript of the conference to Rickover who, using profane language, threatened to recommend to Congress that ONR be abolished unless Van Orden halted such activities.[37] Chief of Naval Operations Elmo R. Zumwalt, who was supporting the effort, sponsored a follow-on conference on lightweight reactors. However, Rickover, by cowing the Navy's civilian secretariat, forced ONR to abandon the effort. There were other efforts by firms directly involved in nuclear propulsion programs to examine possible alternatives to the PWR plant. All were rejected by Rickover.

In this period, an official history of Navy nuclear propulsion summarized the situation:

> The technical competence of Code 1500 [Rickover's group] was unquestionably strong, even outstanding, but was it wise to let one technical group decide what path the Navy would follow in developing nuclear propulsion? . . . Although Rickover's technical judgment in this case seemed correct, the absolute certainty with which he asserted his opinion did not help to convince others that Code 1500 was open-minded on the subject of new reactor designs. It was tempting to conclude that Rickover was simply trying to establish a monopoly to keep himself in power.[38]

Significantly, beyond the *Nautilus* and *Seawolf* plants, back in 1951, Rickover had received authority to begin development of a land-based prototype for a propulsion unit capable of driving one shaft of a major surface warship. A short time later Rickover would also initiate three more reactor programs: a plant one-half the size of the *Nautilus,* that is, 7,300 horsepower; a twin-reactor submarine plant; and a small plant for a hunter-killer submarine. All would use pressurized-water reactors. Thus, a vigorous nuclear propulsion effort was under way by the time the *Nautilus* went to sea, all employing PWR plants. The Navy had earlier considered a gas turbine plant for the third nuclear submarine. The complexity of that plant led to it being discontinued at an early stage. Rather, the third U.S. nuclear submarine—the *Skate* (SSN 578)—would have a PWR derived from the STR/S2W of the *Nautilus.*

## The *Skate* Class

The size and cost of the *Nautilus* and *Seawolf* were considered too great for series-production submarines.[39] The solution was to revert to the basic *Tang*-class design, using a smaller reactor plant with the accompanying reduction in speed. However, a large-diameter pressure hull would be needed for the reactor and its shielding, resulting in a larger submarine. These scaled-down, simplified STR/S2W plants had two different arrangements, hence the first two were designated S3W and the next two S4W.

The *Skate* retained the derivative Type XXI hull design. Like the *Tang* (and Type XXI), the new SSN would have a BQR-2 sonar as well as the SQS-4 active scanning sonar. Forward the *Skate* would have six torpedo tubes, while aft two short torpedo tubes would be fitted for launching torpedoes against enemy destroyers "chasing" the submarine. A total of 22 torpedoes could be carried.

The *Skate* design was essentially completed before the *Nautilus* went to sea, and within six months of that event, on 18 July 1955, the *Skate* and her sister ship *Swordfish* (SSN 579) were ordered. Two months later another pair of SSNs were ordered, the *Sargo* (SSN 583) and *Seadragon* (SSN 584). Before the year was out two more nuclear submarines were ordered—the high-speed *Skipjack* (SSN 585) and the giant *Triton* (SSRN 586). The Navy was thoroughly committed to nuclear submarines the same year that the *Nautilus* went to sea!

Electric Boat designed the *Skate* and would construct her and another of the class, the *Seadragon.* The *Swordfish* was ordered from Portsmouth, marking that yard's entry into nuclear submarine construction, while the *Sargo* introduced the Mare Island shipyard (California) to nuclear construction.

*The USS* Skate *in Arctic ice in April 1959. U.S. nuclear submarines practiced under-ice operations to prepare them for penetration of Soviet home waters for intelligence collection and, later, anti-SSBN missions. Note the submarine's vertical rudder in the foreground.* (U.S. Navy)

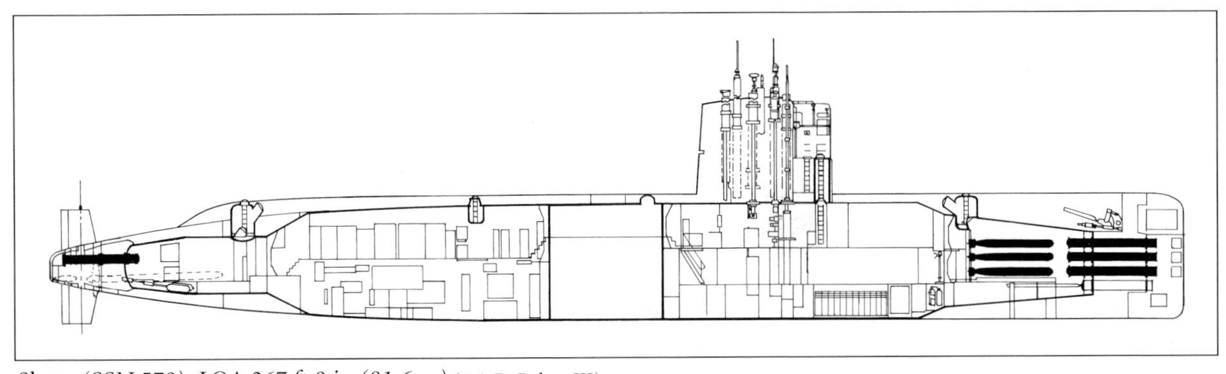

Skate *(SSN 578). LOA 267 ft 8 in (81.6 m)* (©A.D. Baker, III)

The *Skate* was placed in commission on 23 December 1957. She and her three sister ships, completed in 1958–1959, were soon available to help develop nuclear submarine tactics and begin regular overseas deployments by SSNs. The *Skate* undertook under-ice cruises in July 1958. The following month she steamed under the ice pack toward the North Pole, almost simultaneous with the epic cruise of the *Nautilus*. Under Commander James Calvert, the *Skate* reached the top of the world on 11 August. Although unable to surface at the North Pole, the *Skate* did surface nine times in openings in the polar ice during ten days of operations under ice. Subsequently sailing under the ice during the winter season, on 17 March 1959, the *Skate* did surface at the pole while she was spending 12 days operating under the ice pack.[40] For many years the four *Skate*-class submarines were heavily engaged in Arctic research and seafloor mapping as well as fleet operations.[41] The Arctic was an arduous operating area. The Arctic region encompasses 3.3 million square miles (8.5 million km$^2$) of ocean—about the same size as the United States—most of which is covered with pack ice. The extent of the ice varies considerably with the season. Some areas are extremely shallow, although large areas are deeper than 6,000 feet (1,830 m). The surface ice projects downward from about six feet (1.8 m) to more than 20 feet (6.1 m); some ice "keels" of 162 feet (49.4 m) have been observed. On some Arctic transits U.S. submarines have traveled 25 feet (7.6 m) above the seafloor with only seven feet (2.1 m) between the top of the sail structure and the downward-projecting ice.

Acoustic conditions also vary greatly in the Arctic. At times it is remarkably calm. But when the winds blow, the ice begins to move, and there is an incessant creaking noise that can interfere with sonar performance. In some areas the marine life contributes to the Arctic crescendo. Acoustic effectiveness is also affected by the low salinity in the area, as well as the eddies where warm and cold currents further complicate acoustic conditions.

SSN operations under the ice would soon become a significant factor in anti-strategic missile submarine operations. (See Chapter 7.) At the same time, these arctic operations tested, for the U.S. Navy, the concept of operating missile-launching submarines under the Arctic ice—a concept that was never realized.

One other *Skate*-class SSN accomplishment is especially worthy of note: The *Sargo,* shortly after being commissioned, on 21 October 1958 torpedoed and sank the abandoned landing ship *Chittendon County* in the Hawaiian Islands. This was the first ship to be sunk by a nuclear-propelled submarine. (The *Sargo* also suffered one of the earliest nuclear submarine accidents, when, alongside the pier at Pearl Harbor on 14 June 1960, an oxygen fire erupted in her stern compartment; it could only be extinguished by submerging the ship and flooding the space. One crewman was killed.)

"Conventional" SSN construction—based largely on the Type XXI design plus nuclear propulsion—ended with the four *Skate*-class submarines. The fourth *Skate* was followed in the attack submarine series by the *Skipjack,* a very different kind of SSN.

## Special-Mission Submarines

Two other submarine projects are considered to be among the U.S. first-generation nuclear submarines, the radar picket *Triton* and the missile-launching *Halibut* (SSGN 587). Studies of an improved submarine power plant were initiated in 1951, before construction of the *Nautilus* began, with the emphasis on high speed and maximum reliability of operation.[42] From the beginning a two-reactor design was chosen to meet these criteria. In addition to considering pressurized water (as in *Nautilus*) and sodium (as in *Seawolf*) as the coolant/heat transfer fluid, Rickover's staff looked at fused salts, hydrocarbons, lithium, potassium, and various gases. One by one these coolants were discarded from submarine application for technical reasons. By late 1953, after six months of operating experience with the STR/S1W land prototype, the decision was made to proceed with pressurized water. The complexity of a two-reactor propulsion plant required that a land-prototype S3G reactor plant be constructed at West Milton, New York. The similar S4G would be installed in the submarine.

With Rickover exerting his growing influence with Congress, the fiscal year 1956 shipbuilding program included funding for a radar picket submarine (SSRN) with high *surface* speed.[43] The radar picket (SSR) concept grew out of the latter stages of World War II, when Japanese suicide aircraft attacked and sank numerous U.S. picket destroyers that were providing early warning of air attacks against carrier and amphibious forces. The SSR would submerge to escape attack after providing warning of incoming air raids.

The first fleet boat/SSR conversions were undertaken shortly after the war. In all, ten fleet boats were converted, some being lengthened 30 feet (9.1 m) to accommodate electronic equipment and air-plotting spaces. Two new SSRs also were built, the large *Sailfish* (SSR 572) and *Salmon* (SSR 573).[44] However, even as the *Triton* was being ordered from Electric Boat in October 1955, the SSR concept was being overtaken by technology. Land-based aircraft and carrier-based aircraft could provide longer-range and more responsive radar coverage. The few conventional SSRs kept in service after the late 1950s were used primarily in the missile-guidance role to support the Regulus program.

But Rickover justified continuation of the *Triton* project: "The importance of the *Triton* goes beyond the specific military task which has been assigned to her. The *Triton,* in her operations, will test an advanced type of nuclear propulsion plant and will pave the way for the submersible capital ships of the future."[45] A most perceptive U.S. flag officer, Admiral I. J. (Pete) Galantin, a non-nuclear submariner, has written of a discussion in the mid-1950s with Milton Shaw, an early and important member of Rickover's staff:

> In spite of the success being demonstrated by *Nautilus* and the plans to proceed with the single-reactor *Skate* class, [Shaw] said, Rickover shared the uneasiness of senior engineers in the reactor development community about reliance on a single reactor. Difficulties being experienced with the liquid-metal cooled reactor prototype plant for *Seawolf,* and with many other AEC reactor programs, may have been a factor. During this pioneering stage, scientific knowledge and theory were not precise enough to overcome tremendous concern about the unknowns that could arise in the engineering and operation of reactors.[46]

In the event, the U.S. Navy would build only one two-reactor submarine plant, the S4G used in the *Triton.*

Captain Edward L. Beach, President Eisenhower's naval aide, who would become the first commanding officer of the *Triton,* later observed that "Rickover was always engineering-oriented. To him the two-reactor *Triton* was much more important than the follow-on *Skipjack* [and] *George Washington* [SSBN 598], despite all the other exotic stuff the *G.W.* had. We had, of course, a surface ship plant . . . and it's

*The* Triton *was the world's largest submarine when she was completed in 1959. She was the only U.S. nuclear submarine with a two-reactor plant; a superlative undersea craft, her radar picket mission was passé by the time she joined the fleet. Note the fairwater opening for her retracted SPS-26 air-search radar.* (U.S. Navy)

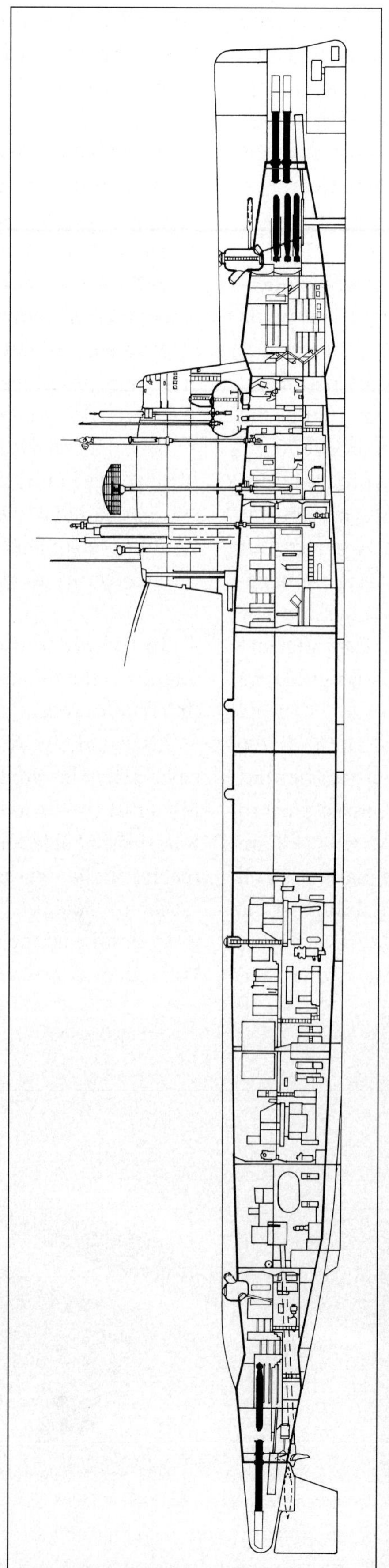

Triton *(SSRN 586). LOA 447 ft 6 in (136.4 m)* (©A.D. Baker, III)

now plain we were the prototype for the surface ship multiple reactor plants then under design."[47]

The *Triton* would also be the world's largest submarine when she was completed in 1959. Previously, the world's largest undersea craft had been the Japanese *I-400* class of submarine aircraft carriers. (See Chapter 15.) The *Triton* was longer, with a slightly greater displacement—the nuclear submarine was 447½-feet (136.4-m) long, displaced 5,662 tons on the surface and 8,500 tons submerged.

The *Triton* had a modified Type XXI hull, optimized for surface operation. Indeed, she probably was the world's only nuclear-propelled submarine to be designed with the same surface and underwater speeds—28 knots. At sea, according to Captain Beach, the *Triton* was "slightly faster surfaced than submerged, though capable of more than 30 knots in either condition, if we slightly exceeded the operating parameters imposed. This we did, with Rickover's approval and under his observation."[48] That event occurred on the *Triton*'s initial sea trials of 27 September 1959, with Rickover on board. The two-reactor plant produced 45,000 horsepower when the "delta T"—the difference between the cold leg and the hot leg of the primary coolant—was increased. Beach believed that the plant could have reached 60,000 horsepower "had that been necessary."[49]

The giant *Triton* retained a classic double hull configuration, with a reserve buoyancy of some 30 percent. The ship had ten compartments: the bow compartment had four torpedo tubes (there were another two tubes aft, with a total of 12 torpedoes being carried); crew's compartment; operations compartment; air control center-"officers' country" compartment; two reactor compartments; No. 1 engine (turbine) compartment; auxiliary machinery compartment; No. 2 engine compartment; and stern compartment. Amidships the ship had three levels.

The sail—the largest of any U.S. submarine in height and length—contained a small conning tower (compartment), the last in a U.S. submarine. The huge sail provided a streamlined housing for an AN/SPS-26 air search radar.[50] It was the first electronically scanned, three-dimension radar, employing frequency scanning for elevation. It could track a high-flying aircraft at some 65 n.miles (120 km) at altitude up to some 75,000 feet (22,866 m). According to Beach, "We very seldom used it, or had a chance to use it. Nearly all of its operating hours were spent in routine testing to keep it functioning."[51] The radar had very low reliability.

The radar antenna and the massive hoisting cylinder took up much of the sail's volume. For stowage the SPS-26 antenna was rotated 90° to the centerline and fully retracted into the sail.

Even before the *Triton* was launched on 19 August 1958, controversy was swirling over the $109-million submarine.[52] Four months earlier the Navy had announced plans to take half of the existing radar picket submarines out of service. The remaining five submarines would follow shortly. Although the subsequent crises in Lebanon and the Taiwan Straits delayed their retirement, the end of the SSR concept was in sight.

The president's wife, Mrs. Mamie Eisenhower, christened the *Triton*. After launching, the *Triton* was completed at Electric Boat, but behind schedule. During 1958–1959 the emphasis at the yard was on Polaris submarines. The *Triton* was to have been commissioned in August 1959, but she did not even get underway on sea trials until 27 September. Belatedly, she was placed in commission on 10 November 1959. She would never serve as a radar picket, and her career would be short.

Speaking at the commissioning, Vice Admiral Bernard L. Austin, the Deputy CNO for Plans and Policy, put a positive perspective on the submarine:

> on this ship will fall the opportunity for demonstrating the practicability of large submarines. Her experience may well point the way for the future course of naval science.
>
> As the largest submarine ever built, her performance will be carefully followed by naval designers and planners the world over.
>
> For many years strategists have speculated on the possibility of tankers, cargo ships and transports that could navigate under water. Some of our more futuristic dreamers have talked of whole fleets that submerge. *Triton* is a bold venture into this field.[53]

Electric Boat workmen added the "finishing touches" to the *Triton*. Then, in late January 1960,

Captain Beach was ordered to a top-secret meeting in Washington on 4 February. There he was informed that the *Triton* was being sent on an underwater, around-the-world cruise, essentially following the course of Ferdinand Magellan's ships (1519–1522). Such a trip would make important contributions to geophysical and oceanographic research and would help to determine the problems of long-duration operations—all important to the Polaris submarine program. However, like the Arctic cruises of the *Nautilus* and *Skate,* the around-the-world cruise was also looked at as a means of demonstrating U.S. technological excellence in the face of continuing Soviet space successes. The highly secret voyage would have the code name "Sandblast," the *Sand,* of course, referring to Beach.

Like the Arctic cruises, an around-the-world voyage would be a dramatic demonstration of nuclear submarine capabilities. Less than two weeks later, on 16 February 1960, the *Triton* departed New London with 184 officers, enlisted men, and civilian technicians on board. Steaming south, the *Triton* was forced to broach her sail above the surface on 5 March to permit a sailor seriously ill with kidney stones to be taken off at Montevideo Harbor and to be transferred to the U.S. cruiser *Macon.*[54] The submarine then rounded Cape Horn, sailed across the Pacific and Indian Oceans, rounded the Cape of Good Hope, and steamed northward. On 2 May Beach again broached the sail to take aboard two officers from a U.S. destroyer off Cadiz. During the cruise the *Triton* was able to maintain continuous radio reception through the use of a buoyant floating cable antenna.

The *Triton* surfaced completely for the first time in 83 days, 19 hours on 10 May, off the coast of Delaware. A helicopter lifted Captain Beach from her deck and flew him to Washington, where, at a White House ceremony, President Eisenhower revealed the voyage. Beach was flown back to the *Triton* and was on board the next morning when the submarine tied up at New London. She had traveled 35,979 n.miles (66,669 km) submerged.[55]

The *Triton*'s achievements, however, were overshadowed in the news by the shootdown of a U-2 spyplane over the Soviet Union on 1 May. Subsequently, the *Triton* carried out routine operations. Without publicity her designation was changed from radar picket to torpedo-attack submarine (SSN) on 1 March 1961. But the *Triton* was too large, carried too few torpedoes, and was too expensive to operate to serve effectively as an SSN. (The SPS-26 radar was not removed.)

As early as 1958—while the *Triton* was still under construction—there were proposals to employ the submarine in roles other than as a radar picket. Lieutenant Commander Walter Dedrick, who would become the first commanding officer of the cruise missile submarine *Halibut,* proposed four possible roles for the *Triton* in an informal paper that he routed to Beach and others: (1) a command ship for a fleet or force commander, (2) advanced sonar scout for the fleet, (3) Regulus cruise missile ship, or (4) minelayer. All but the first would require extensive conversion.

No action resulted from Dedrick's paper nor from several subsequent proposals. The most promising of the latter were to employ the *Triton* as an emergency high-level command ship or an under-ice rescue ship for other nuclear submarines. In the role of a command ship, the Joint Chiefs of Staff or even the president and other senior civilians would be flown by helicopter out to the *Triton* when there was danger of a nuclear conflict. (In the early 1960s the Navy modified a cruiser-command ship and a light aircraft carrier to emergency command posts for presidential use.)

Of more interest to naval planners was the problem of a U.S. nuclear submarine becoming disabled under the Arctic ice. The *Triton*'s twin-reactor, 34,000-horsepower propulsion plant made her the most powerful undersea craft afloat. Should a nuclear submarine become disabled under ice, the *Triton* could bring out spare parts and divers to make repairs, if that were possible. If necessary, the *Triton* also could serve as a tug to tow the disabled craft to open water. At Beach's request, plans were drawn up for this modification, which, he believed, "would have been easy and inexpensive."[56]

In the event, the *Triton* continued to operate in a limited SSN capacity for the next several years until decommissioned on 29 March 1969. She was the world's second nuclear submarine to be taken out of service, the Soviet *K-27* having been the first.

The *Triton,* overtaken by events, was built for the wrong purpose at the wrong time. Her value as a

*The* Nautilus *loading torpedoes. The bulbous dome on the starboard side is the UQS-1 under-ice sonar, a high-frequency sonar normally fitted in minesweepers. The light patch at right covers the tethered rescue buoy, to be released if the submarine is disabled in "rescuable" water.* (U.S. Navy)

development ship for multi-reactor surface ships was small, as was her role in maturing large submarine concepts. If she was a "failure," however, it was an aberration in the U.S. nuclear submarine program. By May 1960, when the *Triton* surfaced from her record underwater cruise, the U.S. Navy had 11 nuclear-propelled submarines in commission and another 17 under construction. At the time the U.S. Navy was far ahead of the Soviet Navy in the production of nuclear submarines. And all evidence available to the U.S. intelligence community indicated that the designs of U.S. submarines were superior.

(The Soviet Navy developed a form of SSR with four Project 613/Whiskey submarines being converted, rejoining the fleet from 1959 to 1963 in the radar picket configuration. These submarines had their conning towers lengthened to provide for installation of a large air-search radar, given the NATO code name Boat Sail, with certain internal equipment and the two stern torpedo tubes and their reloads being removed to provide spaces for electronic gear and aircraft plotting. NATO called the SSR variant Whiskey Canvas Bag, the name derived from the canvas cover sometimes seen over the radar. While the U.S. SSRs were for fleet air defense, the Soviet radar pickets were intended to provide warning of air attacks on Soviet coastal territory.)

**The U.S. Navy had shown an interest in the possibility of nuclear-propelled submarines as early as 1939. During World War II that interest was kept alive by Major General Leslie Groves, head of the atomic bomb project. Immediate postwar proposals for nuclear submarines, while simplistic, demonstrated the continued interest in such ships at all levels of the Navy.**

**By the early 1950s, with solutions being found to the myriad of engineering problems, the construction of the USS *Nautilus* began amidst national publicity and hoopla. By the time the *Nautilus* went to sea in 1955, series production of SSNs as well as the construction of several specialized nuclear-propelled submarines had been initiated by the Navy. Most impressive of the latter was the giant radar picket submarine *Triton;* her construction specifically to demonstrate the feasibility of a twin-reactor submarine in some respects marked the rapidly growing control of H. G. Rickover over the Navy's nuclear propulsion program. While she was a valuable test platform for a two-reactor plant, there was no naval requirement for the *Triton.***

**Rickover, while a technical manager but not a decision maker in developing the first "nukes," was key to ensuring that U.S. nuclear submarines would be relatively safe and would enjoy a high degree of support from Congress. These were both critical issues in the success of the nuclear submarine program in the United States. In that environment total control of the design of submarines was being shifted in the mid-1950s from the Portsmouth and Electric Boat shipyards to the Bureau of Ships in Washington, D.C. The Naval Reactors Branch of the Atomic Energy Commission—soon known as "NR"—and code 08 of BuShips became increasingly powerful, having profound influence on all aspects of nuclear submarine design and construction.**

**The U.S. entry into nuclear propulsion was slow, measured, and cautious.**

TABLE 4-2
**Nuclear-Propelled Submarines**

| | U.S. *Nautilus* SSN 571 | U.S. *Skate* SSN 578 | U.S. *Triton* SSRN 586 | Soviet Project 627 November |
|---|---|---|---|---|
| Operational | 1955 | 1957 | 1959 | 1958 |
| Displacement | | | | |
| surface | 3,180 tons | 2,550 tons | 5,662 tons | 3,087 tons |
| submerged | 3,500 tons | 2,848 tons | 7,781 tons | 3,986 tons |
| Length | 323 ft 8½ in | 267 ft 8 in | 447 ft 6 in | 352 ft 3 in |
| | (98.7 m) | (81.6 m) | (136.4 m) | (107.4 m) |
| Beam | 27 ft 8 in | 25 ft | 36 ft 11 in | 26 ft 1 in |
| | (8.46 m) | (7.62 m) | (11.26 m) | (7.96 m) |
| Draft | 21 ft 9 in | 20 ft 6 in | 23 ft 6 in | 21 ft |
| | (6.68 m) | (6.25 m) | (7.16 m) | (6.42 m) |
| Reactors* | 1 STR/S2W | 1 S3W/S4W** | 2 S4G | 2 MV-A |
| Turbines | 2 steam | 2 steam | 2 steam | 2 steam |
| horsepower | 13,400 | 7,300 | 34,000 | 35,000 |
| Shafts | 2 | 2 | 2 | 2 |
| Speed | | | | |
| surface | 22 kts | 15.5 kts | 28 kts# | 15.5 kts |
| submerged | 23.3 kts | 18 kts | 28 kts# | 30 kts |
| Test depth | 700 ft | 700 ft | 700 ft | 985 ft |
| | (213 m) | (213 m) | (213 m) | (300 m) |
| Torpedo tubes*** | 6 533-mm B | 6 533-mm B | 4 533-mm B | 8 533-mm B |
| | | 2 533-mm S | 2 533-mm S | |
| Torpedoes | 26 | 22 | 12 | 20 |
| Complement | 104 | 95 | 180 | 110 |

Notes: * See Appendix C for U.S. nuclear plant designations; Appendix B for Soviet nuclear plant designations.
** Two submarines built with S3W and two with the similar S4W.
*** Bow; Stern.
# Normal speeds; exceeded on trials. (See text.)

5

# Soviet Nuclear-Propelled Submarines

*Project 627/November was the Soviet Union's first nuclear-propelled submarine design; the lead submarine, the* K-3, *went to sea three and one-half years after the USS* Nautilus. *The sail-mounted and bow sonars of this Project 627A submarine are easily distinguished.* (Malachite SPMBM)

Soviet scientists began considering the possibility of nuclear-propelled ships and submarines soon after World War II. Soviet scientists had been conducting research in the nuclear field as early as 1932, and in 1940 the USSR Academy of Sciences established a senior research committee to address the "uranium problem," including the potential results of nuclear fission. The German invasion of the Soviet Union in June 1941 curtailed nuclear research efforts, with the laboratories that had been conducting research into nuclear physics in Leningrad and Kharkov being evacuated eastward, away from the battles.

Early in the war academicians I. V. Kurchatov and A. P. Aleksandrov, the leading nuclear scientists in the USSR, worked primarily on the protection of ships against magnetic mines at the Leningrad Physico-Technical Institute. Subsequently, Kurchatov went to Sevastopol and Aleksandrov to the Northern Fleet to work on mine countermeasures problems. However, Soviet leader Josef Stalin signed a resolution initiating work on the nuclear weapons program on 11 February 1943, and both Kurchatov and Aleksandrov were ordered to resume their nuclear research activities in Moscow. By that time the Soviet government was aware of the atomic bomb programs in the United States as well as in Germany.

Under the direction of Kurchatov, scientists and engineers were recalled from the military fronts, other research institutes, and industry to work on an atomic bomb. This wartime effort was under the overall supervision of Lavrenti Beria, the head of state security (including the NKVD) and one of Stalin's principal lieutenants. Three months after the atomic bombings of Japan, on 6 November 1945, the anniversary of the October Revolution, Commissar of Foreign Affairs and Stalin sycophant Vlachyslav M. Molotov publicly declared,

> We ought to equal the achievements of modern world technology in all branches of industry and national economy and provide the conditions for the greatest possible advance of Soviet science and technology.... We will have atomic energy too, and much else. [*Stormy prolonged applause. All rise.*][1]

In 1946 P. L. Kapitsa, an internationally renowned physicist, proposed nuclear propulsion for ships. Kapitsa, head of the Institute of Physics Problems, was not involved with nuclear matters, having refused to work on weapons projects. When Beria learned of his interest in nuclear propulsion, he ordered all such considerations to cease. Kapitsa was told to direct his attention to nuclear fuel processing. He continued to refuse to work on weapons. Stalin had him removed from the institute, and Aleksandrov was appointed in his place. (Kapitsa was returned to the post after Stalin's death.)

Aleksandrov, in turn, looked into the issue of nuclear propulsion in 1948. But Beria again demanded that all nuclear efforts go toward weapons development. The first Soviet nuclear weapon was detonated on 29 August 1949 (four years after the first American A-bomb was detonated). By that time Beria and other Soviet leaders were seeking means to deliver nuclear weapons against the United States. Several long-range bomber programs were under way and missile programs were being pursued.

In 1949–1950 the development of a submarine-launched nuclear torpedo was begun. The origins of the T-15 torpedo—to have been the first nuclear weapon employed by the Soviet Navy—being attributed to Captain 1st Rank V. I. Alferov at Arzamas 16, the Soviet nuclear development center.[2] This was one of the most ambitious submarine weapon projects ever undertaken, being intended as a strategic attack weapon to be used against such major Western naval bases as Gibraltar and Pearl Harbor. Western cities were considered to be either too far inland or, if truly coastal cities, their seaward approaches were expected to be too well protected. The submarine carrying a T-15 would surface immediately before launching the torpedo to determine its precise location by stellar navigation and using radar to identify coastal landmarks. The torpedo was to carry a thermonuclear (hydrogen) warhead a distance of some 16 n.miles (30 km).

The T-15 was to have a diameter of just over five feet (1,550 mm) and a length of approximately 77 feet (23.5 m). The 40-ton underwater missile would be propelled to its target by a battery-powered electric motor, providing an undersea speed of about 30 knots.

Obviously, a new submarine would have to be designed to carry the torpedo—*one* torpedo per submarine. The long submerged distances that the submarine would have to transit to reach its targets demanded that it have nuclear propulsion. In 1947 Professor Boris M. Malinin, the dean of Soviet submarine designers, had addressed the concept of nuclear-propelled submarines:

> A submarine must become an underwater boat in the full meaning of the word. This means that it must spend the greater and overwhelming part of its life under water, appearing on the surface of the sea only in exceptional circumstances.... The submarine will remain the most formidable weapon in naval warfare.... If... it is considered that the appearance of superpowerful engines, powered by intranuclear (atomic) energy is probable in the near future—then the correct selection of the direction in which the evolution must go is... the basic condition for the success of submarines.[3]

In the fall of 1952, Soviet dictator Josef Stalin formally approved the development of a nuclear-propelled submarine.[4] His decision came a few weeks after a secret meeting with Aleksandrov. The meeting was the result of a letter that Aleksandrov and Kurchatov had written to Stalin suggesting research into the feasibility of such a craft. On 9 September 1952—at Stalin's direction—the Council of Ministers adopted a resolution for the development of a nuclear submarine, which he formally approved three days later.

The submarine would be capable of carrying a single T-15 torpedo. In addition to the T-15 launch tube, for self-defense the submarine would have two 21-inch (533-mm) torpedo tubes, without reloads.

*Vladimir N. Peregudov*
(Malachite SPMBM)

Design of the submarine—Project 627—began in September 1952 at the Research Institute of Chemical Machine Building (NII-8), which would design the nuclear reactor for the submarine.[5] The following March the SKB-143 special design bureau was reorganized under Captain 1st Rank Vladimir N. Peregudov and was assigned responsibility for design of the submarine.[6] Several hundred engineers and supporting staff from other institutes and bureaus were assigned to SKB-143 (part of its previous staff being reassigned to TsKB-18 to pursue chemical closed-cycle submarines).

Peregudov had been informed of the nuclear submarine project and given his assignment in the fall of 1952 at a meeting in the Kremlin with Viatcheslav Malychev, vice president of the USSR Council of Ministers, one of the most powerful men in the country. Few Navy personnel were to work on the design effort; Peregudov was one of the few exceptions. An engineering specialist, Peregudov was a protégé of Malinin.[7] He had worked on submarine designs since 1927 and had been the initial chief designer of the Project 613/Whiskey submarine.

Peregudov and Malinin had discussed the potential of nuclear propulsion in 1949. As chief designer for Project 627, Peregudov was responsible to the Ministry of Medium Machine Building. The Navy's leadership was not provided with information on the project—it was considered too highly classified.

N. A. Dollazhal would be in charge of designing the reactor for the submarine, under the direction of Aleksandrov, director of the Nuclear Energy Institute, and G. A. Gasanov would design and build the steam generators. Thus Peregudov, Dollazhal, and Gasanov were entrusted with direction of Project 627 and its propulsion plant. Dollazhal had never before even seen a submarine; Peregudov knew nothing of nuclear physics. Their ability to talk to experts was severely restricted by security rules. The participants were warned and warned again not to discuss their work with even their own colleagues unless it was absolutely necessary. Vladimir Barantsev of SKB-143 recalled,

> I was never officially told what we were designing. I had to guess it myself. I didn't know what we were engaged in. The word reactor was never pronounced out loud. It was called a crystallizer not a reactor and it took me about three or even four months to understand what kind of submarine we were designing. . . .
>
> When I looked at the drawings and saw the dimensions of the [propeller] shaft I guessed everything but I was advised not to discuss it with others.[8]

Several types of reactors were considered with various heat exchange fluids for the reactor plant: uranium-graphite (already being used for a power reactor at Obninsk), beryllium oxide, liquid metal (lead bismuth), and pressurized water. The water reactor was selected, although, as in the United States, liquid metal was also regarded as promising, and that effort would be pursued. Pressurized water would require less development time and had the least technological risk. A land prototype of the pressurized-water plant that would propel the submarine was constructed at Obninsk, near Moscow, in the same manner that the prototype plant of the *Nautilus* (SSN 571) was installed at Arco, Idaho.[9]

The facility at Obninsk consisted of three compartments with a single submarine reactor, turbine, and associated heat exchangers, pumps, and related equipment with one propeller shaft. The Obninsk submarine reactor became critical on 8 March 1956. Admiral Pavel G. Kotov recalled the excitement of those days:

> When the . . . reactor was tested, it indicated a speed of 30 knots—a speed which would go on endlessly. It could drive a ship which wouldn't need to surface for a long time. We knew it could reach not only England, it could reach America without surfacing and come back.[10]

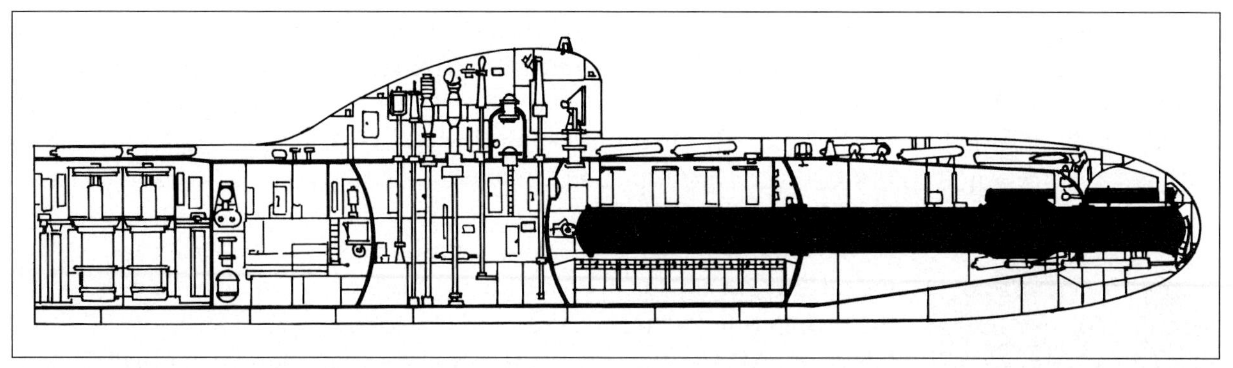

*Project 627 SSN as originally designed to carry a single T-15 nuclear torpedo.* (©A.D. Baker, III)

The Project 627 submarine—named *K-3*—featured a new hull design. Whereas the U.S. *Nautilus* had a modified Type XXI configuration, as used in contemporary U.S. diesel-electric submarines, a totally new hull shape to fully exploit the power of her reactor plant was employed by Peregudov for Project 627. He argued that the submarine should resemble a torpedo in shape, which had been extensively tested in wind tunnels. Barantsev recalled Peregudov's problems:

> The shipyard people were very angry when they realized they had to build a submarine of such a shape. It was a very beautiful shape but very complicated to build. The layers of material which comprised the hull were so thick, and they knew how difficult it was going to be to give such a shape to a submarine when you had to build it with material of such a thickness.[11]

The Project 627 design was based partially on the U.S. *Albacore* (AGSS 569) hull. "We had details of the *Albacore* from magazine photos," according to Russian engineers.[12] The *K-3*'s long, sleek lines of the outer hull were melded into a low, streamlined sail structure—what became known as the "limousine" shape. Aft, Peregudov retained the basic Type XXI stern configuration (which had been rejected by U.S. designers). The "knife" stern featured a single, large rudder, twin propeller shafts penetrating through the horizontal stabilizers, and stern diving planes immediately aft of the propellers.[13]

The classic double hull configuration was retained, providing a reserve buoyancy of about 30 percent, roughly twice that of the *Nautilus*. The Soviets retained "surface unsinkability" in Project 627—and subsequent combat submarines—through the use of double hull, high reserve buoyancy, and internal pressure bulkheads. This meant that the submarine could remain afloat with any one compartment flooded and the loss of two adjacent ballast tanks. This feature would have great significance in later submarine accidents, enabling submarines damaged while underwater to reach the surface. Internally the Project 627 design had nine compartments, three more than the *Nautilus:* torpedo (bow), battery/crew, control, officers, reactor, turbine, two machinery compartments, and the stern compartment. Amidships the reactor compartment contained two VM-A reactors, each providing approximately 70 megawatts thermal power. Two reactors and the related duplication of machinery were deemed necessary for reliability because Soviet designers and engineers were afraid of equipment faults and mistakes by crewmen. The submarine could operate safely with only one reactor on line. Twin steam turbines were provided to generate a total of 35,000 horsepower (almost three times the power of the *Nautilus*).

Electric motors were provided for quiet, low-speed operation, and in an emergency they could propel the submarine at eight knots. In the engineering spaces special efforts were made to reduce vibrations and noise, although shock absorbers were not used for the main turbines.

The double-hull submarine's pressure hull was fabricated of the new AK-25 high-tensile steel, which provided Project 627's increased test depth. The submarine would have a test depth of 985 feet

(300 m), deeper than contemporary U.S. submarines.[14] (The AK-25 steel was first used in the Project 641/Foxtrot-class submarines.)

With the double-hull configuration, an internal hull coating was applied to absorb own-ship machinery noise to reduce the acoustic signature of the submarine. Anechoic coatings on the external hull were applied to reduce the vulnerability to active sonar detection.[15] As a result of the former attribute, the underwater noise level of the submarine when moving at slow and medium speeds was considered to be the same as Project 611 and 613 submarines when using electric motors.

Nuclear submarine noise levels increased considerably at higher speeds. The official history of Russian-Soviet shipbuilding notes that while "the first generation of Soviet nuclear attack submarines, having a good speed quality and an atomic installation two times more powerful than American nuclear ships, significantly lagged behind in stealth."[16]

At the time the issue of acoustic quieting was not regarded by the designers as a high priority. They faced the task of making the submarine independent of the atmosphere and, therefore, safe from detection by surface ships and aircraft. In addition, the methods and means of submarine quieting in the early 1950s were not extensively developed in theoretical or in practical terms.

Meanwhile, the submarine *Narodovolets* (D-2) was fitted as a full-scale research craft for the equipment to evaluate a long-term, sealed atmosphere for nuclear submarines. Her crew lived and worked for 50 days in a sealed environment in support of the nuclear submarine program.[17] For Project 627 a complex system of air-conditioning and oxygen generation was developed for prolonged underwater operations. Chemicals were used to absorb toxic gases in the submarine and regenerate oxygen. Unfortunately, these chemical absorbers were flammable and would be the cause of several fires in first-generation nuclear submarines.

SKB-143 designers worked day and night to complete the design for the *strategic* attack submarine. Wooden scale models for each compartment were made with the latest equipment from the shipbuilding and aviation industries being adopted for the craft. Peregudov monitored each detail of the design. It was intended that a reactor device to produce tritium for the T-15's thermonuclear warhead would be installed in the submarine. The torpedo would be "self-launched," that is, to "swim" out of the tube, a procedure that would increase the length of the torpedo tube by 6½ feet (two meters), but the scheme was significantly simpler than conventional launch systems and would save several tons of weight.

The Project 627 design was completed in 1954, and construction began that year at the Molotovsk shipyard (No. 402) with the formal—and secret—keel laying of the *K-3* taking place on 24 September 1955.[18] The yard is the world's only major shipbuilding facility above the Arctic circle, in the delta of the Northern Dvina River, about 30 miles (48 km) across the delta from the city of Arkhangel'sk. Stalin planned the yard, founded in the late 1930s, as the largest in the world, capable of building battleships. The *K-3* was constructed in the main covered building structure (hall No. 42).[19] Significantly, the first Soviet nuclear submarine was constructed in a horizontal position in contrast to the far less efficient inclined building ways used by the United States and most other nations to build submarines until the 1980s. When ready for launching, the submarine would be pulled from the hall onto a launch dock, which would then submerge to float off the craft.

## Changes in Direction

As Project 627 was getting under way, the country was in a state of political and economic turmoil. When World War II ended, Stalin had allocated major resources to rehabilitate the damaged and destroyed shipyards and initiated the construction of a major oceangoing fleet. Programs were initiated for large battle cruisers, heavy and light cruisers, destroyers, submarines, and even aircraft carriers. In this period series production was begun of advanced diesel-electric submarines as well as chemical closed-cycle undersea craft.

The Commander-in-Chief of the Navy, Admiral Nikolai G. Kuznetsov, was aware that a nuclear-propelled submarine was being developed, but so great was nuclear weapons secrecy in the USSR that he was given no information about its giant nuclear torpedo. When the design was completed and he was briefed on the project in July 1954, he is reported to have declared: "I don't need that kind of boat."[20]

A Navy panel chaired by Rear Admiral Aleksandr Ye. Orël reviewed the project and recommended that the submarine be changed to a torpedo-attack craft for use against enemy ships.[21] Admiral Kuznetsov supported the recommendations of Orël's panel. Although his tenure was to end soon, Kuznetsov's opinions carried great weight. (When he suffered a heart attack in May 1955, Kuznetsov was effectively replaced by Admiral Sergei G. Gorshkov, who would formally relieve him in 1956.)

In 1955—in response to the Navy's objections and recommendations—the Tactical-Technical Elements (TTE) requirement for Project 627 was revised for attacks against enemy shipping, to be armed with conventional (high-explosive) torpedoes. The forward section of the submarine was redesigned for eight 21-inch torpedo tubes, with 12 reloads provided, a total of 20 weapons. (Later nuclear-warhead 21-inch torpedoes were added to the loadout of conventional torpedoes.) Advances in torpedo launching equipment permitted firings to depths of 330 feet (100 m) for the first time.[22] The submarine was fitted with the Arkitka (Arctic) M active/passive sonar in her sail and the Mars-16KP passive sonar in the bow. The latter was an array sonar somewhat similar to the German GHG.

The *K-3* was launched on 9 August 1957. Just over a month later—on 14 September—the *K-3*'s nuclear plant was started up and she went to sea for the first time at 10:30 A.M. on 4 July 1958. Academician Aleksandrov, on board for the trials, wrote in the ship's log, "For the first time in [the] country's history steam was produced without coal or oil."[23] Unlike her American counterpart, which went to sea three and a half years earlier, these milestones in the development of the Soviet nuclear submarine were highly secret.

Under the command of Captain 1st Rank Leonid G. Osipenko, the *K-3*'s trials were highly successful. For safety reasons it had been decided to operate the twin reactor plant at only 60 percent power until the completion of test runs with the Obninsk land prototype. At 60 percent power the *K-3* reached a speed of 23.3 knots, which was "quite a surprise, since the speed obtained was 3 knots higher than expected."[24] That was also the maximum speed of the USS *Nautilus.*

Subsequently, when operated at maximum power, the *K-3* was able to reach approximately 30 knots, although 28 knots is often listed as the maximum operating speed for the class. During her trials the *K-3* reached a depth of 1,017 feet (310 m), a world record for a military submarine.

The senior assistant (executive officer) of the *K-3,* Captain 2d Rank Lev M. Zhiltsov, recalled of the trials:

> When in the tests the reactor drove the submarine to standard speed, everyone on the bridge was shaken . . . by the quietness. For the first time in all my duty on submarines, I heard the sound of the waves near the bow end. On conventional submarines, the sound of the exhaust from the diesel engines covers everything else. But here there was no rattling and no vibration.[25]

The *K-3* carried out extensive trials, albeit with some problems. There were leaks in the steam generators, with the crew having to periodically don respirators while they searched for the leaks. The early generators were found to have an extremely short service life; those initially installed in Soviet nuclear submarines began to leak after some 800 hours of operation. "We felt like heroes," recalled one commanding officer of a Project 627A submarine when his engineers were able to extend the operating time to 1,200 hours. (Tests ashore had demonstrated that the operating time before failure should have been 18,000 to 20,000 hours. The long-term solution was to change the material in the steam generators, the design itself having been found sound and providing benefits over the similar U.S. system, such as higher operating temperatures and hence greater power.)

Still, the trials of the *K-3* were successful trials, and in 1959 Osipenko was honored with the decoration Hero of the Soviet Union, and his crew received other decorations. Designer Peregudov was awarded the title Hero of Socialist Labor, the USSR's highest award for a designer, and SKB-143 and the Molotovsk shipyard were awarded the coveted Order of Lenin for their accomplishments.

*A detailed view of a Project 627A/November SSN. The ship has a streamlined sail typical of Malachite-designed nuclear submarines. This is the* K-8 *in trouble off the coast of Spain in April 1970; she sank soon after this photo was taken. Her forward diving planes are extended.* (U.S. Navy)

The same year that the pioneer *K-3* went to sea, the Soviet government made the decision to mass produce nuclear-propelled submarines. Admiral Gorshkov recalled,

> the major meeting, also dedicated to the future development of the Navy, which took place in Moscow in 1958 was for me very memorable. The fleet commanders, other leading Navy admirals, representatives of the General Staff, the CinCs of component services, and workers of the Party Central Committee and Council of Ministers took part in it. The correctness of the priority construction of nuclear submarines with missile armaments was supported. N. S. Khrushchev spoke in favor of creating about 70 nuclear submarines with ballistic missiles, 60 with antiship cruise missiles, and 50 with torpedoes.[26]

The mass production of nuclear-propelled submarines was approved.

Although the *K-3* was referred to as an "experimental" ship in some Soviet documents, it was, in fact, the lead ship for series production nuclear submarines, unlike the one-of-a-kind USS *Nautilus.* Production was initiated at the Molotovsk/Severodvinsk yard. Minor changes were made to the design, especially improvements to the nuclear power plant, controls, and sonars, and the production submarines—Project 627A—were equipped with improved diving planes and rudders for better control at high speeds. In these submarines the Arkitka M was fitted in the bow (vice sail) and the Mars-16KP was replaced by the MG-10 passive sonar located in the upper portion of the bow.

The first Project 627A submarine, the *K-5,* was laid down in hall No. 42 in August 1956 (a year before the *K-3* was launched). A total of 12 Project 627A submarines were placed in commission from December 1959 through 1964, demonstrating the early and determined support of nuclear-propelled submarines by the Soviet government. On trials—with power limited to 80 percent—the *K-5* reached 28 knots.

The overall success of the *K-3* demonstrated the achievements of Soviet scientists and engineers in the depths as well as in space. In mid-1962 the *K-3* was modified, with her sail structure reinforced, and a special, upward-looking sonar, underwater television cameras, and high-latitude navigation equipment were installed. On 11 July the *K-3,* now

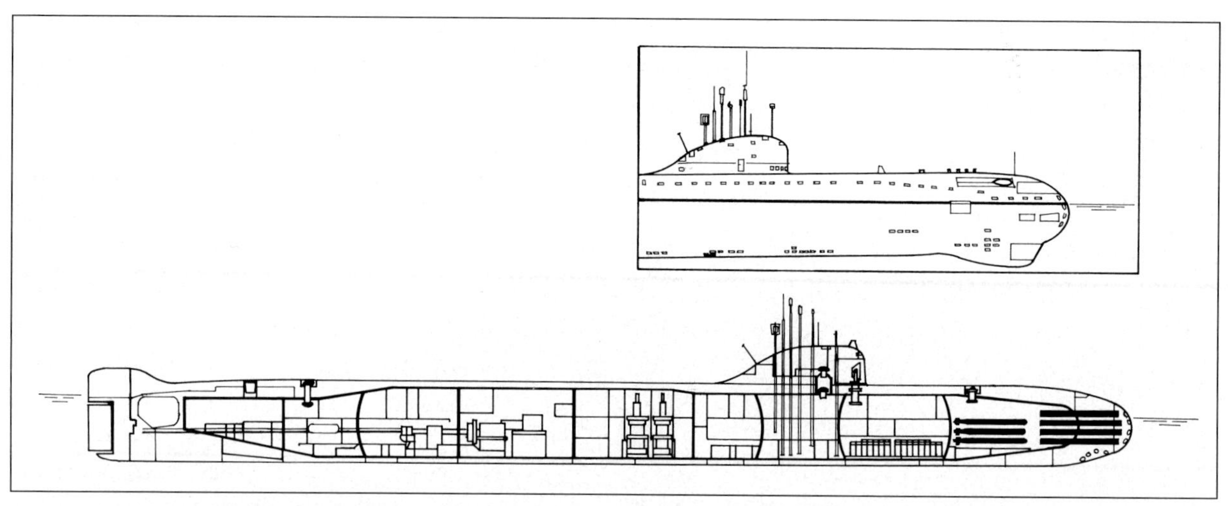

*Project 627A/November SSN. Inset shows* K-3 *with bow-mounted sonar. LOA 352 ft 3 in (107.4 m)* (©A.D. Baker, III)

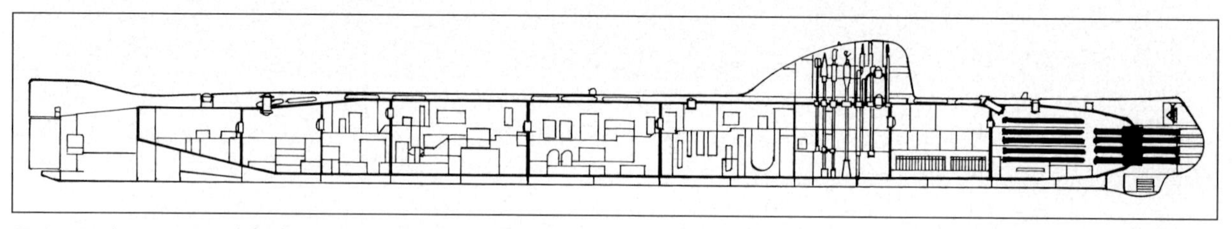

K-27 *Project 645/Modified November SSN. LOA 360 ft 3 in (109.8 m)* (©A.D. Baker, III)

under the command of Zhiltsov, departed her base at Gremikha on the eastern Kola Peninsula and steamed north.[27] The submarine surfaced on 15 July at 84° North latitude, some 360 n.miles (666 km) from the geographic top of the world. Rear Admiral Aleksandr I. Petelin, the submarine flotilla commander, and Zhiltsov went onto the ice and raised the flag of the USSR. The *K-3* reached the North Pole on 17 July but was unable to surface because of the thickness of the ice, estimated at 39 feet (12 m). She did surface several additional times through ice before making a 24-knot run back to Gremikha, arriving on 21 July 1962.

At Gremikha the *K-3* was met by Nikita Khrushchev and other government leaders. They were on a visit to the Northern Fleet, which included viewing extensive displays of the newest naval weapons as well as submarines, and witnessed the underwater launch of a ballistic missile. Khrushchev personally decorated Zhiltsov, his engineer officer Engineer-Captain 2d Rank R. A. Timofeyev, and Rear Admiral Petelin with the nation's highest decoration—Hero of the Soviet Union. (Timofeyev apparently did not receive notice of the award presentation. While at work, wearing dungarees, he was told to report immediately to the meeting hall. Upon his arrival Khrushchev complimented him as a "hero in blue dungarees.") The submarine was honored on 21 July by being given the name *Leninsky Komsomol,* named for the youth group honoring V. I. Lenin. On 29 September 1963 the Project 627A submarine *K-181,* commanded by Captain 2d Rank Yu. A. Sysoyev, surfaced precisely at the North Pole. Those men not on watch went out onto the ice, and the Soviet flag and naval ensign were raised. The crew then engaged in sports on the ice. (Sysoyev also was named a Hero of the Soviet Union for the submarine's polar exploit, while for this and other operations, the *K-181* in 1968 became the first warship since World War II to be awarded the Red Banner Order.) Arctic operations by nuclear submarines became a regular activity.[28] Another submarine of this type, the *K-133,* took part in the February-March 1966, around-the-world Arctic cruise in company with two nuclear-propelled ballistic missile submarines, traveling submerged some 20,000 n.miles (37,000 km) in 54 days.

During their careers the *K-3, K-5,* and *K-11* had the more advanced and reliable VM-AM reactor plants installed. When the submarines were overhauled at the Zvezdochka shipyard in Severodvinsk during the 1960s, their reactor compartments were cut out and the VM-AM plants (built at the Severodvinsk shipyard) were installed.

In 1958, as the *K-3* was being completed, the U.S. Chief of Naval Operations, Admiral Arleigh Burke, observed that "the Russians will soon have nuclear submarines. . . ." In August 1959, shortly after visiting the unfinished nuclear icebreaker *Lenin* at the Admiralty shipyard in Leningrad, Vice Admiral H. G. Rickover, stated,

> I think unquestionably we are ahead of them. I dislike saying this because I am responsible for naval atomic propulsion, but as far as we know, the only marine propulsion plant they have is in the *Lenin*, and it has not yet operated at sea. We have had naval [nuclear] plants operating since 1953. Mr. [F. P.] Kozlov [Soviet First Deputy Premier], when he was in the United States in July, told me that they are building atomic-powered submarines.[29]

Four months after the *Lenin* went to sea, and with the first few Soviet nuclear submarines already completed, in January 1960, Admiral Burke told the U.S. Congress that "there are indications that the Soviet Union is engaged in a nuclear submarine building program" and that "it must be expected that the Soviets will have nuclear-powered submarines in operation in the near future." Accurate U.S. intelligence on the Soviet nuclear submarine program was lacking in the pre-satellite era, with some Western officials questioning at that time whether the Soviets could in fact construct a nuclear submarine with their existing technology base.

Indeed, Project 627 (given the NATO code name November) was the first of several Soviet nuclear-propelled submarines that Western intelligence "got wrong." It was inconceivable to U.S. intelligence and engineering analysts that the Soviets had installed two reactors in the submarine, generating 35,000 horsepower. Thus, in January 1968, when the U.S. nuclear-propelled aircraft carrier *Enterprise* departed San Francisco for Pearl Harbor, Hawaii, there was little concern when intelligence sources (primarily the seafloor Sound Surveillance System [SOSUS]) revealed a November-class SSN closing with the carrier and her escorts.

As the submarine approached the carrier force, the U.S. warships accelerated. The November was believed to have a maximum speed of 23 to 25 knots. The *Enterprise* force accelerated, the carrier being accompanied by a nuclear-propelled and an oil-burning escort ship. Available reports differ as to the speed reached by the task force—some sources say as high as 31 knots. The November kept pace with the carrier. This incident would have a profound effect on the U.S. nuclear submarine program. (See Chapter 17.)

However, propulsion plant reliability problems would plague Project 627/November submarines and cause several major casualties. Soviet nuclear submarines did not deploy to the Western hemisphere during the Cuban missile crisis of October 1962 because of continuing problems with their unreliable steam generators. Only Soviet diesel-electric submarines participated in that superpower confrontation.[30]

On 8 September 1967 the *K-3* was in the Norwegian Sea, returning to her base on the Kola Peninsula, after 56 days at sea. Early that morning a fire flared up in the torpedo compartment, probably from the ignition of hydraulic fluid. The sailors there had no time to use fire extinguishers and fled to the second compartment, which was closed off from the rest of the submarine.

At the time the submarine was steaming at a depth of 165 feet (50 m). Despite the use of respirators, many crewmen suffered carbon-monoxide poisoning and 39 crewmen died. The survivors were able to sail the *K-3* back to port.

Late on the night of 8 April 1970, as Soviet naval forces were conducting a multi-ocean exercise known as Okean (Ocean), the submarine *K-8* suffered an engineering casualty while she was operating submerged at a depth of 395 feet (120 m) in the Atlantic, off Cape Finisterre, Spain.[31] A spark ignited a fire in the flammable chemicals of the air regeneration system. The submarine was able to reach the surface, but smoke and carbon dioxide forced most of the crew onto deck. Soon the ship was drifting without power, because the reactors had been shut down and the auxiliary diesel generators ran for only one hour when they suffered a problem with their cooling system.

After three days on the surface, while the crew attempted to save the *K-8* in the face of strong gales, the submarine sank into the depths on 12 April. Half of her crew of 104 was saved by Bulgarian and Soviet merchant ships; 52 men were lost, among them her commanding officer.[32] This first Soviet nuclear submarine loss occurred after the sinking—with all hands—of the USS *Thresher* (SSN 593) in 1963 and USS *Scorpion* (SSN 589) in 1968.

The casualties in Soviet submarines led U.S. Secretary of the Navy John F. Lehman to declare, "We know there have been some catastrophic health-impairment incidents."[33] One U.S. submarine analyst observed, ". . . if Western submarines had one-tenth of the nuclear reactor casualties and engineering breakdowns that the Soviets have, there would probably be no Western nuclear submarines."[34]

Still, Project 627/November submarines served in the Soviet Navy until the early 1990s. The lead ship, the *K-3*, was decommissioned in January 1988 after almost 30 years of service. She is moored at Gremikha on the Kola Peninsula. (There are plans to preserve the ship as a museum/memorial should funds become available.)

## The Lead-Bismuth Plant

Simultaneous with the development of the pressurized-water plant for Project 627, the Council of Ministers approved parallel development of a torpedo-attack submarine reactor plant employing liquid metal (lead bismuth) as the heat exchange medium. Design of the submarine began under Peregudov at SKB-143 in late 1953, parallel with the design of Project 627. The liquid-metal design was completed in 1956, with A. K. Nazarov having become the chief designer in 1955. It was given the designation Project 645.[35] As with the water-cooled submarine reactor, a land prototype of the Project 645 plant was constructed, also at Obninsk. It became critical in March 1958.

Like U.S. engineers, the Soviets realized that liquid metal offered a higher heat exchange capacity than did pressurized water and thus promised more efficient submarine reactor plants. Also, the liquid-metal plant would operate at far lower pressure, which could simplify some plant components.[36] While the U.S. Navy employed sodium as the heat exchange material, the Soviets selected an alloy of lead and bismuth as the coolant. Although lead bismuth was inferior in thermal properties, it was less chemically active and less dangerous than sodium in the event of an accident.

The Project 645 propulsion system would have two VT reactors, each rated at 73 megawatts thermal power, slightly more powerful than the VM-A reactors of Project 627. The steam generators of the two types of plants had major differences, with the Project 645 design making pipe surfaces accessible for the plugging of individual pipes in the event of failure, that having been a problem that plagued the early Project 627A submarines.

The use of lead bismuth had negative aspects. The alloy had to be continuously heated to keep the coolant from hardening. Thus the plant could not be completely shut down, as could pressurized-water reactors. Upon returning to base the submarine had to be connected to the base's steam-

*Project 645 was a test bed for the lead-bismuth reactor plant, with two VT reactors. Externally, the* K-27 *resembled an enlarged Project 627/November SSN. Like her American counterpart, the USS* Seawolf, *the* K-27 *suffered nuclear plant problems, but the Soviet submarine's problems got out of control.* (Malachite SPMBM)

heating system, and only then could the reactor be shut down. This requirement greatly complicated procedures and raised the cost of providing bases for such a submarine.

In a fit of optimism, the designers of Project 645 dispensed with diesel generators and diesel fuel, depriving the submarine of an auxiliary power plant.

The liquid-metal reactor plant required a larger submarine, although with the same number of compartments (nine) as Project 627, and the internal arrangement was different. Project 645 had several improvements over the Project 627 design: the self-sustaining turbogenerator; an automated, remotely controlled (hydraulic) torpedo reloading system to enable reloading the eight bow torpedo tubes in about 15 minutes; and strengthened flat bulkheads between all compartments, which increased the survivability of the submarine over the combination of four spherical (curved) and four flat bulkheads of Project 627A. These features increased the displacement by 340 tons in comparison with Project 627A. For the first time in Soviet shipbuilding, a low-magnetic steel was used for the manufacturing of the outer hull to reduce the submarine's vulnerability to magnetic mines. This allowed a 50 percent reduction in the deperming (demagnetizing) equipment and related cables fitted in the submarine's hull.

Design work for the submarine began under Peregudov at SKB-143 in late 1953 and was completed in 1956, with Nazarov becoming chief designer in 1955. The keel for the *K-27* was laid down in building hall No. 42 at Severodvinsk on 15 June 1958. Work was slowed because of delays in delivery of the propulsion plant, and the *K-27* was not launched until 1 April 1962. She began sea trials in the summer of 1963 under Captain 1st Rank I. I. Gulyayev.

Only minor problems were encountered in the trials. After delivery to the fleet, from 21 April to 11 June 1964, the *K-27* undertook a cruise of 12,425 n.miles (23,024 km), remaining submerged for 99 percent of the time. Even operating in tropical waters caused no problems for the unique propulsion plant.

In the fall of 1966, after continued and extensive operations, the *K-27* was placed in a dry dock for maintenance and upgrades. There it was discovered that the outer hull—made of low-magnetic steel—had suffered numerous cracks that had not been previously found because of the anechoic coating on the outer hull. Major repairs were required. Subsequently, the twin VT reactors were refueled beginning in 1967. The *K-27* returned to sea in September 1967.

A month later the *K-27* suffered the first of a number of serious engineering problems. On 13 October, while at sea, the lead-bismuth alloy burst into the starboard reactor's primary loop. The cause was slag deposits (oxides of the lead-bismuth alloy) blocking the coolant loop (or piping). As a result, two pumps were flooded with the cooled alloy. The starboard steam-generating system was shut down, and the submarine safely returned to base under her own power.

After repairs the *K-27* again went to sea on 21 May 1968. At about 12 noon on 24 May, one of the ship's reactors suffered an accident. As a result, the core cooling procedure was disrupted and the fuel elements overheated. This, in turn, forced radioactive products into the cooling loops and caused radioactive gas to explode into the reactor compartment.

The submarine came to the surface at 12:15 P.M. The port steam-generating plant was shut down, and the subsequent six-hour return voyage back to Gremikha was accomplished with the starboard reactor driving both turbines. Most of the crew of 124 suffered from radiation exposure and were hospitalized; nine men died in this incident.

In early June a Navy commission came to the conclusion that the plant had to be shut down. Wrote an engineer-historian:

> This was essentially a death sentence for the *K-27*—after the alloy was frozen, it would been almost impossible to restart the reactors. But the commission was forced to take this step because of the serious radiation hazard which still remained on the ship.[37]

By 20 June 1968 the operations necessary to freeze the alloy in the port reactor were completed. The *K-27* was dead. The coolant of the starboard reactor was kept liquid as extensive contamination experiments were carried out on the submarine during this period. Studies also were made of the feasibility of returning the craft to service, replacing the lead-bismuth plant with a pressurized water plant as the U.S. Navy had done with the USS *Seawolf* (SSN 575). But more advanced submarines were becoming available, and rehabilitating the *K-27* was not worth the cost.[38] The *K-27* was the world's first nuclear-propelled submarine to be taken out of service.

Neither the United States nor Soviet Union had been successful in these early efforts to put a liquid-metal reactor plant to sea in a submarine. Metallurgy and other supporting technologies were not yet ready for such reactor plants.

Early in its design efforts, SKB-143 considered a missile variant of the Project 627/November SSN. In the United States the concept of a missile-launching configuration had been considered for the *Nautilus*, but it was immediately rejected by then-Captain Rickover, who wanted no complications for the ship. At SKB-143 design proceeded on the P-627A missile variant under Peregudov and Grigori Ya. Svyetayev. A single, large strategic guided or "cruise" missile was to be housed in a cylindrical hangar behind the submarine's sail. The surface launch of the missile was to take place from a carriage that rolled it from the hangar and elevated it to an angle of 16 degrees for launching. Most of this movement would be automated, with the submarine's time on the surface estimated at 6½ minutes.

The missile was to be the P-20, being developed by Sergei Ilyushin's OKB-240 aircraft design bureau. It was to be a land-attack weapon that could carry a nuclear warhead against targets 1,890 n.miles (3,500 km) from the launching ship. The missile was to attain a maximum speed of Mach 3 and was to fly the last portion of its trajectory at low altitude and zigzag. At that time this weapon held more promise for success against land targets than did potential ballistic missiles.

Development of the P-627A began in the summer of 1956 and was completed by the end of 1957, with the issuing of blueprints following soon after. Construction of the first P-627A at the Severodvinsk shipyard began in 1958. However, military interest in strategic weapons was shifting to land-based ballistic missiles, and in early 1960 work halted on the prototype P-627A submarine as well as the planned follow-on Project 653 cruise missile production submarine. (See Chapter 6.) Next, in 1960–1961 draft designs were prepared for completing the unfinished missile submarine as Project PT-627A, with 25½-inch (650-mm) torpedo tubes as well as a new type of array sonar. These efforts ceased in November 1961.

Almost simultaneous with the construction of the Project 627A/November class, Soviet shipyards began building Project 658/Hotel ballistic missile submarines and Project 659/Echo I cruise missile submarines, both intended for the strategic attack role. Designed by TsKB-18 (later named Rubin), these submarines had very different configurations than Project 627, but retained the same propulsion plant and related machinery. Western intelligence labeled this first generation of Soviet nuclear submarine HEN—for the phonetic code names *H*otel, *E*cho, *N*ovember.

## Political Perspective

When World War II ended and the USSR sought to move into the nuclear era, Josef Stalin had ruled the Soviet Union for almost two decades. In the late 1930s he had begun to construct a "balanced fleet"—battleships, aircraft carriers, cruisers, destroyers, and submarines. This ambitious program, intended primarily to confront Great Britain, was halted with the German invasion of the USSR in June 1941 and the ensuing Great Patriotic War.[39] As World War II concluded in Europe and the Far East in 1945, Stalin again initiated a major fleet construction program. Lesser warships were begun in large numbers—the *Skoryy*-class destroyers, the *Sverdlov*-class light cruisers, and several submarine classes. Preparations were under way for building the *Stalingrad*-class battle cruisers and aircraft carriers.

All major naval decisions during this period had required the personal approval, or at least the acquiescence, of Stalin. Admiral Kuznetsov, Stalin's naval commander-in-chief from 1939 to 1947 and again from 1951 to 1956, wrote in his memoirs: "I well remember the occasion when Stalin replied to a request for more anti-aircraft [guns] on ships in the following words: 'We are not going to fight off America's shores.' "[40]

The additional guns were not installed. Kuznetsov continued:

> On the other hand, Stalin had a special and curious passion for heavy cruisers. I got to know this gradually. At a conference in [A. A.] Zhdanov's office I made several critical remarks on the heavy cruiser project. Zhdanov said nothing, as if he had not heard what I said. When we left his office, one of the leading officials of the People's Commissariat for the Shipbuilding Industry, A. M. Redkin, warned me:
>
> "Watch your step, don't insist on your objection to these ships."
>
> He told me in confidence that Stalin had threatened to mete out strict punishment to anyone objecting to heavy cruisers. . . .[41]

Opposition to Stalin's views on naval matters was minimal. He ordered the execution of eight of the Navy's nine flagmen (admirals) in 1938–1939 as part of his massive purge of the leadership of the Soviet armed forces.[42] Many lesser officers also were shot. In his frightening fictional account of the period, *Darkness at Noon,* Arthur Koestler tells how naval hero "Michael Bogrov" was shot because he differed with "No. 1" on whether to build large, long-range submarines, or small, coastal submarines.[43]

On a regular basis Stalin—and his successors Nikita Khrushchev and then Leonid Brezhnev—met with generals, marshals, and admirals to review and approve (or reject) specific military programs. If a new design or technology was under consideration, the senior designers would join them or, frequently, the leader would meet alone with the designers of aircraft, missiles, and other weapons, including submarines. Khrushchev's memoirs are replete with descriptions of meetings with aircraft, strategic missile, air defense, and other weapon designers, as are those of his son, Sergei, a missile guidance engineer.[44]

Stalin died on 5 March 1953; within days stop work orders were sent to Soviet shipyards to halt the construction of major warships. The *Stalingrad*-class battle cruisers and the planned carrier programs were cancelled, and construction of the *Sverdlov*-class light cruisers was cut back, as were destroyer and submarine programs.

Khrushchev, who succeeded Stalin as the head of the Communist Party and of the Soviet state, in 1957 initiated a defense program that would emphasize cruise and ballistic missiles over conventional air, ground, and naval forces. Some Soviet officials would refer to this shift as a "revolution in military affairs."

Further, Khrushchev believed that submarines would be the only important warship in a future conflict: "We made a decision to convert our navy primarily to submarines. We concentrated on the development of nuclear-powered submarines and soon began turning them out virtually on an assembly line."[45]

He continued,

> Aircraft carriers, of course, are the second most effective weapon in a modern navy. The Americans had a mighty carrier fleet—no one could deny that. I'll admit I felt a nagging desire to have some in our own navy, but we couldn't afford to build them. They were simply beyond our means. Besides, with a strong submarine force, we felt able to sink the American carriers if it came to war. In other words, submarines represented an effective defensive capability as well as reliable means of launching a missile counterattack.[46]

Khrushchev sought a naval commander-in-chief who would support his views. He appointed Admiral Sergei G. Gorshkov as Deputy Commander-in-Chief of the Navy in mid-1955, and Gorshkov officially succeeded Kuznetsov as head of the Navy in January 1956.[47] Khrushchev, who had worked briefly with Gorshkov in the Black Sea area during the war, described him as,

> a former submarine captain. He appreciated the role which German submarines had played in World War II by sinking so much English and American shipping, and he also

appreciated the role which submarines could play for us in the event that we might have to go to war against Britain and the United States."[48]

Gorshkov was not a submarine officer. Rather, he had served in surface ships and small craft during the war, becoming a rear admiral at age 32. Khrushchev directed Gorshkov to scrap the fleet's battleships and cruisers, and to instead build a fleet of smaller, missile-armed ships and submarines that could defend the Soviet Union against Western naval-amphibious attacks. Gorshkov was politically astute and would guide the Soviet Navy into the nuclear-missile era, albeit with an initial emphasis on submarine warfare. Under Khrushchev, who would hold office until October 1964, and Admiral Gorshkov, who would command the Soviet Navy until December 1985 *(29 years),* the USSR embarked on a massive program to produce nuclear-propelled submarines.[49]

**The Soviet Union detonated its first nuclear "device" in 1949, four years after the United States had demonstrated a deliverable nuclear weapon.[50] The Soviet Union sent its first nuclear-propelled submarine to sea less than four years after the United States did. The latter accomplishment was a remarkable achievement in view of the enormous bureaucratic, financial, and technological limitations of the Soviet Union. Further, by the late 1950s the Soviet missile, space, nuclear weapons, and bomber programs all had higher priorities for scarce resources than did submarines.**

**The Soviet *K-3* was, in many respects, superior to the USS *Nautilus,* having significantly greater speed, operating depth, and survivability after suffering damage. Also, greater consideration was given to quieting the *K-3.* However, early Soviet submarines suffered major engineering problems, much more serious than their American counterparts. And the U.S. nuclear submarines stressed safety considerations.**

**While a specialized design bureau was established to produce the first Soviet nuclear submarines, additional design bureaus would soon become engaged in nuclear submarine projects. Even before the *K-3* went to sea, the decision was made to undertake series production of the design, and several missile submarine designs were already under way. This was in contrast to the more measured and cautious U.S. entry into the series production of submarines. This difference was a manifestation, in part, of the Soviet decision to concentrate on naval aviation, submarines, and anti-ship missiles, while the U.S. Navy's missions demanded a more "balanced" fleet with major surface warships, aircraft carriers, and amphibious ships as well as submarines.**

**Two other interesting and important considerations of early U.S. and Soviet nuclear submarine designs were (1) the high level of government support for nuclear submarines in both countries, and (2) the U.S. submarine program was being pursued in the full light of publicity, while the Soviet program was conducted with the greatest of secrecy, as was typical in that society.**

Table 5-1
**Liquid-Metal Coolant Submarines**

| | U.S. *Seawolf* SSN 575 | Soviet Project 645 Mod. November |
|---|---|---|
| Operational | 1957 | 1963 |
| Displacement | | |
| surface | 3,721 tons | 3,414 tons |
| submerged | 4,287 ton | 5,078 tons (approx.) |
| Length | 337 ft 6 in (102.9 m) | 360 ft 3 in (109.8 m) |
| Beam | 27 ft 8 in (8.43 m) | 27 ft 3 in (8.3 m) |
| Draft | 22 ft (6.7 m) | 19 ft 2 in (5.8 m) |
| Reactors | 1 SIR/S2G | 2 VT |
| Turbines | 2 steam | 2 steam |
| horsepower | approx. 15,000 | 35,000 |
| Shafts | 2 | 2 |
| Speed | | |
| surface | 19 knots | 15 knots |
| submerged | 21.7 kts | 29 kts* |
| Test depth | 700 ft (213 m) | 985 ft (300 m) |
| Torpedo tubes** | 6 533-mm B | 8 533-mm B |
| Torpedoes | 22 | 20 |
| Complement | 105 | 105 |

Notes: * A speed of 32.2 knots was achieved on acceptance trials.
** Bow.

6

# Cruise Missile Submarines

*The first and only launch of the Regulus II cruise missile from a submarine was made from the USS* Grayback *in 1958. Here crewmen prepare an unarmed training weapon. Cancellation of the Regulus II deprived U.S. submarines and surface ships of a cruise missile capability for two decades.* (U.S. Navy)

Germany was the first nation to attempt to launch missiles from a submarine. During May-June 1942 the German missile test facility at Peenemünde on the Baltic Sea carried out *underwater* U-boat launches of short-range rockets from the *U-511*.

Six rocket-launching rails were welded to the deck of the *U-511*, and waterproof cables were run from the rockets to a firing switch inside of the submarine. The only modification to the rockets was waterproofing them by sealing their nozzles with candlewax. The firing tests from a depth of some 25 feet (7.6 m) were entirely successful. About 24 rockets were launched from the *U-511*, and additional rounds were fired from a submerged launch frame. The slow movement of the submarine through the water had no effect on the accuracy of the rockets. The 275-pound (125-kg) projectiles had a range of five miles (8 km). The only problem encountered was an electrical ground that caused two rockets to fire simultaneously.[1]

Although these were preliminary experiments, *Generalmajor* Walter Dornberger, the head of the Peenemünde missile facility, presented the findings to the Naval Weapons Department, contending that rocket-firing submarines could attack coastal targets in the United States. The Navy predictably rejected consideration of an Army-designed weapon, the rocket rails were removed from the *U-511*, and in July 1942 the submarine departed on her first war patrol.

Subsequently, as the Type XXI U-boat was being developed, a rocket system was developed for attacking pursuing surface ships. The key to this weapon was a very precise passive, short-range detection system (S-*Analage passir*) to detect propeller noise from ASW ships. The submerged U-boat would then launch a rocket at the target. The echo-sounding gear performed well during trials,

*Sailors aboard the* U-511 *prepare rockets for submerged launching, while naval officers and officials from the Peenemünde missile facility watch from the conning tower. These primitive—and successful—test launchers were the harbinger of today's submarine-launched cruise missiles.* (Imperial War Museum)

but the rockets were still in an early stage of development when the war ended.[2]

In the Pacific near the end of the war, a U.S. submarine commander, Medal of Honor–winner Eugene B. Fluckey, experimented with launching rockets from his submarine while on the surface. At Pearl Harbor, Fluckey had an Army multi-barrel, 5-inch (127-mm) rocket launcher welded to the deck of the fleet submarine *Barb* (SS 220) and took on a store of unguided projectiles.

Early on the morning of 22 June 1945, the *Barb* surfaced off the coast of the Japanese home island of Hokkaido and bombarded the town of Shari. The rockets were launched while the submarine was on the surface, at a range of 5,250 yards (4.8 km). During the next month the *Barb* remained in Japanese waters, attacking ships and carrying out five additional rocket bombardments, some supplemented by gunfire from the submarine's 5-inch and 40-mm cannon.[3] The *Barb*'s rocket attacks were the product of one aggressive commander's action, not part of a formal Navy program.

The initial missiles developed by the Soviet and U.S. Navies for shipboard launching were based on the German V-1. The first V-1 "buzz bombs" had struck London on 13 June 1944, a week after the Allied landings at Normandy.[4] German records indicate that 8,564 V-1 missiles were launched; about 43 percent failed or were diverted or were destroyed by defending fighters, anti-aircraft guns, or barrage balloons. Fighters—piston engine as well as the jet-propelled Meteor—could intercept the missiles, and more than a thousand anti-aircraft guns were positioned in belts near the English coast to intercept the missiles. In addition, under Operation Crossbow, Allied bombers sought out V-1 production facilities and launch sites. (Winston Churchill recorded that almost 2,000 U.S. Army Air Forces and Royal Air Force airmen died in those attacks.) Still, 2,419 missiles fell on England and 2,448 on the Belgian port of Antwerp after it was occupied by the Allies.

The V-1 was powered by a pulse-jet engine that could propel the 27½-foot (8.38-m) missile at 400 miles per hour (644 kmh) for about 150 miles (241 km). It was launched by trolley from a rail 157 feet (48 m) long. When the missile had flown a preset distance, the engine cut off and the missile tipped over and dove to earth. The 1,830-pound (830-kg) high-explosive warhead detonated as it struck the ground, inflicting considerable damage. The total weight of a missile at launch was 4,917 pounds (2,230 kg), providing an impressive 2:7 payload-to-weight ratio.

Within a month of their first use, sufficient V-1 components were in American possession to initiate series production of the missile, given the Army designation JB-2 and called "Loon" by the Navy.[5] Both the U.S. Army and Navy drew up plans for the mass production of the missile for use against the Japanese home islands, to be launched from surface ships and from captured offshore islands. The Army Air Forces had "grandiose" production plans to permit 500 JB-2 sorties per day.[6] But the atomic bomb ended the war in the Pacific before the American copies of the V-1 could be used in combat.

Immediately after the war U.S. submariners—seeking new missions for submarines—expressed interest in the possibility of firing missiles from them, and a proposal for the launching of V-1/Loon missiles was drawn up. On 5 March 1946, Secretary of the Navy James Forrestal approved plans to convert two submarines to conduct experimental launches of the Loon missile.

*A Loon test vehicle—the Americanized test version of the German V-1 missile—blasts off from the submarine* Carbonero. *When Japan surrendered in August 1945, the U.S. Army and Navy were planning a massive assault on the Japanese home island with V-1 missiles.* (U.S. Navy)

The fleet-type submarine *Cusk* (SS 348) was provided with a launching ramp, and on 12 February 1947 she launched a Loon while operating off the coast of California—the world's first launch of a guided missile from a submarine.[7] At this time the U.S. Navy used the term *guided* to describe virtually all missiles, whether they were aerodynamic cruise missiles, or ballistic missiles, and regardless of their means of guidance.

The principal distinction was the missile's means of flight: *cruise* or *ballistic.* A cruise missile (like the V-1) uses continuous propulsion and aerodynamic lift (from wings or fins) to reach its target. A ballistic missile is launched in a specific direction on a ballistic trajectory; it is powered for only the first few minutes of flight. Both types of missiles are "guided" in the sense that they can be aimed at a specific target.

The U.S. Navy continued launching Loon cruise missiles from the surfaced *Cusk* during February and March 1947, with the unarmed missiles flying out to 87 n.miles (160 km). The fleet submarine *Carbonero* (SS 337) also was fitted with a launching ramp in 1949, but she was not fitted with a hangar (as was subsequently provided in the *Cusk*). A major change in the Loons from the original V-1 was the provision for radio-command guidance so that submarines or aircraft could guide the missiles during flight. The *Cusk* and *Carbonero* missile launches with the subsonic (Mach 0.6) and relatively primitive Loon demonstrated the difficulty that surface ships would have detecting and intercepting such weapons.

The Loon was never planned as an operational weapon for use by submarines. Rather, the Navy employed it for training crews and to obtain experience with the problems involved with launching missiles from submarines.[8] With this early Loon experience, in 1947 the U.S. Navy initiated the development of several advanced land-attack missiles for use from surface ships and (surfaced) submarines: The Rigel was to be a Mach 2 missile with ramjet propulsion intended for land attack; ramjets were more powerful than turbojets, which were used to propel existing high-performance aircraft. A follow-on, more-capable ramjet missile named Triton was proposed to have a speed of Mach 3.5 and a range of possibly 2,000 n.miles (3,700 km) in later versions.

But ramjet propulsion was a new, difficult technology. Thus the Regulus Attack Missile (RAM) was initiated as an interim submarine-launched weapon.[9] Regulus would be the U.S. Navy's first operational submarine-launched missile and the first Navy missile to carry a nuclear warhead. It was intended to strike land targets in the Soviet Union and mainland China, complementing the carrier-based nuclear strike capability then being developed.

Developed by Chance Vought Aircraft, the Regulus was initially intended to carry a conventional warhead of 4,000 pounds (1,800 kg), in part because of the limited data on nuclear weapons available to the Navy. In 1949 a nuclear warhead was proposed in place of the conventional warhead.

The Regulus had folding wings and tail fin; subsonic propulsion was provided by a turbojet engine with two rocket boosters fitted for launching, the latter falling away into the sea as they were exhausted. The missile's J33-A-18A engine burned high-octane and highly flammable aviation fuel. Still, this fuel was far less toxic and more stable than the liquid fuels that would be used with submarine-launched ballistic missiles.

Regulus guidance was radio command, initially from the launching submarine (if it kept an antenna above the water), or from an aircraft, or from another submarine. This system—known as TROUNCE—required continuous control until the missile was virtually over the target, at which time it would be given the command to dive onto the target.[10] This meant that an aircraft would have to come within tens of miles of the target; alternatively, a submarine would have to be within radar range of the shore to determine its precise location as it guided the missile against a target; that TROUNCE submarine would control the missile for the final 240 n.miles (444 km) of flight. In a multi-ship test of the TROUNCE radio-control system, on 19 November 1957, the cruiser *Helena* launched a Regulus and guided it for 112 n.miles (207.5 km); the submarine *Cusk* then assumed control for 70 n.miles (130 km); the guidance was then given over to the submarine *Carbonero,* which guided the missile for the last 90 n.miles (167 km) to target—a distance of 272 n.miles (504 km) from launching ship to target. This missile impacted about 150 yards (137 m) from the intended target point.[11]

The first Regulus test vehicle flew on 29 March 1951, being directed by radio control from an aircraft. Fitted with retracting wheels, the Regulus took off under its own power, circled the airfield, and landed safely. The first shipboard launch was made from the missile test ship *Norton Sound* on 3 November 1952; the aircraft carrier *Princeton* launched a Regulus missile on 16 December 1952, the first launch from a warship.

Meanwhile, two fleet submarines were converted to a Regulus configuration, the *Tunny* (SSG 282) and *Barbero* (SSG 317), the latter having served as a cargo submarine (ASSA 317) from 1948 to 1954.[12] Both submarines were fitted with a hangar and launching ramp aft of their conning tower, as well as with TROUNCE guidance equipment. The hangar could accommodate two Regulus missiles with their wings folded. The *Tunny* was recommissioned as an SSG on 6 March 1953 and the *Barbero* on 28 October 1955 following their conversions at Mare Island. The first submarine launch of a Regulus missile occurred on 15 July 1953 from the surfaced *Tunny.*

The Regulus missile—with an Mk 5 nuclear warhead—became operational from surface ships in May 1954. That warhead had an explosive force of 50 to 60 kilotons; from 1956 the Regulus would carry

*The converted fleet submarine* Barbero *prepares to launch a Regulus I missile while operating in Hawaiian waters in 1960. The similar* Tunny *had a streamlined fairwater. Both submarines made combat patrols into the North Pacific with two nuclear-armed Regulus I missiles in their hangars.* (U.S. Navy)

the W27 warhead, slightly lighter, but with an explosive force of one to two megatons. Two megatons was more than 100 times the explosive power of the atomic bomb that was dropped on Hiroshima.

The first overseas deployments of Regulus occurred in 1955, when the heavy cruiser *Los Angeles* (three missiles) and the aircraft carrier *Hancock* (four missiles) deployed to the Western Pacific. Nuclear warheads were provided for all of the missiles. Two purpose-built Regulus submarines, the *Grayback* (SSG 574) and *Growler* (SSG 577), were constructed to a modified *Tang* (SS 563) design. Built at the Mare Island and Portsmouth naval shipyards, respectively, the ships were completed in 1958. They had a streamlined design featuring large, twin missile hangars faired into their bows. Each hangar could hold two of the Regulus I missiles or one of the improved Regulus II weapons. (See below.) Although similar in general characteristics and appearance, the *Grayback* and *Growler* were not identical: the former was 322⅓ feet (98.27 m) long and displaced 2,671 tons on the surface, and the latter was 317½ feet (96.8 m) and 2,543 tons, the differences reflecting the detailed designs of their respective building yards.

*Sailors prepare a Regulus I missile for launching aboard the submarine* Barbero. *The missile's wings (and tail fin) are still folded; the missile rests on the retracted launch catapult. The hangar held two missiles. This unarmed missile was loaded with mail and flown ashore under radio control.* (U.S. Navy)

To launch the Regulus the crew had to surface the submarine, open the missile hangar door, manually extract the missile from the hangar and place it on the launching ramp, extend its wings and tail fin, rotate and elevate the ramp, and plug in appropriate cables before launching the missile. The *Grayback* and *Growler* hangars were significantly larger than the canister-like hangars in the two fleet boat conversions, and in the event of a hangar-flooding accident the submarine probably would have been lost. (This hangar-catapult arrangement reflected the Navy's Bureau of Aeronautics being in charge of the Regulus program and looked at the

*The* Grayback *entering San Diego harbor with a Regulus on the launching rail. Two large hangars are faired into the submarine's bow; four Regulus I missiles could be accommodated. The* Grayback *and* Growler, *although based on the* Tang *design, differed in detail.* (U.S. Navy)

weapon as an unmanned aircraft rather than a true "missile.")

The next Regulus submarine was even more ambitious, being nuclear propelled. There had been brief interest in arming the *Nautilus* (SSN 571) with Regulus cruise missiles. That concept was rejected by then-Rear Admiral H. G. Rickover, who demanded that other than the nuclear plant installation, the submarine employ only existing design features and weapons.

The original Navy shipbuilding program for fiscal year 1956 provided for five conventional submarines as well as for three nuclear submarines. Admiral Arleigh Burke, who became Chief of Naval Operations in August 1955, was a strong proponent of nuclear propulsion and sea-launched missiles. He directed that two additional nuclear submarines be constructed in place of conventional submarines if funding could be made available. Accordingly, the next SSG—to be built at Mare Island—would be nuclear propelled, becoming the USS *Halibut* (SSGN 587). The other nuclear submarine would exploit a radical new hull design, becoming the *Skipjack* (SSN 585; see Chapter 9).

Mare Island was already building nuclear attack submarines and was directed to modify the design for the SSG funded in fiscal year 1956 to employ an S3W reactor plant, the type being installed in that yard's nuclear-propelled *Sargo* (SSN 583). Much larger than the *Grayback* and *Growler*, the *Halibut* would have a single large missile hangar faired into her bow to accommodate four of the later (and larger) Regulus II missiles. However, in practice she would carry only five Regulus I weapons or two of the later Regulus II missiles. The *Halibut*'s unusual hangar configuration and launching ramp was evaluated on board a landing ship before installation in the submarine.[13] (Also considered was a rotary Regulus launcher that could accommodate four Regulus II missiles. This scheme did not progress past the model stage; see drawing.)

The *Halibut* was commissioned on 4 January 1960 under Lieutenant Commander Walter Dedrick. She was the world's *second* nuclear-propelled submarine with a missile armament. The first was the USS *George Washington* (SSBN 598), a ballistic missile submarine that had been commissioned five days earlier. (The first Soviet SSGN, the *K-45* of the Project 659/Echo I design, was placed in commission on 28 June 1960.) But the *Halibut* was intended only as a transitional SSGN design. A larg-

*The* Halibut *as an SSGN. The "bulge" forward is the top of the hangar that could accommodate five Regulus I missiles. The launching rail is in the stowed position, between the hangar and sail. She was the U.S. Navy's only SSGN, pending conversion of Trident submarines in the early 21st Century.* (U.S. Navy)

er and more-capable missile submarine design was on the drawing boards: The *Permit* (SSGN 594), which would have a surface displacement of some 4,000 tons and about 5,000 tons submerged, a length of approximately 350 feet (106.7 m), and the S5W propulsion plant. The *Permit* SSGN was to carry four Regulus II missiles in four hangars—two faired into the forward hull and two outboard of the sail. The four smaller hangars were a major factor in the submarine's design, providing increased survivability in the event of damage or a hangar flooding casualty compared to the *Halibut*.

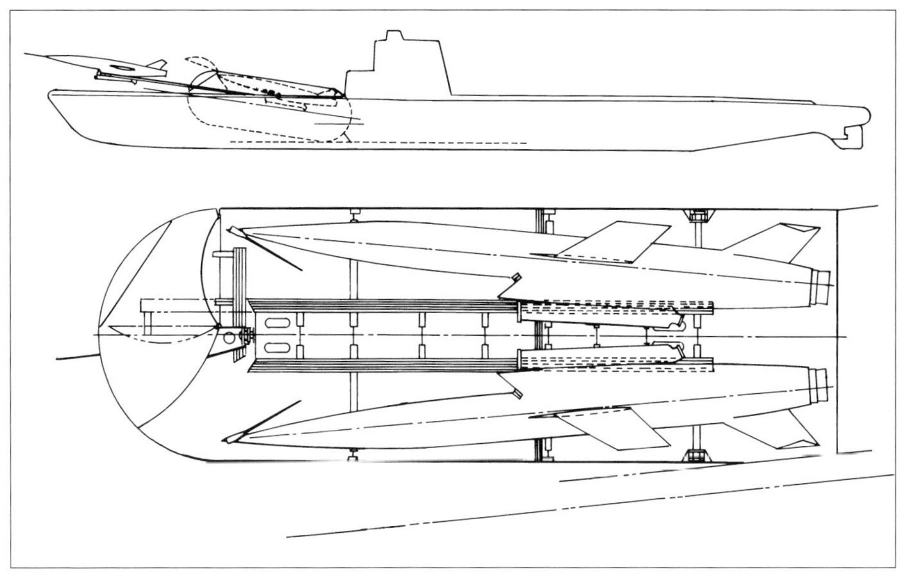

*A rotary launcher holding four Regulus II missiles was proposed for submarine use.* (©A.D. Baker, III)

Early Navy planning had provided for the first three of the *Permit* SSGNs to be funded in fiscal year 1958, a fourth ship in fiscal 1959 and, subsequently, seven more units to be built for a class of 11 submarines. The Navy's long-range plan of 1958 showed an eventual force of 12 SSGNs with Regulus II or later cruise missiles. This was in addition to a force of 40 or more nuclear submarines armed with Polaris (ballistic) missiles.[14]

The weapon, to be carried by these SSGNs, was the successor to the original Regulus, the larger, supersonic Regulus II, which would have a maximum range of 1,000 n.miles (1,853 km), twice that of the earlier weapon. This missile also would carry the W27 two-megaton warhead. The first flight of the definitive Regulus II land-attack missile occurred on 29 May 1956, with a planned 1960 operational date. The first and only launch of a Regulus II missile from a submarine was made from the *Grayback* on 18 September 1958. (The only other Regulus II shipboard launch was from the converted LST *King County* on 10 December 1958.)

Three months after the *Grayback* launch, on 18 December, Secretary of the Navy Thomas S. Gates cancelled the Regulus II program. The decision had been made to accelerate and enlarge the Polaris ballistic missile program, leading to termination of the Regulus II and its planned successors.[15] The Regulus II was considered redundant as a strategic strike weapon, and its cancellation could shift funds to the Polaris effort. Further, the potential of the Regulus for tactical (anti-ship) operations with high-explosive or nuclear warheads was overshadowed by the massive attack capabilities of the U.S. Navy's aircraft carriers.

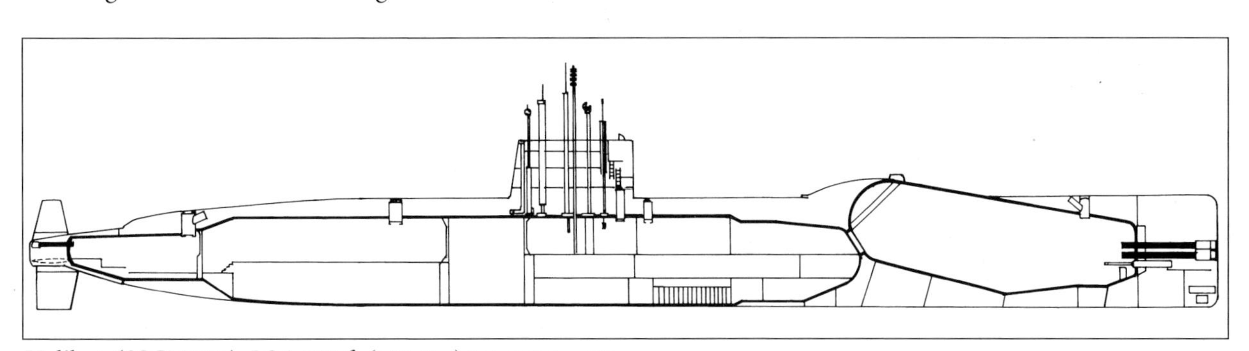

Halibut *(SSGN 587). LOA 350 ft (106.7m)* (©A.D. Baker, III)

With cancellation of the Regulus II, the five existing cruise missile submarines continued to make deterrent patrols in the North Pacific carrying the Regulus I. The four *Permit*-class SSGNs that were in the fiscal 1958–1959 shipbuilding programs were reordered as *Thresher*-class SSNs.[16]

Beyond cancellation of the Regulus II, work also was halted on the more advanced Rigel and Triton submarine cruise missiles. For a brief moment, with the cancellation of the Regulus II missile, the Navy considered using the anti-submarine SUBROC weapon against shore targets with an air-burst warhead. (See Chapter 10.) But that idea was discarded.

*The ultimate Regulus submarine would have been the* Permit *SSGN, with four Regulus II missiles carried in separate hangars. The cancellation of the Regulus program led to completion of the planned SSGNs as torpedo-attack submarines of the* Thresher *design.* (U.S. Navy)

## Regulus on Patrol

The first Regulus-armed submarine to go to sea on a strategic missile or "deterrent" patrol was the *Tunny,* carrying two Regulus I missiles. During the Lebanon crisis of 1958, she was ordered into the North Pacific to substitute for the nuclear strike

TABLE 6-1
**Land-Attack Cruise Missiles**

| | **U.S. Regulus I SSM-N-8** | **U.S. Regulus II SSM-N-9** | **Soviet P-5 (4K95) NATO Shaddock** |
|---|---|---|---|
| Operational | 1954 | cancelled | 1959 |
| Weight* | 13,685 lb (6,207.5 kg) | 22,564 lb (10,235 kg) | 9,480 lb (4,300 kg) |
| Length | 41 ft 6 in (12.65 m) | 57 ft (17.38 m) | 38 ft 10½ in (11.85 m) |
| Wingspan | 21 ft (6.4 m) | 20 ft ½ in (6.1 m) | 8 ft 2½ in (2.5 m) |
| Diameter | 56½ in (1.44 m) | 50 in. (1.27 m) | 29½ in (0.9 m) |
| Propulsion | 1 turbojet 4,600 lbst** 2 booster rockets | 1 turbojet*** 15,000 lbst 2 booster rockets | 1 turbojet 4,960 lbst 2 booster rockets |
| Speed | Mach 0.9 | Mach 2+ | Mach 1.2 |
| Range | 500 n.miles (926.5 km) | 1,000 n.miles (1,853 km) | 300 n.miles (550 km) |
| Guidance | radio control or preset | inertial | preset |
| Warhead | nuclear Mk 5 or W27 | nuclear W27 | nuclear RDS-4 or conventional |

Notes: * Gross takeoff weight; does not include booster rockets.
** lbst = pounds static thrust.
*** Fitted with afterburner.

capability of an aircraft carrier, and the submarine took up station above the Arctic Circle, with her two missiles available to strike targets in the Soviet Far East. This was more than two years before the first U.S. ballistic missile submarine, the *George Washington* (SSBN 598), began the first Polaris missile patrol.

The Regulus-armed *Barbero* operated in the Atlantic from April 1956 to late 1958. She shifted back to the Pacific Fleet as Regulus missiles became a deterrent force. From September 1959 to July 1964, the Regulus-armed submarines were on continuous patrol in the North Pacific, with their missiles aimed at targets in the Soviet Far East. The four diesel boats would make refueling stops at Midway Island or Adak, Alaska, to maintain maximum time on patrol. Forty-one Regulus patrols were conducted in that period by the five missile submarines; one or two submarines carrying a total of four or five missiles were continuously "on station" in the western Pacific. While the numbers were small, the Regulus did permit coverage of specific targets with nuclear weapons without requiring that a carrier task force be kept in the area or basing nuclear-armed aircraft in Japan or South Korea.[17]

In addition to the five submarines, for brief periods up to ten aircraft carriers and four heavy cruisers carried Regulus I missiles. Space and weight were "reserved" for the future installation of the Regulus II in several missile cruisers, including the *Long Beach,* the world's first nuclear-propelled surface warship.[18] There were advocates of a much larger Regulus program with proposals put forward to convert several outdated aircraft carriers to Regulus launching ships.[19]

The demise of the Regulus II at the end of 1958 marked the end of cruise missile development in the U.S. Navy for more that a decade. A later Chief of Naval Operations, Admiral Elmo R. Zumwalt, would write:

> To my mind the Navy's dropping in the 1950s of a promising program for a cruise missile called "Regulus" was the single worst decision about weapons it made during my years of service. That decision was based on the theory that our carriers were so effective that we did not need cruise missiles . . . without cruise missiles practically all our long-range offensive capability was crowded onto the decks of a few carriers.[20]

The retirement of the Regulus I cruise missile in July 1964 occurred five months before the first patrol of a Polaris SSBN in the Western Pacific began in late December 1964.[21] The diesel SSGs *Barbero, Growler,* and *Tunny* were decommissioned, with the ex-fleet boats being scrapped and the *Growler* becoming a museum ship in New York City. The SSG *Grayback* was converted to a transport submarine for swimmers and commandos, serving in that role until 1984 (redesignated APSS 574, subsequently LPSS 574). The *Halibut*—the only U.S. SSGN to be completed—had a most interesting second career as a "spy sub," being converted to carry out clandestine deep-ocean search operations. Redesignated SSN 587, her highly clandestine operations continued until 1976. (See Chapter 10.)[22]

## Soviet Cruise Missile Submarines

Whereas the U.S. Navy initially developed submarine-launched cruise missiles for strategic attack and then shifted to ballistic missiles for the strategic role, the Soviet Navy displayed continuous interest in both types of missiles for the strategic role. The USSR was not the target of German V-1 cruise missiles.[23] In July 1944, however, Great Britain provided a damaged V-1 to the Soviet Union and by the end of the war, work was under way on cruise missiles at the aircraft design bureaus headed by Georgi M. Beriev, Vladimir N. Chelomei, Sergei V. Ilyushin, and Seymon A. Lavochkin. Chelomei would be the most successful in developing ship-launched cruise missiles.

During the war Chelomei, already a recognized mathematician and scientist, had worked on the development of pulse-jet engines, as used in the V-1 missile. He contacted Georgi Malenkov, a member of the State Defense Commission and a close associate of Stalin, telling him that he could build a weapon similar to the German V-1 missile. Malenkov supported the 30-year-old designer, and in 1944 Chelomei was given the resources of the aircraft design bureau previously headed by Nikolai N. Polikarpov, now designated OKB-51.[24] Chelomei created the V-1 analog—with the designation 10X—in 1945. Meanwhile, in 1950 design bureau TsKB-18 began the design of a cruise missile submarine propelled by a closed-cycle steam turbine, that is, the same type of propulsion plant used in the German Type XXVI/Soviet Project 616. This submarine was Project 624; it was to have a submerged displacement of 2,650 tons and to carry cruise missiles—designed by Lavochkin—that would have a range against shore targets of 160 n.miles (300 km). However, development of both the missile and submarine were soon halted.

In 1952–1953 design efforts began on Project 628, an updated Soviet XIV series (K-class) submarine configured to conduct experimental launches of the 10XN Volna (wave) subsonic cruise missile. This missile—developed by Chelomei's design bureau—was powered by twin ramjets; the missile was launched from a ramp with the aid of a single booster rocket. Although Western intelligence reported launchers installed near Leningrad and Vladivostok for this missile, it did not enter ground or naval service. It was rejected for naval service because of guidance limitations, the high fuel consumption of available ramjets, and the ongoing development of supersonic missiles. (A version of the 10XN did enter service with the Soviet Air Forces in 1953.)

A still further refinement of the V-1 design by Chelomei was the 15X missile, based on the availability of the Rolls-Royce Nene centrifugal-flow turbojet, provided by Britain's Labour government for civil use in the Soviet Union. This engine, which also became the powerplant for the famed MiG-15 turbojet fighter, and an advanced point-to-point guidance system gave promise of an effective submarine-launched strategic weapon. The submarine would transit to a preselected launch position with the missile either in a towed launch container or fitted in a container on the deck of the submarine. The former concept had been developed by the Germans to tow V-2 ballistic missiles to an underwater launch position. (See Chapter 7.)

Chelomei's design bureau was disestablished in December 1952 because of political intrigues within the defense industry. The decree closing Chemomei's bureau was one of the last such documents signed by Josef Stalin, who died in March 1953. Chelomei's facilities were taken over by aircraft designer Artyem I. Mikoyan.[25] From January 1953, as a professor at the Bauman Institute of the Moscow Technical University, Chelomei continued to work on missile designs. In 1954 he conceived the idea of ship-launched missiles with wings that automatically extend in flight. This would permit the missiles to be carried in containers essentially the same size as the missile's fuselage. Aerodynamic improvements made possible the elimination of a launch ramp, permitting the missile to be launched directly from the canister. He applied this concept to the 20X missile which, upon provision of more-flexible guidance, become the P-5 (NATO SS-N-3c Shaddock[26]). This deck-mounted canister would be adopted by the Soviet Navy, with the structure providing both the storage and launch functions, simplifying installation in surface ships as well as in submarines. This was in contrast to the U.S. Navy's method, which employed the canister as only a hangar, with the missile having to be manually extracted, placed on launch rails, wings extended, and other manual functions performed before launching.

Chelomei gained the support of Soviet leader Nikita S. Khrushchev. In August 1955 his design bureau was re-established as OKB-52, initially to develop submarine-launched cruise missiles.[27] That year the decision was made to produce both the P-5/Shaddock cruise missile of Chelomei and the P-10 cruise missile being developed by OKB-48 under seaplane designer Beriev. Both missiles were intended for strategic strikes against land targets.

A single Project 611/Zulu design was modified in 1955 to the Project P-611 configuration to test launch the P-10 missile. The P-10 was housed in a hangar, with the missile extracted from the hangar, its wings opened, and then launched (as with the U.S. Regulus). The hangar was on the deck casing, aft of the conning tower, with the missile to fire *forward,* over the bow. The submarine was modified at Molotovsk shipyard (subsequently renamed Severodvinsk). During the fall of 1957 four P-10 missiles were launched from the submarine P-611. But work on this missile was halted because of the successful tests of the P-5 missile, which was supersonic (Mach 1.2), had a range of 300 n.miles (570 km), and incorporated other advantages when compared to the P-10 missile.[28]

In that same year, Project P-613—the modification of the Project 613/Whiskey submarine *S-146*—was undertaken to conduct tests of the P-5 missile. That modification, like the P-611, was under chief designer Pavel P. Pustintsev of TsKB-18. The work was undertaken at the Krasnoye Sovormo yard in Gor'kiy, inland on the Volga River. The missile canister for this submarine was also placed behind the conning tower and, again, the launch took place on the surface, over the bow.

Following P-5/Shaddock tests at Kapustin Yar in 1956, the first P-5 missile was launched from the *S-146* on 22 November 1957 in the White Sea. After extensive tests the P-5 system became operational in 1959 and was installed in operational submarines. Work on competing anti-ship cruise missiles was halted.

Six Project 644/Whiskey Twin-Cylinder submarines were converted in 1960 at Gor'kiy to a Pustintsev design. Each submarine was fitted with paired missile canisters aft of the conning tower, built into the deck casing, which elevated and fired aft, over the stern. Simultaneous with this effort, guided missile submarine Project 646 was developed on the basis of Project 641/Foxtrot. Another design effort of Pustintsev, this craft had a submerged displacement of approximately 3,625 tons and featured several major variants: the basic design was to carry four P-5 missiles or two P-10 missiles; the improved Project 644P had two P-5 missiles. However, further development of these designs was not pursued.

*The Project 644/Whiskey Twin-Cylinder marked the first operational deployment of the Shaddock missile system. The paired canisters elevated to fire aft. The Whiskey-class submarines served as test and operational platforms for several missile systems. Note the blast shields at the forward end of the canisters.*

*The definitive Whiskey SSG was Project 665/Whiskey Long-Bin, the term referring to the enlarged fairwater configured with four forward-firing Shaddock launch tubes. Six of these submarines were produced, their significance soon being overshadowed by the large Soviet SSGN program.*

Instead, the definitive Project 613/Whiskey SSG design was Project 665, known by NATO as the Whiskey Long-Bin (for the enlarged conning tower structure). These were new-construction submarines of a design developed by TsKB-112 under designer B. A. Lyeontyev. Each had four forward-firing P-5/Shaddock missiles installed in the front portion of the large, bulbous conning tower. The launch tubes were fixed at an upward angle of 14 degrees. From 1961 to 1963 the Gor'kiy yard and the Baltic shipyard in Leningrad together delivered six Long-Bin submarines.

The P-5/Shaddock land-attack missile had a range of 300 n.miles (550 km), a terminal speed of mach 1.2, and an accuracy of plus-or-minus two n.miles (four km). While limited in missile range and effectiveness, these submarines provided the Soviet Navy with valuable experience and training in cruise missile submarines. Unlike the

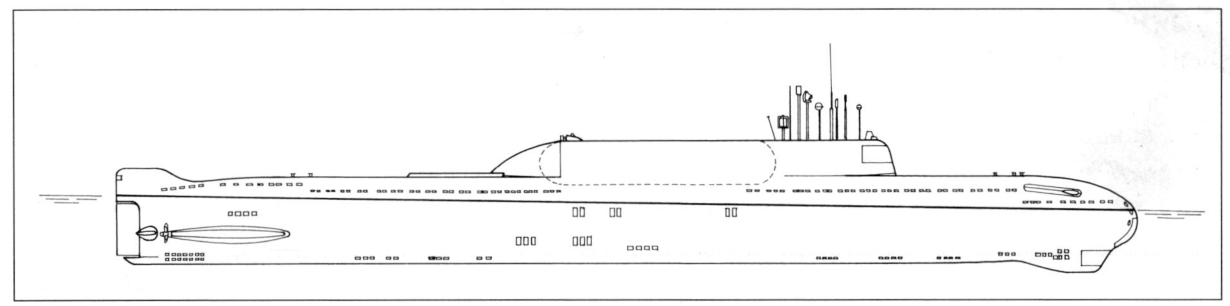

*Project P-627A SSGN showing cruise missile hangar fitted aft of the sail.* (©A.D. Baker, III)

two converted fleet boats used by the U.S. Navy in the Regulus program (*Barbero* and *Tunny*), the Soviet Project 665 submarines did not undertake long-distance missile patrols.

In 1956, simultaneous with the diesel-electric cruise missile submarine program, work began on nuclear submarines with cruise missiles for land attack. Among these was Project P-627A/November —a modification of the first SSN design that would carry one P-20 missile.[29] In the same period work began at SKB-143 on Project 653, a nuclear-propelled submarine of some 7,140 tons submerged displacement that was to carry two of the long-range P-20 cruise missiles. Design work on the submarine was carried out under Mikhail G. Rusanov from 1956 to 1959, when, before the completion of the technical design stage, SKB-143 was directed to complete the working drafts and send them to the Severodvinsk yard. The lead ship was to be completed in 1962, with 18 submarines planned for construction. In early 1960 the P-20 missile was seen to have major flaws, and construction of both the P-627A and 653 submarines was halted.

Also, in 1956 design work was begun on Project 659 (NATO Echo I) under Pustintsev and N. A. Klimov at TsKB-18. This craft would have the same VM-A reactor plant as the Project 627A/November SSN. But Project 659 would not have the advanced hull shape of the Project 627A. Rather, Project 659/Echo I and the subsequent Project 651/Juliett SSG, 675/Echo II SSGN, and 658/Hotel SSBN submarines would have conventional hulls to enhance their stability on the surface while launching their missiles.

The first Project 659 submarine—the *K-45*—was laid down at the Leninsky Komsomol yard (No. 199) at Komsomol'sk-na-Amur on 28 December 1958 and placed in commission on 28 June 1961. This was the second Soviet shipyard to produce nuclear submarines. The shipyard, the largest in Siberia, was begun in 1931, and submarine construction began in the 1930s with components produced in the European USSR. (The largest surface warships built at Komsomol'sk were heavy cruisers.) Through 1963 four more nuclear submarines of this project were built. These submarines had a length of 364¾ feet (111.2 m) and displaced 4,976 tons submerged. Missile armament of the submarines consisted of six P-5/Shaddock cruise missiles in paired launch canisters that were mounted on the deck casing and elevated 15 degrees to launch forward.

The early P-5 cruise missile system had a number of deficiencies, among them low accuracy, limited effectiveness, and a significant amount of time on the surface (20 minutes) being needed to prepare the missiles for launch. Accordingly, from 1958 to 1961 the Chelomei bureau designed a new system based on the P-5 missile and designated P-5D. It had increased range and with a higher probability of penetrating enemy defenses to reach its target. A Whiskey Twin-Cylinder submarine was fitted to test the missile (Project 644D), and the P-5D was accepted for service in 1962.

Based on the P-5 and P-5D missiles, Chelomei's bureau developed the P-7 missile to have twice the range of the earlier weapons, approximately 540 n.miles (1,000 km). Flight tests took place from 1962 to 1964, with a Whiskey SSG again planned as a test platform (Project 644-7). However, the P-7 missile did not become operational in submarines, because the role of cruise missile submarines was shifted from strikes against land targets to the anti-ship role.[30]

The intensive development of ballistic missiles and the problems with land-attack cruise missiles changed Soviet policy toward strategic cruise missiles.[31] The Strategic Rocket Forces was established on 14 December 1959 as an independent military service to control all *land*-based strategic ballistic missiles; at the time Soviet sea-based strategic forces were downgraded, and the construction of submarines with land-attack guided (cruise) and ballistic missiles was halted. Thus, in February 1960, the decision was made to halt the further development of several cruise missiles, including the P-20 for the Projects P-627A and 653 SSGNs. Work on these projects was stopped.

From the array of strategic and anti-ship missiles begun in the 1950s, only the P-5 and P-6 missile systems designed by Chelomei were placed in service on submarines in this period. However, by 1965 the P-5 land-attack missile was taken off all SSG/SSGNs. In place of land-attack weapons, most of the SSG/SSGN force was rearmed with anti-ship missiles.

During the mid-1950s intensive development had began on anti-ship cruise missiles concurrent with the development land-attack missiles. The former weapons were intended primarily to attack U.S. aircraft carriers, which carried nuclear-armed strike aircraft and thus presented a threat to the Soviet homeland. Soviet premier Khrushchev in 1955 announced that "submarines armed with guided missiles—weapons best responding to the requirements indicated for at sea operations—will be deployed at an accelerated rate."[32]

Under the direction of Chelomei at OKB-52, the first practical anti-ship missile was produced, the P-6 (NATO SS-N-3a Shaddock). The liquid-fuel, supersonic (Mach 1.2) missile was surface launched from submarines as well as from surface warships, the latter variant designated P-35. With a range up to 245 n.miles (450 km), the missile could carry a nuclear or conventional warhead. Beyond its onboard guidance with terminal radar homing, the missile could be provided with guidance updates while in flight by a Video Data Link (VDL, given the NATO code name Drambuie). This enabled long-range reconnaissance aircraft—primarily the Tu-20 (NATO Bear-D)—and, subsequently, satellites to identify distant targets and relay the radar picture to the (surfaced) submarine or surface missile-launching ship.[33]

In addition, the technique was developed for the submarine to launch two missiles at an interval of some 90 seconds. Both missiles would climb to their cruise altitude of some 9,840 feet (3,000 m), but as they descended to a lower altitude for seeking the target ship, only the first missile would activate its terminal search radar.

This radar picture would be transmitted up to the Bear aircraft and relayed back to the launching ship, and updated guidance data would be transmitted to the second missile. This provided the second missile with an optimum flight path while preventing the target ship from detecting radar emissions from the second missile. Only in the final phase of the attack would the second missile activate its radar. In this procedure—given the NATO code name Theroboldt—the lead missile's transponder enabled the VDL system to identify the relative position of the in-flight missile as well as the launch platform, thereby providing accurate relative positions to target subsequent missiles. This alleviated the need for the then-difficult task of providing the precise geographic locations of both the launch platform and the over-the-horizon target.

The first successful submarine P-6/SS-N-3a test launches were conducted by a Project 675/Echo II in July-September 1963. Modified Tu-16RT Badger aircraft provided the long-range targeting for the tests. The missile became operational on submarines in 1963—the world's first submarine-launched anti-ship missile.

The missile could carry a conventional warhead of 2,205 pounds (1,000 kg) or a tactical nuclear weapon, with both versions carried in deployed submarines. The first P-6 missiles were installed in submarines of Project 675/Echo II and Project 651/Juliett classes. These submarines also could carry the P-5 land-attack cruise missile, but few if any of the Echo II SSGNs and none of the Juliett SSGs appear to have deployed with the P-5.

The five older submarines of Project 659/Echo I design—each carrying six land-attack missiles—which did not have anti-ship missile guidance systems—were later reconfigured as torpedo-attack submarines (redesignated 659T).

The design work for Project 675/Echo II submarine had begun in 1958 at TsKB-18 under Pustintsev. The Echo II, with a submerged displacement

*A Project 675/Echo II photographed in 1979. These submarines were considered a major threat to Western aircraft carriers. The paired Shaddock launch canisters elevated and fired forward. The recess aft of the sail is for the ultra-high-frequency antenna mast, shown here in the raised position.* (U.K. Ministry of Defence)

The lead submarine, the *K-1,* was built at Severodvinsk and placed in commission on 31 October 1963. Twenty-nine ships were produced through 1968—16 at Severodvinsk and 13 at Komsomol'sk. Components of the terminated Project 658/Hotel ballistic missile submarines may have been shifted to the SSGN program when the construction of strategic missile submarines was halted.

The SSGNs and Juliett SSGs were employed extensively in Soviet long-range operations, deployed to the Mediterranean, and were even deployed as far as the Caribbean. The nuclear-propelled Echo SSGNs, like the contemporary Project 627/November SSNs and Project 658/Hotel SSBNs (i.e., HEN submarines), suffered a large number of engineering problems. For example, in July 1979 the *K-116*—a Project 675/Echo II SSGN—suffered a reactor meltdown because of the accidental shutting off of a main coolant pump, an operator error. Reportedly, radioactivity reached 6,000 rads/hour at the bulkheads of the reactor compartment, far beyond safety limits. The submarine was towed back to Pavlovsk, 40 miles (65 km) southeast of Vladivostok and permanently taken out of service.

of 5,737 tons and length of 378½ feet (115.4 m), was also propelled by a first-generation nuclear plant with two VM-A reactors (i.e., the HEN propulsion plant). Again, the missiles were launched while the submarine was on the surface with eight stowage-launch canisters fitted in the deck casing. The four paired missile canisters elevated 15 degrees before firing. Project 675 had an enlarged sail structure containing a folding radar, which led to the ship's nickname *raskladyshka* (folding bed). When surfaced, the submarine would expose the massive radar antennas by rotating the forward portion of the sail structure 180 degrees. The NATO designation for the radar was Front Door/Front Piece.[34]

On 10 August 1985 another Echo II, the *K-431,* suffered a reactor explosion at Chazuma Bay in the Soviet Far East during a refueling operation.[35] A massive amount of radioactivity was released when the 12-ton reactor cover, blown upward, smashed into the submarine, tore open the hull, and ruptured the reactor compartment. As the *K-431* began to sink alongside the pier, a tug was used to ground the submarine. Subsequently, the radioactive hulk was towed to a "permanent" berth at Pavlovskaya Bay. Of the men in the area at the time and subse-

*An Echo II with the first and third pair of Shaddock canisters in the elevated position. The early Project 659/Echo I cruise missile submarines were suitable only for the land-attack role and hence were converted to SSNs; the later Echo II SSGNs could strike land or naval targets.* (U.S. Navy)

*The Front Door/Front Piece of an Echo II SSGN is exposed in this view. The antenna to track the Shaddock missile in flight and relay mid-course guidance instructions rotates 180 degrees to become the forward end of the sail when the submarine operates submerged.*

*The Front Door/Front Piece radar in a Project 651/Juliett SSG in the stowed position. The arrangement required fitting the bridge in the center of the elongated sail. The same radars also were fitted in several Soviet anti-ship missile cruisers.*

quently involved in the cleanup, 7 developed acute radiation sickness, and a "radiation reaction" was observed in 39 others.

In 1986, while she was moored at Cam Ranh Bay, Vietnam, the *K-175* had a serious reactor accident, also caused by personnel error. She was carried back to the USSR in a transporter dock and was not returned to service. Other submarines of this class had breakdowns at sea and were towed back to Soviet ports.

A few of these submarines also had problems not related to their nuclear powerplants. The *K-122*—an Echo I SSN—suffered a battery-related fire off Okinawa on 21 August 1980 that left the submarine on the surface without power. Nine men died, others were burned, and the boat had to be towed to Vladivostok. There were no radiation problems.

Although by 1965 all land-attack missiles had been removed from these submarines, there still was concern by some U.S. naval officers that these submarines posed a strategic threat to the United States. For example, late in 1968, the director of U.S. Navy strategic systems, Rear Admiral George H. Miller, believed that Soviet submarine-launched cruise missiles could be dual purpose, and used to attack bomber bases of the Strategic Air Command, presenting a different attack profile than ballistic missiles, this complicating U.S. detection and warning of strategic attack.[36] Miller also noted that the cruise missiles could be used to attack Sentinel, the planned U.S. anti-ballistic missile system.

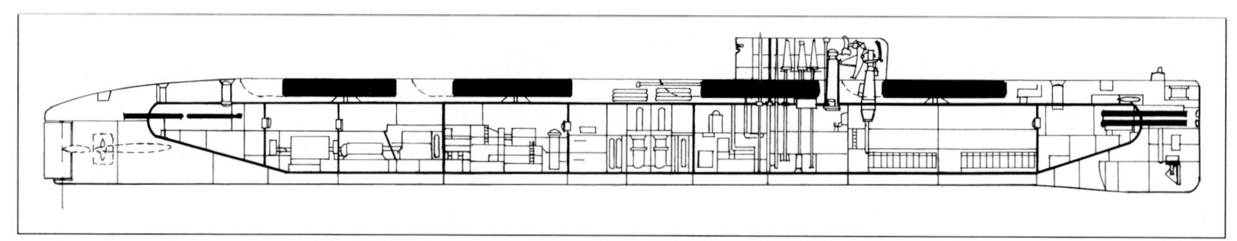

*Project 675/Echo II SSGN. LOA 378 ft 6 in (115.4 m)* (©A.D. Baker, III)

## Non-Nuclear Submarines

Simultaneous with the building of Project 675 nuclear submarines, the Project 651/Juliett diesel-electric submarine was put into production. It had a submerged displacement of 4,260 tons and was 281¾ feet (85.9 m) long. Missile armament consisted of four P-5/P-6 canisters, paired in the same manner as in the SSGNs. This was a TsKB-18 design under chief designer Abram S. Kassatsier.

Seventy-two Project 651/Juliett submarines were planned. In the event, only 16 of these submarines were built at Gor'kiy and the Baltic shipyard in Leningrad from 1963 to 1968. As with Project 675/Echo II submarines, these craft had a large, rotating radar in the leading edge of their sail, with one of these submarines later having a satellite targeting system fitted (Project 651K). The first few submarines were built with low-magnetic steel. These soon suffered significant corrosion damage as well as cracks. The later submarines were of standard steel construction.

The building of diesel-electric SSGs in parallel with nuclear SSGNs occurred because of limitations at that time in producing additional nuclear reactors. Because Juliett SSG construction was contemporaneous to the Echo II SSGN and other factors, some U.S. submarine analysts believed that there was "sufficient evidence to warrant at least consideration of the possibility that, with the advent of nuclear power, the Soviet Union, in addition to continuing the development of conventional diesel-electric submarines, also advanced the development of unconventional nonnuclear [closed-cycle] propulsion systems."[37] But the Project 651/Juliett SSG had conventional propulsion—except for one unit.

A small nuclear reactor—designated VAU-6—was developed to serve as an auxiliary power source in diesel-electric submarines. It was a pressurized-water reactor with a single-loop configuration coupled with a turbogenerator. After land-based trials it was installed in a Juliett (Project 651E) during 1986–1991. The sea trials "demonstrated the workability of the system, but revealed quite a few deficiencies. Those were later corrected."[38]

In 1960, on the basis of Project 651, the Project 683 nuclear-propelled submarine was designed, also to carry the P-5/P-6 missiles. This was to have

*The Project 651/Juliett SSG was an attractive submarine and, being diesel-electric, it demonstrated the scope of the Soviet commitment to the anti-carrier role. An auxiliary nuclear power source was evaluated for these submarines, but appears not to have been pursued.* (U.S. Navy)

been a larger craft with more weapons; two small, 7,000-horsepower reactor plants were to propel the submarine. However, this design was not further developed.

Submarines with anti-ship cruise missiles were developed specifically to counter U.S. aircraft carriers, which since 1950 carried nuclear-capable strike aircraft on forward deployments to within range of targets within the Soviet Union. The high priority of defense of the Soviet homeland against this threat led to major resources being allocated to the SSG/SSGN construction. From 1963 to 1968 a total of 50 "modern" missile-armed submarines of the Juliett, Echo I, and Echo II classes were sent to sea carrying 326 missiles. These submarines, in turn, forced an increase in U.S. tactical anti-submarine warfare. While on the surface the SSG/SSGNs were increasingly vulnerable to Allied detection, although they were immune to the aircraft-carried Mk 44 and Mk 46 ASW torpedoes, which could not strike surface targets.[39] This led to development of the Harpoon anti-ship missile for use by aircraft to attack cruise missile submarines on the surface. (The Harpoon subsequently became the U.S. Navy's principal anti-ship weapon; it was carried by aircraft, surface ships, and, briefly, submarines.)

Cruise missile submarines thus became a mainstay of the Soviet Navy. With both cruise missile and a torpedo armament, these submarines provided a potent threat to Western naval forces. Indeed, some Project 675/Echo II and Project 651/Juliett submarines remained in service into the early 1990s.

---

**Admiral S. G. Gorshkov, Commander-in-Chief of the Soviet Navy from 1955 to 1981, used the term "revolution in military affairs" to describe the period of the late 1950s when cruise and ballistic missiles were integrated into the Soviet armed forces. For the Soviet Navy this led to an intensive period of development of cruise missiles, initially for the land-attack role and then for the anti-ship role, as well as the development and construction of several cruise missile submarine designs. Gorshkov observed that ". . . all the latest achievements in science, technology, and production were utilized in the course of building qualitatively new submarines, surface ships, and [their] armament."[40]**

**Significantly, of the 56 first-generation (HEN) nuclear-propelled submarines built in the Soviet Union, slightly more than half of them (29) were Project 675/Echo II cruise missile submarines. While this proportion was achieved, in part, by the termination of Project 658/Hotel ballistic missile submarine program, when considered together with the 16 Project 651/Juliett SSGs, cruise missiles represented a major Soviet investment to counter U.S. carriers that could threaten the Soviet homeland with nuclear-armed aircraft.**

**This was in sharp contrast to the U.S. Navy. Having aircraft carriers available for the anti-ship as well as for the land-attack roles, the U.S. Navy rapidly discarded cruise missile submarines as the Polaris ballistic missile became available.**

TABLE 6-2
**Cruise Missile Submarines**

| | U.S. *Grayback* SSG 574* | U.S. *Halibut* SSGN 587 | Soviet Project 659 NATO Echo I | Soviet Project 675 NATO Echo II | Soviet Project 651 NATO Juliett |
|---|---|---|---|---|---|
| Operational | 1958 | 1960 | 1961 | 1963 | 1963 |
| Displacement | | | | | |
| surface | 2,671 tons | 3,845 tons | 3,770 tons | 4,415 tons | 3,140 tons |
| submerged | 3,652 tons | 4,755 tons | 4,980 tons | 5,737 tons | 4,240 tons |
| Length | 322 ft 4 in (98.27 m) | 350 ft (106.7 m) | 364 ft 9 in (111.2 m) | 378 ft 6 in (115.4 m) | 281 ft 9 in (85.9 m) |
| Beam | 30 ft (9.15 m) | 29 ft (8.84 m) | 30 ft 2 in (9.2 m) | 30 ft 6 in (9.3 m) | 31 ft 10 in (9.7 m) |
| Draft | 19 ft (5.8 m) | 20 ft (6.1 m) | 23 ft 31⁄2 in (7.1 m) | 23 ft (7.0 m) | 22 ft 8 in (6.9 m) |
| Reactors | — | 1 S3W | 2 VM-A | 2 VM-A | — |
| Turbines | — | 2 | 2 | 2 | — |
| horsepower | — | 7,300 | 35,000 | 35,000 | — |
| Diesel engines | 3 | — | — | — | 2 |
| horsepower | 6,000 m | — | — | — | 4,000# |
| Electric motors | 2 | — | — | — | 2 |
| horsepower | 4,700 | — | — | — | 12,000 |
| Shafts | 2 | 2 | 2 | 2 | 2 |
| Speed | | | | | |
| surface | 20 knots | 20 knots | 15 knots | 14 knots | 16 knots |
| submerged | 17 knots | 20 knots | 26 knots | 22.7 knots | 18 knots |
| Test depth | 700 ft (213 m) | 700 ft (213 m) | 985 ft (300 m) | 985 ft (300 m) | 985 ft (300 m) |
| Missiles | 4 Regulus I or 2 Regulus II | 5 Regulus I or 4 Regulus II | 6 P-5 | 8 P-5/P-6 | 4 P-5/P-6 |
| Torpedo tubes** | 6 533-mm B 2 533-mm S | 4 533-mm B 2 533-mm S | 4 533-mm B 2 400-mm B 2 400-mm S | 4 533-mm B 2 400-mm S | 6 533-mm B 4 400-mm S |
| Torpedoes | 22 | 17 | 4 533-mm 8 400-mm | 10 533-mm 6 400-mm | 6 533-mm 12 400-mm |
| Complement | 84 | 123 | 104 | 109 | 78 |

Notes: * *Growler* (SSG 577) was similar; see text. Their combat systems were the same.
** Bow + Stern torpedo tubes.
*** Several submarines were fitted to carry 18 533-mm and 4 400-mm torpedoes.
# Also fitted with a single 1,360-hp diesel generator for shipboard electric power.

7

# Ballistic Missile Submarines

*A Project 658M/Hotel II SSBN in rough seas. The elongated sail structure housed three R-21/SS-N-5 Serb ballistic missiles. The first Hotel I was completed in 1960, less than a year after the first U.S. Polaris submarine went to sea. The American submarine carried more missiles with greater capability.* (U.S. Navy)

Germany led the world in the development of ballistic missiles, launching V-2 missiles in combat from 6 September 1944 until 27 March 1945.[1] The German Army fired about 3,200 missiles during the 6½-month V-2 campaign, their principal targets being London and the Belgian port of Antwerp. These missiles killed some 5,000 persons, most of them civilians, and injured thousands more.[2] Once launched on its ballistic trajectory the missile was invulnerable to any form of interception.

Although the warheads of all V-2 missiles combined carried less high explosives than a single major 1944–1945 raid against Germany by British and American bombers, and the casualties inflicted were far fewer than many of those raids against Germany and Japan, the V-2 was a remarkable technological achievement as well as an exceptional production accomplishment. The V-2 became the basis for ballistic missile development in Britain, China, France, the Soviet Union, and the United States.

The V-2 was a 13-ton, 46-foot (14.1-m) missile propelled by a rocket engine fueled by liquid oxygen and alcohol. The V-2 carried a high explosive warhead just over one ton (1,000 kg) with an initial range of 185 miles (300 km).[3] It was transported by truck to a forward launch position, where, in a few hours, it was assembled, erected, fueled, and launched. The missile was fitted with preset guidance and its initial accuracy was about four miles (6.4 km). This very poor accuracy led to the V-2 being a "terror" weapon rather than a military weapon used against specific targets. Later V-2 missiles, with beam-guidance during the boost phase, attained an accuracy of approximately one mile (1.6 km).

While the V-2 was a German Army project, a sea-launched version was under development when the war ended. Klauss Riedel of the Peenemünde staff proposed launching V-2 missiles from canisters towed to sea by submarines, a scheme given the code name *Prufstand XII* (Test Stand XII). According to German engineers working on the project, "Its objective was to tow V-2 rockets in containers across the North Sea to continue V-2 launch operations against targets in Britain after, in the wake of the Normandy invasion, the launching bases in Northern France and Holland were lost while V-2 production continued."[4]

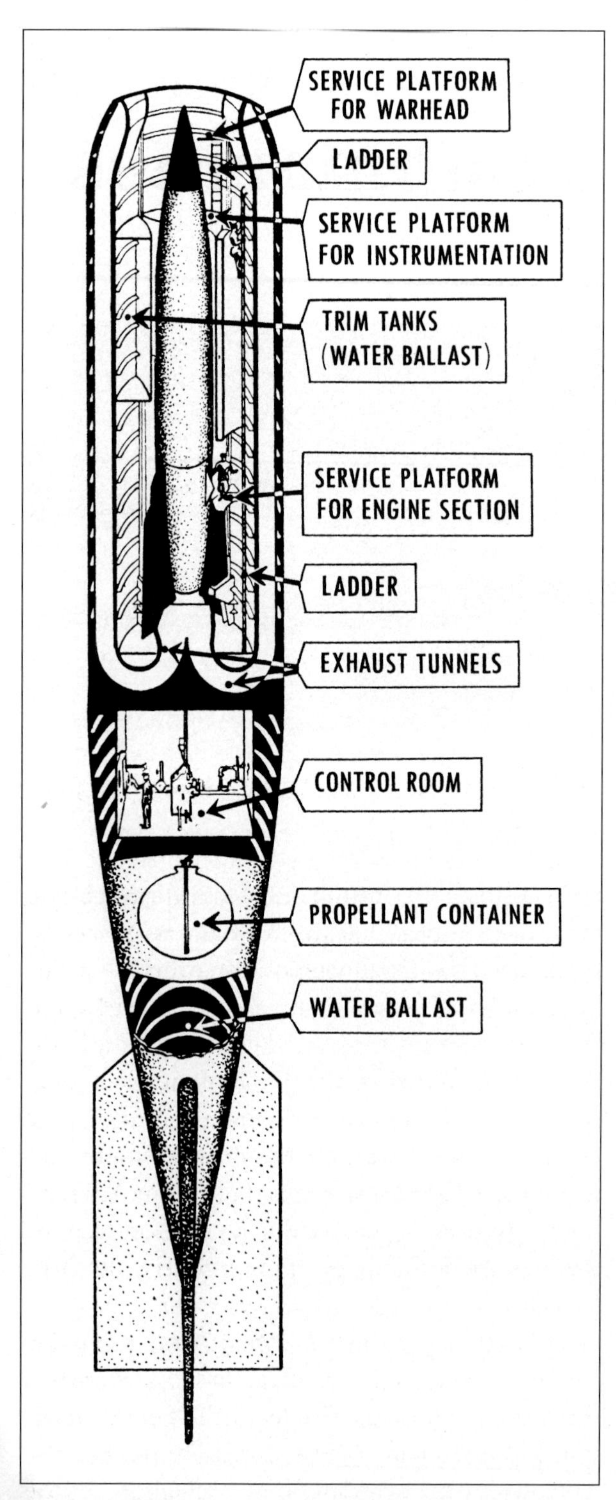


*Submarine-towed V-2 missile launch container.* (*Atomic Submarines*)

The V-2 canisters were approximately 118 feet (36 m) long and 18⅔ feet (5.7 m) in diameter, and displaced about 500 tons. Beyond the missile and the fuel, stored separately, the canisters would have had ballast tanks, a control room, and bunks for a crew of eight to ten men. Obviously, these accommodations would be untenable for long cruises. The underwater voyage across the North Sea to a suitable launch position was estimated to take about 24 hours, with the submarine controlling the submerging and surfacing of the canisters through the towing lines.[5] Studies indicated that a single U-boat could tow three containers for sustained periods. Long transits were considered feasible, in which case the missile crews would be carried in the towing submarine and transferred to the V-2 canisters by rubber raft, a precarious operation. Thus the Germans could bombard American cities, with New York most often mentioned as the principal target.[6]

To launch the missiles the underwater canisters would be ballasted to achieve an upright position, projecting above the surface of the water. The missile would be fueled, the guidance system started, and the V-2 would be launched. A novel exhaust was provided for the canister with ducts to carry the flaming exhaust up through the canister to vent them into the atmosphere. Riedel proposed the sea-launch project early in 1944. Detailed design of the containers was done by the Vulkan shipyard in Stettin with construction of the containers beginning in August 1944. Three of the containers were 65 to 70 percent complete when the war ended. A prototype model was successfully tested ashore at Peenemünde. Discussing the submarine-towed V-2s, *Generalmajor* Walter Dornberger, who commanded the Peenemünde missile center, wrote:

> we did not expect any construction difficulties that could not be overcome, but work on the problem was temporarily suspended because of the A-4 [V-2] troubles . . . at the end of 1944, it was resumed. By the middle of December a full memorandum was being prepared on the preliminary experiments, and we were getting to work on the first design sketches [for series production]. The evacuation of Peenemünde before the middle of February of 1945 put an end to a not unpromising project.[7]

Peenemünde fell to Soviet troops in the spring of 1945, although it had been damaged by British bomber raids. When British, Soviet, and U.S.

troops entered Germany, the V-2 project was high on their lists for intelligence collection. Among their finds were missiles, unassembled components, plans, and related equipment. German scientists and engineers were quickly taken into custody. Many, led by Wernher von Braun, the technical chief of the German mis sile effort, went to the United States, where they became the core of the U.S. Army's missile program. The Soviets initially established the research agency *Institut Rabe* in Nordhausen for German missile specialists to continue their work.[8] Then, without warning, on 22 October 1946 the Soviet NKVD security police took into custody the missile scientists and engineers of *Institut Rabe* and specialists from other military fields—including those working on submarines—a total of some 5,000 men. With their families and belongings, they were sent by train to Russia to continue their work under tighter controls.[9]

*A model of the P-2 combination cruise/ballistic missile submarine, at left are the R-1 ballistic missiles and, beneath the sail, the Lastochka cruise missiles. The problems of such a submarine proved to be insurmountable at the time. This model was photographed at the Central Naval Museum in Leningrad in 1973.* (Capt. E. Besch, USMC Ret.)

In October 1945 the British fired three captured V-2 missiles from Cuxhaven on the North Sea in Operation Backfire. Soviet observers were invited to the third launch. Subsequently, in both the United States and USSR, the Germans helped to assemble and test V-2 missiles. Sixty-eight were test fired in the United States and 11 in the Soviet Union over the next few years. On 6 September 1947 one of the American V-2s was launched from the flight deck of the U.S. aircraft carrier *Midway* (Operation Sandy).[10] This was history's only launch of a V-2 from a moving "platform." In the United States the Navy had some interest in developing a "guided missile" ship to launch missiles of this type.

At that time the Navy was considering both modifying aircraft carriers to launch V-2 missiles (in addition to operating their aircraft) and completing the unfinished battleship *Kentucky* and battle cruiser *Hawaii* as ballistic missile ships. In the event, the U.S. Navy—like the U.S. Air Force—initially concentrated on cruise missile technology for the long-range strike role.[11]

The Soviet and U.S. Navy learned of the V-2 underwater canisters, but there was little interest in the scheme, because available missile resources in both countries initially were concentrated on the cruise missiles for naval use (as described in Chapter 6). In the United States the newly established Air Force sponsored some ballistic missile development, in competition with the Army effort (under von Braun). In the Soviet Union the development of ballistic missiles was sponsored by a complex governmental structure—as were most Soviet endeavors—with the principal direction coming from the NKVD and the Red Army's artillery directorate.

In 1949 a preliminary draft for a missile submarine designated Project P-2—to strike enemy land targets—was drawn up at TsKB-18 (later Rubin) under chief designer F. A. Kaverin. The submarine was to have a surface displacement of almost 5,400 tons and carry 12 R-1 ballistic missiles—the Soviet copy of the V-2—as well as the Lastochka (swallow) cruise missile. But the designers were unable to solve the myriad of problems in the development of such a ship.

The first ballistic missile to be considered for naval use was the R-11.[12] Developed by the OKB-1 design bureau under Sergei P. Korolev, the R-11

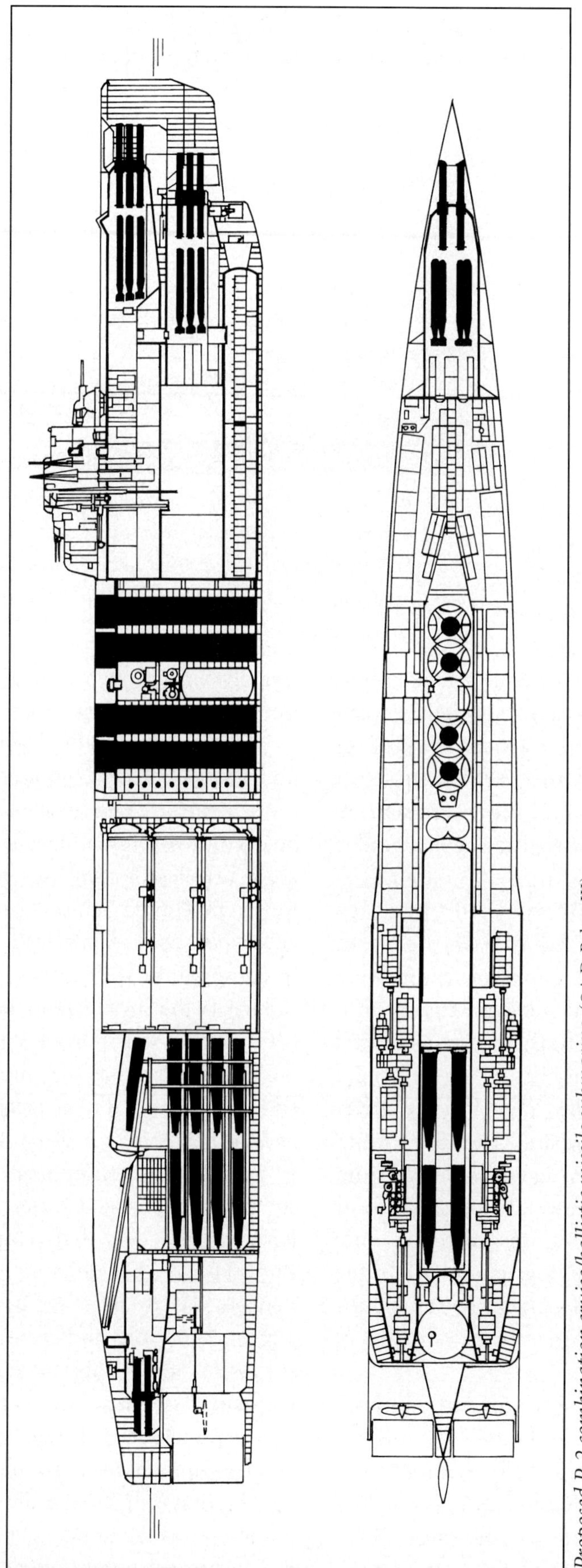

*Proposed P-2 combination cruise/ballistic missile submarine.* (©A.D. Baker, III)

entered Army service in 1956. It was a 4½-ton missile with a storable-liquid propellant that could deliver a 2,200-pound (1,000-kg) conventional warhead to a maximum range of 93 miles (150 km). Later, about 1956, the R-11 was fitted with a nuclear warhead, which initially reduced the missile's range by about one-half.

In 1953–1954 Korolev proposed a naval variant the missile that would be fueled by kerosene and nitric acid in place of the R-11's alcohol and liquid oxygen, which was less stable for long-term storage. Test launches of this R-11FM were conducted at Kapustin Yar in 1954-1955.[13] Three launches from a fixed launch stand were followed by additional launches from a test stand that moved to simulate a ship's motion.

The OKB-1 bureau developed a shipboard launch scheme wherein the R-11FM missile would be housed in a storage canister; for launching, the submarine would surface and the missile would be elevated up and out of the canister. The Soviet Navy had specified a submerge-launch capability for the missile, but Korolev opposed it, believing that his scheme employing the R-11FM could provide a missile capability in less time. Beyond the missile, a multitude of problems confronted the program, including establishing the exact location of the submarine, the bearing to the target, secure communications, and other concerns. All had to be solved.

## The First Ballistic Missile at Sea

The world's first ballistic missile-carrying submarine was the Project 611/Zulu *B-67*. She was modified at the Molotovsk (Severodvinsk) shipyard with two R-11FM missile tubes fitted in the after end of an enlarged sail structure and related control equipment installed.[14] The missile tubes extended down through the fourth compartment, replacing one group of electric batteries and the warrant officers' space. Their accommodations were moved to the bow compartment, from which reload torpedoes were removed. The deck guns were also deleted. The modification—Project V611 (NATO Zulu IV½)—was prepared at TsKB-16 under chief designer Nikolai N. Isanin, who would be associated with other early Soviet missile submarines.[15]

In anticipation of submarine launches, a special test stand was built at the Kapustin Yar test center to simulate the effects of a missile tube in a ship encountering rough seas. Sea trials were conducted hurriedly, and on 16 September 1955 in the White Sea, the submarine *B-67* launched the first ballistic missile ever to be fired from a submarine. The missile streaked 135 n.miles (250 km) to impact in the remote Novaya Zemlya test range. (Later, in the fall of 1957, the *B-67* was employed in an unusual, unmanned underwater test to determine the vulnerability of missile-carrying submarines to depth charge attack. After the tests divers connected air hoses to the submarine to blow ballast tanks to surface the submarine.)

Late in 1955 work began at TsKB-16 on a production ballistic missile submarine (SSB) based on the Zulu—Project AV611, armed with two R-11FM missiles. Four of these diesel-electric submarines, each with a surface displacement of 1,890 tons, were built in 1957 at Severodvinsk. In addition, in 1958, the submarine *B-62* of this type was modified to the SSB configuration at shipyard No. 202 in Vladivostok (also designated AV611).[16]

The R-11FM missile became operational in 1959, giving the Soviet Union the world's first ballistic missile submarines.[17] However, the range of the missiles limited the effectiveness of the submarines to a tactical/theater role. Also, the clandestine operation of the submarines was compromised by the need to surface or snorkel to recharge their batteries.

Meanwhile, in 1955 Korolev's OKB-1 bureau transferred all submarine ballistic missile projects to special design bureau SKB-385 under Viktor P. Makeyev. This shift enabled Korolev to concentrate on longer-range strategic missiles and space programs. Makeyev, who was 31 years of age when he became head of the bureau in June 1955, immediately pursued the D-2 missile system, which would include the new R-13 missile (NATO SS-N-4 Sark) launched from a new series of submarines.[18] The missile, with a storable-liquid propellant, was surface launched and had a range of up to 350 n.miles (650 km).

The submarine to carry this new missile would be Project 629 (NATO Golf), employing the same diesel-electric machinery and other components of the contemporary torpedo submarine of Project

*The Project AV611/Zulu V SSB demonstrated the early Soviet interest in putting ballistic missiles at sea. Two R-11FM/SS-1c Scud-A missiles were fitted in the sail structure. The Soviet Navy carried out the world's first submarine ballistic missile launches at sea.*

*Details of the sail structure of a Project AV611/Zulu V SSB showing the covers for the two ballistic missile tubes. The missile tubes penetrated into the pressure hull. The submarine's periscopes and masts are located between the bridge (forward) and missile tubes.* (U.S. Navy)

*The early Soviet submarine-launched ballistic missiles were elevated out of the tube and above the sail in preparation for launching, as shown here on a Project 629/Golf SSB. This was a difficult procedure in rough seas. Subsequently, Soviet SSB/SSBNs were fitted with submerged-launch missiles.*

641 (NATO Foxtrot).[19] Developed at TsKB-16 under Isanin, Project 629 had a surface displacement of 2,850 tons and could carry three R-13 missiles. The missile tubes penetrated the top *and bottom* of the pressure hull, extending up into the sail. The missiles were raised up out of the sail for launching. The R-11FM missile was fitted in the first five Project 629 submarines, which later would be rearmed with the R-13 missile.

As the first submarines were being built, in 1958–1959 TsKB-16 began the development of a system by which a small nuclear reactor would be fitted in Project 629 submarines as a means of recharging the submarine's battery without running the diesel engine, a form of closed-cycle or air-independent propulsion. The reactor generated

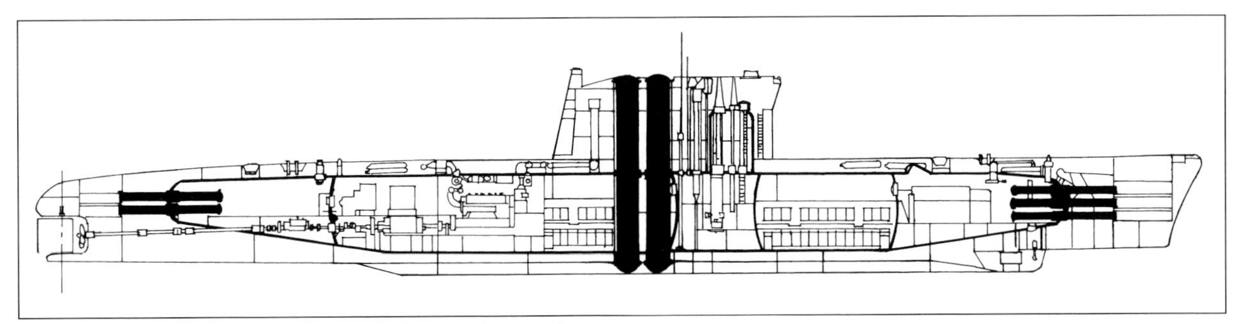

*Project AV611/Zulu V SSB. LOA 296 ft 10 in (90.5 m)* (©A.D. Baker, III)

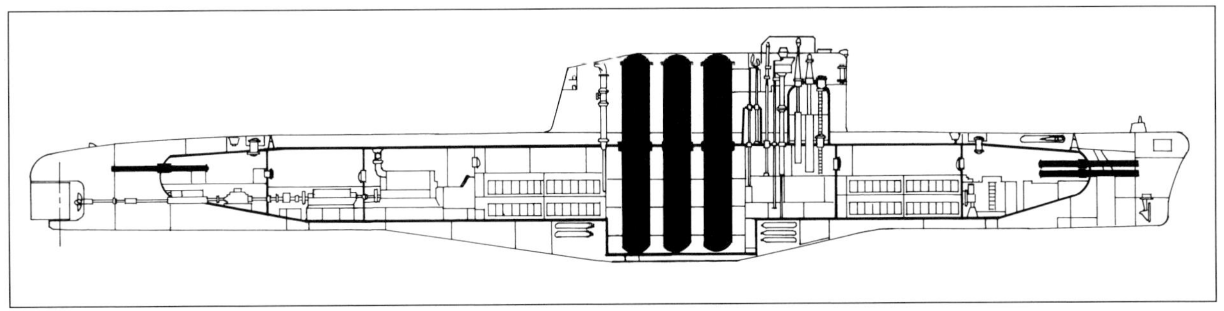

*Project 629A/Golf II SSB. LOA 324 ft 5 in (98.9 m)* (©A.D. Baker, III)

about 600 kilowatts of power. However, this effort was not pursued.

The lead Project 629/Golf submarine was the *B-41* (later *K-79*), built at Severodvinsk and delivered in 1959. From 1959 to 1962, 22 submarines were constructed—15 at Severodvinsk and 7 at Komsomol'sk-on-Amur in the Far East. Sections for two additional submarines were fabricated at Komsomol'sk and transferred to the Peoples Republic of China, with one being assembled at Darien in the mid-1960s (designated Type 035). The second submarine was never completed. The Chinese submarines were to be armed with the earlier R-11FM missile, but no SLBMs were delivered, and the submarine served as a test platform for Chinese-developed SLBMs.

*A Project 629/Golf SSB in heavy seas in the North Atlantic. A large number of these diesel-electric submarines were produced and proved useful as a theater/area nuclear-strike platform. An auxiliary nuclear reactor was also developed for installation in these submarines.* (U.K. Ministry of Defence)

Even though part of the prelaunch preparation of R-11FM

and R-13 missiles was conducted underwater, the submarine had to surface to launch missiles. According to Captain 2d Rank V. L. Berezovskiy: "The preparation to launch a missile took a great deal of time. Surfacing, observation of position, the steadying of compasses—somewhere around an hour and twenty or thirty minutes. This is a monstrously long time. . . . The submarine could be accurately detected even before surfacing [to launch]."[20]

The world's first launch of an SLBM armed with a thermo-nuclear warhead occurred on 20 October 1961 when a Project 629 submarine launched an R-13 missile carrying a one-megaton warhead that detonated on the Novaya Zemlya (Arctic) test range in test "Rainbow." (The first U.S. test launch of a Polaris missile with a nuclear warhead occurred a half year later when an A-2 weapon with a warhead of just over one megaton was fired from the submarine *Ethan Allen*/SSBN 608.) Subsequently, R-13 missiles with nuclear warheads were provided to submarines.

In 1955 experimental work had begun on underwater launching of ballistic missiles from a submerging test stand. Major problems had to be solved for underwater launch. After launches from the submerging test stand, launch tests of dummy missiles were conducted in 1957 from the Project 613D4/Whiskey submarine *S-229.* The submarine had been modified at shipyard No. 444 at Nikolayev on the Black Sea with two missile tubes fitted amidships. The dummy missiles had solid-propellant engines to launch them out of the tubes and a liquid-fuel second stage.

Subsequently, the Project V611/Zulu submarine *B-67,* used earlier for the R-11FM tests, was again modified at Severodvinsk for actual missile tests (changed to PV611). The first launch attempt from the *B-67* occurred in August 1959, but was unsuccessful. On 10 September 1960 an S4-7 (modified R-11FM) missile was launched successfully from the submarine while she was submerged and underway. (Less than two months earlier, the U.S. nuclear submarine *George Washington* [SSBN 598] carried out a submerged launch of a Polaris A-1 missile.)

In 1958—just before completion of the first Golf SSB—work began at the SKB-385 bureau on the liquid-propellant R-21 ballistic missile (NATO SS-N-5 Serb), with a range up to 755 n.miles (1,400 km), twice that of the R-13 missile. The R-21 could be launched from depths of 130 to 165 feet (40–50 m), providing enhanced survivability for the submarine, with the submarine traveling up to four knots at the time of launch.

Initial underwater launches of the R-21 dummy missiles took place in 1961 from the *S-229;* that submarine was fitted with a single SLBM launch tube aft of the conning tower. A year later *two* R-21 launch tubes were installed in the *K-142,* the last Project 629/Golf submarine being built at Severodvinsk (redesignated Project 629B). Completed in 1961, that submarine made the first underwater launch of an R-21 missile on 24 February 1962.

Through 1972, 13 additional submarines of Project 629/Golf and seven Project 658/Hotel submarines were refitted to carry three underwater-launch R-21 missiles, becoming Projects 629A and 658M, respectively. The rearmed Project 629A/Golf II-class submarines were considered theater and not strategic missile platforms. During September-

*The Project 613D4/Whiskey submarine* S-229 *as modified for underwater launch tests of ballistic missiles. The United States led the USSR in the deployment of submerged-launch ballistic missiles. The later Soviet missile submarines could launch their missiles while surfaced or submerged.* (Malachite SPMBM)

October 1976 six of these submarines shifted from the Northern Fleet to the Baltic to provide a sea-based theater nuclear strike capability; the other seven SSBs served in the Pacific Fleet. They remained in Soviet service until 1989 with the last units retired as SSBs after almost 30 years of continuous service. Some units continued briefly in the missile test and communications roles.[21]

The Golf SSBs were relatively successful, although one sank in 1968. This was the *K-129*, a rearmed Project 629A/Golf II submarine, which had departed Avacha Bay at Petropavlovsk (Kamchatka) on 25 February 1968, for a missile patrol in the North Pacific. In addition to her three R-21 nuclear missiles, the submarine was armed with torpedoes, two of which had nuclear warheads. The *K-129* was manned by a crew of 98.

En route to patrol in the area of latitude 40° North and longitude 180° East, on 8 March 1968 the *K-129* suffered an internal explosion and plunged to the ocean floor, a depth of three miles (4.8 km). The U.S. Navy's seafloor Sound Surveillance System (SOSUS) detected the explosion, providing a specific search area of several hundred square miles, and the special-purpose submarine *Halibut* (SSN 587) was sent out to clandestinely seek the remains of the *K-129* with a deep-towed camera. Once the precise location of the *K-129* was determined, the U.S. Central Intelligence Agency used the specially built salvage ship *Glomar Explorer* in a 1974 effort to lift the stricken submarine, which was relatively intact.[22]

The clandestine salvage effort—code name Jennifer—succeeded in raising the submarine, but during the lift operation the hull split, with most of the craft falling back to the ocean floor. The bow section (compartments 1 and 2)—containing two nuclear as well as conventional torpedoes and the remains of six sailors—was brought aboard the *Glomar Explorer*, minutely examined, dissected, and then disposed of. The sailors' remains were recommitted to the sea in a steel chamber on 4 September 1974 with appropriate ceremony.[23] Project Jennifer was the deepest salvage operation ever attempted by any nation. The CIA began planning a follow-up lift of the remainder of the submarine, but that effort was halted when Jennifer was revealed in the press.

Personnel failure probably caused the loss of the *K-129;* of her crew of 98, almost 40 were new to the submarine. Some Soviet officials mistakenly contend that the *K-129* was sunk in a collision with the trailing U.S. nuclear submarine *Swordfish* (SSN 579).[24]

Another Project 629A/Golf II submarine appeared in American headlines when, on 4 May 1972, a U.S. Defense Department spokesmen stated the submarine had entered a port on Cuba's northern coast. This visit occurred a decade after the Soviet government had attempted to clandestinely place ballistic missiles capable of striking the United States in Cuba, leading to the missile crisis of 1962.[25] Another Golf II visited Cuba in 1974.[26] (No nuclear-propelled *ballistic* missile submarines entered Cuban ports, although nuclear-propelled torpedo and cruise missile submarines did so.)

## Nuclear-Propelled Missile Submarines

TsKB-18 also was exploring ballistic missile submarine designs. In 1957–1958 the bureau briefly

TABLE 7-1
**Soviet Submarine-Launched Ballistic Missiles**

| | R-11FM NATO SS-1b Scud-A | R-13 NATO SS-N-4 Sark | R-21 NATO SS-N-5 Serb |
|---|---|---|---|
| Operational | 1959 | 1961 | 1963 |
| Weight | 12,050 lb (5,466 kg) | 30,322 lb (13,745 kg) | 36,600 lb (16,600 kg) |
| Length | 34 ft 1 in (10.4 m) | 38 ft 8 in (11.8 m) | 42 ft 4 in (12.9 m) |
| Diameter | 34⅔ in (0.88 m) | 51 in (1.3 m) | 55 in (1.4 m) |
| Propulsion | liquid-propellant 1 stage | liquid-propellant 1 stage | liquid-propellant 1 stage |
| Range | 80 n.miles (150 km) | 350 n.miles (650 km) | 755 n.miles* (1,400 km) |
| Guidance | inertial | inertial | inertial |
| Warhead** | 1 RV conventional or nuclear 10 KT | 1 RV nuclear 1 MT | 1 RV nuclear 1 MT |

Notes: There were several variants of each missile.
* A later version of R-21 carried an 800-kiloton warhead to a range of 1,600 km.
** RV = Reentry Vehicle.

looked at a closed-cycle propulsion plant using superoxide-sodium (Project 660). This design, with a submerged displacement of about 3,150 tons, was to carry three R-13 missiles. It was not pursued because of the substantial development and testing effort required for the power plant.

Rather, the next Soviet ballistic missile submarine would have nuclear propulsion (SSBN). Design work began in 1956 at TsKB-18, first under chief designer P. Z. Golosovsky, then I. B. Mikhilov, and finally Sergei N. Kovalev. This submarine would be Project 658 (NATO Hotel). The submarine had a surface displacement of 4,080 tons and an overall length of 374 feet (114 m). The submarine initially was armed with three R-13/SS-N-4 missiles. In addition to four 21-inch (533-mm) bow torpedo tubes, four 15.75-inch (400-mm) stern torpedo tubes were provided for defense against pursuing destroyers, as in the Project 675/Echo II submarine.

The nuclear power plant was similar to the installation in the Project 627A/November SSNs and Project 659/675/Echo SSGNs with two VM-A reactors—the HEN installation. And, as did those submarines, the SSBNs suffered major engineering problems. The keel for the lead ship, the *K-19,* was laid down at Severodvinsk on 17 October 1958. She was placed in commission on 12 November 1960. From the outset the *K-19* and her sister ships were plagued with engineering problems.

Less than a year later, while operating in Arctic waters off the southern coast of Greenland on 4 July 1961, the *K-19* suffered a serious accident when a primary coolant pipe burst. The submarine was at a depth of 660 feet (200 m). The entire crew suffered excessive radiation exposure before the submarine was able to reach the surface. Her crew was evacuated with the help of two diesel-electric submarines, and she was taken in tow. Of 139 men in the *K-19* at the time, 8 men died almost immediately from radiation poisoning and several more succumbed over the next few years—a total of 14 fatalities. The reactor compartment was replaced at Severodvinsk from 1962 to 1964, and the original compartment was dumped into the Kara Sea. The cause of the casualty was traced to improper welding.

During 1968–1969 the *K-19* was rearmed with the underwater-launch RSM-40 missile (NATO SS-N-8). Returning to sea, on 15 November 1969 the

*Production of the Project 658/Hotel SSBN was halted because of the Soviet defense reorganization. Design was already under way on the far more capable Project 667A/Yankee SSBN. This is the problem-plagued* K-19, *disabled in the North Atlantic on 29 February 1972.* (U.S. Navy)

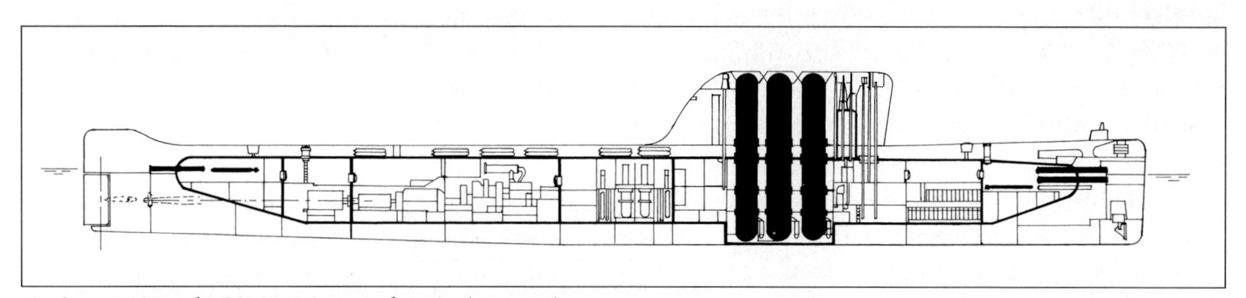

*Project 658/Hotel I SSBN. LOA 373 ft 11 in (114.0 m)* (©A.D. Baker, III)

*K-19* was in a collision with the U.S. nuclear submarine *Gato* (SSN 615), which was observing Soviet submarine operations in the Barents Sea. Both submarines were damaged, but neither was in danger of sinking.

On 24 February 1972 the *K-19* suffered a fire while operating submerged some 600 n.miles (1,110 km) northeast of Newfoundland. The submarine was able to surface. There were fires in compartments 5, 8, and 9. The fires were extinguished, but 28 of her crew died, and the ship was left without propulsion. She was towed across the Atlantic to the Kola Peninsula, a 23-day tow. Twelve survivors were trapped in isolated *K-19* compartments, in darkness, and with cold rations. They were initially forced to drink alcohol from torpedoes. Later rations and water were passed down to them through an opening from the open deck. There was no radiation leakage.

After investigations the Soviet government decided to use the *K-19* to test the mobilization possibilities of the ship repair industry—the ability to repair a warship in a short amount of time. On 15 June 1972 the *K-19* was towed to the Zvezdochka yard at Severodvinsk; with the yard workers laboring at maximum effort, on 5 November the submarine was returned to the fleet.

The surviving crewmen were sent back to the ship. This was done for the "morale and psychological well-being" of the crew. The *K-19* continued in service until 1990. The submarine *K-19*—the first Soviet SSBN—was known in the fleet as the "Hiroshima."[27]

Other Project 658 submarines also suffered problems, some having to be towed back to Soviet ports. As accident after accident occurred, designer Kovalev recalled, "It was literally a disaster—steam generators leaked, condensers leaked and, practically, the military readiness of these submarines was questionable."[28] The problems that plagued these and the other submarines of the HEN-series primarily were caused by poor workmanship and lack of quality control at the shipyards and by component suppliers. Personnel errors, however, also were a factor in these accidents. From 1960 to 1962 eight ships of the Project 658/Hotel design were built at Severodvinsk. Additional SSBNs of this class were planned, but construction was halted by the late 1959 establishment of the Strategic Rocket Forces and cancellation of sea-based strategic strike programs. (See Chapter 11.)

The design of a larger nuclear-propelled submarine—to carry a larger missile—had been started at SKB-143 in 1956.[29] Project 639 was to have a surface displacement of about 6,000 tons and nuclear propulsion with four propeller shafts, and was to be armed with three R-15 missiles. Reactor plants with both pressurized water and liquid-metal coolants were considered. The R-15, designed at OKB-586, which was headed by Mikhail K. Yangel, was a liquid-propellant missile with a range of 540 n.miles (1,000 km). It would have been surface launched.

TsKB-16 also was working on a diesel-electric SSB, Project V629, a variant of the Golf design, that would carry one of the large R-15 missiles. But this was a surface-launch missile, and its performance was overtaken by other weapons in development; consequently, the decision to halt work on the missile was made in December 1958, and both submarine designs were cancelled.[30]

The seven Project 658 submarines that were rearmed with the R-21 missile (658M/Hotel II) served until 1991 in the SSBN role. The problem-plagued *K-19* was one of the last to be retired. In 1979 she was converted to a communications relay ship (renamed *KS-19*). Other units subsequently were employed as research and missile test submarines after being retired from first-line SSBN service.[31]

However, according to the semiofficial history of Soviet shipbuilding, "the characteristics of the Soviet oceanic missile installations and submarines, the carriers of the first generation [missiles], were significantly behind the American submarines and missile installations. Therefore, during the beginning of the 1960s work began on constructing a more modern system for the next generation."[32] The next generation Soviet SSBN would go to sea in the late 1960s.

**The Soviet Union pioneered the development of submarine-launched ballistic missiles and ballistic missile submarines. However, the capabilities of these submarines—both diesel-electric and**

nuclear propelled—were limited, and the latter submarines were plagued with the engineering problems common to first-generation Soviet nuclear submarines. Still, early Soviet SSB/SSBNs provided some compensation for the initial shortfall in Soviet ICBM performance and production. Early planning for more capable ballistic missile submarines was halted because of the preference for ICBMs that followed establishment of the Strategic Rocket Forces.

Further, the Soviets initiated a highly innovative program of developing submarine-launched anti-ship ballistic missiles. While the development of a specific missile for this purpose was successful—with some speculation in the West that an anti-submarine version would follow—that weapon ultimately was abandoned because of arms control considerations. (See Chapter 11.)

TABLE 7-2
**Soviet Ballistic Missile Submarines**

| | Soviet Project AV611 Zulu V | Soviet Project 629 Golf I | Soviet Project 658 Hotel I |
|---|---|---|---|
| Operational | 1957 | 1959 | 1960 |
| Displacement | | | |
| surface | 1,890 tons | 2,850 tons | 4,080 tons |
| submerged | 2,450 tons | 3,610 tons | 5,240 tons |
| Length | 296 ft 10 in (90.5 m) | 324 ft 5 in (98.9 m) | 373 ft 11 in (114.0 m) |
| Beam | 24 ft 7 in (7.5 m) | 26 ft 11 in (8.2 m) | 30 ft 2 in (9.2 m) |
| Draft | 16 ft 9 in (5.15 m) | 26 ft 7 in (8.1 m) | 25 ft 3 in (7.68 m) |
| Reactors | — | — | 2 VM-A |
| Turbines | — | — | 2 |
| horsepower | — | — | 35,000 |
| Diesel engines | 3 | 3 | — |
| horsepower | 6,000 | 6,000 | — |
| Electric motors | 3 | 3 | — |
| horsepower | 5,400 | 5,400 | — |
| Shafts | 3 | 3 | 2 |
| Speed | | | |
| surface | 16.5 knots | 14.5 knots | 18 knots |
| submerged | 12.5 knots | 12.5 knots | 26 knots |
| Test depth | 655 ft (200 m) | 985 ft (300 m) | 985 ft (300 m) |
| Missiles | 2 R-11FM | 3 R-13/SS-N-4 | 3 R-13/SS-N-4 |
| Torpedo tubes* | 6 533-mm B 4 533-mm S | 4 533-mm B 2 533-mm S | 4 533-mm B 2 400-mm B 2 400-mm S |
| Torpedoes | | | |
| Complement | 72 | 83 | 104 |

Notes: *Bow; Stern.

8

# "Polaris—From Out of the Deep . . ."

*The USS* George Washington, *the West's first ballistic missile submarine, on sea trials in November 1959. Despite popular myth, an attack submarine was not cut in half on the building ways for insertion of the ship's missile compartment. Note the high "turtle back" over the 16 Polaris missile tubes.* (U.S. Navy)

Following the dramatic launching of a V-2 from the aircraft carrier *Midway* in 1947, the U.S. Navy gave relatively little attention to ballistic missiles, in part because of other priorities for new weapons, especially those related to aircraft carriers. The ubiquitous Submarine Officers Conference in 1946, while considering a "super-bombardment submarine," reported that it "might be desirable to make exploratory studies of a submarine capable of carrying a V-2 type missile," but there was no follow-up effort.[1]

In 1950 Commander Francis D. Boyle, a World War II submarine commander, proposed a "guided missile" submarine based on the wartime fleet boat. Boyle's proposal provided a "rocket room" and single launch tube aft of the forward torpedo, living, and battery spaces—and forward of the control spaces. Innovative features in the craft included a vertical launcher and a pump-jet, or ducted, propulsion device in place of propellers. But neither the missile nor interest in a ballistic missile program existed in the U.S. Navy at the time.

The proposals to place ballistic missiles in submarines received impetus in the aftermath of the Soviet detonation of a thermo-nuclear (hydrogen) device on 12 August 1953. The fear of Soviet advances in strategic missiles led the U.S. Department of Defense to direct the Navy to join the Army in development of an Intermediate Range Ballistic Missile (IRBM) that could be launched from *surface ships.*[2]

The Navy's leadership had objected strenuously to the joint program, because the Army was developing the liquid-propellant Jupiter missile. The Navy considered liquid propellants too dangerous to handle at sea, and a 60-foot (18.3-m) missile would be troublesome even on board surface ships. In addition, there was general opposition to ballistic missiles at sea within the Navy from the "cultural" viewpoint for two reasons. First, from the late 1940s both the Bureau of Aeronautics and the Bureau of Ordnance were (separately) developing cruise missiles that could be launched from submarines against land targets; neither bureau wished to divert scarce resources to a new ballistic missile program.

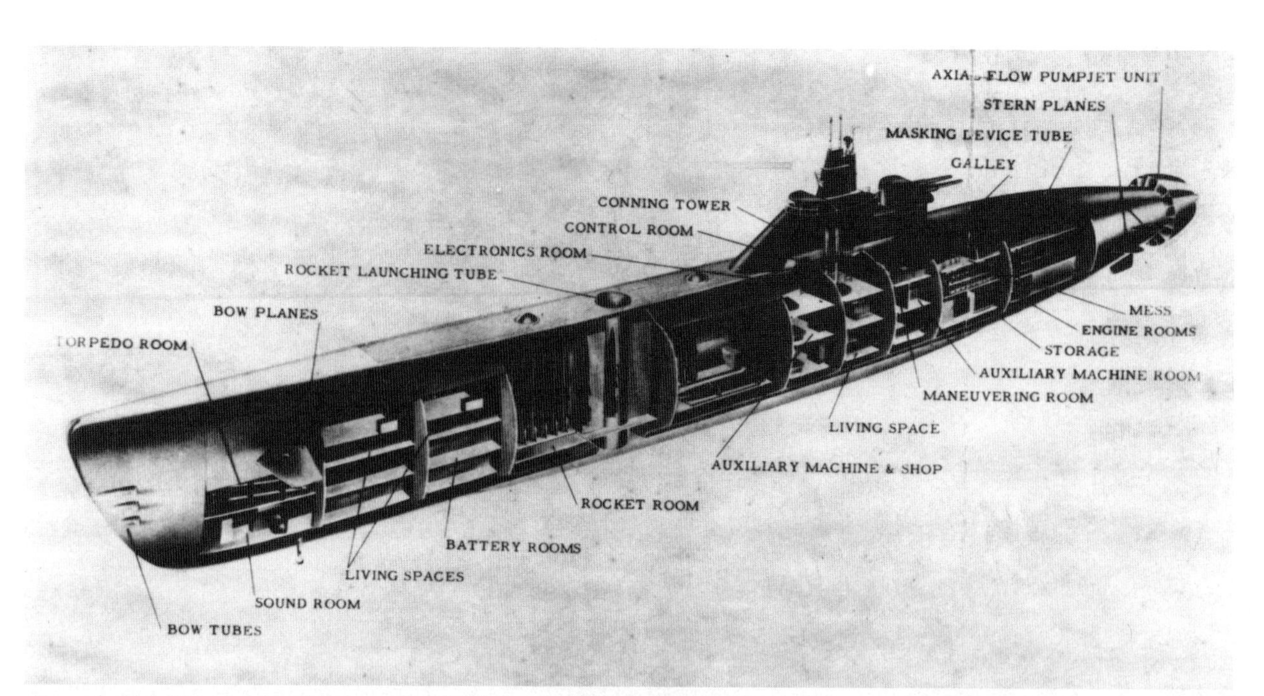


*An early U.S. Navy proposal for carrying ballistic missiles in submarines saw this 1950 design for adapting a fleet boat to carry missiles that would be launched from a single launch tube. The submarine—with a modified conning tower—retains a deck gun; a pump-jet/ducted propulsor is provided.* (U.S. Navy)

Second, the Navy had lost the B-36 bomber-versus-carrier controversy to the Air Force in the late 1940s. That loss had cost the Navy prestige and cancellation of the first postwar aircraft carrier, the *United States*.[3] As a result, the Navy's leadership wanted to avoid another inter-service battle, this time over strategic missiles. Indeed, Admiral Robert B. Carney, Chief of Naval Operations from 1953 to 1955, had placed restrictions on Navy officers advocating sea-based ballistic missiles.

There was another issue that was very real. This was the fear of having to pay for new weapons—such as strategic missiles—out of the regular Navy budget.[4] Navy opposition to developing a sea-based ballistic missile force ended when Admiral Arleigh A. Burke became Chief of Naval Operations in August 1955. According to Admiral Burke's biographer, "Burke's most significant initiative during his first term [1955–1957] was his sponsorship, in the face of considerable opposition, of a high-priority program to develop a naval intermediate-range ballistic missile."[5]

Fearing that the project would be given low priority within the Navy and that it would be doomed to failure if left to the existing Navy bureaucracy, Admiral Burke established the Special Projects Office (SPO) to provide a "vertical" organization, separate from the existing technical bureaus (e.g., Bureau of Ships), to direct the sea-based missile project. Heretofore, all major naval technical developments as well as production had been directed by the technical bureaus, a horizontal organizational structure that dated from 1842. In these actions Burke was strongly supported by the Secretary of the Navy, Charles S. Thomas.[6]

As important as establishing the SPO was selection of its director. Admiral Burke appointed newly promoted Rear Admiral William F. Raborn, a Navy pilot who had considerable experience with guided missiles. Because of the role of Raborn in Polaris development, it is important to cite Burke's criteria for the man to direct this important, controversial,

*William F. Raborn* (U.S. Navy)

and difficult project. Burke selected an officer whom he believed had the ability to direct the project on the basis of his individual qualifications and with minimal concern for his membership in the Navy's warfare specialties or "unions." According to Burke:

I realized that he didn't have to be a technical man. He had to be able to know what technical men were talking about. He had to get a lot of different kinds of people to work. . . . I wanted a man who could get

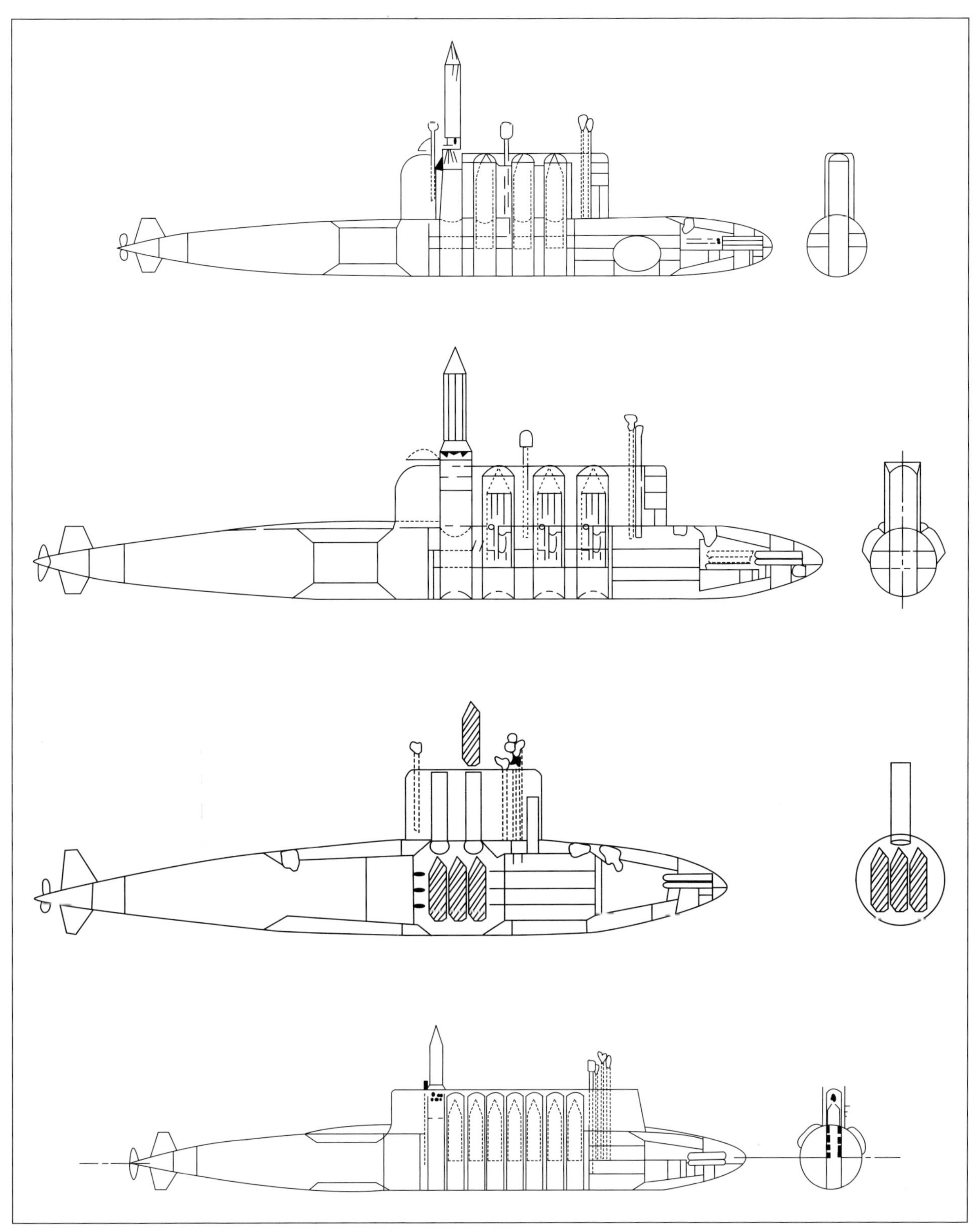

*A variety of Jupiter SSBN configurations were considered, all based on variations of the* Skipjack *design.*

*Forced to join the Army in the quest for a ballistic missile force, the Navy produced several preliminary designs for a submarine to launch the liquid-propellant Jupiter missile. This version has four SLBMs carried in a modified* Skipjack *hull. Satellite navigation is shown being used as the submarine's sail penetrates the surface.* (U.S. Navy)

> along with aviators because this [program] was going to kick hell out of aviators. They were going to oppose it to beat the devil because it would take away, if it were completely successful in the long run, their strategic delivery capability.
>
> It would be bad to have a submariner, in that because it first was a surface ship [weapon]; submariners were a pretty close group and they would have wanted to do things pretty much as submariners had already done . . . besides they were opposed to ballistic missiles.[7]

And, Burke had problems with surface warfare officers as well, because "they don't know much about missiles or strategic [matters]."

Admiral Burke's support for Raborn included telling him that he could call on the best people in the Navy for his project staff but the numbers had to be kept small; and, any time that it looked like the project's goals could not be accomplished, Raborn should simply recommend to Burke that the project be stopped.

On 8 November 1955 the Secretary of Defense established a joint Army-Navy IRBM program. The sea-based Jupiter program, given highest national priority—Brickbat 01—along with the Air Force Atlas ICBM and Army Jupiter IRBM, progressed rapidly toward sending the missile to sea in converted merchant ships. During 1956 a schedule was developed to put the first IRBM-armed merchant ships at sea in 1959. Some studies also were addressing the feasibility of submarines launching the Jupiter IRBM from the surface, with four to eight missiles to be carried in a nuclear-propelled submarine of some 8,300 tons submerged displacement.

The Navy still had severe misgivings about the use of highly volatile liquid propellants aboard ship, and studies were initiated into solid propellant missiles. However, solid propellants had a low specific impulse and hence were payload limited. The major boost for solid propellants came in mid-

1956, when scientists found it feasible to greatly reduce the size of thermo-nuclear warheads. Dr. Edward Teller—"father" of the American hydrogen bomb—is said to have suggested in the summer of 1956 that a 400-pound (181-kg) warhead would soon provide the explosive force of a 5,000-pound (2,270-kg) one.[8] The Atomic Energy Commission in September 1956 estimated that a small nuclear warhead would be available by 1965, with an even chance of being ready by 1963.

This development coupled with the parallel work on higher specific impulse solid propellants permitted (1) a break from the Army's Jupiter program in December 1956, (2) formal initiation of the Polaris SLBM program with a solid-propellant missile, and (3) a shift from surface ships to submarines as the launch platform.[9]

On 8 February 1957, the Chief of Naval Operations (Burke) issued a requirement for a 1,500-n.mile (2,775-km) missile launched from a submarine to be operational by 1965. This range was stipulated to enable a submarine in the Norwegian Sea to target the Soviet capital of Moscow. The February 1957 schedule with a goal of 1965 was followed by a series of revisions and accelerations in the Polaris program.

During this period Admiral Burke was financing the Polaris program entirely from existing Navy budgets. At the time he was fighting for a major naval building program, including the first nuclear surface ships, as well as the Polaris. Burke's task became more urgent when, on 3 August 1957, a Soviet R-7 missile rocketed several thousand miles from its launchpad at Tyuratam to impact in Siberia. This was the world's first long-range ICBM flight test. Five weeks later, on 4 October 1957, the Soviets orbited *Sputnik*, the first artificial earth satellite. On 23 October Secretary of the Navy Thomas S. Gates proposed acceleration of Polaris to provide a 1,200-n.mile (2,225-km) missile by December 1959, with three SLBM submarines to be available by mid-1962, and a 1,500-n.mile missile to be available by mid-1963. A month later the program was again accelerated, to provide the 1,200-n.mile missile by October 1960. Late in December 1957—just as the preliminary plans for the Polaris submarine were completed—the schedule was again changed to provide the second submarine in March 1960, and the third in the fall of 1960.

## Building Missile Submarines

To permit the submarines to be produced in so short a time, on the last day of 1957 the Navy reordered a recently begun nuclear-propelled, torpedo-attack submarine and a second, not-yet-started unit as ballistic missile submarines. Thus the first five Polaris submarines (SSBN 598–602) were derived from the fast attack submarines of the *Skipjack* (SSN 585) class. The *Skipjack* featured a streamlined hull design with a single propeller; propulsion was provided by an S5W reactor plant producing 15,000 horsepower. (See Chapter 9.) Thus the basic missile submarine would be an existing SSN design, which was lengthened 130 feet (39.6 m)—45 feet (13.7 m) for special navigation and missile control equipment, 10 feet (3.0 m) for auxiliary machinery, and 75 feet (22.9 m) for two rows of eight Polaris missile tubes. The number 16 was decided by polling the members of Admiral Raborn's technical staff and averaging their recommendations! Obviously, being much larger than the *Skipjack* with the same propulsion plant, the SSBNs would be much slower.

The first submarine was to have been the *Scorpion* (SSN 589), laid down on 1 November 1957; by this method the Navy was able to use hull material and machinery ordered for attack submarines to accelerate the Polaris submarines. Popular accounts that an SSN "was cut in half" on the building ways and a missile section inserted to create the first Polaris submarine are a myth. The *Scorpion*/SSN 589 was laid down on 1 November 1957 at the Electric Boat yard; she was reordered as a Polaris submarine—designated SSGN(FBM) 598—on 31 December 1957.[10] The "cutting" was done on blueprints with construction using components of both the *Scorpion* and the *Skipjack*. The "new" submarine was named *George Washington*.

Admiral Rickover initially objected to a single screw for the missile submarines, although he had recently accepted that configuration for SSNs. Atomic Energy Commission historians Richard Hewlett and Francis Duncan noted that "under written orders from Admiral Burke [Raborn and other admirals] excluded Rickover from all the

*The* George Washington *at rest, showing the position of the Polaris missile compartment aft of the sail. The reactor compartment is behind the missiles, beneath the square covering. The hull number and name would be deleted before the submarine went on deterrent patrol.* (General Dynamics Corp.)

preliminary studies."[11] And Raborn, as the director of the SPO, "had over-all responsibility for the submarine as well as the missile."[12] Raborn and the other admirals involved in the Polaris project feared Rickover's participation "would lead to domination of the new project" by his office.[13] Burke excluded Rickover by simply directing that the S5W plant would be used. Rickover did gain one concession: The requirement for Polaris submarines to operate under the Arctic ice pack was deleted from the SSBN requirements. (Two years later, however, Rickover would tell a congressional committee that the Polaris submarines "will be able to operate under the polar icecap.")

Spurred on by Soviet strategic missile developments—given the political label "missile gap" in the United States—Polaris was given the highest national priority. Construction of attack submarines and the two-reactor radar picket submarine *Triton* (SSRN 586) were slowed, because Polaris had priority for shipyard workers and matériel. By July 1960 there were 14 Polaris submarines under construction—the five *George Washington* class, five improved *Ethan Allen* (SSBN 608) class, and four of the *Lafayette* (SSBN 616) class.

The first five SSBNs were based on the *Skipjack* design, with a test depth of 700 feet (215 m), except that the first ship, the *George Washington,* with a missile compartment insert of High-Tensile Steel (HTS) rather than the HY-80 steel used in the hull, limited her to about 600 feet (183 m). The improved, five-ship *Ethan Allen* class was larger, incorporating the hull features of the *Thresher* (SSN 593) class with a test depth of 1,300 feet (400 m). These submarines, at 7,800 tons submerged, were larger (compared to 6,700 tons for the *George Washington*); had only four torpedo tubes (vice six); had *Thresher*-type machinery quieting; and were provided with improved accommodations and other features. The subsequent *Lafayette*-class ships—the "ultimate" Polaris design—were still larger, displacing 8,250 tons submerged, with further improved quieting and other features. The propulsion plants in all three classes were similar, with the final 12 submarines of the *Lafayette* class having improved quieting; all would carry 16 Polaris missiles.

The Polaris missiles could be launched while the submarines were completely submerged—approximately 60 feet (18.3 m) below the surface—with a launch interval of about one minute per missile.[14] The major concern for Polaris submarines was communications; how long would the delay be before a submarine running submerged could receive a firing order? Technical improvements, the use of trailing-wire antennas, and, subsequently, satellites and communications relay aircraft would enhance communications to submarines.[15] The one-way communications mode of missile submarines and potential message reception delays meant that they were unsuitable for a first-strike

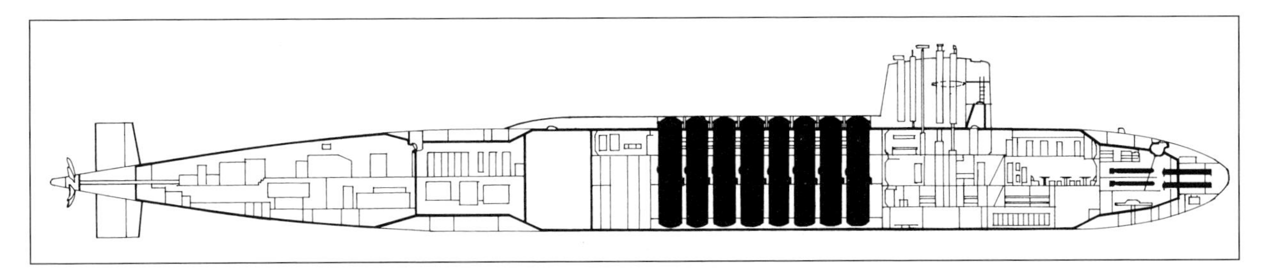

George Washington *(SSBN 598). LOA 381 ft 8 in (116.36 m)* (©A.D. Baker, III)

attack (a factor known to leaders in the Kremlin as well as in Washington). Further, the relatively high survivability of submarines compared to manned bomber bases and land-based ICBMs meant that submarine missiles could be "withheld" for a second-strike or retaliatory attack, making them a credible deterrent weapon.

The *George Washington* was placed in commission on 20 December 1959, under Commanders James B. Osborn *and* John L. From Jr. Polaris submarines introduced a new ship-operating concept, with each submarine being assigned two complete crews of some 135 officers and enlisted men (called Blue and Gold, the Navy's colors). One crew would take a Polaris submarine to sea on a submerged deterrent patrol for 60 days; that crew would then bring the submarine into port for repairs and replenishment, and after a transfer period, the alternate crew would take the submarine back to sea for a 60-day patrol. While one crew was at sea, the other would be in the United States, for rest, leave, and training. Thus, some two-thirds of the Polaris force could be kept at sea at any given time.[16]

The *G.W.* went to sea on 18 June 1960 carrying two unarmed Polaris missiles for the first U.S. ballistic missile launch by a submarine. Rear Admiral Raborn was on board as were both the Blue and Gold crews and a number of technicians; about 250 men crowded the submarine. Numerous minor problems plagued the missile countdown, and, finally, the submarine returned to port without launching the two missiles.

Returning to sea on 20 June, the *George Washington,* after encountering additional minor problems, successfully carried out two submerged launches, several hours apart. Moments after the second missile launch, Raborn sent a message directly from the *G.W.* to President Dwight D. Eisenhower: POLARIS—FROM OUT OF THE DEEP TO TARGET. PERFECT.[17]

The *George Washington* departed on the first Polaris "deterrent" patrol on 15 November 1960, manned by Commander Osborn's Blue crew. The submarine carried 16 Polaris A-1 missiles with a range of 1,200 n.miles and fitted with a W47 warhead of 600 kilotons explosive force.[18] The *George Washington* was at sea on that initial patrol for 67 consecutive days, remaining submerged for 66 days, 10 hours, establishing an underwater endurance record. Before the *George Washington* returned to port, the second Polaris submarine, the *Patrick Henry* (SSBN 599), had departed on a deterrent patrol on 30 December 1960. Additional Polaris submarines went to sea at regular intervals.

Significantly, from the outset Navy planners had prepared for SSBN deployments in the Atlantic and Pacific, broad seas where they would be less vulnerable to Soviet ASW efforts. But the Kennedy administration decided to send three Polaris submarines into the Mediterranean Sea beginning about 1 April 1963, to coincide with the removal of U.S. Jupiter IRBMs from Turkey. Those missiles had been "traded" by the Kennedy administration in the surreptitious negotiations with the Soviet government to end the Cuban missile crisis. The *Sam Houston* (SSBN 609), the first SSBN to enter the "Med," assured that all interested parties knew of the Polaris presence when, on 14 April, the submarine entered the port of Izmir, Turkey.[19] The first Polaris submarine to undertake a deterrent patrol in the Pacific was the *Daniel Boone* (SSBN 629), which departed Guam on 25 December 1964 carrying 16 A-3 missiles.

Initially the U.S. Navy had proposed an ultimate force of about 40 Polaris submarines.[20] Some Defense officials, among them Deputy Secretary of Defense Donald Quarles, felt that many more such

TABLE 8-1
**U.S. Nuclear-Propelled Submarines Completed 1960–1967**

| Shipyard | Ballistic Missile (SSBN) | Torpedo-Attack (SSN) | Cruise Missile (SSGN) |
|---|---|---|---|
| Electric Boat Company, Groton, Conn. | 17* | 4 | — |
| General Dynamics, Quincy, Mass. | — | 1 | — |
| Ingalls Shipbuilding, Pascagoula, Miss. | — | 5 | — |
| Mare Island Naval Shipyard, Vallejo, Calif. | 7 | 3 | 1** |
| Newport News Shipbuilding, Va. | 14 | 4 | — |
| New York Shipbuilding, Camden, N.J. | — | 3 | — |
| Portsmouth Naval Shipyard, Maine | 3 | 4 | — |
| **Total** | **41** | **24** | **1** |

Notes: * Includes the USS *George Washington*, commissioned on 30 December 1959.
** USS *Halibut* (SSGN 587).

craft were possible, predicting "very strong Congressional support for the construction of perhaps as many as 100 such submarines."[21]

Admiral Burke recalled that he had developed a requirement for 39 to 42 Polaris submarines (each with 16 missiles) based on the number of targets for nuclear weapons in the USSR. These were multiplied by two weapons each for redundancy and reliability, multiplied by 10 percent for weapons intercepted or "shot down," and multiplied by 20 percent as the number that would probably malfunction.[22] Burke also had used 50 submarines as his "target" for Polaris.[23] At all force levels it was planned to keep two-thirds of the Polaris submarines at sea through the use of dual crews (see above).

Subsequently, the Navy and Department of Defense programmed a force of 45 submarines—five squadrons of nine ships each. Of those, 29 submarines were to be at sea (deployed) at all times and could destroy 232 Soviet targets (i.e., 464 missiles with a 50-percent pK[24]). But by September 1961, Secretary of Defense Robert S. McNamara had advised President John F. Kennedy that he recommended a final force of 41 Polaris submarines carrying 656 missiles.[25] In January 1962 McNamara told Congress, "Considering the number of Minuteman missiles and other strategic delivery vehicles which will be available over the next few years, it is difficult to justify a Polaris force of more than 41 submarines."[26]

In the event, through 1967 the U.S. Navy took delivery of 41 Polaris submarines (which were organized into four squadrons). Each submarine carried 16 Polaris missiles—a total of 656 SLBMs. In addition, the Royal Navy built four similar Polaris missile submarines, completed from 1967 to 1969; these were armed with American-supplied A-3 missiles carrying British nuclear warheads.[27]

During the seven-year period 1960–1967 U.S. shipyards produced an average of almost 9½ nuclear submarines per year, a construction rate that never again was achieved in the United States, but that would be exceeded by the USSR in the 1970s. (No diesel-electric submarines were added to the U.S. fleet in that period.)

In contrast to the two or three missiles of the first-generation Soviet SSB/SSBNs, the U.S. Polaris submarines each carried 16 SLBMs. Further, the U.S. missiles could be launched while the submarine remained fully submerged and were credited with a greater accuracy than the Soviet weapons.[28] As important as the missiles was the development (at an accelerated rate) of the fire control and navigation systems for Polaris submarines. The latter was particularly critical because of the long ranges of the missile and the need for the submarine to remain submerged (except for masts and antennas raised on a periodic basis). Recalling that this was in the period of primitive navigation satellites such as the Transit, the Ships Inertial Navigation System (SINS) developed for the Polaris program was a remarkable technological achievement. It provided accurate navigation based on movement of the submarine and using external navigation sources only for periodic updates. Finally, providing the needed life support systems, including oxygen and potable water while fully submerged for a crew of 150 to 160 men in a submarine for 60 to 70 days, was also a major achievement. At the time of their construction, the Polaris SLBM submarines were the largest, most complex, and from a viewpoint of weapons payload, the most powerful submarines yet constructed by any nation.[29]

The time to prepare missiles for launching after receipt of a launch order was about 15 minutes; the missiles could then be launched at intervals of about one minute. The submerged Polaris submarine had to be stationary or moving at a maximum of about one knot to launch. The launch depth (from the keel of the submarine) was about 125 feet.

In addition to submarines, from the outset the Navy considered placing Polaris missiles in surface warships as well as specialized strategic missile ships. Particular efforts were made to configure cruisers being built and converted to carry surface-to-air missiles to also have Polaris missiles. Also, extensive analysis was undertaken into the feasibility of placing Polaris missiles in *Iowa*-class battleships, aircraft carriers of the large *Forrestal* class (76,000 tons full load), and smaller *Hancock*-class carriers (42,600 tons). These efforts determined that all of the ships could readily accept Polaris missiles.[30] Indeed, the *Forrestal*s could accommodate up to 30 missile tubes (replacing some 5-inch guns and their accessories) without seriously affecting aircraft numbers or operations. The *Hancock*s similarly could be armed with up to 20 Polaris missiles, albeit with more difficulty because of the lack of weight margin in those ships (built during World War II). The specialized Polaris ships would be merchant designs, armed with a large number of Polaris missiles and manned by Navy personnel. They could steam on the open seas, "hiding" among the world's merchant traffic, or remain in allied harbors, where they would be safe from enemy submarine and possibly air attack. At one point the U.S. government proposed multi-national NATO crews for these ships.[31]

In the event, no surface ships ever were armed with Polaris missiles. The rapid buildup of Polaris submarines coupled with the increasing range of submarine-launched missiles alleviated the need for such ships. The surface ship SLBM concept, however, would be revisited in the future. (See Chapter 12.)

## Improved Polaris Missiles

The 1,200-n.mile A-1 missile was considered an interim weapon from the outset of the solid-propellant program, with longer-range variants in development when the *George Washington* went to sea. The 1,500-n.mile A-2 missile went on patrol in June 1962 in the *Ethan Allen*. A month before she deployed, on 6 May 1962 the *Ethan Allen*, operating near Christmas Island in the mid-Pacific, launched a Polaris A-2 missile with a nuclear warhead that detonated in the only full-systems test of a U.S. strategic missile—land- or sea-based—from launch through detonation (Operation Frigate Bird). The A-2 carried the same W47 warhead of the A-1 missile, but with an increased yield of 1.2 megatons and an increase in range from 1,200 n.miles to 1,500 n.miles.

The 2,500-n.mile A-3 missile was first deployed in the *Daniel Webster* (SSBN 626) in September 1964. All three versions of the Polaris had the same diameter (54 inches/1.37 m). This enabled the submarines to be upgraded to later missiles during normal shipyard overhauls. Thus, all 41 Polaris submarines eventually were upgraded to launch the A-3 variant.[32]

While the A-1 and A-2 carried single warheads, the A-3 carried a Multiple Reentry Vehicle (MRV) that "shotgunned" three warheads onto a single target. Each of its three W58 warheads had a yield of 200 kilotons, with the total MRV weighing some 1,100 pounds (500 kg).[33] The MRV effect compensated for the limited accuracy of the A-3 missile.

The fourth missile of this series to go to sea was the C-3 variant, given the name Poseidon.[34] This was a much larger missile, 74 inches (1.88 m) in diameter, with about the same range as the A-3, that is, 2,500 n.miles. But the Poseidon was the world's first operational strategic missile to carry Multiple Independently targeted Reentry Vehicles (MIRV). Ten to 14 W68 warheads could be fitted, each with an explosive force of 50 kilotons, which could be guided to separate targets within a given area ("footprint").

The impetus for the Poseidon SLBM came from indications that the Soviet Union was developing an Anti-Ballistic Missile (ABM) system that would be

*The* Henry Clay *(SSBN 625) launches a Polaris A-2 missile from the surface off Cape Kennedy, Florida, in 1964. The mast atop the sail is a telemetry antenna used for test launches. This was the first of only two surface SLBM launches from U.S. submarines.* (U.S. Navy)

TABLE 8-2
**U.S. Submarine-Launched Ballistic Missiles**

| **Missile** | **Polaris A-1 UGM-27A** | **Polaris A-2 UGM-27B** | **Polaris A-3 UGM-27C** | **Poseidon C-3 UGM-73** |
|---|---|---|---|---|
| Operational | 1960 | 1962 | 1964 | 1971 |
| Weight | 28,000 lb (12,700 kg) | 32,500 lb (14,740 kg) | 35,700 lb (16,195 kg) | 64,000 lb (29,030 kg) |
| Length | 28 ft (8.53 m) | 31 ft (9.45 m) | 32 ft (9.75 m) | 34 ft (10.37 m) |
| Diameter | 54 in (1.37 m) | 54 in (1.37 m) | 54 in (1.37 m) | 74 in (1.88 m) |
| Propulsion | solid-propellant 2 stage | solid-propellant 2 stage | solid-propellant 2 stage | solid-propellant 2 stage |
| Range | 1,200 n.mi (2,225 km) | 1,500 n.mi (2,775 km) | 2,500 n.mi (4,635 km) | 2,500 n.mi* (4,635 km) |
| Guidance | inertial | inertial | inertial | inertial |
| Warhead** | 1 RV nuclear W47 Y1 600 KT | 1 RV nuclear W47 Y2 1.2 MT | 3 MRV nuclear W58 3 x 200 KT | 10–14 MIRV nuclear W68 10–14 x 50 KT |

Notes: * With reduced payload.
** RV = Reentry Vehicle; MRV = Multiple Reentry Vehicle.
MIRV = Multiple Independently targeted Reentry Vehicle.

able to defeat the planned number of U.S. strategic missiles in a nuclear strike. This innovative Poseidon SLBM was the result of a series of Navy studies that began in March 1964 with the so-called Great Circle Study. Initiated by Secretary of the Navy Paul H. Nitze, this and later studies responded to demands by Secretary of Defense McNamara for more rigorous analysis of U.S. strategic offensive and defensive requirements. At the time the Navy was proposing the Polaris B-3 missile as an eventual follow-on to the A-3. The B-3 missile—with a diameter of 74 inches (1.88 m)—would carry six warheads, each of 170 kilotons, plus penetration aids, to about the same range as the Polaris A-3. But Dr. Harold Brown, Director of Defense Research and Engineering, proposed that the Navy incorporate the latest missile technology.[35] This became the Poseidon C-3 with enhanced payload, range, and separate target attack capability. The first Poseidon patrol began in March 1971, with the *James Madison* (SSBN 627) taking to sea a loadout of 16 missiles. While the "conversion" of the Polaris submarines was required to carry this larger missile, the changes were relatively minor, and all 31 submarines of the *Lafayette* class were converted to launch the Poseidon C-3 missile. A final version of the Polaris/Poseidon was the so-called EXPO (Extended-range Poseidon), which entered advanced development in the early 1970s. This missile was soon renamed the Trident C-4, with 12 of the *Lafayette*-class submarines being further converted to carry the weapon. The C-4 missile, with a diameter of 74 inches (1.88 m), had a theoretical range of 4,000 n.miles (7,400 km) and a MIRV configuration for eight W76 warheads of 100 kilotons each.

The flexibility of the 41 Polaris submarines to be rapidly updated to carry successive SLBMs was one of the several remarkable aspects of the program. From 1967 to 1980 there were 41 Polaris/Poseidon SSBNs in service (including those undergoing overhaul and modernization). The last Polaris patrol by the "original" 41 SSBNs was completed in 1994, bringing to an end a remarkable achievement.

*The missile hatches of the* Sam Rayburn *(SSBN 635) are open, awaiting their deadly cargo of Polaris A-3 missiles. The Polaris configuration permitted relatively simple modifications to launch improved missiles of the Polaris, Poseidon, and Trident C-4 series.* (Newport News Shipbuilding Corp.)

As the SSBNs were retired, there were efforts to employ some of the early Polaris submarines in the torpedo-attack role (SSN), but they were unsuccessful because of their limited sonar, too few weapons, slow speed, and high noise levels. An innovative proposal for retired Polaris/Poseidon SSBNs was to employ them as forward radar tracking platforms as part of a national Anti-Ballistic Missile (ABM) defense system. As proposed in the late 1960s to support the planned Sentinel/Safeguard ABM program, several early SSBNs were to have been rebuilt to mount large, fixed-array radars on an enlarged sail structure. During crisis periods the submarines would surface through the Arctic ice pack to provide early warning/tracking of Soviet or Chinese ballistic missiles. Also under consideration was the possibility of launching ABM intercept missiles from the submarine missile tubes, possibly employing Polaris missiles as the boost vehicle for the interceptor. These proposals were abandoned when the Sentinel/Safeguard ABM program was cancelled in the early 1970s.[36]

More successful, from 1984 onward four Polaris/Poseidon submarines were reconfigured to carry special operations forces (commandos). Each was modified to carry 65 troops or combat swimmers, their equipment, and rubber landing craft.[37] The "very last" of the 41 Polaris submarines in service was the *Kamehameha* (SSN 642), retired in 2002 after serving as a transport submarine.

**The U.S. Polaris system was born as a reaction to indications in the late 1950s of massive Soviet efforts to develop strategic missile and satellite systems. These concerns led the United States to initiate the Polaris SLBM and Minuteman ICBM**

to counter the fear of a strategic "missile gap." (In fact, the missile gap would exist in favor of the United States until the mid-1970s.)

Polaris rapidly became one "leg" of the U.S. Triad—the term coined by the U.S. Air Force to rationalize the "need" for three U.S. strategic offensive forces: manned, land-based bombers, land-based ICBMs, and submarine-launched ballistic missiles.[38] In 1961 Secretary of Defense McNamara described the Polaris as having

> the most survival potential in the wartime environment of any of our long range nuclear delivery systems. Polaris missiles do not have to be launched early in the war, they can be held in reserve and used in a controlled and deliberate way to achieve our wartime objectives. For example, Polaris is ideal for counter-city retaliation.[39]

TABLE 8-3
**Ballistic Missile Submarines**

| | U.S. *George Washington* SSBN 598 | U.S. *Ethan Allen* SSBN 608 | U.S. *Lafayette* SSBN 616 |
|---|---|---|---|
| Operational | 1960 | 1961 | 1963 |
| Displacement | | | |
| surface | 5,900 tons | 6,900 tons | 7,250 tons |
| submerged | 6,700 tons | 7,900 tons | 8,250 tons |
| Length | 381 ft 8 in (116.36 m) | 410 ft 5 in (125.13 m) | 425 ft (129.57 m) |
| Beam | 33 ft (10.06 m) | 33 ft (10.06 m) | 33 ft (10.06 m) |
| Draft | 26 ft 8 in (8.13 m) | 27 ft 6 in (8.38 m) | 27 ft 9 in (8.46 m) |
| Reactors | 1 S5W | 1 S5W | 1 S5W |
| Turbines | 2 | 2 | 2 |
| horsepower | 15,000 | 15,000 | 15,000 |
| Shafts | 1 | 1 | 1 |
| Speed | | | |
| surface | 16.5 knots | 16 knots | 16 knots |
| submerged | 22 knots | 21 knots | 21 knots |
| Test depth | 700 feet (215 m) | 1,300 feet (400 m) | 1,300 feet (400 m) |
| Missiles | 16 Polaris | 16 Polaris | 16 Polaris/Poseidon |
| Torpedo tubes* | 6 533-mm B | 4 533-mm B | 4 533-mm B |
| Torpedoes | 12 | 12 | 12 |
| Complement** | 136 | 136 | 136 |

Notes: * Bow; Stern.
** Alternating Blue and Gold crews.

Polaris submarines were superior to contemporary Soviet missile submarines and were specially configured for being upgraded to launch more advanced missiles as they became available. The Polaris submarines were *completely* invulnerable to Soviet anti-submarine forces of the time. Writing in 1961, a senior spokesman of the Soviet Navy wrote—in a classified publication—"Strictly speaking, we do not yet have finalized methods for combating missile-carrying submarines. Even the main forces for accomplishing this mission have not yet been defined."[40]

In response to Polaris the Soviets would initiate several major ASW programs, but not until the late 1970s could these be considered a credible threat to U.S. strategic missile submarines. Still, in 1978 Secretary of Defense Harold Brown could state, unequivocally, "The critical role of the SLBM force, as the most survivable element in the current TRIAD of strategic forces, both now and in the foreseeable future, is well established."[41]

As a result of its survivability, according to Secretary Brown, "the SLBM force contributes to crisis stability. The existence of a survivable, at-sea ballistic missile force decreases the Soviet incentives to procure additional counterforce weapons and to plan attacks on United States soil since such attacks would not eliminate our ability to retaliate."[42]

The Polaris submarines appear to have been the models for subsequent SSBN development in all countries that developed such ships—China, France, Great Britain, and the Soviet Union.

# 9

# The Quest for Speed

*The Project 661/Papa SSGN was the world's fastest submarine, attaining a speed of 44.7 knots in 1970—which is still unsurpassed. Like several other Soviet submarine designs, she proved to be too expensive and too complex for large-scale production.*

Allied code breaking during World War II revealed that the Type XXI submarine would have a much greater submerged speed than previous U-boats. In response, beginning in 1944 the Royal Navy rapidly modified eight S-class submarines for use as "high-speed" targets for training British anti-submarine forces. The converted submarines could attain 12.5 knots underwater, slow in comparison with the Type XXI, but faster than any other allied submarine.[1]

After the war the British, Soviet, and U.S. Navies began to improve the underwater performance of their submarines with German design concepts and technologies. Britain and the United States rebuilt war-built submarines to improve their submerged speeds, and all three navies incorporated German technology in their new designs.

As early as 1945 a U.S. Navy officer, Lieutenant Commander Charles N. G. Hendrix, drew up a single page of "Ship Characteristics" that proposed a radical, high-speed submarine design, incorporating such features as:[2]

reduced-cavitation propeller
relocation of propeller [aft of rudder]
radical changes in hull design
flood valves on main ballast tanks
double or triple hull design
automatic depth control
automatic steering
two vertical rudders or an upper rudder
high-speed bow and stern planes
improved pitot (speed) log
abolish conning tower compartment
provide combat information center
reduce hatches in pressure hull

This single page—in the opinion of the authors of this book—represents the origins of a remarkable submarine, the USS *Albacore* (AGSS 569). The *Albacore* was a revolutionary design, because it began with a "clean sheet of paper." On 8 July 1946 the Bureau of Ships requested that the David Taylor Model Basin in nearby Carderock, Maryland, undertake tests of a high-speed hull form, soon known as Series 58.[3] A few days later, on 26 July, Dr. Kenneth S. M. Davidson, chairman of the hydrodynamics panel of the Committee on Undersea Warfare of the National Academy of Sciences, wrote to the commanding officer of the model basin, Captain Harold E. Saunders, calling for a completely new approach to submarine design—"a rational design" instead of "ceaseless modification and juggling" of existing designs, yielding "a second rate answer." During the next few years several Navy offices began supporting the development of a high-speed undersea research craft, especially Rear Admiral Charles B. (Swede) Momsen, the Assistant Chief of Naval Operations for Undersea Warfare.[4]

The *Albacore* was intended primarily "To provide information required for future submarine design by acting as a test vessel in regard to hydrodynamics [and] . . . To act as a high-speed target during the evaluation of anti-submarine systems and weapons and to determine the anti-submarine requirements necessary to defeat this type of submarine." The secondary task was to train personnel to operate high-speed submarines.[5]

Addressing the need for submarine speed, another admiral declared: "the next step in the submarine will be to make speeds so that they will be able to outrun any type of homing weapon. We consider that as the most important submarine development that we have in our [shipbuilding] program."[6]

Dr. Davidson, professor of engineering at the prestigious Stevens Institute of Technology in Hoboken, New Jersey, and John C. Niedermair, in the preliminary design section of the Bureau of Ships, produced a radical hull design for the *Albacore,* using results of model basin tests and modeling of early, turn-of-the-century submarines (e.g., John Holland's *Plunger* of 1897). There also was input from aerodynamic analyses in that, while air and water differ in density, both are fluids and key parameters apply to both. Early papers on the *Albacore* refer to her inverted blimp shape or the "Lyon-form," an apparent reference to British scientist Hilda Lyon, who helped develop the streamlined shape of the airship *R-101.* This hull form was a starting point for the *Albacore,* with a 30-foot (9.15-m) submarine model that was "flight tested" in the National Advisory Committee for Aeronautics (NACA) aerodynamic wind tunnel at Langley Air Force Base in Virginia.[7]

To reduce the turbulence generated as water flowed over the hull, all projections were removed (guns, bits, cleats, railings, etc.) except for a streamlined fairwater and the control surfaces. As designed, the submarine was planned to achieve a submerged speed of 27.4 knots with an electric motor producing only 4,000 shaft horsepower.[8] The submarine would have a 500-cell battery, enabling her to maintain that speed for 30 minutes, or 21.5 knots, for one hour. Two General Motors 16/338 "pancake" diesel engines—from the *Tang* (SS 563) class—were fitted for charging batteries. Those problem-plagued engines were used because they were readily available and, being smaller and lighter than conventional diesel engines, seemed suitable for the small research submarine. The single electric motor produced 7,500 horsepower.

*An* Albacore *model undergoing wind-tunnel testing at Langley Air Force Base (Virginia) to determine flow/drag characteristics. High-speed submarine shapes were also tested in water tanks at the David W. Taylor model basin in Carderock, Maryland.* (U.S. Navy)

The early *Albacore* designs provided for a submarine with a length of 150 feet (45.73 m), a beam of 30 feet (9.15 m), and a surface displacement of some 1,600 tons. In time the length grew to 204 feet (62.2 m), but the beam was reduced to 27 feet (8.23 m), and the surface displacement to 1,517 tons. The final *Albacore* design featured a low length-to-beam ratio of 7.5:1, rounded bow, tapered stern, and "body of revolution," that is, a circular cross section along the entire hull. A double hull was provided to facilitate the radical outer shape of the hull as well as surviving inert torpedo hits in the secondary training-target role. The inner hull was 21 feet (6.4 m) in diameter at its widest point, four feet (1.2 m) greater than in the *Tang* design. This enabled the *Albacore* to have the equivalent of almost three deck heights amidships.

No weapons were provided in the submarine, although the issue was continually raised, even after the *Albacore* was completed. Admiral I. J. (Pete) Galantin later explained, "If the ship was given a torpedo-firing capability we would gain one attack boat of very limited capability and lose flexibility and operational time needed for the exploration of new design concepts."[9]

The consideration of the *Albacore* for a secondary role as a target for ASW weapons led to the submarine being listed as SST in early planning documents. The designation soon was changed to AG(SST) and, subsequently, to AGSS 569.[10]

Construction of the *Albacore* was authorized in the fiscal year 1950 shipbuilding program, and the submarine was ordered on 24 November 1950 from the Portsmouth Naval Shipyard (Kittery, Maine). Portsmouth was the Navy's lead submarine construction yard and had a highly innovative design staff.

The *Albacore* was commissioned on 5 December 1953 and quickly demonstrated that she was the most agile and maneuverable and the fastest undersea craft yet constructed. The *Albacore* was fitted with a fully automated control system featuring aircraft-type "joy sticks" for high-speed maneuvering. Before tests of the automated ship control system, the *Albacore*'s officers went to Lakehurst, New Jersey, to fly blimps to better understand one-operator, automated control systems similar to that in the submarine. An early commanding officer, Jon L. Boyes, wrote:

> One test trial was with *Albacore* traveling at a certain depth at speeds in excess of twenty-five knots; go into a thirty degree dive, and when a specified rate of descent was reached, to reverse propulsion or controls. At times in this maneuver, we took heels of over forty degrees with down angles around fifty degrees—learning from this important lesson about ship equipment and human performance.[11]

And,

> Later, in other tests and exercises at sea, we found that in automated flight *Albacore* was frequently quieter than when in a manual-operator control mode. We found also that when operating at high speeds the ship's sonar detection range of certain surface and submerged targets was somewhat longer than when in human operator control mode.[12]

The *Albacore*'s hull was fabricated of HY-80 steel, but because her piping, hull penetrations, and other components were made of High-Tensile Steel (HTS), she was credited with a test depth of only 600 feet (185 m). However, Boyes has stated that 800 feet (240 m) was a normal operating depth, and on one occasion the *Albacore* descended to 1,400 feet (430 m) in trials. According to Boyes:

> They called for us to go to maximum flight speed, then roll her over in a direct descent with a one-pilot override. He [the pilot] was mesmerized by the instrumentation and we were at 1,200 feet before we caught it.[13]

Boyes also stated that a later dive reached 1,600 feet (480 m).

A large number of modifications and changes were made to the *Albacore* to test and evaluate various design concepts during her 19 years of service. The major changes were made under four "phases" and several lesser projects. (Phase I was the configuration as constructed.)

*1956–1957* (Phase II): The stern control surfaces were moved forward of the propeller; *Aquaplas* sound-dampening plastic—to absorb vibration and dampen water-flow noise—was

*The* Albacore *in her original configuration. The* Albacore *established the hull form for future U.S. submarines, both conventional and nuclear. However, the Navy did not pursue all of the performance-enhancing features developed with the* Albacore. (U.S. Navy)

applied to the submarine's outer structure and ballast tank interiors.

*1958:* The small bow diving planes were removed to reduce flow noise and drag; these planes had been useful only when operating near the surface and at slow speeds.[14] And, during 1958–1959 the *Albacore* was fitted with a towed-array sonar mounted in her sail, the first time that hydrophones were towed behind a U.S. submarine.[15]

*1959* (Phase III): The original 11-foot (3.35-m) propeller was replaced by a 14-foot-diameter (4.27-m) propeller, which turned more slowly and was quieter.

*1960–1961:* The stern control surfaces were configured in an "X" configuration; dive brakes were fitted; a new bow configuration, a BQS-4 sonar, an auxiliary rudder at the trailing edge of the sail structure, and a drag parachute from a B-47 bomber were installed.[16]

The X-tail consisted of two essentially equal pairs of control surfaces, one forward of the other to leave space for yokes, tillers, and stocks as well as the propeller shafting within the tapered stern. The dive brakes—ten large, hydraulically operated "doors" arranged around the hull, aft of the sail—and the sail rudder were among several schemes evaluated to counter excessive pitch or dive angles during high-speed maneuvers.

*1962:* A Digital Multi-beam Steering (DIMUS) sonar was installed, providing an improvement in passive acoustic detection.

*1962–1965* (Phase IV): The submarine was lengthened to 210½ feet (64.18 m), with installation of contra-rotating propellers mounted on a smaller shaft within an outer shaft, with a second, 4,700-horsepower electric motor being provided.[17] The spacing between the contra-rotating propellers initially was set at ten feet (3.05 m); in 1965, after sea trials, the spacing was reduced to 7½ feet (2.29 m) and later five feet (1.52 m). The forward, seven-blade propeller had a diameter of 10⅔ feet (3.25 m) and the after, six-blade propeller was just over 8¾ feet (2.68 m).

Silver-zinc batteries, a new, 3,000 psi (208 kg/cm$^2$) ballast blow system, and an improved sonar also were installed. While the silver-zinc batteries provided more power than conventional lead-acid batteries did, they required 22 hours to charge while using both of the *Albacore*'s diesel generators.

*1969–1971* (Phase V): A polymer ejection system to reduce drag was installed. Called Project SURPASS, the system had tanks, pumps, and piping installed in the bow compartment for mixing and distributing the polymer. It also had ejection orifices set in the hull and around the bow and sail. Three "soft" tanks were used to carry 40,000 gallons (151,400 liters) of polymer mixed with fresh water.[18] During SURPASS trials that began in November 1971 the *Albacore* demonstrated a 9 percent speed increase using polymer, at a sustained speed of 21 knots with only 77 percent of her normal horsepower. At that expenditure rate her 40,000 gallons of

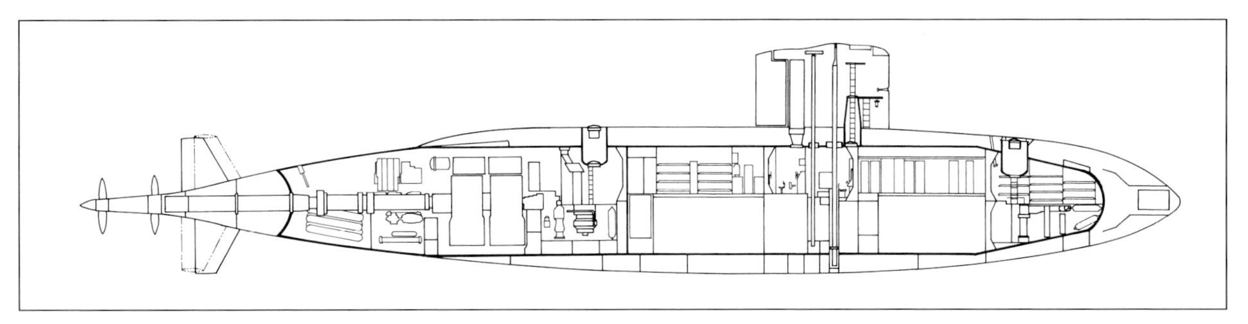

Albacore *(AGSS 569) in 1964 configuration with X-tail and contra-rotating propellers. LOA 210 ft 6 in (64.18 m)* (©A.D. Baker, III)

*The* Albacore *in dry dock, showing her advanced X-stern and contra-rotating propellers. The two propellers had a different number of blades of different diameters. This aspect of the* Albacore *development effort was continued in the nuclear-propelled submarine* Jack. (U.S. Navy)

*A close-up view of the dive brakes on the* Albacore. *These and other techniques were evaluated in an effort to improve response to the snap-role phenomena during high-speed maneuvers. The* Albacore *also was fitted—briefly—with a B-47 bomber's drag parachute for the same purpose.* (U.S. Navy)

mixed polymer would last for 26 minutes. Thus, the use of polymer successfully reduced drag and hence increased performance although only a limited amount of the polymer mixture could be carried. Other problems were encountered with that early system. The Phase V tests ended in June 1972, shortly before the submarine was taken out of service.

The *Albacore* purportedly attained the highest speed achieved to that time by a submarine of any nation in the Phase IV/contra-rotating propeller configuration. With that configuration the *Albacore* was expected to achieve 36 knots.[19] In fact she reached 37 knots. Senior submarine designer Captain Harry Jackson explained that, "The *Albacore* obtained her high speed from a combination of developments and not just an increase in power. The doubling of [electric motor] power was indeed a large factor but without the others, it would not have happened."[20] Those developments were:

(1) contra-rotating propellers (driven by separate electric motors, with the shafts on the

same axis, turning in opposite directions).

(2) "X" configuration of the after control surfaces.

(3) absence of forward control surfaces.

(4) attention to detail in the smoothing of the outer hull and minimizing the size of holes into the free-flood spaces.

(5) closing the main ballast tank flooding holes with smooth fairing plates.

(6) attention to detail with a goal of making high speed.

(7) keeping the submarine to the smallest size possible.

More research and development activities were planned for the *Albacore,* but continued problems with her "pancake" diesel engines caused repeated delays in her operations. No replacement engines were available; plans were drawn up to install conventional diesel engines, but that would add a parallel mid-body of 12 feet (3.66 m) amidships, which would have increased drag. Accordingly, the decision was made to deactivate the *Albacore,* and she was decommissioned on 1 September 1972 after 19 years of pioneering speed-related submarine development.

## Combat Submarines

As early as 1954 the Bureau of Ships had considered combat derivatives of the *Albacore.* There were major differences in views about whether or not combat submarines should have a single screw: some engineers wanted two screws for reliability; some wanted a single screw for increased efficiency and speed. Also, the tapered stern of a single-shaft submarine would prevent tubes being fitted aft for torpedoes to be launched against a pursuing enemy warship.

In November 1954 the Bureau of Ships developed a set of submarine options based on the *Albacore* design (table 9-1). The Electric Boat yard (Groton, Connecticut) had designed the successor to the *Tang* class and the lead ship was under construction. This would be the *Darter* (SS 576), similar in design to the *Tang* with improvements in quieting. However, her underwater speed was rated at only 16.3 knots compared to 18 knots for the earlier submarine.

Meanwhile, the Portsmouth Naval Shipyard, which had largely designed as well as built the *Albacore,* had proposed a combat variant in late 1953. BuShips approved the project, and Portsmouth produced the *Barbel* (SS 580) design, with some input from the BuShips, especially with respect to the arrangement of sonars and torpedo tubes.

*The* Albacore *in 1966 with her X-tail and contra-rotating propellers. The* Albacore's *life as a research platform was shortened by the use of "pancake" diesel engines that had been procured for the* Tang*-class attack submarines. They were difficult to use and maintain, and were short-lived.* (U.S. Navy)

TABLE 9-1
**BuShips Combat Submarine Options**

| | ***Darter* SS 576** | **Twin-Screw *Albacore* design** | **Single-Screw *Albacore* design** |
|---|---|---|---|
| Displacement | | | |
| surface | 2,102 tons | 2,033 tons | 2,020 tons |
| submerged | 2,369 tons | 2,300 tons | 2,285 tons |
| Length | 268 ft 7 in (81.88 m) | 208 ft (63.41 m) | 205 ft (62.5 m) |
| Beam | 27 ft 2 in (8.28 m) | 29 ft (8.84 m) | 29 ft (8.84 m) |
| Shafts | 2 | 2 | 1 |
| Speed | | | |
| surface | 15.7 knots | 14 knots | 14.1 knots |
| submerged | 16.3 knots | 17.4 knots | 18.6 knots |
| Torpedo tubes* | 6 533-mm B<br>2 400-mm S | 6 533-mm B<br>2 400-mm S | 6 533-mm B |

*Notes:* * Bow + Stern.

Because of her combat role, the *Barbel* required forward diving planes to better control depth while near the surface and for slow-speed operations. Initially, the planes were fitted on the bow, but they subsequently were refitted to the sail structure (bridge fairwater position).

*The* Blueback *(SS 581), one of the three* Barbels, *the ultimate U.S. diesel-electric attack submarine. These modified* Albacore-*hull submarines later had their forward diving planes moved to the sail structure. The three* Barbels *had long service lives—an average of 30 years per ship.* (U.S. Navy)

Forward, the *Barbel* had six 21-inch (533-mm) torpedo tubes with 16 reloads carried; no stern tubes were fitted because of the tapered stern. The submarine's speed and maneuverability were considered sufficient to alleviate the need for stern tubes. Her combat system—torpedoes, BQR-2/BQS-4/SQS-4 sonars, and fire control equipment—with the related crew increase, made the *Barbel* significantly larger than the *Albacore,* with a surface displacement of 2,150 tons compared to 1,242 tons for the *Albacore.* The *Barbels* were fitted with three diesel engines and two relatively small electric motors; the latter generated only 3,125 horsepower (less than half of the *Albacore*'s power). Accordingly, the *Barbel* was slower, capable of only 18.5 knots submerged.

The broad diameter (29 feet/8.84 m) of the *Barbel* provided three full deck levels amidships, greatly improving the placement of control spaces and accommodations.

The *Darter,* completed in October 1956, was inferior to the potential *Barbel* design. Hence only one submarine of the *Darter* design was built. Series production was planned for the *Barbel* design, with the lead ship ordered from Portsmouth in 1955. The following year one additional submarine was ordered from Ingalls Shipbuilding in Pascagoula, Mississippi, and one from New York Shipbuilding in Camden, New Jersey. Neither yard had previously built submarines; opening them to submarine construction would facilitate the production of large numbers of diesel-electric units while Electric Boat and the naval shipyards concentrated on producing nuclear submarines.

Three *Barbels* were completed in 1959. No more were built for the U.S. Navy because of the decision made by 1956 that all future combat submarines would be nuclear propelled. The fiscal year 1956 shipbuilding program (approved the previous year) was the last U.S. program to provide diesel-electric combat submarines. Significantly, that decision was made little more than a year after the pioneer *Nautilus* (SSN 571) had gone to sea, demonstrating the Navy's readiness to adopt an all-nuclear submarine fleet. (Subsequently, the *Barbel* design was adopted for diesel-electric submarines by the Dutch and Japanese navies.)

Meanwhile, the Bureau of Ships and Electric Boat were moving to combine the two American submarine "revolutions"—nuclear propulsion and the *Albacore*'s teardrop hull. In October 1955, six weeks after the *Barbel* was ordered from Portsmouth, the Navy awarded a contract to Electric Boat for construction of the *Skipjack* (SSN 585).

The *Skipjack* was half again as large as the *Barbel* in surface displacement, being longer and having a greater diameter, a requirement for her nuclear propulsion plant. Further, her hull shape was a body-of-revolution design, whereas the *Barbel* design had the vestiges of a flat superstructure deck. The *Skipjack*'s forward diving planes were mounted on the sail rather than on the bow. The principal motivation for this change was to reduce the level of flow-induced noise as well as the mechanical noises of the diving planes near the bow-mounted sonar. The penalty in reducing the effectiveness of the planes by moving them closer to the submarine's center of buoyancy was compensated for in part by increasing their size. This resulted in more drag, more noise (albeit away from the sonar), reduced maneuverability, and placing the heavy plane-moving equipment higher in the ship. The larger planes and their equipment dictated larger sail structures, which generally created more drag, more self-noise, and greater disturbance of water flow into the propeller. But all of these penalties—some not appreciated at the time—were considered to be acceptable in the face of enhanced sonar performance. (Also, sail-mounted diving planes could not be retracted, preventing the submarine from surfacing through ice, although a later scheme solved this problem.)

Being based on the *Albacore* hull also meant that the *Skipjack* would be a single-hull submarine, the first U.S. SSN with that configuration. The single-hull design was favored because of the minimum total structural weight as well as smaller outside dimensions for a given pressure-hull volume.[21] This resulted in less wetted surface area and, therefore, with all other factors being equal, improved underwater speed for a given horsepower. The continuous outer surface of single-hull design also minimized hydrodynamic flow noise. The transition to single-hull designs involved trade-offs:

- reserve buoyancy was reduced from 25 to 35 percent in early U.S. nuclear submarines to 9 to 13 percent of surface displacement, which limits the amount of flooding a submarine can overcome.
- the pressure hull no longer has the collision or weapons damage protection provided by the stand-off distance between the inner (pressure) and outer hulls.
- ballast tanks became primarily located in the bow and stern sections, limiting longitudinal stability should tanks rupture or suffer damage.
- hydrodynamic shaping became more costly when pressure hull plates were used in the outer hull in place of lighter, outer plates.
- the number of watertight compartments would be reduced to five (in the *Skipjack*) and subsequently to three in later submarines, reducing the possibility of recovery from serious flooding casualties.

The *Skipjack* was propelled by an S5W reactor plant, an improved pressurized-water system employing the same basic process as all previous U.S. reactor plants except for the liquid-sodium reactor originally fitted in the *Seawolf* (SSN 575). The S5W was rated at 15,000 horsepower, slightly more than the horsepower of the nuclear plant installed in the *Nautilus,* but was more efficient and easier to maintain than the earlier reactor plants.[22] Its similarity to the *Nautilus* S1W/S2W design made it unnecessary to build a land prototype.[23]

Rear Admiral H. G. Rickover opposed the adoption of a single propeller shaft for nuclear submarines. Captain Donald Kern, in the Bureau of Ships at the time, recalled "great arguments with Admiral Rickover about going to a nuclear submarine with a single screw and even more severe arguments about going to a nuclear submarine with a single screw [operating] under the ice, and he just wanted no part of a submarine under the ice with a single screw."[24]

Rickover argued that a submarine going under the ice with a single screw could be lost if the screw were damaged. He asked Kern: "Suppose you had to go under the ice. Which submarine would you choose?" According to Kern:

> The obvious answer he was looking for was that I'd choose the twin screw because I had one screw to get back on if I lost one. I said, "I'm not sure I'd ever want to go under the ice with a twin-screw submarine. . . . I'd only go under the ice in a single-screw submarine," which was not the answer he wanted at all.

Rickover: "Why would you do an idiotic thing like that?"

Kern:

> In this case my answer, to me, was perfectly straightforward and understandable from an engineering point of view, that the twin-screw submarine has a very delicate set of shafts that are exterior to the hull, are out there on struts, and you got to have struts to keep the propeller off the hull and hold the end of the shaft. A single-screw submarine has no unprotected shaft line. It has an

> extremely rigid protection on the hull around the shaft line. It has a much bigger screw and a much tougher screw.

Kern continued:

> . . . the *Sea Dragon* [SSN 584] . . . hit the whale and we lost the prop. All we did was run a mushy old whale through the propeller and we bent the propeller, we bent the shaft, we bent the strut. We wrecked the bloody submarine by hitting a whale. I said suppose you hit a chunk of ice. You would have really some bad damage. If you put a submarine up there and start running ice through the blade, you're going to lose both propellers in a hurry.

Rickover finally acquiesced, probably thinking, said Kern: "Okay, you dummies, remember, if we lose the submarine because of the single screw, it was your idea, not mine." (In the event, until 1967 only the early, twin-screw U.S. submarines would operate under the Arctic ice pack.)

To provide an emergency "come-home" capability in the event of a shaft or propeller casualty, U.S. single-screw submarines would be provided with an electric motor-propeller pod that could be "cranked down" from the engineering spaces. The pod could provide emergency propulsion.

Even with a single-shaft propulsion plant the Navy was not prepared to adopt all of the *Albacore*'s features "as they were considered too complicated," explained Captain Jackson.[25] Thus, the *Skipjack* had the teardrop hull shape and single shaft turning a single, five-blade, 15-foot (4.57-m) propeller, but with conventional, cruciform stern control surfaces, albeit mounted forward of the single propeller.[26] Also, hull coatings and polymers were eschewed. (The later USS *Jack*/SSN 605, completed in 1967, was built with contra-rotating propellers on a single shaft.)

The *Skipjack*'s weapons and sonar were similar to the *Barbel* configuration—the BQS-4 bow-mounted sonar and six 21-inch torpedo tubes; 18 torpedo reloads could be carried for a total of 24 weapons.

The submarine was placed in commission on 15 April 1959 with Lieutenant Commander William W. (Bill) Behrens Jr. in command. The *Skipjack* immediately demonstrated the success of her design. She achieved just under 33 knots submerged with her original, five-blade propeller. This

*The* Skipjack *on sea trials, showing her modified* Albacore-*type hull with her sail-mounted diving planes. This design combined nuclear propulsion and the* Albacore *hull. Several of her periscopes and masts are raised. She has a small, crowded bridge.* (U.S. Navy)

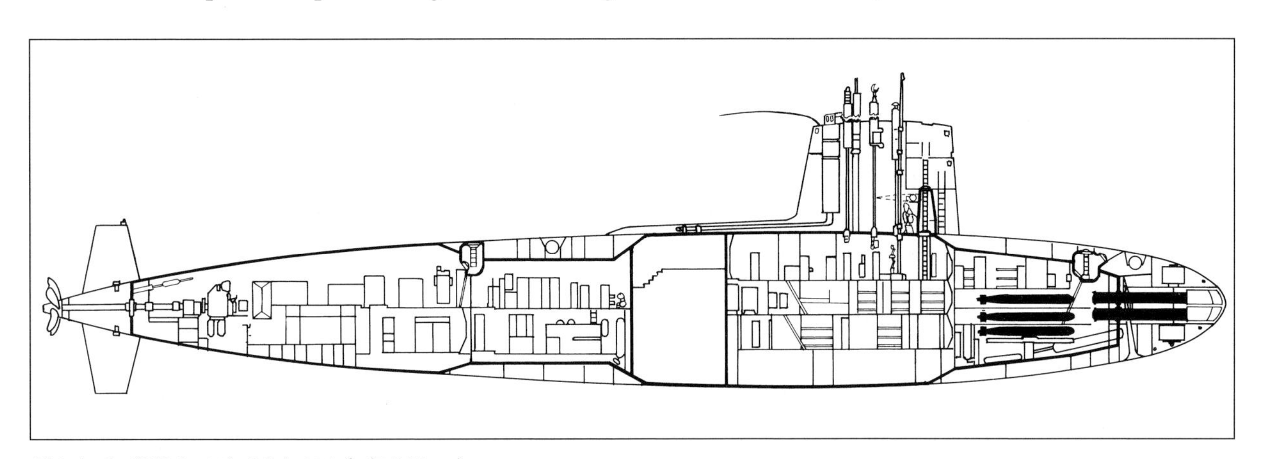

Skipjack *(SSN 585). LOA 252 ft (76.83 m)* (©A.D. Baker, III)

was 15 knots faster than the previous *Skate*-class SSNs. However, the *Skipjack* had minimal quieting features. A later change in propeller—to a seven-blade, slower-turning propeller to reduce cavitation as well as shaft noise—reduced her speed.

Like the conventionally propelled *Barbel,* the *Skipjack* had a test depth of 700 feet (215 m), with an estimated 1.5 safety margin, that is, 1,050 feet (320 m). She exceeded the latter depth in the spring of 1960 when the U.S. Navy was demonstrating the submarine's prowess for the British First Sea Lord, Admiral Sir Caspar John. With Sir Caspar and two British submarine officers on board, during a hard dive and starboard turn at a speed in excess of 20 knots, the controls jammed and the submarine plunged downward at an angle of almost 30 degrees. After some difficulties the crew was able to recover, and the craft returned safely to the surface.[27] (The Royal Navy procured a *Skipjack*/S5W power plant for the first British nuclear submarine, HMS *Dreadnought,* completed in 1963; she was a hybrid design—forward she was British, aft American with the transition point called "Check-Point Charlie."[28])

Admiral Arleigh Burke, the Chief of Naval Operations, had described the *Skipjack* as having "unmatched speed, endurance and underwater maneuverability."[29] That endurance was demonstrated in 1962, when the *Scorpion* (SSN 589) of this class remained at sea, submerged with a sealed atmosphere for 70 consecutive days.[30]

## The Superlative Papa

Speed was important to Soviet attack/anti-ship missile submarines because of the requirement to counter U.S. and British aircraft carriers. By the end of 1958—the same year that the first Soviet nuclear-propelled submarine went to sea—work began in Leningrad and Gor'kiy at the submarine design bureaus on the next generation of submarines. Both conservative and radical designs were contemplated for the next-generation torpedo-attack (SSN), cruise missile (SSGN), and ballistic missile (SSBN) submarines. These efforts would produce two high-speed submarines—Project 661 (Papa SSGN) and Project 705 (Alfa SSN).

A resolution was issued by the Council of Ministers on 28 August 1958 "On the creation of a new fast submarine, new types of power plants, and the development of scientific-research, experimental-design and design work for submarines." The resolution called for:

a twofold increase in speed

an increase of one and one-half times in diving depth

creation of smaller nuclear power plants

creation of smaller turbine plants

creation of small-dimension missile systems with long range and submerged launch

development of automation to ensure control and combat use of future submarines at full speed

improvement of submarine protection against mines, torpedoes, and missiles

decrease in the overall displacement and dimensions of submarines

improvement in crew habitability

use of new types of materials

In 1958 design bureau TsKB-16 began to address these ambitious goals in an experimental cruise missile submarine—Project 661.[31] The ship's chief designer was N. N. Isanin, who had been in charge of the design of early Soviet ballistic missile submarines. Project 661 was Isanin's first totally new design submarine, and he would design a superlative submarine with new materials, propulsion plant, weapons, and sensors. (At the end of 1963 Isanin was appointed head of SDB-143/Malachite and until 1965 he remained the chief designer of Project 661; he was succeeded in both roles by N. F. Shulzhenko.)

The goal of Project 661—achieving all of the desired features in a single submarine—would be extremely difficult. Fourteen pre-preliminary designs were developed. At the same time, according to Dr. Georgi Sviatov, at the time a junior naval engineer working for Isanin, "It was desired that the submarine be totally 'new': it must have an advanced design, new materials, new powerplant, and new weapons system—it must be superlative."[32]

No single design option could meet all TTE

goals, which required a submerged speed of 35 to 40 knots. In July 1959 a proposed design was submitted to the Navy and the State Committee for Shipbuilding. This was one of the smallest design options but also the one estimated to have the highest speed—up to 38 knots. During the pre-preliminary stage three alternative hull materials were addressed: steel, titanium alloy, and aluminum. The unsuitability of aluminum was soon established; of course, the submarine could be built of steel, which had been used in submarine construction for more than a half century. Titanium alloy evoked the most interest of Soviet designers and naval officials because of its high strength with low weight, and resistance to corrosion and because it was non-magnetic. Applied to a submarine titanium could provide a reduction in structural weight and increase in hull life, while enabling a deeper test depth for the same relative pressure hull weight. And, the non-magnetic properties afforded increased protection against magnetic mines and Magnetic Anomaly Detection (MAD) devices.

Taking into consideration the significance of titanium for future submarine effectiveness, the State Committee for Shipbuilding adopted the decision to use this alloy for Project 661. *The high cost of titanium alloys compared to steel was not taken into account.*[33] The use of titanium in submarine construction was approved in 1959 and required the creation of special facilities for producing titanium rolled sheets, stampings, packings, castings, and piping. Up to that time nothing as large as a submarine had been fabricated of titanium alloy. At the time the metal was used in small quantities in aircraft and spacecraft. There were many problems in using titanium, in particular the special conditions required for welding the metal. Initially titanium was welded in an argon-rich atmosphere, with workers wearing airtight suits to prevent contamination of the welds. Special facilities had to be provided for handling and welding titanium in building hall No. 42 at the Severodvinsk (formerly Molotovsk) shipyard. Atmosphere-controlled areas were required, with special paints, floor coverings, and air purification measures being provided in the building hall.[34]

Another important issue was the type of nuclear power plant to be used in Project 661. Both water and liquid-metal (lead-bismuth alloy) were considered as the coolant medium. The second option potentially gave some benefit in reducing displacement; however, there were significant drawbacks (discussed below with project 705/Alfa). The decisive argument in favor of a pressurized-water plant was the additional time required to develop the liquid-metal plant. The two VM-5 pressurized reactors would enable the twin turbines to produce 80,000 horsepower by a great margin more than any previous submarine propulsion plant.

As a result of these decisions, it was calculated that a submarine could be constructed with launchers for 10 to 12 anti-ship cruise missiles and a submerged speed of up to 38 knots. The displacement with a two-shaft propulsion plant exceeded the specified size. A single-shaft design was seriously considered; twin shafts were selected to provide reliability through redundancy and led to the deletion of the auxiliary diesel-generator plant found in other nuclear submarines. Subsequent operations of the submarine would demonstrate the error of that decision.

Based on these decisions, work on the preliminary design began at TsKB-16 in February 1960. At the same time, the Navy assigned three submarines to serve as test beds for Project 661 components: The Chernomorsky shipyard (No. 444) at Nikolayev on the Black Sea modified the Project 613AD/Whiskey for trials of the P-70 Amethyst cruise missile; the Dalzavod yard at Vladivostok modified the Project 611RU/Zulu for the trials of the Rubin sonar system; and the repair yard at Murmansk modified Project 611RA/Zulu for the trials of the Radian-1 mine-hunting sonar.[35] In addition, the submarine *K-3*—the first Project 627/November submarine—was modified for shipboard trials of the advanced ship control (Shpat and Turmalin) system and the *K-181*, a Project 627A submarine, was fitted with the navigation (Sigma) system intended for Project 661. The Rubin system was the most advanced sonar yet developed in the Soviet Union and would provide long-range detection of surface ships, especially multi-screw aircraft carriers. Its large size and shape required a new, rounded bow configuration. The Radian-1 sonar was designed specifically for the detection of moored mines.

The Project 661 submarine—given the NATO code name Papa—had a double-hull configuration with nine compartments. The forward section of the pressure hull consisted of two horizontal cylinders, 19 feet, 4½ inches (5.9 m) in diameter; in cross section this resembled a figure eight. The upper cylinder was the first compartment (torpedo room) and the lower cylinder was the second compartment (sonar and storage batteries). Behind them was the third compartment. Beginning with the fourth compartment the pressure hull had a cylindrical form with a diameter of 29½ feet (9 m) in the area of the center frames. The narrow pressure hull forward permitted the installation of five angled launchers for the Amethyst missile along each side of the forward pressure hull and within the outer hull. The launchers had an upward angle of 32.5 degrees. The torpedo armament consisted of four 21-inch torpedo tubes in the upper bow compartment with eight reloads. This would be the smallest number of torpedo tubes fitted in a Soviet nuclear submarine.

The principal armament would be ten Amethyst anti-ship missiles (NATO designation SS-N-7). Development of the Amethyst missile was initiated in April 1959 by V. N. Chelomei's OKB-52. It was the world's first anti-ship cruise missile that could be launched from underwater. A solid-propellant missile, the Amethyst could be launched from depths down to 98 feet (30 m). Upon leaving the launch tube, the missile's wings automatically extended, and the missile's rocket motor ignited underwater. The missile streaked to the surface, where the second motor ignited, followed by the third (cruise) motor. The flight toward the target was at an altitude of about 200 feet (60 m) at subsonic speed. On trials a range of 38 n.miles (70 km) was achieved.

The Amethyst missile had an autonomous control system that later became known as "fire and forget." Radar terminal-homing guidance was intended to analyze returns of geometric indicators of target disposition to determine the location of the aircraft carrier within a formation of warships.

On par with Amethyst's many attributes, Russian writers list a number of shortcomings, primarily its relatively short range and not being standardized, that is, it could be launched only from submarines and only while submerged. These and other shortcomings led to the Amethyst being provided only in Project 661 and in 12 Project 670/Charlie I submarines.[36] Still, its underwater launch was a milestone in submarine development.

The launch trials of the Amethyst missile began on 12 December 1962 from the Project 613AD/Whiskey. Development and launch trials continued through September 1967, at which time the missile became operational.

Meanwhile, construction of the submarine—given the tactical designation *K-162*—encountered difficulties. Plates of titanium with a thickness up to 2⅓ inches (60 mm) would be used in the craft's pressure hull. Two half-scale submarine compartments were built with the titanium alloy 48-OT3V.[37] One was tested in a pressure chamber at the building yard, and the other was tested for shock strength on Lake Ladoga (Ladozhskoye) near Leningrad. While these compartments were being constructed, cracks appeared in the welded seams, which propagated into the basic structural material. This occurred not only during the cooling of the structure immediately after welding, but also afterward. The destruction of the test compartment under hydraulic testing caused grave doubts among Navy officials and designers—even the enthusiasts of using titanium.

Concern over this situation led the State Committee for Shipbuilding employing the expertise of the nation's top naval engineers and technical experts to solve the titanium problems. At the end of 1961, the first set of titanium alloy construction plates arrived at Severodvinsk, with the submarine's official "keel laying" taking place on 28 December 1963.

Low-quality titanium sheets were delivered by the Kommunar Metallurgical Plant for the light (outer) hull of the submarine. These sheets had an increased content of hydrogen and accordingly had a tendency to form cracks. During the on-ways construction period, about 20 percent of the surface of the outer hull was replaced because of cracks. When the submarine was launched, it was found that ten ballast tanks were not watertight in spite of careful examination of the shell of the ballast tanks. Further, corrosion appeared where non-titanium components were not properly isolated from the titanium hull.

The many advanced features of Project 661—the hull material, reactor plant, control systems, sonar, and missiles—created a multitude of con-

*The Project 661/Papa at rest. She has a circular hull form with a relatively small sail. There are ten Amethyst/ SS-N-7 anti-ship missiles fitted forward, between the pressure hull and the outer hull. The free-flood, or limber, holes amidships and aft indicate her double-hull construction.*

struction challenges. Thus construction of the *K-162* progressed slowly and with difficulty. She was rolled out of the building hall in December 1968, and the launch basin was flooded on the 21st. Work continued for another year.

Under the command of Captain 1st Rank Yu. F. Golubkov, on 13 December 1969 the *K-162* was placed in commission. It soon became evident that the submarine was a speedster. The speed trials were very demanding, because the depth of water at the test range did not exceed 655 feet (200 m). Above was ice. The slightest mechanical failure or error in control of the diving planes and, in seconds, the submarine would slam either into the ice or into the mud. The test depth of the *K-162* was 1,300 feet (400 m).

On these trials a speed of 42 knots was achieved with 90 percent power, instead of the 37 to 38 knots "guaranteed" by the specifications. This was faster than any previous (manned) undersea craft had traveled. During the 12-hour full-speed run, the fairings on the access hatches, the emergency signal buoy in the superstructure, and several grills at the water intakes were ripped off. Portions of these grills entered the ship's circulating pumps and were ground up. But the submarine continued. The noise was horrific; a Russian account reads:

> the biggest thing was the noise of the water going by. It increased together with the ship's speed, and when 35 knots was exceeded, it was like the noise of a jet aircraft. . . . In the control room was heard not simply the roar of an aircraft, but the thunder of "the engine room of a diesel locomotive." Those [present] believed the noise level to be greater than 100 [decibels].[38]

Later, in 1970, with the reactor plant at maximum power, the submarine attained a speed of 44.7 knots—the fastest ever traveled by a manned underwater vehicle.

Trials and modifications continued through December 1971, when the submarine became operational with the Northern Fleet. During 1970 the issue of constructing additional Project 661 submarines was considered by the Navy's leadership. The tactical shortcomings of the missile armament, high submerged noise levels, insufficient service life of the basic mechanisms and ship's equipment, the long construction period, and the high cost of the submarine led to the decision not to produce additional submarines of this design. Indeed, the high cost led to the *K-162* being referred to as the "golden fish" (*Zolotaya Rybka*), as were submarines of the subsequent Project 705. Russian engineers joked that the *K-162* cost more being made from titanium than if she had been made from gold.

The *K-162* made operational cruises and continued research and development work. Late in the

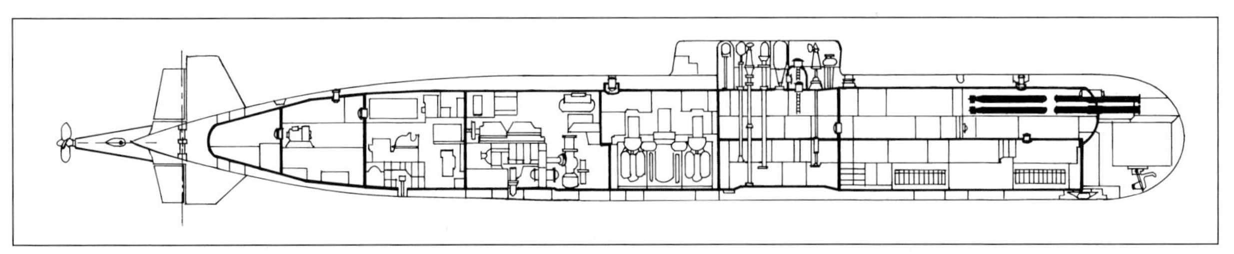

K-162 *Project 661/Papa SSGN. LOA 350 ft 8 in (106.9 m)* (©A.D. Baker, III)

1970s, as the submarine was being refueled (recored), a workman dropped a wrench into reactor-related machinery. The reactor had to be refueled to correct the damage and, in the haste to complete the work, control rods were improperly installed. The problems soon were further compounded, but corrections were made and the submarine continued to operate for another decade.[39] She was decommissioned and laid up in reserve in 1988. The next Soviet titanium-hull, high-speed submarine would be series produced.

## The Remarkable Alfa

The direct successor to the Project 627/November SSN would be the second-generation Project 671 (NATO Victor). This was a *relatively* conservative design intended primarily for the ASW role. The near-simultaneous Project 705 (NATO Alfa) was a major step forward in submarine development.

The design team for Project 705 at SDB-143/Malachite was led by Mikhail G. Rusanov, who had just completed work on Project 653, a missile submarine intended to carry the P-20 missile. Project 705 was to be a high-speed ASW submarine, intended to seek out and destroy Western missile and attack submarines in Soviet defensive areas.[40] The endurance of the ASW submarine was to be 50 days.

From the outset Project 705 was to have had minimal manning. Periodically perceptive submariners had realized that numerous submarine functions could be automated. In his classic story *Das Boot,* journalist Lothar-Günther Buchheim quoted a wartime U-boat commander:

> Actually we ought to be able to get along with a lot fewer men. I keep imagining a boat that would only need a crew of two or three. Exactly like an airplane. Basically, we have all these men on board because the designers have failed to do a proper job. Most of the men are nothing but links in a chain. They fill the gaps the designers have left in the machinery. People who open and close valves or throw switches are not what you'd call fighting men.[41]

By the late 1950s available automation technology could permit a very small SSN crew. Early proposals for Project 705 provided for a crew of 12, and subsequently 18, by using aircraft crew concepts. As completed the crew consisted of 25 officers, 4 warrant officers, and 1 petty officer (cook)—a total of 30 men. Major crew reduction was possible, in part, by providing the world's first integrated combat information-control system fitted in a submarine. The system provided integration of navigation, tactical maneuvering, and weapons employment; it automatically developed optimum decisions and recommended them to the commanding officer.[42]

By reducing the crew and going to a single-hull design with a high-density nuclear propulsion plant, a very small submarine was envisioned—possibly with a submerged displacement of 2,500 tons. However, after the design was reviewed by Navy and shipbuilding leaders in 1963 following the loss of the USS *Thresher* (SSN 593), SDB-143 was directed to extensively revise the design to provide a double-hull configuration and to build an "automated fast [SSN] with torpedo armament."

Project 705 would be a "small submarine." SDB-143 estimated that going from a single-hull to a double-hull design cost about 5 percent (two knots) in speed.[43] And a 30-man crew would still be a dramatic departure from conventional manning practices. For example, the contemporary USS *Skipjack* and Project 671/Victor SSNs had crews three times as large. The spectacular crew reduction was achieved by having ship control, propulsion plant, weapons, sensors, and communications controlled from the main control center. All crew functions and living spaces were located in a single compartment; other compartments, including the torpedo and propulsion spaces had no permanent watch standers and were entered only for maintenance and limited repair work. Crew concentration in a single compartment, according to SDB-143, "provided the prerequisites to ensure their safety at a heretofore unachieveable level."[44]

Overall, the submarine would have six compartments (the original design having only three). Above the control compartment an escape chamber that could accommodate the entire crew was provided in the sail—another submarine first.[45] In an emergency situation the crew would enter the chamber through a hatch within the submarine; the chamber would then be released upon command, and, with part of the sail structure attached for buoyancy, float to the surface. The entire crew could thus leave the submarine together, without being exposed to cold or pressure. Upon reaching the surface the chamber would provide protection from the elements until rescue forces could reach the area. The only disadvantage was an increase in submarine weight.

Another means of reducing ship size was the use of 400 Hz electrical power in place of the 50 Hz systems used in previous Soviet submarines. This change permitted smaller electrical equipment. As in the near-contemporary Project 661/Papa, Project 705 would have a titanium-alloy, double-hull design. Compared to steel, titanium provided a number of advantages in Project 705:

- 30 percent lower mass
- 25 percent lower displacement
- 10 percent increase in speed
- reduction of magnetic field
- significantly lower operating costs as a result of the corrosive resistance of titanium alloy

The titanium hull, advanced fittings and ballast system, and other features would give the submarine a test depth of 1,300 feet (400 m), comparable to U.S. second-generation SSNs of the *Thresher* and later classes. While titanium would lower the hull's magnetic field, the submarine's acoustic signatures were higher than contemporary U.S. submarines. A streamlined sail was "blended" into the advanced hull shape. The submarine would prove to be highly maneuverable.

Armament of the submarine consisted of six 533-mm torpedo tubes with 12 reloads. In addition to standard torpedoes, the submarine could carry the RPK-2 Vyuga (blizzard) ASW missile, given the U.S.-NATO designation SS-N-15 Starfish. This was a ballistic missile carrying a nuclear warhead, similar to the U.S. Navy's SUBROC (Submarine Rocket). The torpedo complex, including rapid reloading, was totally automated. Advanced sonars and fire control system were provided to support the armament.

In some respects the most futuristic aspect of Project 705 was the propulsion plant. Although the Project 645/modified November SSN as well as the USS *Seawolf* (SSN 575) had liquid-metal reactor plants, neither submarine was considered successful. The Project 705 reactor plant would use a lead-bismuth alloy as the heat exchange medium. This would provide increased efficiency (i.e., a more dense power plant) with a single reactor and single OK-7 turbine providing 40,000 horsepower.

Two types of nuclear plants were developed simultaneously for the submarine: the OK-550 for Project 705 was modular, with branched lines off the first loop, having three steam lines and circulating pumps; the BM-40A for Project 705K was a modular, two-section reactor, with two steam lines and circulating pumps.

There were significant drawbacks to both plants because the use of a liquid-metal heat carrier required always keeping the alloy in a liquid (heated) condition of 257°F (125°C) and, in order to avoid it "freezing up," the plant could not be shut down as was done on submarines with a pressurized-water plant. At Zapadnaya Litsa on the Kola Peninsula, the base for these submarines, a special land-based complex was built with a system to provide steam to keep the liquid metal in the submarine reactors from solidifying when the reactors were turned off. In addition, a frigate and floating barracks barge supplied supplemental steam to the submarines at the piers. Because of the inherent dangers of using these external sources of heat, the submarine reactors were usually kept running when in port, albeit at low power.

Like previous Soviet submarines, the craft would have a two-reactor plant, but with a single propeller shaft. The decision to provide a single shaft in this design followed considerable deliberations within the Soviet Navy and at SDB-143 (similar to the discussions within the U.S. nuclear program). The single screw was adopted for Project 705/Alfa almost simultaneous with the Project 671/Victor second-generation SSN design. To provide these submarines with emergency "come-home" propulsion and, in some circumstances, low-speed, quiet maneuvering, these submarines additionally were fitted with small, two-blade propeller "pods" on their horizontal stern surfaces. (See Chapter 10.)

SDB-143 completed the conceptual design for Project 705 in 1962 and the technical design the following year. The first submarine of the class to be completed—the *K-64*—was laid down at the Sudomekh/Admiralty yard in Leningrad on 2 June 1968. That ceremonial "keel" laying was a political event. The *K-64* was more than 20 percent complete at the time, with several hull sections already on the building way. This was the second shipyard to qualify in the construction of titanium-hull submarines and was the fifth Soviet shipyard to construct nuclear submarines:[46]

| Shipyard | Location |
|---|---|
| No. 112 Krasnoye Sormovo | Gor'kiy (Nizhny Novgorod) |
| No. 194 Admiralty | Leningrad (St. Petersburg) |
| No. 196 Sudomekh | Leningrad (St. Petersburg) |
| No. 199 Komsomol'sk | Komsomol'sk-on-Amur |
| No. 402 Severdovinsk | Severdovinsk |

The Project 705/Alfa made use of processes and technologies previously developed for Project 661/Papa. The *K-64* was launched at Sudomekh on 22 April 1969 (V. I. Lenin's birthday anniversary). She was then moved by transporter dock through the Belomor-Baltic Canal system to Severodvinsk, where her fitting-out was completed and her reactor plant went critical. The *K-64* was accepted by the Navy in December 1971, with Captain 1st Rank A. S. Pushkin in command.

The submarine underwent trials in the Northern Fleet area beginning in mid-1972. That same year the *K-64* suffered a major reactor problem when the liquid metal in the primary coolant hardened, or "froze." She was taken out of service, and her hull was towed to Sverodvinsk, where she was cut in half in 1973–1974. The forward portion of the submarine, including control spaces, was sent to Leningrad for use as a training device. The reactor compartment was stored at the Zvezdochka yard in Severodvinsk. (Because of the distance between the two hull sections, Soviet submariners joked that the *K-64* was the world's longest submarine!)

As a result of this accident, in 1974 Rusanov was relieved of his position as chief designer, although he remained at Malachite. The project was continued under his deputy, V. V. Romin.[47] Solving the engineering problem that plagued the *K-64* delayed the completion of the other submarines of the class. Seven additional units were ordered—four from Sudomekh (705) and three from Severodvinsk (705K), with six units laid down from 1967 to 1975.[48] They were launched from 1969 to 1981, and the first to commission was the *K-123*, built at Severodvinsk, on 26 December 1977. The remaining five operational units were completed in 1978–1981.

*A Project 705/Alfa SSN photographed in the Barents Sea in 1983. The Alfa had clean lines, revealing her underwater speed potential. Several of her masts—including her Quad Loop RDF antenna—and periscopes are raised in this photo, as was her bridge windshield.* (U.S. Navy)

The appearance of Project 705/Alfa was a shock to Western naval officers. Initially there was wide disbelief that the submarine had a titanium hull. This view persisted even after evidence was presented: First, ground-level and satellite photography of the Sudomekh yard revealed sections of a submarine hull outside of the closed construction hall. Herb Lord, an analyst at the naval intelligence center in Suitland, Maryland, a suburb of Washington, D.C., believed that the sections were too highly reflective to be steel and, over time, showed no signs of oxidation or corrosion. He concluded that the Soviets were building titanium-hull submarines. (In London a British intelligence analyst, Nick Cheshire, reached a similar conclusion about the same time.)

*An Alfa on the surface, showing how her sail blends into her hull. A mast is raised forward of the windshield. When the masts were retracted they were covered over to minimize water flow disturbance over the sail structure. Although a titanium-hull submarine, the Alfa—like the Papa SSGN—was not a deep diver.* (U.S. Navy)

Second, Commander William (Bill) Green, an assistant U.S. naval attaché, while visiting Leningrad in the winter of 1969–1970, at great personal risk, retrieved some debris outside of the Sudomekh yard—a scrap of machined titanium that "fell" from a truck leaving the yard.[49]

Third, in the mid-1970s, two U.S. submarine analysts, Richard Brooks of the Central Intelligence Agency, and Martin Krenzke, with the Navy's research center in Carderock, Maryland, visited a scrap yard in Pennsylvania that was buying exotic metal scrap from the USSR. Looking at "every piece of metal," they made a singular discovery: a piece of machined titanium with numbers scratched into the surface. The first three digits were "705."[50]

With this evidence the U.S. Navy submarine community *began* to accept that titanium was being used in Soviet submarine construction. (Not until later did Western intelligence ascertain that Project 661/Papa also was built of titanium.)

The Alfa SSN's speed of 41 knots also took Western intelligence unaware. In a 1981 congressional colloquy between a U.S. senator and the Deputy Chief of Naval Operations (Surface Warfare), Vice Admiral William H. Rowden, the senator noted that the Alfa could travel at "40-plus knots and could probably outdive most of our anti-submarine torpedoes." He then asked what measures were being taken to redress this particular situation.

Admiral Rowden replied, "We have modified the Mk 48 torpedo . . . to accommodate to the

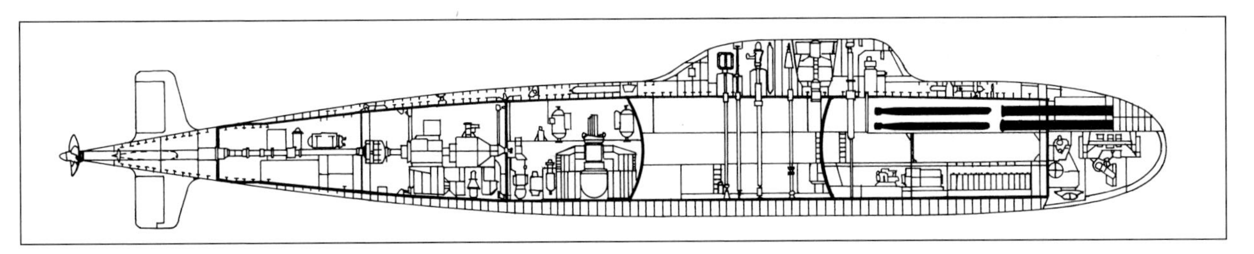

*Project 705/Alfa SSN. LOA 261 ft 2 in (79.6 m)* (©A.D. Baker, III)

increased speed and to the diving depth of those particular submarines." The admiral was less confident of the Mk 46 ASW torpedo used by aircraft, helicopters, and surface ships: "We have recently modified that torpedo to handle what you might call the pre-Alfa. . . ."[51]

In private U.S. naval officers were more vocal in their consternation over the ineffectiveness of existing Western ASW weapons against the Alfa.[52] While U.S. officials reacted to press inquiries by citing the high noise levels of the Alfa at high speed, at lower speeds the submarine would have been difficult to detect. (The submarine could not operate at depths of 2,100 feet [640 m] or more, as estimated by the principal Western intelligence agencies.[53])

In the event, the eighth submarine of the class was not completed. The six operational units entered the fleet during 1978–1981 and served in the Northern Fleet for more than a decade. SDB-143 considered variants of the design, including an elongated submarine (Project 705D) with a second torpedo room and a payload of 30 torpedoes plus four cruise missiles in the sail, and a cruise missile variant (Project 705A) with six cruise missiles. (See Chapter 10.)

Project 705 was impressive, but it did suffer problems; according to Malachite designers:

> First, submarine systems had been complicated in quality and quantity from the moment of conception of the Alfa Class. That was a consequence of increased requirements for such combat qualities as stealth, search capabilities, strike power, reliability, and so forth. The full automation of such complex and ramified systems had led to unacceptable levels of complications in the control and automatic systems. As a result, it was not possible to guarantee their reliability.
>
> Secondly, operational experience demonstrated the lack in small crews of the reserves necessary for non-routine conditions of operation, such as during a struggle for survivability in an emergency, for repair work on equipment that breaks down at sea, for replacement of ill crew members, etc. The drastic reduction in crew numbers also exacerbated several social problems, including the need to train highly skilled specialists . . . the need to provide their interchangeability, difficulties in the execution of duties, and the promotion of the officers [of] these complex automated ships.[54]

Designer Radiy Shmakov of Malachite concluded:

> But in looking back now, it should be admitted that this submarine was the "design project of the 21st Century." She was many years ahead of her time and so turned out to be too difficult to master and operate.[55]

Beyond the lead submarine, other Project 705/Alfa SSNs suffered problems. In the lead operational submarine, the *K-123,* liquid metal from the primary cooling circuit leaked and contaminated the reactor compartment with almost two tons of metal alloy. The compartment was removed in 1982 and a new reactor installed. It took almost nine years to replace the reactor, with the submarine being launched again in 1990 and placed back in commission the following year.

The other Project 705/705K submarines operated until the end of the Cold War, being decommissioned in 1990–1991. The *K-123* served for several more years, being finally decommissioned on 31 July 1996.[56]

Two other unusual submarines were approved for development in this period. In the summer of 1965, Admiral Sergei G. Gorshkov, Commander-in-Chief of the Soviet Navy, approved the development of designs for Projects 696 and 991, both by SDB-143/Malachite. Project 696 was to be a high-speed nuclear submarine with a single turbine and propeller shaft, contra-rotating propellers, advanced drag reduction features—boundary layer control and polymers—and a reactor plant producing approximately 100,000 horsepower. [57] (See Chapter 10.)

Project 991 was to produce an ultra-quiet nuclear submarine. A team led by designer G. N. Cherynshev employed advanced sound mitigation systems (but not polymers). The design work was highly successful, but it was not felt necessary to build a specialized quiet SSN. Rather, in 1972 the

*Five of the six operational Project 705/Alfa SSNs moored at Zapadnaya Litsa on the Kola Peninsula. A special base complex was built to support the use of lead-bismuth alloy as the heat-exchange medium in their reactors. The remote areas of most Soviet submarine bases caused privations for crews and their families.* (Malachite SPMBM)

decision was made to transfer the technologies to Project 945 (NATO Sierra) for series production.

**The U.S. and Soviet Navies initially developed high-speed submarines for different reasons. The U.S. Navy—like the Royal Navy—became interested in high submerged speed to train ASW forces to counter "enemy" submarines based on Type XXI technology. Subsequently, the Soviet Navy sought speed to enable attack submarines (SSN/SSGN) to close with Western aircraft carrier groups, a part of their homeland defense strategy to counter nuclear strike aircraft from the carriers.**

**The USS *Albacore* was a revolutionary undertaking, the next major step in submarine hull form and supporting systems developed after the Type XXI. This test craft reportedly attained an underwater speed of 37 knots. The subsequent USS *Barbel* marked the rapid transition of the *Albacore* hull form to a combat submarine. But the *Albacore* was too revolutionary for the U.S. Navy's leadership; although her basic hull design was adopted for the *Barbel,* and that submarine was more capable than the previous *Tang* design (based on the Type XXI), many performance features of the *Albacore* were not applied to the *Barbel.***

**The subsequent step in the development of advanced submarines—from the American perspective—was the USS *Skipjack.* That submarine combined two underwater revolutions: the *Albacore* hull form with nuclear propulsion. The *Skipjack* went to sea in 1959, the same year the three submarines of the *Barbel* class were completed. Earlier, the Navy's leadership had made the decision to produce only nuclear-propelled combat submarines.**

**That historic decision was made by 1956—only a year after the *Nautilus* went to sea. This commitment contradicts the later myth that "the Navy" opposed nuclear submarines and, even after the *Nautilus* proved the efficacy of nuclear submarines, persisted in wanting to construct diesel boats.**

**Similarly, the early Soviet government decisions to procure advanced submarines so soon after completion of the first Project 627/November (*K-3*) must be applauded. The use of titanium, very-high-power density reactor plants, advanced hull forms, and improved weapons and sensors made Projects 661/Papa and 705/Alfa significant steps in submarine development. Although titanium was used in the construction of these submarines, they did not**

have a greater-than-normal test depth for their time (a characteristic that was attributed to them by Western intelligence).

Further, while Project 661 was a one-of-a-kind submarine, Project 705 was produced in numbers. Despite major problems with the prototype Project 705 submarine and a major reactor problem with another submarine of that class, Project 705 was in many ways a highly successful design.

Essentially simultaneous with the advanced Projects 661 and 705 high-speed submarines, the Soviet design bureaus and shipyards were producing "conventional" SSN/SSGN designs.

TABLE 9-2

**High-Speed Submarines**

| | U.S. *Albacore** AGSS 569 | U.S. *Barbel* SS 580 | U.S. *Skipjack* SSN 585 | Soviet Project 661 NATO Papa | Soviet Project 705 NATO Alfa |
|---|---|---|---|---|---|
| Operational | 1953 | 1959 | 1959 | 1969 | 1977# |
| Displacement | | | | | |
| surface | 1,517 tons | 2,150 tons | 3,070 tons | 5,280 tons | 2,324 tons |
| submerged | 1,810 tons | 2,640 tons | 3,500 tons | 6,320 tons | <3,210 tons |
| Length | 203 ft 10 in (62.14 m) | 219 ft (66.77 m) | 252 ft (76.83 m) | 350 ft 8 in (106.9 m) | 261 ft 2 in (79.6 m) |
| Beam | 27 ft 4 in (8.31 m) | 29 ft (8.84 m) | 32 ft (9.75 m) | 37 ft 9 in (11.5 m) | 31 ft 2 in (9.5 m) |
| Draft | 18 ft 7 in (5.66 m) | 25 ft (7.62 m) | 25 ft (7.62 m) | 26 ft 3 in (8.0 m) | 23 ft 3½ in (7.1 m) |
| Reactors | — | — | 1 S5W | 2 VM-5 | 1 OK-550/BM-40A |
| Turbines | — | — | 2 | 2 | 1 |
| horsepower | — | — | 15,000 | 80,000 | 40,000 |
| Diesel engines | 2 | 3 | — | — | — |
| horsepower | 2,000 | 4,800 | — | — | — |
| Electric motors | 1 | 2 | — | — | — |
| horsepower | 4,700 | 3,125 | — | — | — |
| Shafts | 1 | 1 | 1 | 2 | 1 |
| Speed | | | | | |
| surface | 15 knots | 14 knots | 15 knots | 16 knots | 12 knots |
| submerged | 27.4 knots | 18.5 knots | 33 knots*** | 42 knots## | 41 knots |
| Test depth | 600 feet (185 m) | 700 feet (215 m) | 700 feet (215 m) | 1,300 feet (400 m) | 1,300 feet (400 m) |
| Torpedo tubes** | none | 6 533-mm B | 6 533-mm B | 4 533-m B | 6 533-m B |
| Torpedoes | none | 22 | 24 | 12 | 18### |
| Complement | 37 | 77 | 90 | 75 | 29 |

*Notes:* * 1953 configuration.
** Bow.
*** Reduced with subsequent propeller change.
# The prototype *K-64* was completed in 1971.
## 44.7 knots achieved on trials.
### Torpedoes and ASW missiles.

10

# Second-Generation Nuclear Submarines

*The* Thresher *on sea trials in July 1961. Her antecedents in the* Albacore-Skipjack *designs are evident. But the* Thresher *design was deeper diving, significantly quieter, better hardened against explosive shock, and introduced new sensors and weapons to the fleet.* (U.S. Navy)

The United States and Soviet Union began developing their second-generation nuclear submarines even as their prototype nuclear submarines were going to sea. In the U.S. Navy the first second-generation nuclear submarine was the high-speed *Skipjack* (SSN 585), which went to sea in 1959. Her principal features—the S5W reactor plant and *Albacore* (AGSS 569) hull form—gave her excellent underwater performance. But she retained essentially the same weapons, sensors, and operating depth of her immediate predecessors, both nuclear and conventional. And the *Skipjack*—like her nuclear predecessors—was a very noisy submarine.

This last factor was becoming increasingly significant. The U.S. seafloor Sound Surveillance System (SOSUS) was confirming the high noise levels of U.S. submarines. In 1961 the USS *George Washington* (SSBN 598) was tracked continuously across the Atlantic by SOSUS arrays along the East Coast. The high noise levels of U.S. submarines—conventional and nuclear—had long been realized. A decade earlier Rear Admiral Charles L. Brand had stated,

> There have been practically no developments in silencing of [submarine] machinery. In fact, our submarines are noisier now than during the war due to the fact that we have not improved the machinery and the personnel [have] deteriorated and they don't take as much interest in maintaining low level of noise. The crews are less experienced and actually the noise level of our present submarines is not as good as during the war.[1]

Little changed in the subsequent decade, and early U.S. nuclear submarines had high noise levels. The high speed of the *Skipjack* had exacerbated the situation. A submarine has five noise sources:

*Machinery operation*—main propulsion and auxiliary machinery, especially pumps in nuclear propulsion plants that circulate coolant fluids; unbalanced rotating machinery and poorly machined gears are significant contributors.

*Propeller*—flow-induced propeller blade vibrations and cavitation (formation and collapse of bubbles produced by the propeller).

*Hydrodynamic flow*—the flow of water around the submarine's hull, sail, and control surfaces as the submarine moves through the water.

*Transient perturbations*—pulselike noises created by the movement of torpedo tube doors or shutters, control surfaces, and other submarine activities.

*Crew*—crewmen opening and closing hatches and doors, using and dropping tools, and so forth.

Quieting nuclear submarines would be particularly difficult because of their primary and secondary loop pumps, turbines, and propulsor. Dick Laning, first commanding officer of the *Seawolf* (SSN 575), recalled Admiral H. G. Rickover saying that "making nuclear submarines quieter is a bigger problem than nuclear propulsion."[2] In reality, Rickover had little interest in quieting submarines; rather, he sought innovation in propulsion plants. His attention to the actual submarine was limited, especially if the next SSN used the same S5W plant of the *Skipjack*. After the loss of the *Thresher* (SSN 593), he told a congressional committee that he had no responsibility for the *Thresher*, "because it was a follow-on ship. . . . The *Thresher* was essentially a *Skipjack*-type submarine except that all the equipment was mounted on resilient mounting. . . ."[3]

The next two second-generation U.S. nuclear submarines would be highly innovative in several ways, especially in a dramatic reduction of machinery noises. These were the *Thresher* and *Tullibee* (SSN 597). The six *Skipjack*-class SSNs funded in fiscal years 1955–1957 were joined in the last year by a new-design SSN—the *Thresher*, which would be the last combat submarine designed largely by the Portsmouth Naval Shipyard. This submarine would incorporate the S5W reactor plant in a modified *Albacore* hull, but with major advances in weapons, sensors, operating depth, and quieting.

The design staff at Portsmouth was given relatively wide latitude in the *Thresher* design, in part because of employing the existing S5W reactor plant, and in part because of the parallel development of the Polaris submarines of the *Ethan Allen* (SSBN 608) class, for which the Special Projects Office sought improved quieting and operating depth. The *Thresher* design marked major advances in several areas:

- depth
- quieting
- sonar
- shock hardening
- weapons

The dominating feature in the *Thresher*'s internal arrangement was the AN/BQQ-2 sonar "system," the replacement for the long-serving BQR-2 and BQR-4 passive sonars. The BQQ-2 was centered on a 15-foot (4.6-m) diameter sphere that mounted 1,241 hydrophones. Compared to the older sonars, the BQQ-2 could achieve greater passive detection ranges through the improved ability to distinguish between target noise and own-ship and background sea noises, and its ability to electronically search in three dimensions, enabling it to take maximum advantage of sound energy propagated by bouncing off the ocean bottom. The BQQ-2 also had an active acoustic ("pinging") component, although by this time U.S. submarine sonar doctrine had shifted to passive-only operations.

Locating the BQQ-2 in the bow placed it in an optimum search position and as far as possible from machinery and propeller noises. This location required that the 21-inch (533-mm) torpedo tubes be moved farther aft and angled outboard at approximately 15 degrees, two per side. Amidships torpedo tubes—*firing aft*—had been proposed by the Germans in World War II for the Type XXIB, XXIC, and XXVI submarines. The Bureau of Ships was concerned that firing at an angle would limit the speed at which a submarine could launch torpedoes, but the feasibility of angled launch was demonstrated up to a speed of 18 knots.[4] The reduction in torpedo tubes was caused by hull constraints (i.e., two per side) and was accepted because of the perceived effectiveness of U.S. torpedoes against Soviet undersea craft. It was believed that the Mk 37 torpedo—in service since 1956—would have a probability of kill against Soviet submarines of almost 1.0, that is, a kill for every torpedo fired.[5] This sonar-torpedo tube arrangement was adopted for all later U.S. SSNs as well as ballistic missile submarines of the *Ohio* (SSBN 726) class.

In addition, the *Thresher*-class SSNs would be fitted to launch the SUBROC (Submarine Rocket), operational from 1964.[6] This weapon provided a means of attacking targets detected by sonar at distances beyond the range of Mk 37 and Mk 45 ASTOR (nuclear) torpedoes. After being launched from a standard torpedo tube, SUBROC streaked to the surface, left the water, and traveled on a ballistic trajectory for a predetermined distance out to about 25 n.miles (46.3 km). At that point the expended rocket booster fell away and the W55 nuclear warhead re-entered the water, to detonate at a preset depth. The warhead could be selected to detonate with an explosive force of from one to five kilotons. Before launching a SUBROC the submarine had to transmit an active acoustic "ping" to determine the range to the target, initially detected by passive sonar. This use of active sonar—however brief—was eschewed by U.S. submarine captains, who throughout the Cold War embraced passive acoustic tactics.

The *Thresher* also had a greater operating depth than previous U.S. nuclear submarines. The

*The BQQ-2 sonar sphere before being installed in the* Thresher. *The 15-foot-diameter sphere was fitted with 1,241 hydrophones. All subsequent U.S. attack and strategic missile submarines have had similar sonar configurations.* (U.S. Navy)

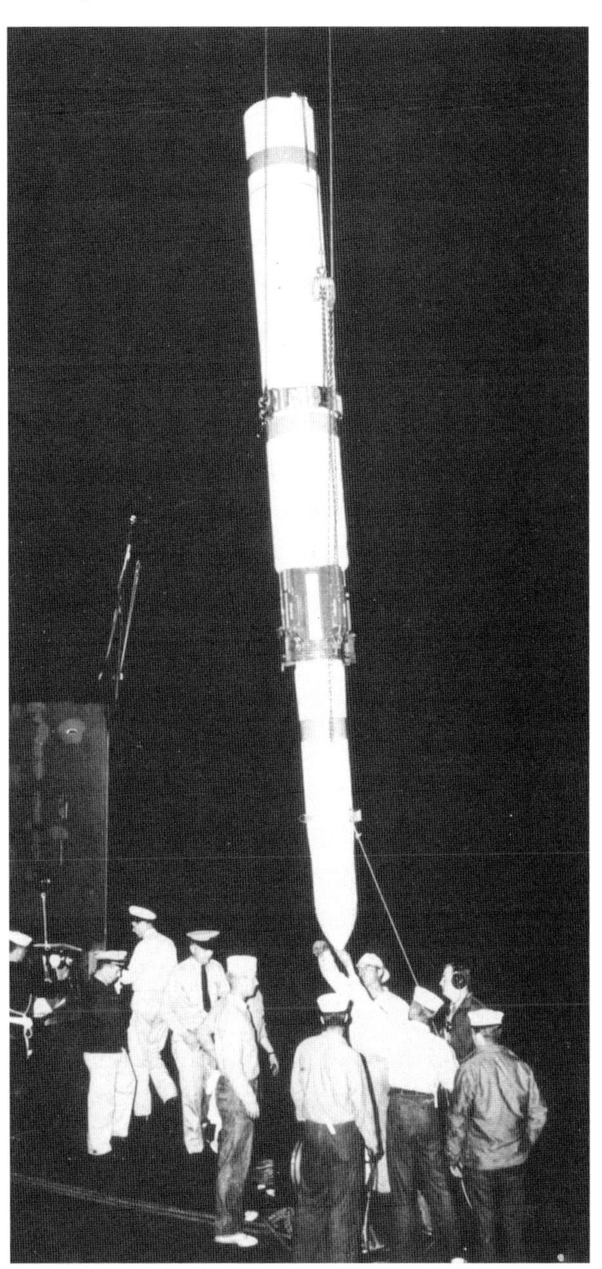

*A SUBROC anti-submarine rocket being lowered to the submarine* Permit. *Produced by Goodyear Aerospace Corp., the SUBROC was the first U.S. underwater-launched tactical missile. The anti-submarine weapon was fitted with a nuclear warhead.* (Goodyear)

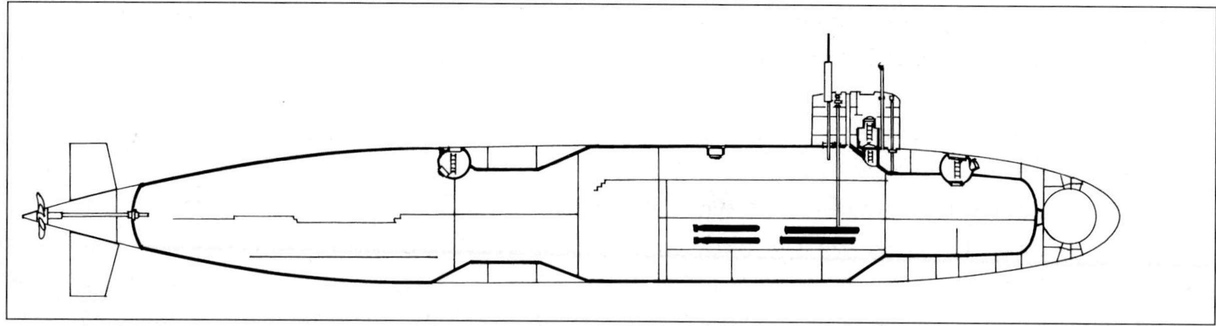

Thresher *(SSN 593). LOA 278 ft 6 in (84.9 m)* (©A.D. Baker, III)

*Thresher*'s pressure hull was fabricated of HY-80 steel, which had been used in the previous *Skipjack* class, but that class was rated at a test depth of 700 feet (215 m), the same as other U.S. post–World War II submarines.[7] Improved welding techniques, piping, and related improvements gave the *Thresher* a test depth of 1,300 feet (400 m). Beyond the operational advantages of greater depth (e.g., going below thermal layers to escape sonar detection), greater depth increased the margin for recovery from control error or malfunction during high-speed maneuvers.

The increase in depth became the most controversial aspect of the *Thresher* design. Captain Donald Kern, head of preliminary design in the Bureau of Ships, recalled Admiral Rickover's view:

> He wanted no part of the deep diving. He thought we were wasting our time and that this was foolish going to the deeper depth, and he fought that tooth and nail, too. I had knock-down, drag-out battles with him on that.[8]

While the *Thresher* was under construction, the following exchange occurred between Admiral Rickover and Representative George Mahon, a member of the House Appropriations Committee:

> *Mahon.* Are you over designing these ships? I am talking now mostly about submarines. Are you putting on refinements that are really not necessary? You spoke of the *Thresher* diving to a very great depth.
> *Rickover.* Yes, sir.
> *Mahon.* How deep are you going?
> *Rickover.* The World War II submarines were designed for [400] feet. Right after World War II we developed the present [700]-foot submarines. Now we are going to [1,300] feet. The reason is that the deeper a ship goes, the less it is possible to detect it. It can take advantage of various thermal layers in the ocean. It also is less susceptible to damage by various types of depth charges and other anti-submarine devices. The greater depth gives it greater invisibility, greater invulnerability. We would like to go deeper if we could, but a point comes where existing hull steel may not be able safely to withstand the greater pressure. . . . However, there is considerable military advantage, Mr. Mahon, to be able to go deeper; it is somewhat analogous to having airplanes which can fly higher.

Three years later—after the *Thresher* was lost at sea—the issue of going deep again was discussed in Congress. At the 1964 hearing before the Joint Committee on Atomic Energy, Vice Admiral Lawson P. Ramage, a much-decorated submariner and at the time a Deputy Chief of Naval Operations, explained the advantage of going deep, including enhanced safety in certain maneuvering situations. Rickover—after he had repeated to the committee the advantages of going deep—changed his viewpoint:

> Sure, you can intuitively say—as Admiral Ramage said in the comparison he made—you would like to go deeper. It is good to have a machine that can perform better. However, I claim we have to be realistic and find out how important this is first, because right now we are incurring considerable expenses in building these ships.[9]

In many respects, the most important aspect of the *Thresher* design was quieting her S5W propulsion plant, with special efforts being made to reduce those narrow-band noises produced by machinery. Previously, little effort had been made to reduce machinery noises in U.S. nuclear submarines—the noise of coolant pumps, fluids in the coolant piping, and turbine noises.

The Royal Navy had developed the concept of "rafting" propulsion machinery in the Ton-class minesweepers built in the 1950s and 1960s, by isolating the machinery from the hull to reduce vulnerability to acoustic mines. This arrangement did not affect the noise-generating machinery, but decreased sound transmission through the hull into the water. This concept was selected for the *Thresher* class in April 1957; the lead submarine was ordered from the Portsmouth Naval Shipyard nine months later.[10] But rafting was not a simple process; British submarine commander John Hervey wrote:

> Submarine sound isolation is an activity in which the good intentions of the designer are very soon frustrated by a badly briefed, or confused, construction worker. At worst, quite serious errors may remain concealed until the first noise trials. For instance, when [the SSN] *Warspite* was [sound] ranged in 1968, the trials officer reported, quite early on, that unlocking the raft did not seem to alter the signature being recorded, news almost guaranteed to ruin the day![11]

In the *Thresher* the twin steam turbines and related gearing were mounted on a sound-isolating raft and not directly attached to the hull. This method increased the volume required for machinery and hence the overall submarine size, contributing to the two-foot (0.6-m) increase in hull diameter over the previous *Skipjack* design. The *Thresher* had a surface displacement of 3,750 tons—some 22 percent larger than the *Skipjack*, with a parallel midbody section inserted in the *Albacore*-type hull. However, with a sail structure about one-half the size of the *Skipjack*'s sail, and several other drag-reducing features (such as smaller hull openings), and improvements to the steam plant, the *Thresher* had a speed of about 28 knots.[12]

*The* Thresher *showing her modified-*Albacore *hull configuration. The design had a short sail structure, which was found unsatisfactory from a perspective of space for masts and electronic surveillance equipment. Three later submarines of this class were extensively modified to remedy these shortfalls.* (U.S. Navy)

The smaller sail was possible, in part, by deleting some masts and electronic intelligence collection capabilities from these submarines. There were some advocates of doing away completely with the sail to further reduce drag, but it was considered necessary for mounting the forward diving planes, housing masts and periscopes, and providing a navigation bridge for entering and leaving harbor.

The *Thresher* was commissioned at Portsmouth on 3 August 1961, with Commander Dean L. Axene as the first commanding officer. The success of the *Thresher* design was reflected by all subsequent U.S. attack submarines having the same basic configuration. Through fiscal year 1961 the Navy ordered 14 SSNs of this design; these ships included four SSGNs reordered as attack submarines after cancellation of the Regulus II program (SSGN/SSN 594–596, 607).

Of these 14 submarines, only two beyond the *Thresher* were constructed at Portsmouth, the *Jack* (SSN 605) and *Tinosa* (SSN 606). The design team at the yard, expanding on experience with the contra-rotating propellers in the *Albacore,* gained approval to build the *Jack* with a similar propulsion configuration. The *Jack* was fitted with a large-diameter, sleevelike outer propeller shaft housing a smaller-diameter inner shaft, each fitted with a propeller. The *Jack*'s machinery spaces were ten feet (3.0 m) longer than in other units of the class because of the turbine arrangement and the use of slow-speed, direct-drive turbines with the elimination of the reduction gear. There was an increase in turbine efficiency with speed increased a fraction of a knot.

This plant was looked on as a potentially quieter propulsion arrangement, in part as a means of providing a quieter plant than the standard S5W plant as fitted in the *Thresher.* In the event, the *Jack*'s two turbines developed problems, mainly because of difficulties with seals and bearings in the concentric shafts, and the arrangement was not repeated. Still, in the opinion of Portsmouth engineers, contra-rotating propellers held promise for improved performance.[13] The *Jack* also was used to test polymer ejection, apparently to reduce flow noises that could interfere with sonar performance.

(In the USSR, the Malachite bureau considered a single-turbine, single-shaft, contra-rotating propeller arrangement for Project 696, a large, high-speed submarine design. Project 696 was envisioned as having a submerged displacement of 5,500 to 6,800 tons, a nuclear plant producing 100,000 shp, and capable of speeds up to 45 knots, increased to more than 50 knots with polymer ejection. The design, begun in the late 1960s, encountered numerous problems, and this project—under designer A. K. Nazarov—was cancelled in 1975. Project 696 also would employ polymer ejection. That design was not pursued.[14])

The USS *Thresher* sank on 10 April 1963 while on postoverhaul trials, the world's first nuclear submarine to be lost. The *Thresher* sank with all on board—112 Navy personnel and 17 civilians, the most casualties ever of a submarine disaster.[15]

This disaster led to an extensive examination of submarine propulsion emergency procedures, because the first "event" in her demise was apparently a reactor scram (shutdown). A scram could occur for many reasons, among them a control system failure, reactor error, stuck control rod, or a too-rapid withdrawal of rods.[16]

Admiral Rickover ordered a reduction in the time needed to restart a reactor in an emergency.[17] The Navy ordered careful examination of all piping and welding in submarines of this class. The ballast blowing systems, which were found incapable of bringing the submarine to the surface from deep depths, were modified; and other changes were made, many collectively known as SUBSAFE (Submarine Safety).[18]

Rickover's criticism of Navy submarine construction and overhaul methods—made to the Joint Committee and widely publicized—led to his procedures and standards being forced on the Navy for the "forward end" of the submarine. This significantly increased his influence and even his control over U.S. submarine design and construction.

The unfinished *Thresher*s were delayed for SUBSAFE modifications, and three, the SSN 613, 614, and 615, were lengthened 13¾ feet (4.2 m), with a section inserted forward of the bulkhead for the reactor compartment. The section provided space for SUBSAFE features, including additional buoyancy, but the primary reason for their delay and for the added section was to provide improved accommodations and intelligence collection equipment, and for a larger sail to house additional masts. (Table 10-1 shows the principal variations of the *Thresher* class.)

Soviet submarine designers carefully monitored the Navy Department's statements about the loss of the *Thresher,* primarily through congressional hearings. As a senior Soviet designer wrote,

> The Admiralty officials presented tens of proposals to make monitoring more strict, to institute new methods of checking pressure [watertight] structures and systems. Tens and hundreds of people participated in checking the work already completed, of the metal and hull parts and systems currently in work. Everybody understood—there were no small details in a submarine.[19]

TABLE 10-1
***Thresher*-class Variations**

| Ship | Surface Displacement | Submerged Displacement | Length Overall |
|---|---|---|---|
| *Thresher* | 3,750 tons | 4,310 tons | 278 ft 6 in (84.89 m) |
| *Jack* | 3,990 tons | 4,467 tons | 297 ft 4 in (90.63 m) |
| SSN 613—615 | 4,260 tons | 4,770 tons | 292 ft 3 in (89.08 m) |

## Nuclear Hunter-Killers

Almost parallel with the design of the *Thresher,* an effort was under way to develop a small, specialized hunter-killer submarine—the *Tullibee.* The previous *K1*-class killer submarines had failed, in part, because of their limited operating range; a nuclear-propelled SSK could reach its deployment area and remain on station to the limit of her crew or until her torpedoes were expended. A small, nuclear SSK would be less expensive than an SSN and thus could be produced in large numbers to counter the increasing Soviet submarine threat.

The proposed "SSKN" had a design goal of 900 tons surface displacement and would be powered by a new pressurized-water reactor. The Naval Reactors Branch developed the S1C/S2C reactor plant that produced 2,500 horsepower, one-sixth the power of the S5W plant in the *Skipjack* and *Thresher.* The *Tullibee* had turbo-electric drive, which dispensed with the gears of steam turbine drive and gave promise of being the quietest nuclear propulsion system possible. Further, the propulsion system would provide excellent response, being able to go from full ahead (200 revolutions per minute) to full astern in seconds. The submarine had a single propeller shaft, as did other second-generation U.S. nuclear submarines. But unlike the other single-shaft submarines, the *Tullibee* did not have an emergency, "come-home" propeller pod that could be lowered to provide limited propulsion in the event of a main propulsion plant or shaft failure.[20] The prototype S1C reactor plant was installed at the Atomic Energy Commission's Windsor, Connecticut, test site; the similar S2C was installed in the *Tullibee.*

Electric Boat was chosen to construct the lead SSKN. The *Tullibee*'s reactor and other considerations made her grow during design until her surface displacement was 2,177 tons. The *Albacore* hull form was corrupted with a lengthy parallel mid-body, topped by a small sail. However, the *Tullibee* was built with HTS (High Tensile Steel), and not HY-80, limiting her test depth to 700 feet (215 m). The *Tullibee* was provided with the AN/BQQ-1 sonar system as well as four angled torpedo tubes. The BQQ-1 was the Navy's first integrated sonar system composed of the BQR-7 low-frequency, passive array, which was wrapped around the first spherical active sonar, the BQS-6, this combination being fitted in a bow sonar sphere.

The *Tullibee* was commissioned on 9 November 1960. She was designated SSN 597 vice the SSKN that had been used during her design and development. Her crew numbered six officers and 50 enlisted men, 32 fewer personnel than initially assigned to a *Thresher*-class SSN. (The *Thresher*-class manning soon increased to about 100.) The *Tullibee*'s first skipper, Commander Richard E. (Dick) Jortberg, recalled:

> . . . We could stand watch and man battle stations. We were weak in in-port maintenance. We solved that problem by adding a training allowance. In effect, we had a four-section crew—of which only three sections went to sea at one time. The section remaining in port took care of schools, training, and leave. . . . During in-port periods, all personnel worked on the ship—all four sections. The crew loved it—we had a 100% reenlistment rate for three years.[21]

The submarine was employed in research and, subsequently, undertook standard SSN deployments. Operationally, the *Tullibee* suffered major limitations, including low speed for ocean transits to deployment areas, Commander Jortberg observed: "In the operational area, people often questioned the slower speed of the *Tullibee*—16 knots. In my experience with the ship, I never found that to be a real problem."[22] As an ASW platform the *Tullibee* was unsurpassed at the time.

The construction of additional submarines of this design was not undertaken, the decision being made by the Navy to instead procure only the

*The small hunter-killer submarine* Tullibee *in October 1960, also on sea trials. The Navy constructed only one ship of this design, preferring instead to employ larger, multi-purpose SSNs in the killer role. The "fins" forward and aft were part of the PUFFS fire control system.* (U.S. Navy)

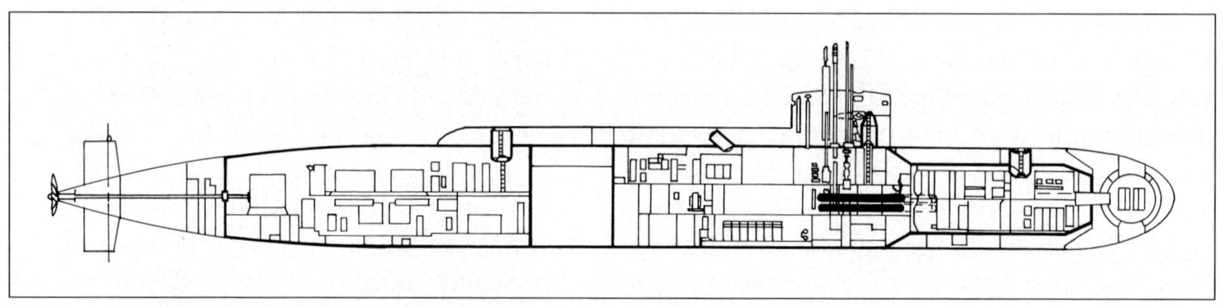

Tullibee *(SSN 597). LOA 272 ft 9½ in (83.16 m)* (©A.D. Baker, III)

larger and more versatile *Thresher* class for all "attack" submarine roles.

The 14 *Thresher*-class SSNs were followed by the similar but larger submarines of the *Sturgeon* (SSN 637) class. Again powered by an S5W reactor plant, these later submarines had increased quieting features and, significantly, a larger sail structure. This permitted the return of the electronic intelligence capability deleted from the short-sail *Threshers* and an arrangement whereby the sail-mounted diving planes could be rotated 90 degrees to facilitate the craft penetrating through Arctic ice. The cost was a submarine 510 tons larger than the original *Thresher* design and several knots slower. The last nine *Sturgeons* were lengthened ten feet (3.1 m) during construction to provide additional space for intelligence collection equipment and technicians, and to facilitate their carrying dry-deck shelters or "hangars" for swimmer operations. And, several units of this class received (minor) modifications to serve as "mother" submarines for rescue submersibles. (See Chapter 13.)

One *Sturgeon,* the *Parche* (SSN 683), was extensively modified, to perform ocean-engineering and other "special missions." The submarine underwent modifications at the Mare Island Naval Shipyard for the ocean engineering role; she was refueled and extensively modified at Mare Island from January

*The* Whale *was the second submarine of the* Sturgeon *class. These submarines were essentially enlarged* Threshers *with enhanced quieting, an under-ice capability, and added electronic intelligence collection capabilities (note the taller sail). The* Whale *undertook several under-ice missions.* (General Dynamics Corp.)

1987 to May 1991. These later modifications included the addition of a 100-foot (30.5-m) section forward of the sail to accommodate special search and recovery equipment. This is reported to have included a clawlike device lowered by cable to recover satellites and other equipment from the ocean floor.

In a program known as Ivy Bells, beginning in 1971 the extensively converted *Halibut* (SSN 597) and, from 1976, the *Parche* were employed to install recording devices for tapping into an underwater communications cable at a depth of some 400 feet (122 m) in the Sea of Okhotsk, between Soviet bases on the Kamchatka Peninsula and the Soviet Far Eastern coast. The submarines carried divers that could "lock out" of special living chambers at depth to install and service the seafloor wiretaps.[23] (This operation was revealed to the Soviets in January 1980 by the U.S. traitor Ronald Pelton; the Soviets quickly shut down the American spy operation.)

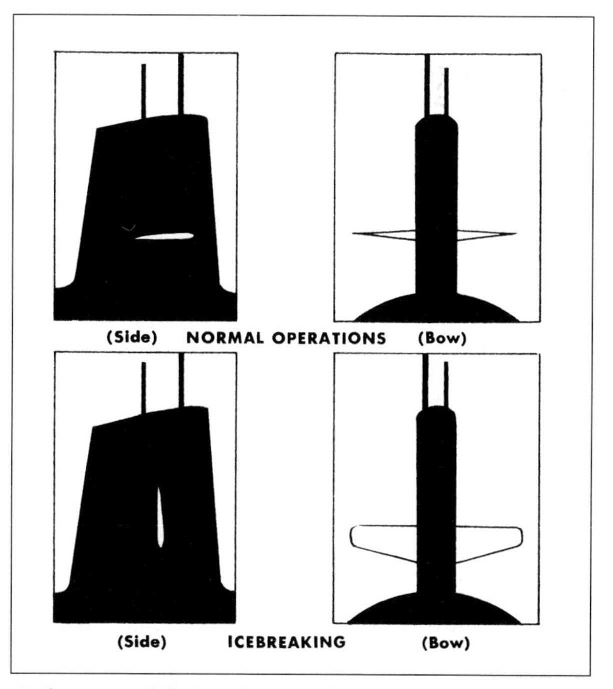


*Sail-mounted diving planes in the* Sturgeon *class could be rotated 90 degrees to reduce ice damage when breaking through the Arctic ice pack.* (*Atomic Submarines*)

With the end of the 41-submarine Polaris program in sight, the Congress provided funds to accelerate the production of SSNs. A total of 37 *Sturgeon* SSNs were completed from 1967 through 1975. The building rate—an average of four SSNs per year—occurred as U.S. shipyards were completing the Polaris program (with the last SSBN completed in 1967).

One modified *Sturgeon* design was built, the enlarged *Narwhal* (SSN 671). She had an S5G reactor plant configured for natural circulation, wherein convection moved pressurized water at low speeds, alleviating the use of the noisy circulating pumps. Slow-speed, direct-drive turbines were employed in the *Narwhal* (as they were in the *Jack*). Completed in 1969, the *Narwhal* quickly demonstrated the efficacy of this natural circulation, which was adopted for future U.S. submarine reactor plants.

One additional SSN would be built with a modified S5Wa plant, the *Glenard P. Lipscomb* (SSN 685; see Chapter 17). The S5W represented exactly what Rickover had sought in nuclear plants—"practicality and simplicity."[24] Further, the *Thresher* served as the design model for the *Ethan Allen* (SSBN 608) class of Polaris missile submarines in terms of interior design, systems, and depth capability, in the same manner that the *Skipjack* had been the model for the *George Washington* (SSBN 598) class. The United States built 98 submarines with the S5W reactor plant—57 second-generation SSNs, the *Lipscomb,* and 41 Polaris SSBNs. Britain built its first SSN—HMS *Dreadnought*—with an S5W plant provided by the United States.

With the availability of the *Sturgeon*-class SSNs, U.S. nuclear submarines returned to under-ice Arctic operations. They had abstained from those voyages following the loss of the *Thresher* because of limitations imposed on SSN operations until the SUBSAFE program was completed. The *Queenfish* (SSN 651) was the first of several submarines of this new class to conduct extensive operations under the Arctic ice pack and was the first single-screw submarine of any nation to surface through ice, in February 1967 in Baffin Bay. The *Whale* (SSN 638) and, subsequently, *Pargo* (SSN 650) were the first submarines of this class to surface at the North Pole, both doing so in April 1969.[25] Operating under Arctic ice was not without problems, including difficulties in firing acoustic homing torpedoes because of under-ice reverberations.

Increasingly, U.S. submarine operations in northern latitudes were for the purpose of observing and tracking the rapidly growing fleet of Soviet nuclear-propelled submarines. By 1970 the Soviet Navy outnumbered the U.S. Navy in numbers of nuclear-propelled submarines.

## Soviet Second-Generation Submarines

Like the U.S. Navy, even as its first-generation nuclear submarines were going to sea, the Soviet Navy was already engaged in developing the second-generation "nukes." And both navies were developing multiple nuclear submarine classes, with the Soviet second-generation effort being much broader.

In the late 1950s, shortly after the first Project 627/November SSN went to sea, the three design bureaus put forward proposals for several advanced submarine designs. Beyond the advanced Project 661/Papa SSGN and Project 705/Alfa SSN, in 1958 the State Committee for Shipbuilding initiated a competition among three bureaus—TsKB-18/Rubin, TsKB-112/Lazurit, and SDB-143/Malachite—to design four second-generation nuclear submarines:

Project 667: nuclear ballistic missile submarine (SSBN)

Project 669: large nuclear attack submarine with heavy torpedo armament (SSN)—follow-on to Project 627/November

Project 670: small nuclear submarine for mass production with torpedo armament (SSN)

Project 671: nuclear ASW submarine with torpedo armament and large sonar (SSKN)

The last was to have a surface displacement of not more than 2,000 tons to permit series construction at inland shipyards; it was to have an armament of four torpedo tubes and carry eight torpedoes, have a speed of about 30 knots, and a test depth of 985 feet (300 m). By Western criteria this would be an SSKN.

As early as 1961 a chief spokesman for the Soviet Navy, Admiral Vasiliy I. Platonov, wrote in a highly classified publication:

> Anti-submarine submarines armed with the most improved sonar and hydroacoustic equipment can be the only real forces for combating missile-carrying submarines. Underwater sonar search, underwater patrolling, and underwater patrols and ambushes must become their tactical methods. Active combating of missile-carrying submarines and all maneuvering connected with hunting and destroying them must now be carried on deeply underwater instead of on the surface. There is no other way.[26]

It soon became evident that Project 671 would be the only SSN that would be built with a torpedo armament, and the design took on a more "universal" or multi-purpose character. Its displacement soon reached the critical mark of 3,500 tons (surface), which was the limit for construction in a Leningrad shipyard and being carried by transporter dock on the Baltic-Belomor Canal to Severodvinsk for completion.[27] The design of Project 671 (NATO Victor) was initiated at SDB-143 in 1958 under Leonid A. Samarkin. A year later, as Project 671 was formalized, George N. Chernyshev was named chief designer with Samarkin as his deputy.

Project 671 introduced a single propeller shaft in Soviet combat submarines, that feature being accepted on the basis of providing lower noise levels, lower displacement, and higher speed in comparison to two-shaft propulsion plants of the same power. According to the deputy chief designer of Project 671,

> The firm position of the Chief Designer and the bureau leadership on this issue, the support of the leadership of the [State Committee for Shipbuilding] (B. E. Butoma), the bold report of the Chief Navy Overseer V. I. Novikov and the Navy [Commander-in-Chief] S. G. Gorshkov gave his agreement "as an exception" to use one drive shaft on this design. This is the way the "breach" was made on this issue.[28]

The single propeller was fitted 14¾ feet (4.5 m) aft of the stern control surfaces. The submarine was provided with twin propeller pods, one on each of the horizontal supports for her stern planes, to be used both as a "come-home" capability in the event of a main propulsion problem and for slow-speed maneuvering. (In an emergency propulsion mode, the propeller pods would be powered by the ship's battery, which could be recharged by the nuclear plant or, if the submarine could raise a snorkel, an auxiliary diesel engine.)

Project 671/Victor had a modified body-of-revolution configuration (i.e., *Albacore*-type hull), providing a large-diameter pressure hull that permitted side-by-side installation of the twin VM-4P reactors and twin OK-300 turbines. The submarine was fitted with a streamlined ("limousine") sail, further contributing to underwater performance. The two-reactor plant had a single steam turbine rated at 31,000 shaft horsepower.

The single-shaft propulsion plant increased the propulsive characteristics of the submarine by about 30 percent relative to a twin-shaft plant and permitted placing the turbine reduction gear and the two independent turbogenerators in a single compartment with all of their auxiliary systems and equipment. This, in turn, reduced the relative length of the submarine, thus decreasing submerged displacement and wetted surface, which additionally increased the ship's speed. Although Project 671 had a 30 percent greater displacement than the earlier Project 627/November SSN, the wetted surface of the two designs was practically the same.[29] With similar horsepower, Project 671 had a maximum speed of some 33 knots.

Most second-generation nuclear submarines had a pressurized-water reactor of improved design, having the two-reactor configuration similar to that of the earlier HEN-series submarines. The nuclear plants for Project 671/Victor SSN and Project 667/Yankee SSBN were similar, with the smaller Project 670/Charlie SSGN having a single reactor of similar design. The introduction of an alternating-current, 380V 50Hz electrical system in Project 671 provided a major improvement in the reliability of the ship's electrical network.[30]

There was a major effort in Project 671 to reduce self-generated noise, including limber hole covers that close automatically with depth to reduce flow noise, over the hull. The second and subsequent submarines of Project 671 had anechoic (anti-sonar) coating on the outer hull and dampening coating on the inner (pressure) hull to reduce the transmission of machinery noises.

With a double-hull configuration, the Project 671 pressure hull has seven compartments—torpedo/berthing/battery, command center, reactors, turbine, auxiliary equipment, berthing and diesel generators, and electric motors. The double-hull configuration, with ballast tanks along the hull as well as forward and aft, continued the Soviet practice of "surface unsinkability." The design also

*A Project 671/Victor SSN being moved from the building hall at the Admiralty shipyard in Leningrad onto a launching dock. The Soviet Union led in the horizontal construction of submarines, an advantage to building on inclined ways. The camouflage nets provide concealment from photo satellites and onlookers.* (Malachite SPMBM)

The *K-38*, the first Project 671 submarine, was laid down at the Admiralty shipyard in Leningrad in April 1963. She was the first nuclear submarine to be constructed at that yard, which had built the world's first nuclear surface ship, the icebreaker *Lenin*, completed in 1959. The *K-38* was launched into the Fontanka River from the building hall via a launching dock in July 1966. That fall, during a test of the nuclear plant, an accident occurred as a result of an overpressure in a steam generator. Repairs were made and construction continued. In July 1967 the submarine was placed in a transporter dock and moved through the inland canal-river system to Severodvinsk for completion and trials in the White Sea.

During trials the *K-38* set three Soviet records: Underwater speed, test depth, and weapons firing depth. Also noted were the submarine's excellent maneuvering characteristics. With Captain 2d

introduced the stronger AK-29 steel to submarine construction, the equivalent of the U.S. Navy's HY-100 steel. This increased test depth by 33 percent to 1,300 feet (400 m) over first-generation submarines, the same as the Project 661/Papa and Project 705/Alfa, and the USS *Thresher.*

The Project 671 armament consists of six 21-inch bow torpedo tubes with 12 reloads. All weapons handling is automated with torpedoes loaded aboard (through a bow loading chute), moved within the torpedo compartment, placed on racks, and loaded into tubes with a hydraulic system. Torpedoes can be launched at depths down to 820 feet (250 m). The torpedo tubes are fitted in the upper portion of the bow in two horizontal rows—two tubes above four tubes. This arrangement permits torpedoes to be launched at higher submarine speeds than possible with the angled tubes of U.S. submarines. The lower bow contains the large Rubin MGK-300 sonar system (as in the Project 661/Papa SSGN); later units are fitted with the Rubikon MGK-503 sonar.

*A Victor SSN is eased into a transporter dock at the Admiralty shipyard. The dock will transport the submarine through inland waterways to the Severodvinsk shipyard for fitting out and for trials in the White Sea. The transporter docks also provide concealment from Western eyes.* (Malachite SPMBM)

Rank E. D. Chernov in command, the *K-38* was placed in commission on 5 November 1967. Project 671 was a *relatively* conservative design, a direct successor to the Project 627/November SSN. The success of the design led to the construction of a class of 15 submarines completed through 1974, all built at the Admiralty yard.

Subsequently, the design was enlarged in Project 671RT/Victor II, with the ship lengthened 30½ feet (9.3 m) and the diameter of the pressure hull increased up to about 1½ feet (0.5 m), most of the addition being forward to provide an enhanced weapons/sonar capability and quieting.[31] Two of the six 21-inch torpedo tubes were replaced by two 25½-inch (650-mm) tubes for the large Type 65-76 torpedoes.[32] Total weapons load would be 18 21-inch torpedoes and six 25½-inch torpedoes (or tube-launched missiles). In addition, the improved Skat sonar system replaced the Rubin, and there were enhanced quieting features.

The Project 671RT introduced "rafting" to Soviet submarines, significantly reducing machinery-produced noise. According to one Russian reference work, "The turb[ine]gear assembly and turbogenerators with their mechanisms are mounted on one unit-frame support with . . . shock-absorbing layout to reduce the acoustic field."[33] The U.S. intelligence community estimated that Project 671RT submarines had the same noise levels as the U.S. *Sturgeon* class (completed in 1967—five years earlier).

Only seven submarines of Project 671RT were constructed in 1972–1978, three at Admiralty and four at the Krasnoye Sovmoro yard in Gor'kiy. Project 671RT construction was truncated because of the decision to pursue a further improvement of the design, Project 671RTM/Victor III.[34] Lengthened a further 15¾ feet (4.8 m), this final variant was quieter, was fitted with tandem propellers (fitted to the same shaft), had improved electronics (including the Skat-K sonar), and carried more advanced weapons than did the earlier submarines. These weapons included the RPK-55 Granat torpedo tube-launched cruise missile (NATO SS-N-21),

*A Project 671RT/Victor II at sea. The free-flood, or limber, holes have spring-loaded covers that close as the submarine dives to improve the submarine's hydrodynamic qualities, reducing drag and noise. Her forward diving planes are fully retracted into the hull, forward of the sail.* (U.K. Ministry of Defence)

*A Victor SSN is backed out of a transporter dock at Severodvinsk. The extensive use of inland waterways, including canals built by tens of thousands of prisoners in the Soviet-era Gulag, permitted submarines and smaller warships to be built at inland as well as coastal shipyards.* (Malachite SPMBM)

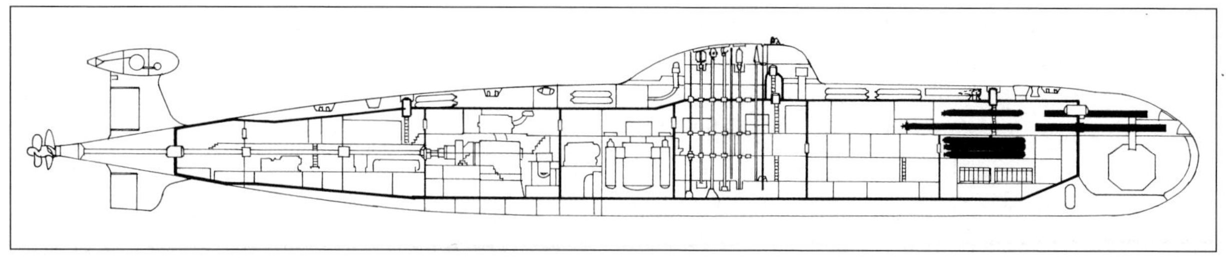

*Project 671RTM/Victor III SSN. LOA 351 ft 4 in* (©A.D. Baker, III)

a weapon similar to the U.S. Tomahawk. (See Chapter 18.)

The 671RTM improvements include a towed-array sonar (hydrophones) streamed from a massive pod mounted atop the upper tail fin/rudder. This pod is 25 7/12 feet (7.8 m) long and 7 1/6 feet (2.2 m) in diameter, about the size of a Volkswagen minibus. Several alternative purposes of the pod were suggested by Western observers when it first was observed—a housing for a towed communications cable, a torpedo decoy system or an auxiliary propulsion system for extremely quiet underwater operations at speeds below 12 knots, or for high, "burst" speeds. The theorized propulsion concepts included Magnetohydrodynamic (MHD) or Electromagnetic Thrust (EMT) drive in which seawater is accelerated through or around the structure, somewhat comparable to waterjet propulsion. For a number of technical reasons, the use of the pod for MHD or EMT propulsion was not practical, among them the very high electrical requirements and the lack of a large frontal opening for water intake.[35] (Similar towed-array pods were fitted in the later Projects 945/Sierra and 971/Akula SSNs and one Project 667A/Yankee conversion.) The enlarged hulls of the 671RT and 671RTM, however, reportedly resulted in a loss of speed with the latter submarine estimated to have had a maximum speed of about 31 knots.

The Admiralty yard in Leningrad constructed the first Project 671RTM submarine, the *K-524,* completed in 1977.[36] The Admiralty yard and Komsomol'sk in the Far East each produced 13 Project 671RTM submarines through 1992. Thus a total of 48 Project 671/Victor SSNs of three variants were produced, the largest SSN series constructed by the USSR, and exceeded in the West only by the near-simultaneous U.S. *Los Angeles* design, with 62 SSNs.

The primary role of these submarines was ASW. According to a director of U.S. Naval Intelligence, a major task of Project 671 SSNs was "operating in conjunction with [Soviet SSBNs], trying to assure that we are not trailing."[37] Indeed, the adoption of Project 671/Victor as the principal SSN design marked the shift of the Soviet attack submarine force away from an open-ocean, anti-shipping campaign, to an emphasis on countering U.S. strategic missile submarines as well as aircraft carriers.

Another Director of Naval Intelligence, discussing the Victor III variant, noted that by the early 1980s, "These submarines often are assigned ASW missions off our coasts."[38] A Victor III, the *K-324,* operating off the U.S. East Coast, fouled the towed hydrophone array of the U.S. frigate *McCloy* in November 1983. The submarine was towed to Cuba for repairs—and recovery of a portion of the U.S. towed array, an intelligence "find" for the Soviets.

A rare, detailed report of a Victor III patrol was published in the Soviet Navy journal *Morskoy Sbornik* in 1993.[39] The article discussed problems of the two-month patrol:

> [Day 23] The shortage of spare parts increases the demand placed on all the equipment, as a result of which wear and tear is growing threateningly. Although the decision to extend the use of the equipment is made by a commission on the basis of required calculations, it is left to me to demand the maximum out of our equipment, literally running it until complete failure.

Successes were also recorded:

> [Day 54] The "contact" this time is a highly prized target, a *Los Angeles*-class submarine.

*A Project 671RTM/Victor III, showing the large sonar-array pod atop her vertical fin and the tandem propellers fitted to a single propeller shaft. The submarine's electronic surveillance mast is raised (NATO Brick Pulp). When this photo was taken in 1983, the submarine had fouled a U.S. frigate's towed sonar array.* (U.S. Navy)

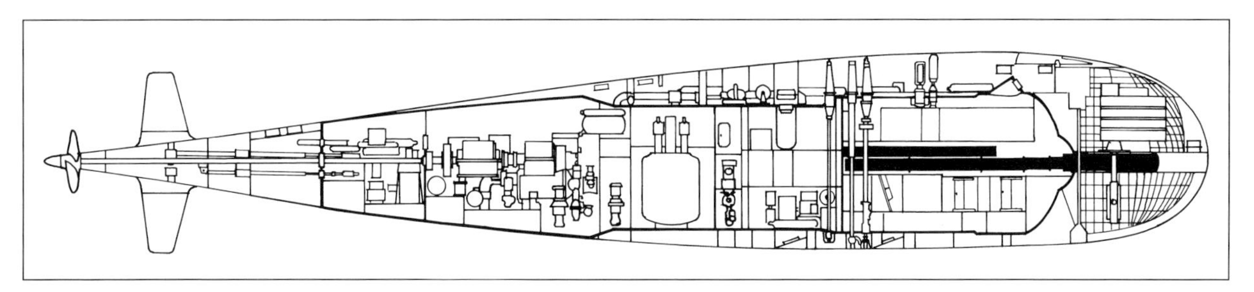

*Project 673—a "sailless" SSN.* (©A.D. Baker, III)

> The boat is like new, "clean," quiet. The Americans have been very successful in this area, which is why they primarily use passive sonar as their means of detection, so as not to give themselves away.
>
> We are no slouches either. This time the trick was that we picked them up on a piece of equipment the equivalent of which, as far as I know, the Americans don't have. We held the hiding contact for a day and a half. Then we lit off our active sonar, which of course illuminated us.

As Project 671 evolved from the original SSKN concept into a large, multi-purpose SSN, the decision was made not to pursue other torpedo-attack submarine designs, among them Projects 669 and 673. The first was conceived as the follow-on to the Project 627/November SSN and was to be armed with eight bow torpedo tubes and carry 30 weapons. SBD-143 was selected to design Project 669, which was to have a single reactor and a displacement of 4,000 to 5,000 tons.

Project 673, developed by SBD-112 in 1960, was envisioned as a smaller SSN with a surface displacement possibly as small as 1,500 tons with a single reactor producing up to 40,000 horsepower for speeds of 35 to 40 knots. The faster variant would have no sail structure (see drawing). Six torpedo tubes and 12 to 18 torpedoes were to be provided.

## Project 670 Charlie SSGN

The high cost of the Project 661/Papa SSGN led the Navy's leadership to consider adopting the P-70 Amethyst (NATO SS-N-7) missile to a smaller, less costly SSGN.

Project 672 was developed by SDB-143 during 1958–1960 as a small SSGN with a surface displacement of less than 2,500 tons, but with a nuclear plant one-half the specific weight of the Project 627 (VM-A) and Project 645 (VM-T) plant that would generate 40,000 horsepower for a speed of 30 to 35 knots. A titanium hull was to provide a test depth of 1,640 feet (500 m). Armament was to consist of torpedo tubes and Amethyst or longer-range cruise missiles. To provide that capability in so small a submarine was found too difficult, and the project was not pursued.

Instead, the Soviets developed Project 670 as a cruise missile submarine. Project 670 originally was intended as a small, mass-production SSN. From May 1960 the design was completely revised by SKB-112/Lazurit in Gor'kiy to carry the Amethyst (SS-N-7) missile. The bureau previously was involved only with diesel-electric submarines. The chief designer of Project 670 was V. K. Shaposhnikov from 1958 to 1960, when V. P. Vorobyev succeeded him. SKB-112 developed Project 670 specifically to attack Western aircraft carriers and other "high value targets."

As an SSGN, Project 670 (NATO Charlie) was a relatively small submarine with a surface displacement of only 3,574 tons while carrying eight Amethyst missiles.[40] It was a double-hull submarine with seven compartments; the sail was fabricated of aluminum to save weight. Project 670 was the first Soviet nuclear submarine to have only one reactor, the VM-4. The OK-350 steam turbine provided 18,800 horsepower to turn a single screw. The reactor plant was "one-half" of the plant in Project 671, although capable of producing some 60 percent of the horsepower of the twin-reactor plant. The SSGN had a submerged speed of 26 knots and a test depth of 1,150 feet (350 m).

Although given a streamlined hull similar in many respects to Project 671, the SSGN would have a large, blunt bow section with four Amethyst missile tubes on each side, between the inner and outer hulls, angled upward at 32.5 degrees. Torpedo armament consisted of four 21-inch and two 15.75-inch (400-mm) torpedo tubes in the bow, with 12 and four torpedoes carried, respectively. The Kerch-670 sonar—intended to track aircraft carriers—with a large hydrophone antenna fitted in the forward section of the sail.

Eleven Project 670/Charlie I SSGNs were completed at the Krasnoye Sovmoro yard in Gor'kiy from 1967 to 1972. These were the first nuclear sub-

*A Project 670/Charlie SSGN steams at high speed in the South China Sea in 1974. These submarines had the advantage of submerged missile launch in comparison with earlier Soviet SSG/SSGNs. The eight Amethyst/SS-N-7 missiles are fitted in the bow; the forward diving planes retract into the hull, immediately forward of the sail.* (U.S. Navy)

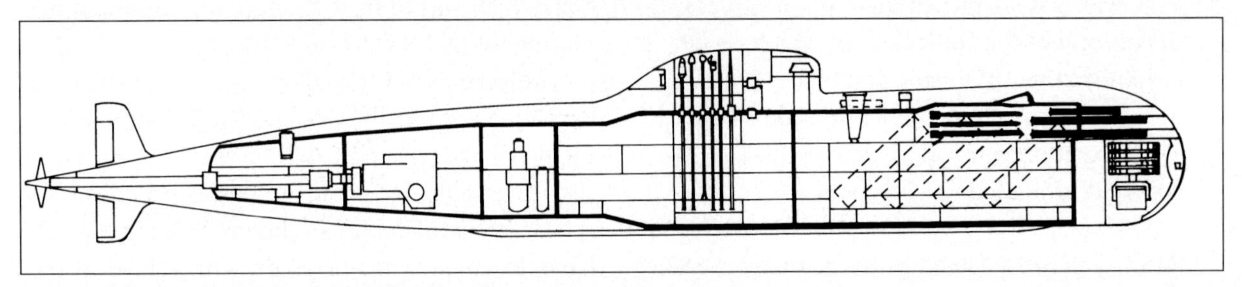

*Project 670/Charlie I SSGN. LOA 309 ft 4 in (94.3 m)* (©A.D. Baker, III)

marines to be constructed at the yard, some 200 miles east of Moscow on the Volga River. The lead ship, the *K-43,* was placed in commission on 27 November 1967, the world's first submarine with an underwater-launch cruise missile capability.

Along with its attributes, the Amethyst missile had a number of shortcomings: It had a relatively short range (38 n.miles/70 km), a limited ability to overcome defensive countermeasures, and required a complex submarine control system. In addition, the missile was not "standardized" in that it could be launched only from submerged submarines. Accordingly, in 1963 advanced development had begun on the P-120 Malachite missile system (NATO SS-N-9).[41] Major differences between the new solid-propellant missile, with a submerged launch, and the Amethyst were a 1.5 increase in range and a more capable guidance/control system. Updated electronics and sonar (Skat-M) were provided in the submarine.

Consequently, SKB-112 designed an enlarged Project 670M/ Charlie II submarine, 29⅙ feet (8.9 m) longer and 680 tons (surface) larger than her predecessor. The larger size would cost about two knots in speed.

The Krasnoye Sovmoro yard produced six Project 670M SSGNs from 1973 to 1980. While Project 670/Charlie submarines were superior to their predecessors of the Project 675/Echo II SSGNs because of their submerged missile launch, they still lacked the underwater speed to effectively track and attack Western aircraft carriers. This would have been corrected in the Project 661/Papa, and was corrected in the third-generation SSGN, the Project 949/Oscar design.

After 20 years in Soviet service, the lead ship of Project 670, the *K-43,* was leased to the Indian Navy, the first nuclear submarine of any nation to be transferred to another.[42] Indian sailors began training at Vladivostok in 1983, both ashore and aboard the *K-43.* The Indian colors were raised on the submarine on 5 January 1988, and she was renamed *Chakra.* The *K-43* was operated by that navy until January 1991 and then returned briefly to Soviet colors, being decommissioned in July 1992. The *K-43/Chakra* trained some 150 Indian naval personnel for nuclear submarine service. In the event, no additional nuclear submarines have been operated by India.

One of her sister ships had a less commendatory career: She sank twice! The *K-429,* a Charlie I SSGN, sank in water 165 feet (50 m) deep off Petropavlovsk in the Far East on 24 June 1983, with the loss of 16 of the 90 crewmen on board. Rescue forces located the sunken submarine, and the next day the surviving men were able to escape to the surface. A series of personnel errors caused the loss.[43] She was salvaged in August 1983. While being repaired, the *K-429* sank again, at her moorings, on 13 September 1985 (without casualties). She was again salvaged but not returned to service. A sister ship, the *K-314,* suffered a reactor meltdown while at Pavlovsk in the Far East, again as the result of personnel error. No crewmen died, and no radiation escaped from the submarine in the December 1985 accident.

Other cruise missile submarine designs were considered in this period, notably Project 705A, an SSGN variant of the Alfa SSN. The preliminary sketches for this variant showed an elongated sail structure, with six fixed-angle tubes for launching the Amethyst/SS-N-7 anti-ship missile. The launchers were fitted aft of the periscopes, masts, and the escape chamber of the standard Project 705 SSN. Internally Project 705A had significant arrangement changes, with approximately the same length as the torpedo-attack submarine (267 feet/81.4 m) and a slightly greater surface displacement (2,385

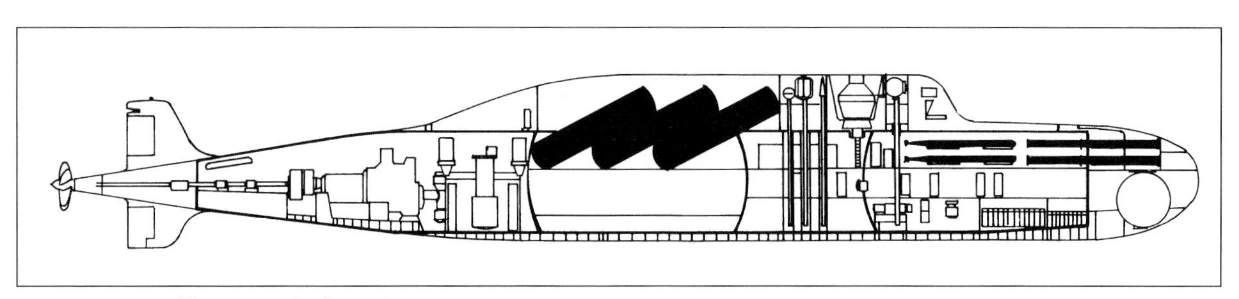

*Project 705A/Alfa cruise missile variant.* (©A.D. Baker, III)

tons versus 2,300 tons). The number of torpedo tubes remained the same. But in this period only the Project 670/Charlie SSGN was series-produced for the anti-ship role.

While the design and construction of the lead second-generation submarines were under way, in October 1964, Nikita Khrushchev was ousted in a peaceful coup. His successor as head of the Communist Party and subsequently also head of the government was Leonid Brezhnev. Khrushchev had generally held back the development and production of conventional weapons, instead stressing strategic nuclear weapons and—for the Navy—submarines.

Under Brezhnev the armed forces underwent a massive infusion of funds; the development and production of conventional as well as strategic weapons was accelerated. Admiral Gorshkov had dutifully followed Khrushchev's dictum to dispose of outdated surface warships and concentrate available resources on submarines, missile ships, and missile-armed aircraft. He had gained approval to construct combination helicopter carriers-missile cruisers of some 19,000 tons full load. These ships had the designation *protivolodochnyy kreyser,* or anti-submarine cruiser, reflecting their specialized ASW role.[44]

Under the Brezhnev regime, Admiral Gorshkov put forward an expanded program of naval construction. Large missile cruisers and aircraft carriers were constructed, the latter program cumulating in the nuclear-propelled *Ul'yanovsk* of some 75,000 tons full load, which was laid down in 1988 but which was never completed because of the demise of the USSR. Significantly, along with this massive increase of investment in surface warships and naval aviation, Gorshkov continued to stress submarine construction.

**While the United States produced relatively few first-generation "nukes"—eight SSN/SSGN/SSRNs—the Soviet Union built 56 SSN/SSBN/SSGNs of the HEN series. The United States initiated the second-generation nuclear submarines much earlier than did the USSR and—employing the S5W reactor plant almost exclusively—built 100 SSN/SSBNs during a 17-year period.[45] The Soviet Union produced more nuclear submarines of this generation—137 SSN/SSBN/SSGNs during 26 years.**

**The Soviet effort was more remarkable for the broad approach in design. U.S. nuclear submarines of the second generation mostly were based on the highly capable *Thresher.* The Soviet efforts included the high-speed, highly automated Project 705/Alfa SSN, the most innovative submarine of the era, while the Project 661/Papa and Project 670/Charlie SSGNs introduced submerged-launch cruise missiles to submarines. The Alfa and Papa also introduced titanium to submarine construction. And, while U.S. submarines remained acoustically quieter, the Soviets were making progress in this area.**

**From the viewpoint of roles and missions, this was a most significant period, because the Victor SSN marked the shift of Soviet naval strategy from open-ocean, anti-shipping to primarily ASW operations against U.S. strategic missile submarines coupled with the still-important anti-carrier role. Both roles sought to protect the Soviet homeland against U.S. nuclear strikes.**

**The Project 670/Charlie SSGN overcame many of the shortcomings of the earlier Projects 651/Juliett SSG and 675/Echo II SSGN and poised a major threat to Western warships: The Charlie's propulsion plant was more reliable, its missiles were launched while submerged, and the shorter-range missiles eliminated the need for mid-course guidance and flew faster and at lower altitudes than earlier anti-ship missiles. (The submarine's major shortfall was its limited speed, which made it difficult to operate against high-speed carriers.)**

**The second-generation nuclear submarines of both nations provided the majority of the submarines that served in the U.S. and Soviet Navies into the 1990s, carrying the burden of undersea warfare for the second half of the Cold War.**

TABLE 10-2
**Second-Generation Nuclear Submarines**

| | U.S. *Thresher* SSN 593 | U.S. *Tullibee* SSN 597 | U.S. *Sturgeon* SSN 637 | Soviet Project 671 Victor I | Soviet Project 671RTM Victor III | Soviet Project 670 Charlie I | Soviet Project 670M Charlie II |
|---|---|---|---|---|---|---|---|
| Operational | 1961 | 1960 | 1967 | 1967 | 1977 | 1967 | 1973 |
| Displacement | | | | | | | |
| surface | 3,750 tons | 2,177 tons | 4,250 tons | 3,650 tons | 4,750 tons | 3,580 tons | 4,250 tons |
| submerged | 4,310 tons | 2,607 tons | 4,780 tons | 4,830 tons | 5,980 tons | 4,550 tons | 5,270 tons |
| Length | 278 ft 6 in (84.9 m) | 272 ft 9½ in (83.16 m) | 292 ft (89.0 m) | 305 ft (93.0 m) | 351 ft 4 in (107.1 m) | 309 ft 4 in (94.3 m) | 344 ft 1 in (104.5 m) |
| Beam | 31 ft 8 in (9.65 m) | 23 ft 4 in (7.09 m) | 31 ft 8 in (9.65 m) | 34 ft 9 in (10.6 m) | 34 ft 9 in (10.6 m) | 32 ft 6 in (9.9 m) | 32 ft 6 in (9.9 m) |
| Draft | 26 ft (7.93 m) | 19 ft 4 in (5.88 m) | 28 ft 10 in (8.8 m) | 23 ft 7 in (7.2 m) | 26 ft 3 in (8.0 m) | 24 ft 7 in (7.5 m) | 26 ft 7 in (8.1 m) |
| Reactors | 1 S5W | 1 S2C | 1 S5W | 2 VM-4P | 2 VM-4P | 1 VM-4 | 1 VM-4 |
| Turbines | 2 steam | 1 steam | 2 steam | 1 steam | 1 steam | 1 steam | 1 steam |
| horsepower | 15,000 | 2,500 | 15,000 | 31,000 | 31,000 | 18,800 | 18,800 |
| Shafts | 1 | 1 | 1 | 1 | 1 | 1 | 1 |
| Speed | | | | | | | |
| surface | 15 knots | 13 knots | 15 knots | 12 knots | 18 knots | | 12 knots |
| submerged | 27–28 knots | 16 knots | 26–27 knots | 33 knots | 31 knots | 26 knots | 24 knots |
| Test depth | 1,300 ft (400 m) | 700 ft (215 m) | 1,300 ft (400 m) | 1,300 ft (400 m) | 1,300 ft (400 m) | 1,150 ft (350 m) | 1,150 ft (350 m) |
| Torpedo tubes* | 4 533-mm A | 4 533-mm A | 4 533-mm A | 6 533-mm B | 4 533-mm B<br>2 650-mm B | 4 533-mm B<br>2 400-mm B | 4 533-mm B<br>2 400-mm B |
| Torpedoes | 23 | 12 | 23 | 18 | 18+6 | 12+4 | 12+4 |
| Missiles | Harpoon/ SUBROC | nil | Harpoon/ SUBROC | nil | various*** | 8 SS-N-7 | 8 SS-N-9 |
| Complement | 88** | 56 | 99** | 76 | 82 | 90 | 90 |

Notes: * Angled; Bow.
** Increased during service life.
*** Torpedo-tube launched weapons.

11

# The Ultimate Weapon I

*A Project 667BD/Delta III ballistic missile submarine after surfacing in Arctic ice pack. The Arctic Ocean—the backyard of the USSR—was a major operating area for Soviet submarines. SSBNs practiced surfacing through the ice to launch their missiles. Massive ice chunks rest on the submarine's deck.* (Rubin CDB ME)

The Soviet Union had sent ballistic missile submarines to sea before the United States had. However, the U.S. Polaris submarines that went on patrol from the fall of 1960 were superior in all respects to contemporary Soviet ballistic missile submarines. Coupled with the limitations of Soviet anti-submarine forces, the Polaris SSBNs represented an invulnerable strategic striking force.

By 1967 the Polaris force was complete, with 41 SSBNs in commission, carrying 656 missiles. The Soviet Union had earlier constructed eight nuclear and 29 diesel-electric submarines, carrying a total of 104 missiles.[1] In comparison with the Soviet systems, all U.S. missiles were carried in "modern" nuclear-propelled submarines; the U.S. Polaris missiles had greater range, were more accurate, and were capable of underwater launch. The Poseidon missile, introduced in 1971, carried the world's first Multiple-Independently targeted Re-entry Vehicles (MIRVs), up to 14 warheads per missile that could be directed to separate targets within a specific geographic "footprint." (See Chapter 7.)

The Soviet second-generation SSBNs sought to alter this U.S. advantage. Design of the Soviet second-generation SSBNs had begun in the late 1950s at TsKB-18 (Rubin). But in 1959 the Soviet government ordered a halt to production of SSBNs as defense policy underwent a reappraisal at the direction of Nikita Khrushchev. Based on this reappraisal, the Strategic Rocket Forces (SRF) was established on 14 December 1959.[2] This was a separate military service, on par with the Army, Navy, Air Forces, and Air Defense of the Country (PVO), that last having been established in 1954.[3] The SRF was given responsibility for development and operation of all Soviet land-based intermediate and intercontinental-range ballistic missiles.[4]

Academician and chief designer Sergei N. Kovalev attributed the delay in new SSBN construction to a combination of factors, including the

*Sergei N. Kovalev* (Rubin CDB ME)

high priority given to construction of Project 675/Echo II SSGNs at Severodvinsk; the influence of Vladimir N. Chelomei, head of special design bureau OKB-52, who urged the deployment of strategic cruise missiles rather than ballistic missiles for land attack; and difficulties encountered in development of the D-4 missile system (R-21/NATO SS-N-5 Sark), the first underwater-launch missile.[5]

With the halt in the construction of strategic missile submarines, some components for unfinished Hotel SSBNs reportedly were shifted to the Echo SSGN program.[6] Nevertheless, as a result of shortcomings in the Soviet ICBM program and the Cuban missile crisis, in 1962–1963 the decision was made to resume the construction of SSBNs. The lead submarine of the Project 667A (NATO Yankee) was laid down at the Severodvinsk shipyard on 4 November 1964.

## Modern Ballistic Missile Submarines

The first second-generation SSBN design was Project 667, a large submarine intended to carry *eight* R-21/SS-N-5 ballistic missiles.[7] Developed under the direction of chief designer A. S. Kassatsier, the Project 667 design was highly complicated because of the proposal that the R-21 missile undergo final assembly in the submarine and provision of a rotating launcher for the missile.[8] As usual, a number of designs were considered; one design model, in the Central Naval Museum (Leningrad), shows a cruise-missile launcher firing forward through the sail with ballistic missiles fitted amidships.

However, a new generation of smaller Submarine-Launched Ballistic Missiles (SLBMs) was under development that were to have increased range and accuracy as well as improved launch readiness. Development began in 1958 on the first solid-propellant SLBM system, the D-6. Major difficulties were encountered and work on the D-6 system was discontinued in 1961.

At that time work began on a naval version of a land-launch missile with a solid-propellant first stage and second stage, the RT-15M.[9] The submarine Project 613D7 (Whiskey *S-229*) was fitted for test launching the RT-15M; but this missile also proved to be too difficult and was cancelled in mid-1964. The Soviet missile design bureaus were unable to produce a missile comparable to the U.S. Polaris SLBM, which had become operational in late 1960.

In place of solid-fuel SLBMs, the SKB-385 bureau, directed by V. P. Makeyev, proposed the liquid-propellant R-27 missile (NATO SS-N-6 Serb) of the D-5/RSM-25 missile system for the Project 667A/Yankee submarine. Smaller than the R-21, the R-27 was a single-stage, submerged-launch missile with a range of 1,350 n.miles (2,500 km). The missile was test launched from the modified Project 613D5 (Whiskey) submarine, which had been fitted with two launch tubes in 1963–1967; further trials took place with the *K-102,* a modified Project 629/Golf SSB that was redesignated Project 605 when fitted with four launch tubes.

In the early 1960s the possibility of rearming the earlier Project 658 (Hotel) SSBNs with the D-5/R-27 system was considered. In addition, in 1964–1965 in TsKB-16, preliminary designs were considered for Project 687, a high-speed submarine based on Project 705/Alfa with a submerged displacement of about 4,200 tons carrying D-5 SLBMs—the R-27 or R-27K anti-ship variant (see below). At the same time, in SKB-143 work began on the Project 679 SSBN based on Project 671/Victor, also to be armed with the D-5 missile system, schemes that somewhat imitated the U.S. conversion of the *Skipjack* (SSN 585) design to a Polaris configuration. Work ceased on these preliminary SSBN designs before any came to fruition.

Rather, based on the promise of the D-5/R-27 missile, in 1962 TsKB-18 shifted SSBN design efforts to Project 667A. Chief designer for the project was Kovalev, who had been the final chief designer of the Project 658/Hotel SSBN. As construction began at Severodvinsk, preparations also were being made for series production at the Komsomol'sk-na-the-Amur shipyard.

The Project 667A/Yankee had a surface displacement of almost 7,760 tons and an overall length of 420 feet (128 m). Internally the double-hull submarine had ten compartments. Later, after engineering problems cut off men trapped in the after section from the forward portion of a submarine, a two-man escape chamber that could be winched up and down was fitted in the stern of these submarines.[10]

Superficially similar to the early U.S. Polaris SSBN design, the Project 667A/Yankee also had 16 missiles, fitted in two rows aft of the sail; however, the 16 missile tubes were fitted in two compartments (No. 4 and 5), unlike the one-compartment missile arrangement in U.S. submarines. And—a first in Soviet submarines—the SSBN had sail-mounted diving planes; the sail position was closer to the ship's center of gravity to help hold her at the proper depth for missile launching.[11]

Other major differences between the U.S. Polaris and Soviet Yankee SSBNs included the latter having twin VM-2-4 reactors providing steam to OK-700 turbines turning twin propeller shafts. The Soviet submarine also had a greater diving depth and, significantly, could launch missiles at a faster rate, from a greater depth, at higher speeds than could Polaris SSBNs. The 667A/Yankee could launch from depths to 165 feet (50 m), compared to less than half that depth for Polaris launches,[12] and the submarine could be moving at three to six knots, while U.S. missile submarines were required to move considerably slower or—preferably—to hover while launching missiles.[13] The time for prelaunch preparations was approximately ten minutes; the launch time for a salvo of four missiles was 24 seconds. However, there were pauses between salvoes so that at least 27 minutes were required from the launching of the first and last missiles. (The later Project 667BDRM/Delta IV SSBN could launch all 16 missiles within one minute while steaming up to five knots.)

Still, Soviet officials candidly admitted that the Yankee's missiles were inferior to the contemporary Polaris A-3 missile, while their submarines were noisier and had other inferiorities in comparison to U.S. SSBNs. Kovalev observed,

> The submarines of Project 667A from the very beginning were good except for their noisiness. It was not that we did not pay attention to this problem, but simply that in the scientific and technical field we were not prepared to achieve low levels of noise. In the scientific field we poorly appreciated the nature of underwater noise, thinking that if a low-noise turbine reduction gear was made all would be in order. But then, in trials we were convinced that this was not altogether so.[14]

The first Project 667A submarine was launched on 28 August 1966, and the ship was commissioned

*The Project 667A/Yankee was the Soviet Navy's first "modern" strategic missile submarine. While outwardly resembling the U.S. Polaris SSBNs, in actuality the Yankee design was very different. The open "hatch" at the forward end of the sail was for a telescoping satellite antenna device, given the NATO code name Cod Eye.*

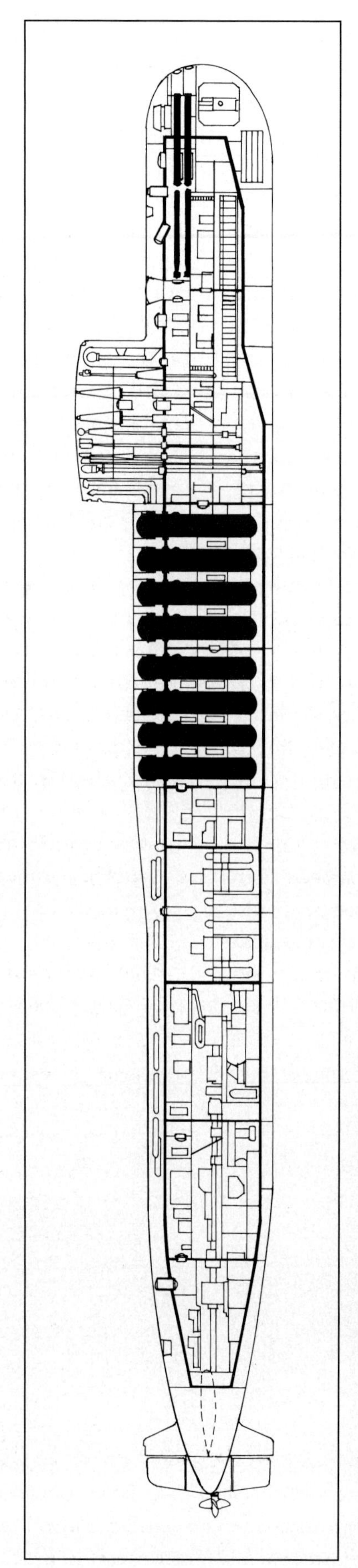

*Project 667A/Yankee SSBN. LOA 420 ft (128.0 m)* (©A.D. Baker, III)

the *Leninets* (K-137) on 5 November 1967. Thirty-four of these submarines were built through 1972—24 at Severodvinsk and 10 at Komsomol'sk, a building rate that exceeded the U.S. Polaris program—that is, approximately 5.5 Polaris per annum compared to 6.8 Yankee SSBNs. And while the U.S. construction of ballistic missile submarines halted at 41 units, Soviet SSBN construction continued.

The first Project 667A/Yankee SSBNs went on combat patrol in the Atlantic in June 1969. Sixteen months later, in October 1970, SSBNs of this class began patrols in the Pacific. These submarines came within missile range—1,350 n.miles—of American coasts. Although their missile range was significantly less than the contemporary Polaris A-3 (i.e., 2,500 n.miles/4,630 km), principal U.S. cities and military bases were along the coasts, whereas the most important Soviet targets were far from the open seas. By 1971 there were regularly two and on occasion up to four Project 667A SSBNs in that western Atlantic within missile range of the United States and one in the eastern Pacific.[15]

While construction of these submarines continued, existing units were upgraded from 1972 to 1983 with the D-5U missile system and the R-27U ballistic missile. This weapon increased range to 1,620 n.miles (3,000 km), with one or three nuclear warheads that would strike the same target, the latter similar to the Multiple Re-entry Vehicle (MRV) warhead of the Polaris A-3 missile. All but one of the 34 ships were refitted with the D-5U and advanced navigation systems during the course of construction or modernization. (The submarines were redesignated 667AU.)

From 1973 to 1976 one of the earlier Project 667A submarines, the *K-140,* was modernized at the Zvezdochka yard at Severodvinsk, being refitted with the first solid-propellant missile, the D-11 system. This was a two-stage RSM-45/R-31 missile (NATO SS-N-17 Snipe). The SLBM had a range of 2,100 n.miles (3,900 km) and demonstrated the Soviet ability to put solid-fuel missiles to sea. As modified, the *K-140*—designated 667AM—accommodated 12 missiles in place of the standard 16. (From 1969 to 1973 a ship design was begun that would carry 16 RSM-45/R-31 missiles; this was Project 999 at TsKB-16 (Volna) under S. M. Bavilin. The missile system was not pursued, and Project 667AM became the only ship fitted with this weapon—the only operational Soviet SLBM with solid-propellant until the RSM-52/SS-N-20 Sturgeon entered service in 1983.)

During the 1970s and early 1980s, there was concern among some Western defense leaders that the Yankee SSBNs on patrol off the U.S. coasts would launch their missiles on "depressed trajectory" or reduced-flight-time profiles. Such a technique could reduce flight times substantially, making strategic bomber bases, command centers, and other time-urgent targets significantly more vulnerable to submarine-launched missiles because of shorter radar warning times.

However, no evidence has been forthcoming from Soviet sources that depressed-trajectory techniques were employed. The U.S. Central Intelligence Agency expressed doubt that such techniques were being developed, noting that while "Theoretically, depressed trajectories can reduce flight times substantially below those normally expected. . . . We believe that the current Soviet SLBM force would require extensive modifications or, more likely, redesign to fly a reduced-flight-time profile."[16] Further, Western intelligence had not observed any Soviet SLBM flights that demonstrated a deliberate attempt to minimize flight time.[17]

Liquid missile propellant continued to cause problems in SSBNs. In 1979 the *K-219* suffered a missile fuel leak, a fire, and an explosion while at sea. The ship survived and a well-trained crew brought her back to port. She was repaired, although the damaged missile tube was not rehabilitated. While carrying 15 nuclear missiles, the *K-219* suffered another missile fuel accident while submerged on patrol some 600 n.miles (1,110 km) off Bermuda on 3 October 1986. The ship came to the surface, "but the crew could not handle the fire. Moreover, unskilled actions led to the sinking of the ship and the death of [four] men."[18] A Soviet merchant ship had tried—unsuccessfully—to tow the submarine. She then took on board the survivors when the *K-219* sank on 6 October. Four crewmen went down with the *K-219.*[19]

A few months later, in March 1987, a strange event transpired in which five submarines sailed from bases along the Kola Peninsula to the area of the Bermuda Triangle in Operation Atrina. Some were SSBNs, others were attack submarines. Although some of the submarines were detected en route, U.S. naval intelligence did not detect the assembly until the submarines reached the objective area; then all were (incorrectly) identified as strategic missile submarines. At the same time, Tu-20 Bear-D reconnaissance-missile guidance aircraft flying from bases on the Kola Peninsula and from Cuba coordinated their operations with the submarines.

The White House was informed of the assembly, and a "crisis" condition briefly existed. The Soviets claimed a "victory."

The SSBN missile patrols continued.

## U.S. Strategic ASW

The appearance of the Project 667A/Yankee SSBN had a profound impact on the U.S. Navy's anti-submarine strategy.[20] Heretofore Western naval strategists looked at the Soviet submarine force as a reincarnation of the U-boat threat of two world wars to Anglo-American merchant shipping.

From the late 1940s, for two decades the U.S. Navy contemplated an ASW campaign in which, in wartime, Soviet submarines would transit through "barriers" en route to attack Allied convoys in the North Atlantic and then return through those same barriers to rearm and refuel at their bases. These barriers—composed of maritime patrol aircraft and hunter-killer submarines guided or cued by the seafloor Sound Surveillance System (SOSUS)—would sink Soviet submarines as they transited, both going to sea and returning to their bases.[21] Also, when attacking Allied convoys, the Soviet submarines would be subjected to the ASW efforts of convoy escorts.

In reality, by the mid-1950s the Soviets had discarded any intention of waging an anti-shipping campaign in a new Battle of the Atlantic. The U.S. Navy's development of a carrier-based nuclear strike capability in the early 1950s and the deployment of Polaris missile submarines in the early 1960s had led to defense against nuclear strikes from the sea becoming the Soviet Navy's highest priority mission. New surface ship and submarine construction as well as land-based naval and, subsequently, Soviet *Air Forces* aircraft were justified on the basis of destroying U.S. aircraft carriers and missile submarines as they approached the Soviet homeland.

When the Project 667A/Yankee SSBNs went to sea in the late 1960s, the Soviet Navy was given another high-priority mission: Strategic (nuclear) strike against the United States *and* the protection of their own missile submarines by naval forces.

The Yankee SSBNs severely reduced the effectiveness of the U.S. Navy's concept of the barrier/convoy escort ASW campaign. These missile submarines—which could carry out nuclear strikes against the United States—would be able to pass through the barriers in peacetime and become lost in the ocean depths, for perhaps two months at a time. Like the U.S. Polaris SSBNs, by going slow, not transmitting radio messages, and avoiding Allied warships and shipping, they might remain undetected once they reached the open sea.

If the Soviets maintained continuous SSBN patrols at sea (as did the U.S. Navy) there would always be some ballistic missile submarines at sea. During a period of crisis, additional Soviet SSBNs would go to sea, passing through the barriers without Allied ASW forces being able to attack them.

Efforts to counter these submarines required the U.S. Navy to undertake a new approach to ASW. A variety of intelligence sources were developed to detect Soviet submarines leaving port, especially from their bases on the Kola Peninsula. These included High-Frequency Direction Finding (HF/DF) facilities in several countries, Electronic Intelligence (ELINT) intercept stations in Norway and, beginning in the 1950s, Norwegian intelligence collection ships (AGI) operating in the Barents Sea.[22] Commenting on the AGI *Godoynes,* which operated under the code name Sunshine in 1955, Ernst Jacobsen of the Norwegian Defense Research Establishment, who designed some of the monitoring equipment in the ship, said that the *Godoynes*—a converted sealer—was "bursting at the seams with modern American searching equipment, operated by American specialists."[23] The Central Intelligence Agency sponsored the ship and other Norwegian ELINT activities. The Norwegians operated a series of AGIs in the ELINT role in the Barents Sea from 1952 to 1976. In the Pacific, there was collaboration with Japanese intelligence activities as well as U.S.

HF/DF and ELINT stations in Japan to listen for indications of Soviet submarine sorties.

From the early 1960s U.S. reconnaissance satellites also could identify Soviet submarines being prepared for sea. Once cued by such sources, SOSUS networks emplaced off the northern coast of Norway and in the Greenland-Iceland-United Kingdom (GIUK) gaps would track Soviet SSBNs going to sea. Presumably, SOSUS networks in the Far East were cued by similar ELINT and other intelligence sources.

Directed to possible targets by SOSUS, U.S. attack submarines would attempt to trail the ballistic missile submarines during their patrols. These SSBN trailing operations were highly sensitive and until the late 1970s were not referred to in even top secret U.S. Navy documents. Navy planning publications—highly classified—began to discuss trailing operations at that time as the U.S. understanding of the Soviet submarine roles in wartime began to change.

Beginning in the late 1960s, the Soviet Union gained an intelligence source in the U.S. Navy that could provide details of U.S. submarine operations, war plans, communications, and the SOSUS program. This source was John A. Walker, a Navy communications specialist who had extensive access to U.S. highly classified submarine material. Based on Walker's data and other intelligence sources, the Soviets restructured their own naval war plans. The previous American perception was that the U.S. Navy would win "easily, overwhelmingly," according to a senior U.S. intelligence official.[24] "From the late 1970s . . . we obtained special intelligence sources. They were available for about five years, until destroyed by [Aldrich] Ames and others." Based on those sources, "we learned that there would be more holes in our submarines than we originally thought—we had to rewrite the war plan."[25]

In the mid-1980s U.S. officials began to publicly discuss the Western anti-SSBN strategy. Probably the first official pronouncement of this strategy was a 1985 statement by Secretary of the Navy John Lehman, who declared that U.S. SSNs would attack Soviet ballistic missile submarines "in the first five minutes of the war."[26] In January 1986, the Chief of Naval Operations, Admiral James D. Watkins, wrote that "we will wage an aggressive campaign against all Soviet submarines, including ballistic missile submarines."[27] Earlier Watkins had observed that the shallow, ice-covered waters of the Soviet coastal seas were "a beautiful place to hide" for Soviet SSBNs.[28]

Only in 2000 would the U.S. Navy reveal some of the details of trailing Soviet SSBNs. In conjunction with an exhibit at the Smithsonian Institution's Museum of American History commemorating one hundred years of U.S. Navy submarines, heavily censored reports of two U.S. trailing operations were released: the trail of a Yankee SSBN in the Atlantic,[29] and that of a Project 675/Echo II SSGN in the Pacific by SSNs.[30] This particular Yankee trailing operation—given the code name Evening Star—began on 17 March 1978 when the USS *Batfish* (SSN 681) intercepted a Yankee SSBN in the Norwegian Sea. The *Batfish,* towing a 1,100-foot (335-m) sonar array, had been sent out from Norfolk specifically to intercept the SSBN, U.S. intelligence having been alerted to her probable departure from the Kola Peninsula by the CIA-sponsored Norwegian intelligence activities and U.S. spy satellites. These sources, in turn, cued the Norway-based SOSUS array as the Soviet missile submarine sailed around Norway's North Cape.

After trailing the Soviet submarine for 51 hours while she traveled 350 n.miles (650 km), the *Batfish* lost contact during a severe storm on 19 March. A P-3 Orion maritime patrol aircraft was dispatched from Reykjavik, Iceland, to seek out the evasive quarry. There was intermittent contact with the submarine the next day and firm contact was reestablished late on 21 March in the Iceland-Faeroes gap.

The trail of the SSBN was then maintained by the *Batfish* for 44 continuous days, the longest trail of a Yankee conducted to that time by a U.S. submarine.[31] During that period the Yankee traveled 8,870 n.miles (16,440 km), including a 19-day "alert" phase, much of it some 1,600 n.miles (2,965 km) from the U.S. coast, little more than the range of the submarine's 16 RSM-25/R-27U missiles.

The *Batfish* report provides day-to-day details of the Yankee's patrol and the trailing procedures. Significantly, the SSBN frequently used her MGK-100 Kerch active sonar (NATO designation Blocks of Wood).[32] This sonar use and rigidly scheduled maneuvers by the Soviet submarine, for example, to clear the "baffles," that is, the area behind the submarine, and to operate at periscope depth twice a day continuously revealed her position to the trailing SSN.[33] The *Batfish* ended her trailing operation as the Yankee SSBN reentered the Norwegian Sea.

The routine repetitiveness of the "target" was used to considerable advantage by the *Batfish.* Certain maneuvers indicated a major track change or impending periscope depth operations. But would such predictable maneuvers have been used in wartime? The repeated use of her sonar in the *Batfish* operation was highly unusual for a Yankee SSBN on patrol. Would the missile submarine have employed countermeasures and counter-tactics to shake off the trailing submarine during a crisis or in wartime? "You bet they would change their tactics and procedures," said the commanding officer of the *Batfish,* Commander Thomas Evans.[34]

There are examples of tactics being employed by Soviet submarines to avoid U.S.-NATO detection. Among them have been transiting in the proximity of large merchant ships or warships in an attempt to hide their signatures from Western sensors, and reducing noise sources below their normal level when transiting in areas of high probability of SOSUS detection.[35] When the Russian cruise missile submarine *Kursk* was destroyed in August 2000, a Russian SSBN, believed to be a Project 667BDRM/Delta IV, may have been using the fleet exercise as a cover for taking up a patrol station without being detected by U.S. attack submarines in the area. (Another Delta IV, the *Kareliya* [K-18] was participating in the exercise at the time.)

Not all U.S. trailing operations were successful. Periodically Soviet SSBNs entered the Atlantic and Pacific without being detected; sometimes the trail was lost. A noteworthy incident occurred in October 1986 when the U.S. attack submarine *Augusta* (SSN 710) was trailing a Soviet SSN in the North Atlantic. The *Augusta* is reported to have collided with a Soviet Delta I SSBN that the U.S. submarine had failed to detect. The *Augusta* was able to return to port; but she suffered $2.7 million in damage. The larger Soviet SSBN suffered only minor damage and continued her patrol.

(U.S. and Soviet submarines occasionally collided during this phase of the Cold War, many of the incidents undoubtedly taking place during trail operations. Unofficial estimates place the number of such collisions involving nuclear submarines at some 20 to 40.)

It is believed that the limited range of the Yankee's RSM-25/SS-N-6 missile forced these submarines to operate relatively close to the coasts of the United States. Under these conditions, and upon the start of hostilities, the trailing U.S. submarines would attempt to sink the Soviet SSBNs as they released their first missiles (or, under some proposals, when their missile tube covers were heard opening). If feasible, the U.S. submarines would call in ASW aircraft or surface ships, and there were proposals for U.S. surface ships to try to shoot down the initial missiles being launched, which would reveal the location of the submarine to ASW forces. These SLBM shoot-down proposals were not pursued.[36]

U.S. anti-SSBN efforts again were set back in 1972 when the first Project 667B/Delta I ballistic missile submarine went to sea. This was an enlarged Yankee design carrying the RSM-40/R-29 (NATO SS-N-8 Sawfly) missile with a range of 4,210 n.miles (7,850 km). This missile range enabled Delta I SSBNs to target virtually all of the United States while remaining in Arctic waters and in the Sea of Okhotsk. In those waters the SSBNs could be defended by land-based naval aircraft as well as submarines and (in ice-free waters) surface warships. These SSBNs were equipped with a buoy-type surfacing antenna that could receive radio communications, target designations, and satellite navigational data when the ship was at a considerable depth.

Further, communications with submarines in Arctic waters were simplified because of their proximity to Soviet territory. The use of surface ships and submarines for communications relay were also possible. It was possible that *civilian* nuclear-propelled icebreakers—which were armed on their sea trials—were intended to provide such support to submarines in wartime.[37]

Also, having long-range missiles that would enable SSBNs to target the United States from their bases or after short transits fit into the Soviet Navy's procedure of normally keeping only a small portion of their submarine fleet at sea, with a majority of their undersea craft held in port at a relatively high state of readiness. These submarines—of all types—would be "surged" during a crisis.

This procedure was radically different than that of the U.S. Navy, which, for most of the Cold War, saw up to one-third of the surface fleet and many SSNs forward deployed, with more than half of the SSBN force continuously at sea—at a cost of more personnel and more wear-and-tear on ships.

The Soviet SSBN operating areas in the Arctic and Sea of Okhotsk—referred to a "sanctuaries" and "bastions" by Western intelligence—were covered by ice for much of the year and created new challenges for Western ASW forces. Attack submarines of the U.S. *Sturgeon* (SSN 637) class were well suited for operating in those areas, being relatively quiet and having an under-ice capability.[38] However, the Arctic environment is not "ASW friendly": communications—even reception—are extremely difficult under ice; passive sonar is degraded by the sounds of ice movement and marine life; and under-ice acoustic phenomena interfere with passive (homing) torpedo guidance. Also, the Arctic environment, even in ice-free areas, is difficult if not impossible for Allied ASW aircraft and surface ship operations.

The Soviet SSBN force thus became an increasingly effective strategic strike/deterrent weapon, especially when operating in the sanctuaries or bastions.

## The Delta Series

In 1963 work began on an advanced, long-range SLBM, the D-9 system with the R-29 liquid-propellant missile (NATO SS-N-8 Sawfly). That same year, at TsKB-16, A. S. Smirnov and N. F. Shulzhenko developed the preliminary proposals for Project 701, a strategic missile submarine for which the bureau initially examined variations of *diesel-electric* as well a nuclear propulsion. The proposed Project 701 SSBN would have a surface displacement of some 5,000 tons and carry six R-29 SLBMs.

As a result of their work, in 1964 the Navy decided to rearm a Project 658/Hotel SSBN as a test platform for the new missile. That submarine, the *K-145*, was lengthened and fitted to carry six R-29 missiles (designated Project 701/Hotel III). She was followed by the Project 629/Golf SSB *K-118*, which also was lengthened and rearmed with six R-29 missiles (designated Project 601/Golf III). The conversions were undertaken at Severodvinsk and at the adjacent Zvezdochka yards, respectively.

A completely new-design SSBN to carry the R-29 was rejected by the Navy, and in 1965 work began at TsKB-18 under the direction of S. N. Kovalev on Project 667B (NATO Delta I). This was an enlarged Project 667A/Yankee SSBN to carry the R-29. Although only 12 missiles were carried, their initial range of 4,210 n.miles (7,850 km) would permit a major revision in Soviet naval strategy and operations.[39]

As Delta I SSBN production was succeeding the Yankee SSBN in the building halls at Severodvinsk and Komsomol'sk, the Soviet and U.S. governments concluded the world's first nuclear weapons agreement—SALT I.[40] In 1974 the Soviet SSBN construction program overtook the U.S. SSBN force of 41 "modern" submarines; the following year the number of Soviet SLBMs in modern submarines overtook the 656 missiles carried in U.S. submarines.[41] Although the United States retained the lead in independently targetable SLBM warheads—primarily because the Poseidon missiles could carry up to 14 MIRVs—and in missile accuracy, the larger Soviet SLBM warheads led to the gross warhead yield of Soviet SLBMs, also exceeding that of U.S. SLBMs by the early 1970s.[42]

In addition, older Project 629/Golf and 658/Hotel submarines remained in service, although their missiles were targeted against regional threats; for example, six Golf SSBs were shifted to the Baltic Fleet. (Several of these older submarines also were used as test beds for newer SLBMs.)

Under the SALT I agreement, signed in 1972, both nations would accept limitations on the number of nuclear delivery "platforms" as well as on nuclear-armed strategic missiles. The United States, already planning the Trident missile program, accepted a limit of 44 SSBNs carrying 710 SLBMs. The Soviet Union accepted a limit of 62 "modern" SSBNs—the number reported to be in service and under construction at the time—carrying 950 SLBMs. However, the Soviet SLBM number could exceed 740 only as equal numbers of land-based ICBMs or missiles on older submarines (deployed before 1964) were dismantled.

*The* K-118 *was a Project 629 submarine that was lengthened and rearmed with six R-29/SS-N-8 Sawfly ballistic missiles, becoming the single Project 601/Golf III. Several early Soviet SSB/SSBNs were modified to test and evaluate advanced missile systems.* (Malachite SPMBM)

Thus, a finite limitation was placed on the strategic submarine forces of both nations. This was the first international agreement intended to limit warship construction since the London naval arms conference of 1930. Although SALT I was applicable only to Soviet and U.S. strategic missile submarines, Soviet officials called attention to the fact that Britain, China, and France—all potential enemies—also were developing SSBNs.

*Project 667B/Delta I was an enlarged Yankee SSBN, armed with a longer-range and more capable missile. Note the increased height of the "turtle back" over the missile compartment. The Delta I had only 12 missile tubes; all other 667-series SSBNs had 16 tubes, except for the single Project 667AM/Yankee II conversion.*

Meanwhile, construction of the Project 667B/Delta I submarines proceeded at a rapid rate. The lead ship, the *K-279,* was laid down at Severodvinsk in 1971; construction was rapid, and she was placed in commission on 27 December 1972. Ten ships were built at Severodvinsk and eight at Komsomol'sk through 1977. These submarines had a surface displacement of some 9,000 tons and had been lengthened 36 feet (11 m) over the 667A to accommodate the larger missiles (albeit only 12 being fitted). In most other respects the two designs were similar.

Even as these Project 667B submarines were being built, Severodvinsk additionally constructed four enlarged Project 667BD/Delta II submarines, again slightly larger to accommodate 16 of the R-29/SS-N-8 missiles. These submarines were too large to be built at Komsomol'sk, located some 280 miles (450 km) inland from the mouth of the Amur River. The *K-182* and three sister SSBNs were completed in 1975. No additional units were built because of the shift to the longer-range RSM-50 missile system.

A Delta II, the *K-92,* launched two SLBMs while operating under the Arctic ice pack in 1983. This marked the world's first under-ice ballistic missile launch.

*The Project 667BD/Delta II SSBN had a more gentle slope from her "turtle back" to the after-hull section compared to the "steps" in the Delta I. All 667-series submarines had sail-mounted diving planes, the only Soviet nuclear submarines with that feature. The tracks along the deck are safety rails for men working on deck.*

*A Project 667BDR/Delta III showing the massive missile structure. There are three rows of free-flood holes along the base of the missile structure, speeding the flooding of the area for diving and draining when surfacing. The Delta III was the oldest Russian SSBN in service at the start of the 21st Century.*

In all, during the decade from late 1967 to 1977, the two Soviet SSBN building yards produced 56 submarines:

34 Project 667A/Yankee
18 Project 667B/Delta I
4 Project 667BD/Delta II

This average of 5½ ballistic missile submarines per year was contemporaneous with a large SS/SSN/SSGN building rate.

The early Project 667B submarines were used to develop tactics for the launching of ballistic missiles from pierside (the first experiments being conducted in 1975), missile launching from a bottomed position in shallow water, and surfacing through the Arctic ice and subsequent launch of missiles. These tactics significantly increased survivability of Soviet SSBNs; with the increase in the missile range, these missile ships became "the least vulnerable strategic force component."[43]

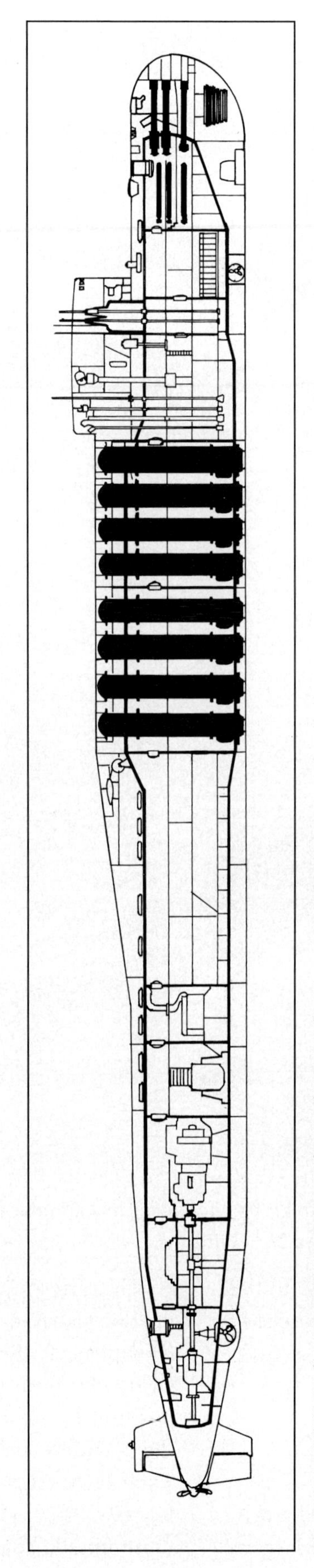

*Project 667BDR/Delta III SSBN. LOA 508 ft 6 in (155.0 m)* (©A.D. Baker, III)

The Soviet modus operandi for these submarines, according to U.S. intelligence sources, was described as:

> The day-to-day disposition of Soviet SSBNs is governed by the wartime requirement to generate maximum force levels on short notice. The Soviet Navy seeks to maintain 75 percent of its SSBNs in operational status, with the remaining 25 percent in long-term repair. . . . Every operational SSBN could probably be deployed with three weeks' preparation time. To maintain this high state of readiness, a relatively small portion of the modern SSBN force—typically about 25 percent, or 14 units—is kept at sea. However, additional [Delta] and [Yankee] class units are probably kept in a high state of readiness in or near home port in order to be ready to fire their missiles on short notice.[44]

At TsKB-18 work began in 1972 on Project 667BDR/Delta III, again under the direction of the indefatigable Kovalev. This submarine would carry the first Soviet SLBM with MIRV warheads for striking individual targets, the two-stage, liquid-propellant RSM-50/R-29R (NATO SS-N-18 Stingray). This missile could deliver a single warhead to a range of 4,320 n.miles (8,000 km) or three or seven MIRV warheads to a range of at least 3,500 n.miles (6,500 km). Sixteen missiles were carried.

The Project 667BDR—which would be the penultimate manifestation of the basic design—would have a surface displacement of almost 10,600 tons with a length of 508½ feet (155 m). Although still propelled by a twin-reactor plant with twin screws, the stern configuration was extensively modified to provide separate propeller bodies for each shaft, a move intended to improve quieting and increase speed. Indeed, each successive class of SSBNs was quieter, albeit at significant expense.[45] And, as nuclear submarines—including SSBNs—went into the yards for overhaul and refueling, efforts were made to further quiet their machinery and in other ways reduce their acoustic signature. According to a senior U.S. flag officer, "The Soviets are incorporating quieting technology in every new and overhauled submarine, and this has seriously reduced our ability to detect them."[46]

Also, with regard to habitability, the Delta III and all subsequent Soviet nuclear submarines were provided with a specially ventilated "smoking room" for those men addicted to tobacco. Historically all navies forbade smoking in submarines because of atmospheric contamination.

From 1976 to 1982 the Severodvinsk yard, at that time the only Soviet yard capable of constructing "big" SSBNs, built 14 of these ships. These brought to 70 the number of modern Soviet SSBNs carrying a total of 1,044 missiles.

## Tactical Ballistic Missiles

Soviet ballistic missile submarines initially were developed for attacking strategic land targets. The early patrol areas for the Project 629/Golf and Project 658/Hotel ballistic missile submarines in the Atlantic were generally believed to be "holding areas" from which the submarines would move to within missile range of U.S. coastal targets in time of crisis or war.

Table 11-1
**Submarine-Launched Ballistic Missiles**

| Missile | R-27 NATO SS-N-6 | R-29 NATO SS-N-8 Sawfly | R-29R NATO SS-N-18 Stingray |
|---|---|---|---|
| Operational | 1968 | 1974 | 1979 |
| Weight | 31,300 lb (14,200 kg) | 73,400 lb (33,300 kg) | 77,820 lb (35,300 kg) |
| Length | 31 ft 6 in (9.61 m) | 42 ft 8 in (13.0 m) | 46 ft 3 in (14.1 m) |
| Diameter | 59 in (1.5 m) | 71 in (1.8 m) | 71 in (1.8 m) |
| Propulsion | liquid-propellant 1 stage | liquid-propellant 2 stage | liquid-propellant 2 stage |
| Range | 1,350 n.miles (2,500 km) | 4,240 n.miles (7,800 km) | 4,300 n.miles* (8,000 km) |
| Guidance | inertial | inertial | inertial |
| Warhead** | 1 RV*** 1 MT | 1 RV | 1 or 3 or 7 MIRV |

Notes: * Range with single RV; range with 3 to 7 RVs is 3,510 n.miles (6,500 km).
** RV = Re-entry Vehicle.
*** The R-27U variant carried three RVs (200 KT each).

However, these holding areas were located on the great circle routes used by U.S. aircraft carriers to steam from East Coast ports to the Norwegian Sea and the Mediterranean Sea for forward deployments. Further, these SSB/SSBN holding areas were regularly overflown by Tu-20 Bear-D aircraft that provided targeting data to Soviet submarines (see page 97).[47]

Some Western analysts argued that the presumed *strategic submarine* holding areas were in fact anti-carrier positions. Their R-13/SS-N-4 and R-21/SS-N-5 Sark missiles fitted with nuclear warheads could be used as area bombardment weapons against carrier task forces.[48] Also there were instances when Project 675/Echo II cruise missile submarines sometimes occupied those SSB/SSBN holding areas. In the event, in 1962 the Soviet government approved the development of SLBMs for use against surface ships, especially U.S. aircraft carriers that threatened nuclear strikes against the USSR. At the time the improved R-27/SS-N-6 SLBM was under development in Makeyev's SKB-385 bureau. The decision was made to provide the nuclear missile with a terminal guidance for the anti-ship role, designated R-27K.[49] The R-27 and R-27K were similar with the 27K having a second stage containing the warhead and terminal guidance.

After preliminary research, the TsKB-16 design bureau was directed to develop a suitable platform—Project 605—based on the reconstruction of a Project 629/Golf-class SSB to test R-27K missiles. V. V. Borisov was the chief designer for this and other Project 629 upgrades. The design was not approved by the Navy until October 1968, and the following month the submarine *K-102* (formerly *B-121*) entered the Zvezdochka shipyard. There the submarine was refitted with four launch tubes, interchangeable for the R-27 and R-27K missiles. To accommodate the missiles as well as improved target acquisition, fire-control, navigation, and satellite link equipment, the *K-102* was lengthened to 380½ feet (116 m), an increase of 56 feet (17.1 m), and her surface displacement was increased by almost 780 tons to 3,642 tons. (Submerged displacement was approximately 4,660 tons.)

Submarine trials began on 9 December 1972, but were curtailed on the 18th because the "Record-2" fire control system had not yet been delivered. Other aspects of the *K-102* trials continued. Even after the control system was installed, there were several malfunctions that delayed completion of the missile-firing trials. From 11 September to 4 December 1973 the *K-102* belatedly carried out test firings of the R-27K missile, using both the Record-2 and the Kasatka B-605 target acquisition system, which used satellite tracking data. Trials of the weapon system continued with the *K-102* firing R-27 strategic (land-attack) as well as R-27K anti-ship missiles. The submarine and missile system were accepted for service on 15 August 1975.

The R-27K/SS-NX-13 would be targeted before launch with data provided from aircraft or satellites tracking U.S. and British aircraft carriers and other surface ships. Western intelligence credited the missile with a range of 350 to 400 n.miles (650–740 km) and a homing warhead that had a terminally guided Maneuvering Re-entry Vehicle (MaRV) to home on targets within a "footprint" of about 27 n.miles (50 km). An accuracy of 400 yards (370 m) was reported. Warheads between 500 kilotons and one megaton could be fitted.

The anti-ship weapons system was fitted only in the Project 605/Golf SSB. Although the missile could be accommodated in the Project 667A/Yankee SSBNs, those submarines were not fitted with the requisite Kasatka B-605 target acquisition system. The R-27K missile did not become operational, because the Strategic Arms Limitation Talks (SALT) agreements of the 1970s would count every SLBM tube as a strategic missile regardless of whether it held a land-attack or anti-ship (tactical) missile.[50] Western intelligence had expected the R-27K to become operational about 1974.

Some Western analysts also postulated that the missile also might be employed as an *anti-submarine* weapon.[51] This evaluation fit with the U.S. shift of sea-based strategic strike forces from carrier-based aircraft to missile-armed submarines, and the related shift in the Soviet Navy's emphasis from anti-carrier warfare to anti-submarine warfare in the 1960s. Kovalev would say only that ASW calculations for the R-27K were made.[52] Other submarine designers would say only that the ASW concept "was not real."[53] (When discussing the use of land-based ballistic missiles against submarines, the source called that proposal "science fiction," although, he observed, "very serious people believed in this [role for ICBMs].") In the event, the

R-27K was not deployed. The Chairman of the U.S. Joint Chiefs of Staff observed:

> The SS-NX-13 is a tactical ballistic anti-ship missile. It may have been intended for deployment in Yankee class SSBNs. It has not been tested since November 1973 and is not operational. However, the advanced technology displayed by the weapon is significant and the project could be resurrected.[54]

The Yankee and Delta SSBNs continued to operate in the strategic role, making regular patrols. Although some Yankees were withdrawn from the SSBN role for conversion to specialized cruise missile (SSGN) and research configurations, many remained in the SSBN role for the duration of the Soviet era. Even after Delta SSBNs took over the majority of strategic submarine patrols, Yankee SSBNs periodically patrolled in the Atlantic and Pacific, moving closer to U.S. coasts during periods of tension between the two super powers. For example, in June 1975 a Yankee came within 300 n.miles (555 km) of the U.S. East Coast. A U.S. national intelligence report noted:

> Because the warning time for missiles launched from a submarine at such a close-in location would be very short, some U.S. targets—such as strategic bombers—would become quite vulnerable unless relocated farther inland or maintained at a higher state of readiness.[55]

Significantly, with the SSBN operations in bastions, the U.S. Navy increased torpedo-attack submarine operations in those waters, primarily to determine Soviet patrol patterns, operating techniques, submarine noise signatures, and other characteristics. As noted above, operations in those waters were difficult for U.S. submarines. Periodically there were "incidents," with several collisions between U.S. submarines and Soviet SSNs, some of the latter operating in support of missile submarines.

The Soviet SSBN patrol rates continued at about 15 percent of the force during the latter stages of the Cold War.[56] The majority of the submarines not on patrol were pierside at bases along the Kola Peninsula, in the Far East at Pavlovsk near Vladivostock, and at Petropavlovsk on the Kamchatka Peninsula, and were kept at a relatively high state of readiness for going to sea.[57] Although their reliability and readiness were significantly less than comparable U.S. strategic missile submarines, a U.S. admiral who visited a 21-year-old Delta I SSBN at Petropavlovsk in February 1997 stated:

> Clearly the capabilities of this ship were very limited and it will soon be decommissioned. But, the crew and her captain seemed committed and proud of what they were doing with the tools they [had] been given. As we all know, the Russian submarine force has some very impressive platforms and ships like those in the hands of a well trained crew will remain a force to be reckoned with for years to come.[58]

**Soviet development of "modern" ballistic missile submarines had been delayed for a variety of reasons—establishment of the Strategic Rocket Forces as a separate military service, the influence of cruise missile designer Chelomei in the decade after the death of Stalin, and difficulties in producing solid-propellant and underwater-launch missiles. Thus, the first modern Soviet SSBN—Project 667A/Yankee—went to sea in late 1967, eight years after completion of the first U.S. Polaris submarine.**

**Like the United States, once the decision was made and the technology made available to produce a modern SSBN, the Soviet Union undertook mass production of these ships. Similarly, the basic Project 667A design—as in the U.S. Polaris program—periodically was updated, evolving into the Project 667B-series/Delta SSBN classes.**

**However, whereas the United States halted Polaris SSBN production at 41 ships, the Soviet construction program appeared to be open-ended and to be limited in the near-term only by strategic arms agreements. (In the longer term, with overwhelming numerical superiority in SLBMs as well as ICBMs, the Soviets probably would have emphasized other nuclear as well as conventional military forces.)**

Further, the geography of the USSR—long considered a limitation to naval operations because of the remote locations of major bases—became an asset in the deployment and protection of SSBNs carrying long-range missiles. Their longer-range SLBMs also permitted the Soviets to keep more of their force in port at a high state of readiness rather than on deterrent patrols at sea, as were U.S., British, and French SSBNs. Still, some in the West—including senior submarine flag officers—would contend that the U.S. Navy's attack submarines "chased" the Soviet Navy into the Arctic and Sea of Okhotsk "bastions."[59]

The decision to retain liquid propellants for SLBMs and to exploit their range flexibility made it obvious that Soviet missile submarines would operate where they could be more readily protected while making it difficult for Allied anti-submarine forces to reach them. (Indeed, a similar logic led to continuing U.S. efforts to increase SLBM range.) Further, according to a U.S. CIA analysis, long-range missiles also provided increased accuracy (because the submarine could operate in home waters where more navigational aids were available) as well as a higher degree of strategic readiness by reducing the time submarines required to reach launch positions, thereby reducing the number of SSBNs needed to be kept at sea.[60]

TABLE 11-2
**Strategic Missile Submarines**

| | Project 667A Yankee | Project 667B Delta I | Project 667BD Delta II | Project 667BDR Delta III |
|---|---|---|---|---|
| Operational | 1967 | 1972 | 1975 | 1976 |
| Displacement | | | | |
| surface | 7,760 tons | 8,900 tons | 10,500 tons | 10,600 tons |
| submerged | 9,600 tons | 11,000 tons | 13,000 tons | 13,000 tons |
| Length | 420 ft (128.0 m) | 456 ft (139.0 m) | 506 ft 9 in (154.5 m) | 508 ft 6 in (155.0 m) |
| Beam | 38 ft 4 in (11.7 m) | 38 ft 4 in (11.7 m) | 38 ft 4 in (11.7 m) | 38 ft 4 in (11.7 m) |
| Draft | 26 ft (7.9 m) | 27 ft 6 in (8.4 m) | 28 ft 2 in (8.6 m) | 28 ft 6 in (8.7 m) |
| Reactors | 2 VM-2-4 | 2 VM-4B | 2 VM-4B | 2 VM-4S |
| Turbines | 2 | 2 | 2 | 2 |
| horsepower | 52,000 | 52,000 | 52,000 | 60,000 shp |
| Shafts | 2 | 2 | 2 | 2 |
| Speed | | | | |
| surface | 15 knots | 15 knots | 15 knots | 14 knots |
| submerged | 28 knots | 26 knots | 24 knots | 24 knots |
| Test depth | 1,475 ft (450 m) | 1,475 ft (450 m) | 1,475 ft (450 m) | 1,300 ft (400 m) |
| Missiles | 16 R-27/SS-N-6 | 12 R-29/SS-N-8 | 16 R-29/SS-N-8 | 16 R-29R/SS-N-18 |
| Torpedo tubes* | 4 533-mm B 2 400-mm B | 4 533-mm B 2 400-mm B | 4 533-mm B 2 400-mm B | 4 533-mm B 2 400-mm B |
| Torpedoes | 12 + 8 | 12 + 6 | 12 + 6 | 12 + 6 |
| Complement | 120 | 120 | 135 | 130 |

Notes: *Bow.

**12**

# The Ultimate Weapon II

*The* Ohio*—the first submarine built to carry the Trident SLBM—approaching Bangor, Washington, the West Coast base for strategic missile submarines. This foreshortened perspective belies the submarine's huge size. This design carries 24 ballistic missiles, more than any other SSBN design.* (U.S. Navy)

The Soviet buildup of strategic nuclear weapons in the 1960s—both ICBMs and SLBMs—became a major concern to U.S. defense officials. Beyond the numbers of Soviet missiles, the development of improved missile guidance and multiple warheads, officials feared, could place the U.S. ICBMs at risk to a surprise first strike by Soviet land-based missiles. In addition, there were indications of Soviet development of an Anti-Ballistic Missile (ABM) system that would reduce the effectiveness of U.S. strategic missiles that survived a Soviet first strike. At the same time, the U.S. Navy was developing the Poseidon SLBM to succeed the Polaris missile, while the Air Force was promoting a very large ICBM, at the time known as WS-120A.[1]

In response to these concerns and the lack of coordination of the new strategic missile programs, Secretary of Defense Robert S. McNamara established the so-called Strat-X study in late 1966 to

> characterize U.S. alternatives to counter the possible Soviet ABM deployment and the Soviet potential for reducing the U.S.

> assured-destruction-force effectiveness during the 1970s. It is desired that U.S. [missile] alternatives be considered from a uniform cost-effectiveness base as well as from solution sensitivity to various Soviet alternative actions.[2]

The Start-X study, which included Air Force and Navy officers as well as civilian scientists and engineers, marked the first time that U.S. strategic weapon requirements had been addressed in an integrated and analytical manner. The study took a half year in which time about 125 candidate missile systems were seriously considered—of which *only two were sea-based.*[3] (Some of the land-based options were *waterborne,* with proposals to mount ICBMs on barges on existing waterways or on a new series of canals to make them mobile and hence more survivable.)

The survivability of U.S. missiles from a Soviet first-strike attack was given particular consideration by the Strat-X participants, hence mobile ICBMs—moved by truck, railroad, and even in aircraft—were particularly attractive.[4] Two of the most expensive methods of enhancing ICBM survivability were to provide ABM interceptors to defend the weapons and to provide hardened missile silos that could survive a nearby detonation of a Soviet warhead.

The two sea-based candidates proposed carrying missiles in surface ships and in a new generation of nuclear-propelled submarines. ("Suboptions" of Strat-X looked into fitting the new Poseidon missile in up to 31 of the Polaris submarines plus building 20 to 25 new submarines to carry Poseidon. This probably would have been the least-costly method of reaching the Strat-X warhead weight objective with a high degree of weapon survivability.)

The final Strat-X recommendations to Secretary McNamara proposed the consideration of four new missile systems, which were determined as much by interservice politics as by analysis. Two were land-based and two were sea-based: (1) ICBMs in hardened underground silos, (2) land-mobile ICBMs, (3) a Ship-based Long-range Missile System (SLMS), and (4) an Undersea Long-range Missile System (ULMS). Thus, rather than providing the *best* strategic options—as Secretary McNamara had requested—Strat-X proposed a set of options balanced between land-based (Air Force) and seagoing (Navy) alternatives.

The SLMS option consisted of specialized, merchant-type ships carrying long-range missiles that could operate in U.S. coastal waters, at sea in company with other naval ships, or in merchant shipping lines. They also could launch their missiles while in ports or harbors.[5] Although surface ships were far more vulnerable than were submarines, the argument could be made that they were less vulnerable than fixed, land-based ICBMs and were the lowest cost of the proposed four options for attaining a fixed amount of missile payload.

Significantly, the ULMS submarine option was the only Strat-X candidate to be developed. From its inception, ULMS was envisioned as the successor to the Polaris and Poseidon. At the time the oldest SSBNs were scheduled to begin being retired in the early 1980s. The ULMS submarine—with missile ranges of more than 6,000 n.miles (11,100 km)—would have a very high at-sea to in-port time ratio, would be designed for rapid missile modernization, and would have enhanced noise-reduction features.

The last was particularly significant. Although U.S. Navy officials steadfastly denied that the Polaris-Poseidon submarine force was vulnerable to detection or attack by Soviet forces, there was continued concern over possible Soviet advances in ASW.

## Soviet Strategic ASW

By 1967 U.S. spy satellites had identified Soviet second-generation submarines under construction, including new SSNs intended primarily for the anti-submarine role. Within a year the Soviets were testing their first ocean surveillance satellites for detecting surface ships.[6] Satellites offer many advantages for ASW, including high search rates and autonomy from land bases. Obviously, there are difficulties in employing satellites for submarine detection beyond the limited ability of their sensors to discriminate between natural ocean disturbances and those produced in the ocean, on the surface, or even in the atmosphere by a submerged submarine. Other difficulties include data links to cue ASW forces to bring weapons to bear against submarines that could be hundreds or thousands of miles from the nearest Russian ASW ship or land base.

Still, in this period there was increasing concern within the U.S. Department of Defense and even among Navy officials over the possibility of a Soviet "ASW breakthrough." The Chairman of the U.S. Joint Chiefs of Staff, Admiral Thomas H. Moorer, in 1972 told Congress:

> the Soviets are believed to be working on a number of new ASW developments which could significantly improve their anti-submarine warfare capability in those waters in which our [missile] submarines are now required to operate. Therefore, as the Soviet anti-submarine warfare capability improves, we need to expand our area of operations in our constant effort to maintain our survivability.[7]

Historically the Soviet Navy has sought a much broader anti-SSBN effort than that of the U.S. Navy. (See Chapter 11.) This wide-ocean search evolved because of the inferior state of Soviet electronics (e.g., passive sonar), the increasing ranges of U.S. submarine-launched missiles, and the evolving Soviet doctrine of "combined arms" operations that provides for the cooperation of combat arms—and multiple military services—to a far greater degree than within the U.S. defense establishment.

Soviet wide-area search efforts were manifested in the development of ocean surveillance satellites to detect, identify, and track *surface ships*. This program subsequently was expanded into a major effort to detect SSBNs, using various combinations of optical, electronic intercept, radar, infrared, and other sensors. It became a most controversial issue in the West.

Two knowledgeable Soviet Navy officers stated in 1988 that space reconnaissance "is accomplishing many missions, including the detection of submerged submarines," and that radars deployed on aircraft and satellites were being used to "detect the wakes of submarines." These specific statements, by Captain 1st Rank Ye. Semenov and Rear Admiral Yu. Kviatkovskii, the latter the head of Soviet naval intelligence, apparently referred to systems for submarine detection that had *already* been deployed.[8] A 1993 report in the Russian general staff journal *Voennaia mysl'* declared:

> All-weather space reconnaissance and other types of space support will allow detecting the course and speed of movement of combat systems and surface and subsurface naval platforms [submarines] at any time of day with high probability, and providing high-precision weapons systems with targeting data in practically real time.[9]

Probably as significant as space-based radar in Soviet ASW efforts were space-based radio-location and optical detection systems. The most comprehensive U.S. analyses of these Soviet ASW efforts undertaken on an unclassified basis are the works of Hung P. Nguyen, who at the time was at the Center for Naval Analyses.[10] His conclusion: A third-generation satellite detection system for use against submarines was ready for deployment by the early 1990s. The Soviets, Nguyen wrote, believed that "mobile strategic targets" (i.e., submarines) at sea could be destroyed with high confidence by a single nuclear warhead if (1) the warhead is a nuclear reentry vehicle that can be guided to the target and (2) the requisite global, near-real-time detection systems are available. Nguyen's analysis cites a 1992 article by Vladislav Repin, a Russian official responsible for designing Soviet early warning systems, stating that the first is already a "routine" technological task and the second is a technological feat deemed fully possible within the "sufficiently near future" (five to ten years).[11]

U.S. officials were long aware of Soviet efforts. The testing of radars for submarine detection on the Soviet *Salyut-7* satellite in 1983–1984 was reported in U.S. newspapers. In 1985 Admiral James Watkins, the U.S. Chief of Naval Operations, said that scientific observations from an American space shuttle (orbital) flight the year before had perhaps revealed some submarine locations. A Navy oceanographer on the flight "found some fantastically important new phenomonology *[sic]* that will be vital to us in trying to understand the ocean depths," the admiral explained. While not releasing details of these observations, which were called "incredibly important to us," a U.S. Navy spokesman implied that "internal waves"—left by a submarine's underwater transit—were involved while Watkins acknowledged that the technology "is clearly opening some doors on submarine tracking."[12]

Beginning in the early 1980s the U.S. Central Intelligence Agency undertook a series of studies in the field of non-acoustic ASW at the request of Congress.[13] Results of those efforts—Project Tsunami—remain classified although the magnitude of the non-acoustic ASW problem was considered so great that a broader, independent program subsequently was established in the Office of the Secretary of Defense. Congressional support was vital to these non-Navy efforts, because the Navy's submarine leadership protested strongly against other agencies looking into their domain of antisubmarine warfare.[14] In the words of one participant in the effort the studies demonstrated:[15]

> Radar detections of ocean subsurface objects [submarines] are feasible and possible.
>
> - We gained significant insight into the possibility of submarine detections from space.
> - The United Kingdom, Norway, [West] Germany and Russia have made significant progress in understanding ASW phenomenology, submarine signature characteristics, flow field and other hydrodynamics effects such as vorticity control—and in some instances are well ahead of the U.S.
> - We have a much better understanding of multi-spectral sensor capabilities and potential.
> - It would be interesting for the [Senate] committee to evaluate the expenditures and return on investment of the [Department of Defense] program with the Navy's programs—both special access and unclassified.[16]

Several types of non-acoustic phenomena have been associated with submarine detection. Most of these could best be exploited some of the time—but not always—by aircraft or submarines; none could be exploited from space according to one American involved in much of this effort.[17]

While some of these phenomena could be exploited some of the time, a key question is how much could be exploited at a given time and place to indicate the presence of a submerged nuclear-propelled submarine. Participants in these studies have said little in public, but one, Dr. Richard Twogood of the Lawrence Livermore National Laboratory, observed, "We have discovered new phenomena that are not fully understood, nor explained by any known models, that appear to be very important to the sensing of surface effects produced by undersea disturbances. . . ," and

> These discoveries bring into question the validity of all previous assessment that were based on models that did not include these effects. In addition, the nature of our results also raises the possibility that certain claims by Russian scientists and officials that they have achieved non-acoustic ASW success with radars merits serious consideration.[18]

One ASW area of great interest to the Soviet Navy has been submarine wake detection. Dr. John Craven, former chief scientist of the U.S. Polaris-Poseidon project and the first director of the Navy's deep submergence project, observed as early as 1974 that the Soviets were conducting research "meant to study the turbulence characteristics of natural wakes associated with storms and to differentiate between those wakes and the wakes of submarines."[19] By the late 1980s several Project 671/Victor SSNs and subsequent ASW submarines had been fitted with wake sensor devices. Earlier a disarmed Project 658M/Hotel SSBN had been fitted as a multi-sensor test bed. Coupled with the introduction of wake-homing torpedoes, it is evident that the Soviets placed considerable importance on wake detection and other in-situ phenomena for hunting and trailing submarines.[20]

The Soviets also developed seafloor acoustic arrays, although they are believed to have been far less effective than SOSUS. These arrays had a limited capability to detect quiet Western submarines, in part because of their deployment in relatively shallow waters, and they were not as widely positioned as were the U.S. arrays. But the Soviet arrays purportedly could detect the unique sounds related to the underwater launch of ballistic missiles, possibly providing both "early warning" and "counterfire" targeting for anti-SSBN forces. Such arrays in the Arctic and Sea of Okhotsk areas were intended to contribute to the security of Soviet SSBNs by tracking intruding Western SSNs. There has additionally been interest by the USSR in bottom-mounted, non-acoustic submarine detection of the "electromagnetic" field of a ship or submarine passing overhead. These are usually considered short-range systems, but they could be used in narrow straits transited by Western SSNs. Research also has been conducted into Extremely Low Frequency (ELF) electromagnetic signatures radiated by surface ships and submarines.[21]

The Soviet approach to countering SSBNs also called for the targeting of U.S. strategic submarine bases, their communications stations, and fixed sensors (e.g., SOSUS terminals). Similarly, the U.S. Navy's Transit navigation satellites, necessary to provide accurate navigation for early U.S. SSBNs, appear to have been the target of the initial Soviet anti-satellite program.[22] (Later, as U.S. SSBN navigation improved and became less dependent on external data, such as the Transit satellites, Soviet anti-SSBN strategy became more focused on the detection of submarines.)

The Soviet anti-SSBNs strategy also included the early observation of missile submarines leaving port. These attempts to detect Western SSBNs as they departed their bases included stationing intelligence collection ships (NATO designation AGI) and oceanographic research and surveying ships near submarine bases in the United States and overseas.[23] The detection capabilities of these ships included visual, electronic intercept (e.g., radio transmissions of pilot boats and escorting Coast Guard craft), radar (against surfaced SSBNs), and acoustic. In some situations these surface ships could point Soviet SSNs toward their targets; these AGIs, research, and surveying ships periodically were observed operating with Soviet SSNs (and sometimes air and surface forces) off of Western SSBN bases.[24] A U.S. intelligence appraisal of the Soviet Ocean Surveillance System (SOSS) stated:

> The Soviet Navy views all combatants, auxiliaries, research ships, naval aircraft, merchant ships, and fishing ships as potential contributors to SOSS. Every bit of data, regardless of source, is validated, entered into the SOSS data base, and correlated with other information in an effort to provide a near-real-time picture of the locations and projected locations of Western targets of interest, including submarines.[25]

Still another indication of operating areas of Western strategic missile submarines apparently came from the intercept of *U.S. radio transmissions* containing SOSUS tracking reports. These communications were compromised from about 1968 by John A. Walker, the U.S. Navy warrant officer who sold communications secrets to the Soviets. As a senior U.S. intelligence officer later remarked, "Remember the uncanny 'coincidence' of Victor [SSNs] being sent to the Med[iterranean] coincidental with a U.S. SSBN coming in? If you know where they [SSBNs] are going and [then you] can man natural barriers or choke points while waiting for them. . . ."[26]

Indeed, Walker's espionage efforts generally are credited with the vulnerability of U.S. strategic missile submarines that may have been at the heart of Mikhail Gorbachev's comment to President Ronald Reagan at Geneva in November 1985 when the Soviet leader declared that the USSR would have won a war between the two countries.[27]

Beginning in 1967 the Soviet Navy completed numerous "large ASW ships," in reality multi-purpose cruisers and aviation ships. Western intelligence considered those ships to have limited open-ocean ASW capabilities because of the perceived need to support their long-range, 30-n.mile (55-km) SS-N-14 ASW missiles and helicopters with off-board sensors. However, the ships were fitted with the 850-megahertz Top Sail radar (Russian MR-600) and, subsequently, later radars that could apparently detect submarine-generated convection cells that can extend up to the local temperature inversion altitude (up to 1.86 miles/3 km) at intervals along the track of a submerged submarine. The convection cells result from rising vortex rings created as the submarine passes through the ocean.[28] At the same time, long-range ASW aircraft and hunter-killer submarines were too few in number to constitute an effective anti-submarine force. The specialized Project 671/Victor SSNs and Project 705/Alfa SSNs were introduced in limited numbers as they competed on the building ways with other types of submarines. (See Chapter 10.) Not until the 1980s were large numbers of SSNs available.

Beyond the use of aircraft, surface ships, and submarines to attack Western SSBNs, the Soviets also considered the use of *land-based* ballistic missiles in the anti-SSBN role. A 1983 U.S. intelligence analysis cited "vague references in Soviet writings to the possible use of land-based ballistic missiles against submarines in the open ocean."[29] The report continued, "Exploring such a technique would be consistent with past Soviet interest in innovate solutions to naval problems. . . . It would also be consistent with Soviet doctrinal emphasis on a multi-service approach to the accomplishment of wartime tasks."

The U.S. analysis also noted, "We are skeptical that such problems [related to the use of land-based missiles] could be overcome, at least during the period of this estimate and believe the Soviets would be unlikely to pursue seriously such a course unless they had high confidence that the initial detection problem would soon be solved."

Articles in Soviet professional journals continued to discuss the use of land-based missiles against Western SSBNs. And there have been periodic activities that indicated interest in this concept. For example, on 2 April 1984, six SS-20 Intermediate-Range Ballistic Missiles (IRBM) were fired into the Barents Sea, reportedly in conjunction with a large-scale naval exercise taking place in the same area.[30]

Another U.S. SSBN security concern was "pin-down"—the fear that during a conflict the Soviets would launch ICBMs/IRBMs to detonate over U.S. missile submarine operating areas to employ Electromagnetic Pulse (EMP)—the radiated electromagnetic field from high-altitude nuclear detonations—to destroy or damage SLBM guidance systems and warheads. The U.S. Navy, responding to this threat, sought to harden missile components and provided SSBNs with a receiver on a trailing-wire antenna to detect EMP pulses (Project Look). The crews of U.S. SSBNs were to monitor onboard radio receivers to determine the pattern of EMP bursts and launch their missiles between them.[31]

Also, ballistic missile defense systems in the USSR would attempt to intercept surviving SLBMs and ICBMs launched against the Soviet Union.[32] An anti-SSBN strategy is exceedingly difficult. But, as one U.S. naval officer observed, to a Soviet war planner working out his strategy for waging nuclear war, an SSBN attrition of "10 or 20 percent may be sufficient to justify the costs, on grounds that it could make the difference between victory (that is, survival) and defeat (obliteration)."[33] He concluded,

> There can be no clear-cut conclusion as to the likely effectiveness of such measures, but for the advocates of a counterdeterrent strategy, the anti-Polaris tasks are perhaps not as *impossible* as conventional Western thinking tends to assume.[34]

## From ULMS to Trident

The expanding Soviet strategic offensive and defensive capabilities—especially ASW—formed the backdrop for development of the Undersea Long-range Missile System (ULMS). During the Strat-X discussions, a generic missile size was used to simplify comparisons between the various missile-basing options. This generic weapon was 50 feet (15.2 m) long and 80 inches (2 m) in diameter, significantly larger than the Poseidon missile.[35]

The use of such a large weapon led to preliminary ULMS studies envisioning a submarine with a surface displacement of 8,240 tons and a length of 443 feet (135 m). The missiles—24 was a nominal number—would be carried horizontally, *external* to the pressure hull in protective canisters. With this scheme, "Missiles may be released from the submarine at all speeds and at depths up to the [submarine's] maximums, and missile firing may be delayed to avoid the backtracking of trajectory so that submarine survivability is not inhibited by missile launch constraints."[36]

The ULMS submarine, of course, would be nuclear propelled. However, it would not be fast, probably capable of some 25 knots at most. At faster speeds it would be more vulnerable to acoustic detection, and a large missile submarine certainly could not "outrun" a Soviet SSN.

The U.S. Navy's Special Projects Office (SPO), which had managed the Polaris and Poseidon missiles and submarines, was redesignated Strategic Systems Projects Office (SSPO) in July 1968 and the director, Rear Admiral Levering Smith, was made project manager for ULMS. Smith, who had been with the Polaris project almost since its inception, had became director of SPO in 1965. He was an ordnance engineering specialist and, like Admiral Raborn—the "father of Polaris"—he had no submarine experience.[37]

Many observers envisioned that ULMS would be developed much like Polaris had been, rapidly and efficiently outside of the standard Navy bureaucracy. But as historian Graham Spinardi wrote: "in marked contrast to Polaris, the development of the next generation FBM submarine was to be a divisive and contested battle. . . ."[38]

The S5G natural-circulation reactor being developed for the submarine *Narwhal* (SSN 671) was proposed for the ULMS submarine. Adopting an existing plant would alleviate or at least reduce Vice Admiral H. G. Rickover's role in the project (much the same as the S5W plant had been used in the Polaris-Poseidon SSBNs with minimal participation by Rickover). Because it was questionable whether a single S5G reactor of some 17,000 horsepower could provide the speed desired for ULMS, the submarine's propulsion soon became a major issue.

Meanwhile, the design concept for ULMS underwent a major revision in early 1969 when Admiral Smith decided against external encapsulated missiles; instead, the missiles would be launched from internal missile tubes, as in the Polaris-Poseidon submarines. This approach would save time and reduce development risks.

The next major change occurred early in 1970 as SSPO requested technical data on the S5G reactor plant, which had gone to sea the previous year in the *Narwhal*. Rickover argued that the ULMS submarine must be able to attain at least 24 knots—and preferably more. He wanted to employ a reactor plant that would produce as much as 60,000 horsepower. This was the plant he was developing for a high-speed cruise missile submarine, the design of which had also started in 1970. Interestingly, 60,000 horsepower was one of the four propulsion plant options proposed by the Bureau of Ships in 1950! (See table 4-1.)

Rickover increasingly took control of the ULMS design. In March 1971, in response to pressure from Rickover, the Chief of Naval Operations established a separate ULMS project office, under Rear Admiral Harvey E. Lyon, a submarine officer, to manage development of the submarine. (Smith's SSPO would retain control of missile development.) A short time later Electric Boat/General Dynamics was awarded a contract for design of the submarine.

Rickover pushed for a very large submarine to accommodate his proposed reactor plant as well as 24 large missiles; his proposals ran afoul of the Department of Defense as well as Admiral Elmo R. Zumwalt, who had become Chief of Naval Operations on 1 July 1970.[39] Zumwalt objected to the construction of an all "high-end" fleet with all aircraft carriers and major surface warships being nuclear propelled, as demanded by Rickover. And Zumwalt opposed the development of the large cruise missile submarine proposed by Rickover, which was to be powered by the 60,000-horsepower reactor plant.

Rickover was forced to scale back his proposals for the SSBN, finally agreeing to provide a new reactor plant, the S8G that generated some 35,000 horsepower through two steam turbines to a single propeller shaft. This would be the most powerful single-reactor plant placed in a U.S. submarine. Still, with 24 internal missile tubes and the S8G reactor plant, ULMS would have a conventional SSBN design that would, ultimately, have a surface displacement of almost 17,000 tons—twice the displacement initially estimated for ULMS—with a length of 560 feet (170.7 m). The submarine would be longer than the height of the Washington Monument in Washington, D.C.

The submarine would have a submerged speed of approximately 25 knots. Special emphasis was placed on quieting the propulsion plant and, reportedly, the submarine exceeded design goals for quieting at low speeds, that is, when using natural circulation (convection) rather than pumps for the circulation of pressurized water in the primary loop.

The ULMS would follow the U.S. third-generation nuclear submarine design, with the BQQ-6 spherical passive sonar array in the bow, four 533-mm torpedo tubes mounted farther aft, and three full deck levels for most of the submarine's pressure hull. The submarine had four major compartments: forward (torpedo, control, living spaces); missile; reactor; and engineering. These SSBNs had the most comfortable level of accommodations yet provided in a U.S. submarine.

The submarine would carry 24 missiles, half again as many as the Polaris-Poseidon SSBNs. The increased weapons load was looked upon as a more efficient means of taking missiles to sea. The potential implications of strategic arms agreements were not considered.

Admirals Rickover and Zumwalt collaborated on obtaining congressional approval for ULMS, which was renamed Trident on 16 May 1972.[40] Several members of Congress as well as anti-nuclear organizations lobbied against the Trident. The SALT agreement of 1972 was used by both advocates and opponents of Trident to justify their positions. A subsequent factor complicating the Trident effort was President Jimmy Carter's desire for drastic nuclear arms reduction. After he became president in January 1977, he asked the Joint Chiefs of Staff for an opinion on the possibility of eventually reducing the entire U.S. strategic offensive force to some 200 submarine-launched missiles.

But Harold Brown, Carter's Secretary of Defense, in January 1980 announced that for the next few years the nation would build one "SSBN" per year and then increase the rate to three ships every two years. He used the designation "SSBN" in part to indicate that these ships might not be submarines of the existing Trident design; he noted,

> Funds are programmed to support concept and design studies leading to a follow-on, less expensive SSBN. This SSBN could be a re-engineered Trident design or a new design of a 24-tube SSBN with tubes the same size as the Trident SSBN.[41]

Dubbing the design the SSBN-X, the Carter administration proposed spending more than $106 million for research and development of a smaller, lower-cost strategic missile submarine. To many observers, this meant that the end of the existing Trident program was in sight.

However, President Carter was defeated by Ronald Reagan in the November 1980 election. As part of the Reagan administration's strategic and naval buildup, the Trident program was reinstated at the building rate of one submarine per year, although the Trident program—as well as the attack submarine program—continued to be plagued by problems at the Electric Boat yard, attributed to the Navy's poor program management and to rapidly increasing submarine costs.

Meanwhile, these problems and delays had led Admiral Smith's SSPO to propose an interim Trident missile, initially called EXPO (for Extended-range Poseidon). This was put forward as a weapon that could be available by 1972, several years before the Trident missile albeit with a lesser range—a maximum of 4,000 n.miles (7,410 km); eight MIRV warheads of 100 kilotons each could be delivered to lesser ranges. That missile, eventually designated Trident C-4, also could be backfitted into existing Polaris-Poseidon SSBNs as well as being used in the Trident submarines before the full-range, Trident D-5 missile would be available.[42]

The Lockheed Corporation was selected to develop both the C-4 and D-5 missiles, building on the firm's experience with Polaris and Poseidon. Indeed, the first two stages of the EXPO/Trident C-4 was the two-stage Poseidon missile.

The C-4 missile was designed to alternatively carry the Mk 500 Evader, a *Maneuvering* Reentry Vehicle (MaRV) warhead; this re-entry system was intended to overcome ballistic missile defenses but was not developed.

## Trident Submarines

After extensive debate, the lead Trident submarine had been funded in fiscal year 1974. An initial schedule was put forward to build ten Trident SSBNs at an annual rate of 1-3-3-3, to be completed from 1977 to 1982. But when the first Trident submarine was ordered from Electric Boat in July 1974, the planned delivery had slipped to 30 April 1979. The shipyard agreed to attempt to make delivery in December 1977 because of the high priority of the program. Like Polaris two decades earlier, Brickbat—the highest priority—was assigned to Trident.[43] Subsequent delays caused by Navy mismanagement of the project, design changes, and chaotic problems at the Electric Boat yard resulted in the late deliveries of the early submarines, with

*The* Ohio *on sea trials in 1981, showing the long "turtle back" that faired into her hull lines. Like second-generation and later SSNs, the Trident missile submarines are single-hull ships. Efforts are under way to convert up to four Trident SSBNs to combined cruise missile/special operations submarines.* (U.S. Navy)

authorizations for the first ten submarines covering a ten-year period instead of four years. The problems at Electric Boat included construction difficulties with the new *Los Angeles* (SSN 688) attack submarines. Thus, with the start of the Trident program, Electric Boat's capabilities were overloaded.[44] Newport News Shipbuilding in Virginia also was encountering difficulties with the *L.A.* class; from 1972 onward these were the only U.S. shipyards constructing nuclear submarines.

The disarray in submarine construction and, especially, the rapidly deteriorating Navy relationship with the Electric Boat yard, a component of General Dynamics Corporation, had another effect on Cold War submarines. Admiral Rickover had favored EB over other shipyards since building the *Nautilus* (SSN 571). Now he spoke out against the yard. And his relationship with Newport News Shipbuilding had deteriorated to the point that the yard's management told the Navy all work would stop if he again entered the yard unannounced.[45]

Rickover had reached the statutory retirement age of 62 in 1962. He was retained on active duty as the head of nuclear propulsion by presidential order—and congressional political pressure. The Reagan administration realized Rickover's role in the chaotic submarine situation and on 13 November 1981—two days after the commissioning of the first Trident submarine—Secretary of the Navy John Lehman announced that Admiral Rickover would not be retained on active duty after January 1982.[46]

In the past Rickover's supporters in Congress had frustrated such efforts to "fire" him. Now, most of his supporters had retired, had died, or had come to acknowledge that Rickover was "part of the problem" and not part of the solution.[47] His successors as head of nuclear propulsion, submarine admirals appointed for eight-year terms, would have less influence in Congress and the Navy.

The lead Trident submarine, the *Ohio* (SSBN 726), was launched at EB on 7 April 1979.[48] She was the largest submarine built by any nation up to that time. The *Ohio* was commissioned on 11 November 1981 and went to sea on her first Trident missile patrol—in the North Pacific—on 1 October 1982, carrying 24 C-4 missiles. The C-4 had already gone to sea in October 1979 on board the *Francis Scott Key* (SSBN 657), a Polaris-Poseidon submarine that had been modified to carry the C-4 missile. Eleven more of these older SSBNs were so modified to carry 16 of the weapons, again demonstrating the flexibility of the Polaris submarine design.

The ultimate U.S. submarine-launched ballistic missile was the Trident D-5. The missile introduced greater range and more accuracy than previous SLBMs, with D-5 accuracy generally credited to be

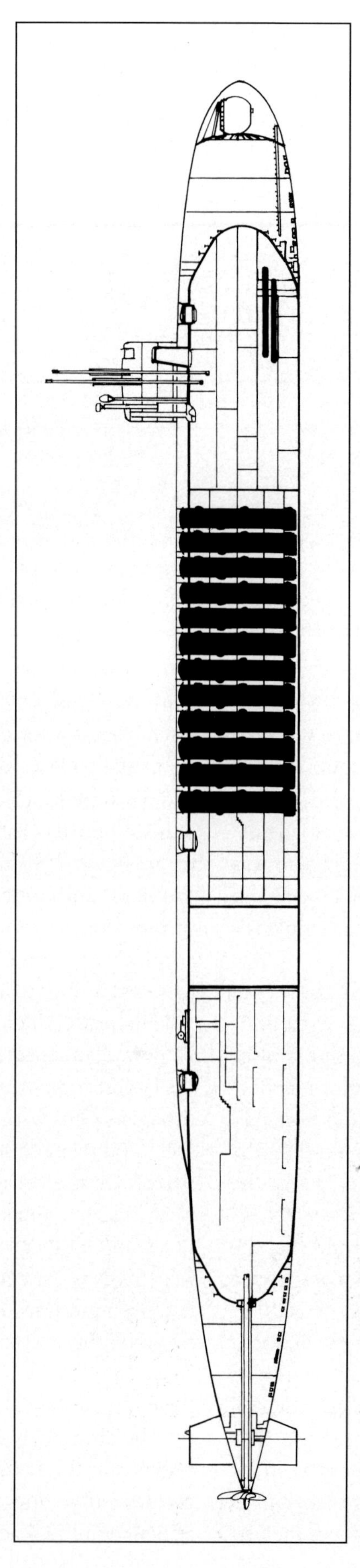

Ohio *(SSBN 726). LOA 560 ft (170.7 m)* (©A.D. Baker, III)

*The world's largest submarine is the Project 941/Typhoon SSBN. The design departed from the missile arrangement used in previous U.S., Soviet, British, and French strategic missile submarines. The location of SLBMs forward—between twin pressure hulls—pushed the massive sail structure aft.* (U.K. Ministry of Defence)

superior to contemporary U.S. land-based ICBMs. Reportedly, the missile's guidance could place the eight warheads within a circle 560 feet (170.7 m) in diameter at a range of 4,000 n.miles (7,400 km).[49] The D-5 missile normally carries eight MIRV nuclear warheads, each having an explosive force of between 100 and 475 kilotons.[50] After a difficult development period, with several major launch failures, the D-5 missile became operational in the USS *Tennessee* (SSBN 734) in March 1990. The *Tennessee* was the ninth Trident submarine to be completed.

By the early 1980s the Navy was planning a class of 24 *Ohio*-class submarines (with 576 SLBMs). These were to replace the 41 Polaris-Poseidon submarines (with 656 SLBMs). The SALT I strategic arms agreement of 1972 required the decommissioning of the Polaris A-3 submarines *Theodore Roosevelt* (SSBN 600) and *Abraham Lincoln* (SSBN 602) to compensate for the *Ohio* entering service; those were the first U.S. SSBNs to be taken out of service.

The fall of the Soviet Union in 1991 also marked the end of the Trident construction. The eighteenth and final Trident SSBN was authorized in 1990 (fiscal year 1991 program). When that submarine was completed in 1997, all 41 Polaris-Poseidon-Trident C-4 submarines had been discarded except for two former SSBNs employed as special operations/transport submarines. (See Chapter 14.) Thus, through the end of the Cold War, U.S. submarine construction programs had produced three cruise missile submarines (1 SSGN, 2 SSG) and 59 ballistic missile submarines (SSBN).

Of the 18 Trident SSBNs, the first eight were completed with the C-4 missile, the next ten with the D-5 missile.[51] Initially the Navy intended to rearm all early submarines with the D-5 missile, but that plan ran afoul of Soviet-U.S. strategic arms agreements.

*A Typhoon at high speed, demonstrating the power of her large twin-reactor plant. The 20 large SLBM tubes are arranged in two rows, forward of the sail. A built-in escape chamber is located just outboard of the sail on each side. The retracted bow diving planes are visible in this view.* (U.K. Ministry of Defence)

*The massive "sail" of a Project 941/Typhoon SSBN. A single periscope is raised; the other scopes and masts are recessed and protected. The anechoic tiles are evident as is (bottom center) the outline of the top of the starboard-side escape chamber (with exit hatch).* (U.K. Ministry of Defence)

V. P. Makeyev's SKB-385 design bureau. In 1972 work had began on the missile submarine, Project 941 (NATO Typhoon), and the following year on the solid-propellant RSM-52/R-39 missile (later given the NATO designation SS-N-20 Sturgeon).[53]

The Project 941/Typhoon SSBN would be the largest undersea craft to be constructed by any nation. The submarine was designed by S. N. Kovalev at the Rubin design bureau. Kovalev and his team considered numerous design variations, including "conventional" designs, that is, a single elongated pressure hull with the missile tubes placed in two rows aft of the sail. (See sketch.) This last approach was discarded because it would have produced a submarine more than 770 feet (235 m) long, far too great a length for available dry docks and other facilities.[54]

Instead, Kovalev and his team developed a unique and highly innovative design—the 441st variant that they considered. The ship has two parallel main pressure hulls to house crew, equipment, and propulsion machinery. These are full-size hulls, 488¾ feet (149 m) long with a maximum diameter of 23⅔ feet (7.2 m), each with eight compartments.[55] The 20 missile tubes are placed between these hulls, in two rows, *forward* of the sail. The amidships position of the sail led to the submarine's massive diving planes being fitted forward (bow) rather than on the sail, as in the Project 667 designs.

The control room-attack center and other command activities are placed in a large, two-

## The World's Largest Submarine

The American pursuit of the Trident program caused an acceleration of the third-generation Soviet SSBN.[52] During their November 1974 meeting at Vladivostok in the Soviet Far East, General Secretary Leonid Brezhnev and President Gerald Ford agreed on a formula to further limit strategic offensive weapons (SALT II). In their discussions Brezhnev expressed his concern over the U.S. Trident program and declared that if the United States pursued deployment of the Trident system the USSR would be forced to develop its *Tayfun* strategic missile submarine.

Although Soviet naval officials and submarine/missile designers believed that liquid-propellant missiles "had irrefutable merit," and those SLBMs were used in the first- and second-generation SSBNs, research continued on solid-propellant SLBMs at

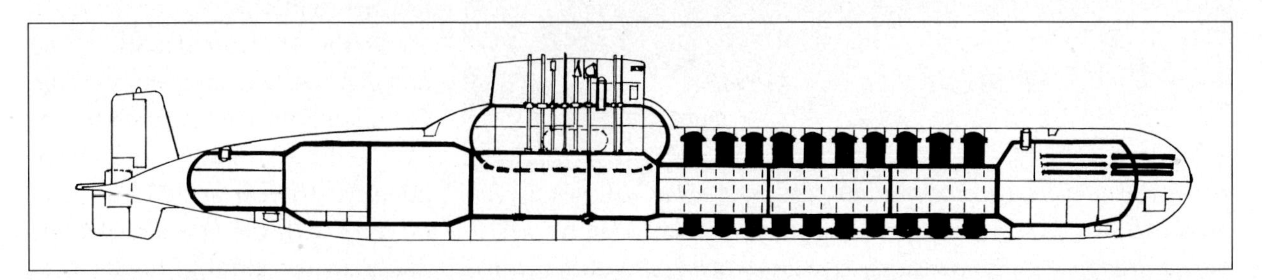

*Project 941/Typhoon SSBN. LOA 564 ft 3 in (172.0 m)* (©A.D. Baker, III)

compartment pressure hull between the parallel hulls (beneath the sail). This center hull is just over 98 feet (30 m) in length and 19⅔ feet (6 m) in diameter. The sail structure towers some 42⅔ feet (13 m) above the waterline. An additional compartment was placed forward, between the main pressure hulls, providing access between the parallel hulls and containing torpedo tubes and reloads. Thus the Typhoon has a total of 17 hull compartments, all encased within a massive outer hull 564⅙ feet (172 m) long. This arrangement gives the Project 941/Typhoon a surface displacement of 23,200 tons. Although the submarine is approximately as long as the U.S. *Ohio* class, it has a beam of 74¾ feet (23.3 m). With a reserve buoyancy of some 48 percent, the submerged displacement is 48,000 tons, about three times that of the *Ohio* (which has approximately 15 percent reserve buoyancy).

The large reserve buoyancy helps decrease the draft of the ship. In addition, it contributes to the ability of the submarine to surface through ice to launch missiles. (On 25 August 1995 a Typhoon SSBN surfaced at the North Pole, penetrating about eight feet [2.5 m] of ice, and launched an RSM-52 missile with ten unarmed warheads.)

Beneath the forward hull is the ship's large Skat sonar system, including the MGK-503 low-frequency, spherical-array sonar (NATO Shark Gill).

There are crew accommodations in both parallel hulls, and in the starboard hull there is a recreation area, including a small gym, solarium, aviary, and sauna. The crew is accommodated in small berthing spaces; the large number of officers and warrants have two- or four-man staterooms. Above each hull there is an escape chamber; together the chambers can carry the entire crew of some 160 men to the surface.

Within each of the parallel hulls, the Typhoon has an OK-650 reactor plant with a steam turbine producing about 50,000 horsepower (190 megawatts) and an 800-kilowatt diesel generator. The twin propellers are housed in "shrouds" to protect them from ice damage. The ship also has two propulsor pods, one forward and one aft, that can be lowered to assist in maneuvering and for hovering, although missiles can be launched while the Typhoon is underway.

The design and features of the Typhoon SSBN were evaluated in a one-tenth scale model built at Leningrad's Admiralty shipyard. The model was automated and provided an invaluable design and evaluation tool.

The keel of the lead Project 941/Typhoon—the *TK-208*—was laid down on 30 June 1976 at Severodvinsk, by that time the only Soviet shipyard constructing SSBNs.[56] A new construction hall—the largest covered shipway in the world—was erected at Severodvinsk, being used to build the Typhoon SSBNs and Project 949/Oscar SSGNs. Most U.S. intelligence analysts had been confused by Brezhnev's reference to a *Tayfun* missile submarine. Not until 1977—when U.S. reconnaissance satellites identified components for a new class of submarines at Severodvinsk—was it accepted that an entirely new design was under construction.[57] The *TK-208* was put into the water on 23 September 1980; trials began in June 1981, and she was commissioned in December 1981. Series production followed.

The D-19 missile system, however, lagged behind schedule with failures of several test launches of the RSM-52/R-39 missile. The Project 629/Golf submarine *K-153* was converted to a test platform for the RSM-52/R-39, being provided with a single missile tube (changed to Project 619/Golf V). That missile became operational in 1984. It carried a larger payload, had greater accuracy than any previous Soviet SLBM, and was the first Soviet solid-propellant SLBM to be produced in quantity.[58] The missile is estimated to have a range of 4,480 n.miles (8,300 km) carrying up to ten MIRV warheads of approximately 100 kilotons each. Their firing rate is one missile every 15 seconds (the same firing rate as the U.S. Trident submarines[59]). Still, the use of solid propellant in the R-39 led to a sharp increase in the dimensions of the missile, with the launch weight reaching 90 tons.

Six Typhoon SSBNs were completed through 1989. Eight ships had been planned, with the seventh, the *TK-210*, having been started but then abandoned while still in the building hall. The six-Typhoon SSBN division was based at Nerpichya, about six miles (ten km) from the entrance to Guba Zapadnaya Litsa on the Kola Peninsula, close to the border with Finland and Norway. The Typhoon base was distinguished by the extremely poor facilities ashore for the crews as well as for the base workers and their families.[60]

*A Typhoon at rest with crewmen on deck. The twin propellers are housed in protective, circular shrouds, outboard of the upper rudder. The stern "diving planes" are fitted behind the propellers, attached to the shrouds.* (Russia's Arms Catalog)

refueled and rearmed with the RSM-52V missile, although such planning was considered tentative in view of the continuing Russian fiscal problems. The three modernized Typhoon SSBNs would be expected to remain in service at least until 2010–2012.[62]

There have been proposals to convert some of the giant submarines to cargo carriers; this could be done expeditiously—albeit at considerable cost—by replacing the missile tubes with a cargo section. Ironically, an earlier analysis of the Typhoon by the Central Intelligence Agency addressed the possibility of using submarines of this type to (1) carry mini-submarines; (2) support other submarines in the *milch cow* replenishment role; (3) carry troops for special operations; and (4) serve as major command ships.[63]

Discussing Project 941/Typhoon on the macro level, Viktor Semyonov, the deputy chief designer of the Typhoon, stated that the program had encountered "No technical problems—the problems are all financial."[64] But Soviet views of the submarine and her D-19 missile system were not unanimous. One submarine designer wrote:

> To my mind the creation of the D-19 missile system and the Project 941 was a great mistake.
>
> A solid-propellant SLBM had no appreciable advantages [over] a liquid-propellant SLBM. . . . Such an expensive project like the D-19 missile system and the Project 941 which had been [developed] parallel with D-9RM and Project 667BDRM were the ruin of the USSR. Such ill-considered decisions, which were lobbied by the definite industrial circles, undermined the economy of the USSR and contributed to the loss [of] the Cold War.[65]

Almost parallel with Project 941/Typhoon, another SSBN was developed under Academician Kovalev at Rubin. Project 667BDRM/Delta IV—as its designation denotes—outwardly was a further develop-

At sea the Typhoon SSBN had some difficulties with control and seakeeping.[61] Still, the ships could be considered highly successful and provided a highly capable strategic striking force. Their Arctic patrol areas made them immune to most Western ASW forces, and simplified Soviet naval forces providing protection, if necessary. On 9 September 1991, during a missile test launch by one of these SSBNs of this type, a missile did not exit the tube and exploded and burned. The submarine suffered only minor damage.

The *TK-208* entered the Severodvinsk shipyard in October 1990 for refueling of her reactors and for modernization to launch the improved R-39M missile (NATO SS-N-28 Grom). The other ships were to follow, but the end of the Cold War brought an end to the Project 941/Typhoon program. The *TK-208,* which was renamed *Dmitri Donskoi* in 2000, did not emerge from the Severodvinsk yard (renamed Sevmash) until 2002. She sailed for her home base of Guba Zapadnaya Litsa on 9 November 2002. She had been refueled but instead of the R-39M missile, which had encountered development problems, she was refitted to carry the smaller, solid-propellant RSM-52V Bulava, a variant of the land-based Topol-M (SS-27).

Meanwhile, of the five other Typhoon SSBNs, the *TK-12* and *TK-202* were taken out of service in 1996, and the *TK-13* in 1997. They are being scrapped. When this book went to press the *TK-17* (renamed *Arkhangel'sk*) and *TK-20* were also to be

ment of the Yankee-Delta series but in large part was an entirely new, third-generation nuclear submarine. The submarine was significantly larger than its predecessors, had enhanced quieting features, and carried a larger missile, the three-stage, liquid-propellant RSM-54/R-29RM (NATO SS-N-23 Skiff). Depending on the number of warheads, the missile could reach 4,480 n.miles (8,300 km). This long-range missile was the last submarine weapon developed by Makeyev.

Like its predecessors, the Delta IV had twin reactor plants (VM-4SG reactors) and twin propellers, but with an improved stern configuration that enhances propeller efficiency and quieting. One Soviet analyst addressing the Project 667BDRM/Delta IV observed,

> The second generation of Soviet submarines was less noisy, but progress in decreasing the SSBN signature was achieved by the Soviet shipbuilding industry only during the 1980s with the appearance of the strategic submarine design 667BDRM. During this period new technologies were introduced which resulted in an order of magnitude improvement in the accuracy of manufacturing gear assembly, shafts and propellers. A significant decrease in noise level was also achieved with the application of active noise suppression methods for submarines.[66]

The last was a notable innovation by the Soviets. This technique of active sound and vibration control is a remarkably simple method of removing an unwanted disturbance. The disturbance can take the form of either a pressure wave (sound) or structural motion (vibration). It is detected as a signal using suitable sensors, and the detected signal is then used to control actuators in such a way that the disturbance is essentially suppressed by cancellation (insertion of an out-of-phase signal) or by interference. To achieve this, it is necessary to reproduce not only the temporal characteristics of the disturbance, but also its spatial propagation and distribution. This often requires the use of high-speed computing and signal processing techniques. With such techniques, it is possible to achieve substantial reductions in sound or vibration intensity. Depending on the circumstances, reductions of over 40 decibels are possible. The enlarged Project 667BDRM/Delta IV submarine also stressed crew habitability for long patrols, with an exercise space, solarium, and sauna.

The lead submarine was the *K-51,* laid down in 1981, launched in 1984, and commissioned on 29 December 1984. Thus construction of this class lagged several years behind the Typhoon schedule. The Severodvinsk shipyard completed seven Delta IV submarines through 1992. At least two additional units had been started when President Boris Yeltsin directed a halt to SSBN construction. The Soviet Navy accordingly took delivery of 83 "modern" SSBNs during the Cold War (compared to 59 U.S. SSBNs). Because of SALT agreements, the total number of SSBNs in service probably did not exceed 63 ships.[67]

*Four hatches of this Typhoon's 20 missile tubes are open. The safety tracks that are fitted over the tubes are evident in this photograph. The Soviets developed a scheme to rearm SSBNs from supply ships. The U.S. Navy had provided that capability for its Polaris submarines.* (Russia's Arms Catalog)

*The Project 667BDRM/Delta IV submarine had a bizarre "turtle-back" structure accommodating 16 SLBMs. The lines of limber holes along the base of the missile housing familiar in previous Deltas appears to be absent in this design. Rubin designers stated that they encountered no major problems with this configuration.* (U.K. Ministry of Defence)

The Delta IV submarines joined other SSBNs in combat patrols in the Arctic and Sea of Okhotsk areas, with the *K-18* surfacing at the North Pole in August 1994.[68] During firing demonstrations in 1990, a Delta IV launched an entire loadout of 16 missiles in a single, sequential salvo.

When the Cold War ended in 1991, the U.S. and Soviet Navies each provided a significant portion of their nations' strategic offensive forces:[69]

| | Soviet Union | United States |
|---|---|---|
| Modern SSBNs | 62 | 34 |
| Missiles | 940 | 632 |
| Percent of offensive warheads | 27%[70] | 45% |

The U.S. Trident D-5 missile had the same or better accuracy than did land-based ICBMs, providing the sea-based missiles with a hard-target-kill capability, that is, Soviet ICBMs in underground silos and other "hard" targets. At the same time, according to U.S. officials, the accuracy of both of the missiles carried by Typhoon and Delta IV SSBNs could "eventually have a hard-target-kill potential" against U.S. ICBM silos.[71] Further, despite massive efforts to develop effective anti-SSBN capabilities, these sea-based strategic forces continue to enjoy a high degree of survivability.

**As the latest generation SSBNs emerged from the Electric Boat yard in Groton, Connecticut, and from Shipyard 402 at Severodvinsk, the submarine-launched ballistic missile was acknowledged as a principal component of U.S. and Soviet strategic offensive forces. Further, these SSBNs undoubtedly had the highest survivability level of all strategic offensive forces.**

**The U.S. Trident program required more than 11 years from initiation of the program (1970) until completion of the USS *Ohio* (1981). That contrasts to three years from the decision to develop the solid-propellant Polaris SLBM (December 1956) until completion of the *George Washington* (December 1959). The Trident submarine was an adaptation of advanced SSN designs; the Trident C-4 (née EXPO) and D-5 mis-**

siles were improvements of the Polaris-Poseidon missile systems.

The disparity in the development period of Polaris and Trident was caused by several factors: (1) the more complex development procedures and regulations of the Department of Defense in the 1970s; (2) the lack of comparative emphasis on Trident development by the Navy's leadership; (3) the chaotic state of the U.S. submarine construction industry in the late 1960s and into the 1970s; and (4) the extensive involvement of Admiral Rickover in the Trident program (in comparison with his limited role in the Polaris effort).

The U.S. Navy did not launch a strategic missile submarine from 1966 (*Will Rogers*/SSBN 659) until 1979 (*Ohio*), a 13-year building "holiday." This was possible because of the effectiveness of the 41 Polaris submarines and the feasibility of updating their missile system. The Soviet construction of second- and third-generation SSBNs was continuous from the launching of the *K-137* in 1966 until the end of the Cold War.

The Soviet government sought a strategic offensive force—including ballistic missile submarines—as large as those of the United States and all other nations combined.[72] Some Soviet Navy and civilian officials believed that the large size of the SSBN program was unnecessary. Academician Igor Spassky, head of the Rubin design bureau, said that "there were more [SSBNs] than we really needed to fulfill the mission."[73]

In both countries the third-generation SSBN building program was superimposed on large "attack" submarine construction programs as well as impressive surface warship programs.

TABLE 12-1
**Submarine-Launched Ballistic Missiles**

| | Trident C-4 | Trident D-5 | RSM-52/R-39R NATO SS-N-20 Sturgeon | RSM-54/R-29RM NATO SS-N-23 |
|---|---|---|---|---|
| Operational | 1979 | 1990 | 1983 | 1986 |
| Weight | 73,000 lb (33,115 kg) | 130,000 lb (58,970 kg) | 198,420 lb (90,000 kg) | 88,850 lb (40,300 kg) |
| Length | 34 ft (10.37 m) | 44 ft (13.4 m) | 52½ ft (16 m) | 48½ ft (14.8 m) |
| Diameter | 74 in (1.9 m) | 83 in (2.1 m) | 94½ in (2.4 m) | 74¾ in (1.9 m) |
| Propulsion | solid-propellant 3 stage | solid-propellant 3 stage | solid-propellant 3 stage | liquid-propellant 3 stage |
| Range | 4,000 n.miles (7,400 km) | >4,000 n.miles (>7,400 km) | 4,480 n.miles (8,300 km) | 4,480 n.miles (8,300 km) |
| Guidance | inertial | inertial | inertial** | inertial** |
| Warhead | 8 MIRV (100 KT each) | 8 MIRV * | 10 MIRV (100 KT each) | 4 MIRV*** |

Notes: * The Trident D-5 carries either eight W88 (100-KT) or eight W76 (300–475 KT variable yield) re-entry bodies.
** Possibly with stellar inflight update.
*** Up to eight MIRVs can be carried; under arms agreements four are carried.

Table 12-2
**Strategic Missile Submarines**

| | **U.S. *Ohio* SSBN 726** | **Soviet Project 941 NATO Typhoon** | **Soviet Project 667BDRM NATO Delta IV** |
|---|---|---|---|
| Operational | 1981 | 1981 | 1984 |
| Displacement | | | |
| surface | 16,764 tons | 23,200 tons | 11,740 tons |
| submerged | 18,750 tons | 48,000 tons | 18,200 tons |
| Length | 560 ft (170.7 m) | 564 ft 3 in (172.0 m) | 547 ft 9 in (167.0 m) |
| Beam | 42 ft (12.8 m) | 76 ft 1 in (23.2 m) | 38 ft 4 in (11.7 m) |
| Draft | 36 ft 3 in (11 m) | 36 ft (11 m) | 28 ft 10 in (8.8 m) |
| Reactors | 1 S8G | 2 OK-650 | 2 VM-4SG |
| Turbines | 2 | 2 | 2 |
| horsepower | ~35,000 | ~100,000 | 60,000 |
| Shafts | 1 | 2 | 2 |
| Speed | | | |
| surface | | 12 knots | 14 knots |
| submerged | ~25 knots | 25+ knots | 24 knots |
| Test depth | 985 ft (300 m) | 1,300 ft (400 m) | 1,300 ft (400 m) |
| Missiles | 24 Trident C-4/D-5 | 20 RSM-52/R-39R | 16 RSM-54/R-29RM |
| Torpedo tubes* | 4 533-mm A | 6 533-mm B | 4 533-mm B<br>2 650-mm B |
| Torpedoes | 24 | 22 | 12 |
| Complement | 165** | 160 | 135 |

Notes: * Angled + Bow.
** Alternating Blue and Gold crews.

13

# "Diesel Boats Forever"

*Project 641/Foxtrot was an attractive and highly capable torpedo-attack submarine. The successor to the long-range Project 611/Zulu design, these submarines operated far and wide during the Cold War with several units being in the Western Atlantic during the Cuban missile crisis of 1962.* (Japanese Maritime Self-Defense Force)

The U.S. Navy had made the decision to stop constructing diesel-electric submarines by 1956. The Navy's long-term blueprint for the future fleet—developed in 1957—provided for 127 submarines by the 1970s:[1]

> 40 Polaris ballistic missile submarines
> (1,500-n.mile/2,780-km missile range)
>
> 12 cruise missile submarines
> (1,000-n.mile/1,850-km missile range)
>
> 75 torpedo-attack (ASW) submarines

All but ten of the attack submarines were to have nuclear propulsion. Those exceptions were the existing six submarines of the *Tang* (SS 563) class, the *Darter* (SS 576), and the three *Barbel* (SS 580) class; they would serve many years in the U.S. Navy. The last of these non-nuclear submarines to be retired would be the *Blueback* (SS 581), decommissioned in 1990 after 30 years of service. The Navy's leadership realized that nuclear propulsion provided the long-range capabilities needed to support the nation's worldwide military operations.

Of the other nations that have constructed nuclear submarines, Britain's Royal Navy, faced with severe fiscal constraints, also decided, in the early 1990s, to operate an all-nuclear submarine force.[2] Subsequently France reached a similar decision. Thus, only China and Russia have continued to operate diesel-electric as well as nuclear submarines. And, when possible, all have sold new or surplus diesel-electric submarines to other nations.

During the Cold War a U.S. Navy intelligence officer, then-Captain Thomas A. Brooks, observed,

> The Soviets see a continuing utility of the diesel submarine. It is excellent for confined waters such as those in the Mediterranean; it makes a superb "mobile minefield" in Soviet

> parlance; for purposes of forming submarine barriers, it can be more effective; and it can serve quite successfully for delousing high-value units, reconnaissance, sealing off choke points and many traditional submarine missions where the speed and endurance of a nuclear submarine are not required. . . . the Soviets clearly have a commitment to diesel boats forever.[3]

In the Cold War era, that commitment began with the massive submarine construction programs initiated immediately after World War II—the long-range Project 611/Zulu, the medium-range Project 613/Whiskey, and the coastal Project 615/Quebec classes. Not only did these craft serve as the foundation for the Soviet Navy's torpedo-attack submarine force for many years, but converted Zulus and Whiskeys were also the first Soviet submarines to mount ballistic and cruise missiles, and several other ships of these designs were employed in a broad range of research and scientific endeavors.

These construction programs were terminated in the mid-1950s as part of the large-scale warship cancellations that followed dictator Josef Stalin's death in March 1953. But the cancellations also reflected the availability of more-advanced submarine designs. Project 641 (NATO Foxtrot) would succeed the 611/Zulu as a long-range torpedo submarine, and Project 633 (NATO Romeo) would succeed the 613/Whiskey as a medium-range submarine. There would be no successor in the coastal category as the Soviet Navy increasingly undertook "blue water" operations. Early Navy planning provided for the construction of 160 Project 641/Foxtrot submarines.

Designed by Pavel P. Pustintsev at TsKB-18 (Rubin), Project 641 was a large, good-looking submarine, 299½ feet (91.3 m) in length, with a surface displacement of 1,957 tons. Armament consisted of ten 21-inch (533-mm) torpedo tubes—six bow and four stern. Project 641/Foxtrot had three diesel engines and three electric motors with three shafts, as in the previous Project 611/Zulu (and smaller Project 615/Quebec). Beyond the increase in range brought about by larger size, some ballast tanks were modified for carrying fuel. Submerged endurance was eight days at slow speeds without employing a snorkel, an exceptional endurance for the time. The Foxtrot introduced AK-25 steel to submarines, increasing test depth to 920 feet (280 m). The large size also provided increased endurance, theoretically up to 90 days at sea.

*A Foxtrot at speed, showing the clean lines of these submarines. The red-and-white buoy recessed into the deck forward of the sail is the tethered rescue buoy. The bow diving planes retract into the hull almost level with the long row of limber holes, which provided free flooding between the double hulls.* (Soviet Navy)

The lead ship, the *B-94,* was laid down at the Sudomekh yard in Leningrad on 3 October 1957; she was launched—64 percent complete—in less than three months, on 28 December. After completion and sea trials, she was commissioned on 25 December 1958. Through 1971 the Sudomekh-Admiralty complex completed 58 ships of this design for the Soviet Navy.[4]

Additional units were built at Sudomekh from 1967 to 1983 specifically for transfer to Cuba (3), India (8), and Libya (6). The Indian submarines were modified for tropical climates, with increased air-conditioning and fresh water facilities.[5] Later, two Soviet Foxtrots were transferred to Poland. The foreign units brought Project 641/Foxtrot production to 75 submarines, the largest submarine class to be

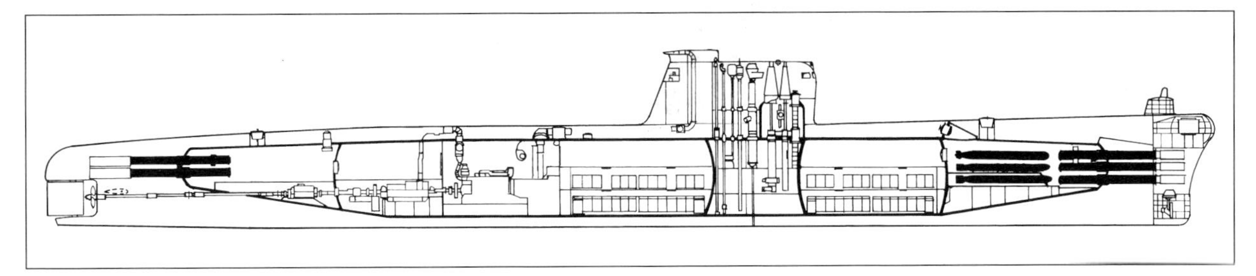

*Project 641/Foxtrot SS. LOA 299 ft 6 in (91.3 m)* (©A.D. Baker, III)

constructed during the Cold War except for the Project 613/Whiskey and Project 633/Romeo programs.

(Two Project 641 submarines are known to have been lost, the *B-37* was sunk in a torpedo explosion at Polnaryy in 1962 and the *B-33* sank at Vladivostok in 1991.)

The Soviet units served across the broad oceans for the next three decades. They operated throughout the Atlantic, being deployed as far as the Caribbean, and in the Pacific, penetrating into Hawaiian waters. And Foxtrots were a major factor in the first U.S.-Soviet naval confrontation.

## Cuban Missile Crisis

The Cuban missile crisis of 1962 demonstrated that Soviet submarines could operate at great distances and could have an influence on world events. On 1 October 1962 four Project 641/Foxtrot submarines departed bases on the Kola Peninsula en route to the Caribbean to support the Soviet weapons buildup in Cuba. In addition to conventional torpedoes, each of these submarines had one nuclear torpedo on board.[6] Another Foxtrot and one Project 611/Zulu submarine were already in the Western Atlantic.

Earlier, Soviet leaders had considered sending surface warships to escort the merchant ships carrying weapons to defend the island and ballistic missiles capable of striking the United States. That proposal was dropped because of the overwhelming superiority of U.S. naval and air forces in the region. Further, at one point it was proposed that the nuclear warheads being sent to Cuba—almost 200—would be carried in submarines. The submarines would have been able to carry only two or three warheads at a time, with protection of the crew from warhead radiation being a major problem.[7] Accordingly, it was decided to transport them in two Soviet merchant ships that would "blend in" with the stream of shipping.

During the Cuban missile crisis, all Soviet submarines in the Western Atlantic were detected by U.S. ASW forces—one Project 611/Zulu and four Foxtrots. One other Foxtrot (pennant No. 945) suffered mechanical difficulties; she was sighted on the surface, en route back to the Kola Peninsula in company with the naval salvage tug *Pamir.*[8] The Zulu was also sighted on the surface, refueling from the naval tanker *Terek.*

The United States and Canada mounted a major effort to locate the four other submarines, which had departed the Kola Peninsula on 1 October.[9] Argus, P2V Neptune, and P3V Orion maritime patrol aircraft, flying from Argentia, Newfoundland, and bases in Canada and the United States, searched for the submarines. U.S. Air Force RB-47 and RB-50 reconnaissance aircraft—without ASW equipment—briefly joined in the search. The quarantine (blockade) naval forces that were to stop merchant ships believed to be carrying offensive arms to Cuba included the specialized ASW aircraft carrier *Essex,* later joined by two additional ASW carriers.

The Cuban crisis reached a critical point a few minutes after 10 A.M. (Washington, D.C. time) on Wednesday, 24 October. At the time Secretary of Defense Robert S. McNamara advised President Kennedy that two Soviet freighters were within a few miles of the quarantine line, where they would be intercepted by U.S. destroyers. Each merchant ship was accompanied by a submarine—"And this is a very dangerous situation," added McNamara.[10]

Robert F. Kennedy, the attorney general and brother of the president, recalled, "Then came the disturbing Navy report that a Russian submarine had moved into position between the two [merchant] ships."[11]

The decision was made at the White House to have the Navy signal the submarine by sonar to surface and identify itself. If the submarine refused, small explosives would be used as a signal. Robert Kennedy recalled the concern over that single Foxtrot submarine: "I think these few minutes were the time of gravest concern for the President. . . . I heard the President say: 'Isn't there some way we can avoid having our first exchange with a Russian submarine—almost anything but that?' "[12]

At 10:25 that morning word was received at the White House that the Soviet freighters had stopped before reaching the blockade line.

No one participating in the White House discussions nor anyone else in the U.S. government knew that each of the four Soviet submarines in the area had a nuclear torpedo on board in addition to 21 conventional weapons. Rear Admiral Georgi Kostev, a veteran submarine commander and historian, described the views of Nikita Khrushchev in approving the sending of submarines to support the clandestine shipment of ballistic missiles to Cuba (Operation Anadyr): "Khrushchev believed that it was proper for the submarines, if the convoy would be stopped, to sink the ships of the other party."[13] When asked how he believed a nuclear war could have been sparked at sea, Admiral Kostev replied:

> The war could have started this way—the American Navy hits the Soviet submarine [with a torpedo or depth charge] and somehow does not harm it. In that case, the commander would have for sure used his weapons against the attacker.

Some of the Soviet submarine commanders believed that they were under attack as U.S. destroyers and aircraft harassed them, dropping small explosive signal charges. To some commanders these were simply small depth charges.

TABLE 13-1
**Soviet Submarines Identified by the U.S. Navy During the Cuban Missile Crisis**

| First Sighting* | Position | Type | Notes |
|---|---|---|---|
| 22 Oct | 1024Z42°55'N 39°30'W | 611/Zulu | returning to USSR |
| 24 Oct | 1929Z25°25'N 68°10'W | 641/Foxtrot | returning to USSR |
| 25 Oct | 2211Z27°30'N 68°00'W | 641/Foxtrot | northeast of Cuba |
| 26 Oct | 1048Z21°31'N 69°14'W | 641/Foxtrot | east of Cuba |
| 26 Oct | 1908Z24°40'N 72°15'W | 641/Foxtrot | northeast of Cuba |
| 26 Oct | 2105Z18°05'N 75°26'W | 641/Foxtrot | south of Cuba |

Notes: * Greenwich Mean Time (Z).

One of the submarine commanding officers, Captain 2d Rank Aleksey F. Dubivko of the *B-36*, recalled, "I had my orders to use my weapons and particularly [the] nuclear torpedo only by instructions from my base. Absolutely."[14]

Dubivko added: "As far as I'm concerned, having received an order to use my nuclear torpedo, I would surely have aimed it at an aircraft carrier, and there were plenty of them there, in the area."

Intensive U.S. search operations reported six Soviet submarines, albeit two of them on the surface and returning to the USSR. The four Foxtrot-class submarines in the Cuban area during the crisis were the *B-4*, *B-36*, *B-59*, and *B-130*. (Early Soviet plans had proposed major surface and submarine groups to support the missile deployment to Cuba. The latter was to consist of four Project 658/Hotel SSBNs and seven diesel-electric torpedo submarines—all with nuclear weapons.)

One Foxtrot was forced to the surface after 34 continuous hours of sonar contact and harassment by the U.S. destroyer *Charles P. Cecil* and maritime patrol aircraft on 30–31 October.[15] The submarine/ASW aspects of the Cuban missile crisis provided considerable insight into Soviet submarine capabilities and tactics as well as problems. However, as the official U.S. analysis of the ASW operations—given the code name CUBEX—states:[16]

> The reliability of results of CUBEX evaluation is affected by small sample size and biased by two major artificialities. The factitious aspects of the operation included the non-use of destructive ordnance and the priority scheduling of aircraft during daylight hours for the visual/photographic needs of the Cuban quarantine force. The unnatural case of not carrying a contact through "to the kill" affected the tactics of both the hunter and hunted, as did the unbalanced day/night coverage.[17]

The Soviet submarines sought to evade detection by short bursts of high speed, radical maneuvering—including backing down and stopping, taking advantage of thermal layers, turning into the wakes of ASW ships, and releasing "slugs" (bubbles) of air and acoustic decoys. But knowing that they probably were safe from attack with lethal weapons, and not being required to carry out attacks against U.S. ships, the Soviet submarine captains were not realistically tested in a conflict situation. Also, the submarines made extensive use of radar, which they might not employ to the same extent in wartime. Extensive snorkeling also occurred with durations of one-half hour to 11 hours being detected by U.S. forces. Undoubtedly in wartime the submarines would have practiced less frequent and shorter-duration snorkel operations.

The experience for both Soviet submariners and U.S. ASW practitioners was invaluable.

The success of Project 641/Foxtrot, led to several cruise missile (SSGN) variants being proposed. (See Chapter 6.) In 1956–1957 preliminary designs were developed at TsKB-18 for Project 649, a more advanced diesel-electric submarine with a surface displacement of some 2,560 tons, a submerged displacement of 3,460 tons, and an underwater speed up to 20 knots. The project was not pursued.

Instead, from 1965 a TsKB-18 team under Yuri Kormilitsin developed Project 641B (NATO Tango). This was a major improvement of the Foxtrot design with respect to hull form and battery capacity, the latter providing an increase in submerged endurance. Also, the sonar was linked directly to the torpedo fire control system, enabling more rapid handling of firing solutions and the automatic transmission of data to torpedoes before firing.

Eighteen submarines of this long-range design were delivered from 1973 to 1982 by the Krasnoye Sormovo shipyard in Gor'kiy (now Nizhny Novgorod). Construction of these submarines, in turn, was halted with development of the improved Project 877 (NATO Kilo) design.

Meanwhile, with the completion of the truncated Project 613/Whiskey program—the last submarine being delivered in 1958—work was under way on a successor medium-range submarine, Project 633 (NATO Romeo). Design work had begun at SKB-112 (Lazurit) in 1954 under chief designers Z. A. Deribin and, after 1959, A. I. Noarov.[18] The new design had enhanced armament, sensors, and performance over Project 613.

The number of bow torpedo tubes was increased to six, in addition to the two stern tubes, while sonar, habitability, and food storage were improved, the last increasing practical endurance from 30 days for the Whiskey to 60 days for the Romeo. The latter submarine was viewed as the ultimate undersea craft of its type by the Soviet Navy's leadership with preliminary plans being made for the production of 560 ships!

The first Project 633/Romeo, the *S-350*, was laid down on 22 October 1957 at the Krasnoye Sormovo shipyard in Gor'kiy, which had been the lead yard for the Whiskey class. The submarine was delivered to the Navy in December 1959 with a total of 20 units being built through 1962. The early cancellation of large-scale production and the subse-

*The simplistic lines of the Project 941B/Tango are evident in this view. The Tango was intended as the successor to the Foxtrot class, but, in the event, few were built. The Tango's principal advantage over her predecessor was increased submerged endurance.* (U.K. Ministry of Defence)

*The Project 633/Romeo had the same sonars as the larger Foxtrot. The Romeo's sail was distinguished by the "top hat" extension that housed the submarine's periscopes. The massive Romeo construction program was cut back in the USSR because of the shift of emphasis to nuclear propulsion and long-range submarines.*

quent transfer of 15 of these ships to other countries reflected the Soviet Navy's emphasis on nuclear propulsion and long-range conventional undersea craft. The recipients were Algeria (2), Bulgaria (4), Egypt (6), and Syria (3), with the Egyptian units being transferred upon completion.

China was given plans and documentation for Project 633/Romeo and initiated mass production construction of these submarines at four shipyards. According to Chinese sources, slightly more than 160 units were completed in China from 1962 to 1987, which would make this the second largest class of undersea craft constructed after World War II.[19] (Western sources place Chinese production of the Romeo at 92 units, completed between 1960 and 1984; eight of those were transferred to Egypt (4) and North Korea (4).[20]

*Project 690/Bravo was a specialized target submarine for Soviet ASW forces. The reinforced outer hull has a "hump" aft of the sail, the "aim point" for practice torpedoes. The U.S. Navy constructed two smaller training submarines in the early 1950s, the* T1 *(SST 1) and* T2 *(SST 2). They were of limited use.* (U.K. Ministry of Defence)

One Soviet unit of this class, the *S-350,* sank in 1962 at the naval base of Polyarnyy on the Kola Peninsula as a result of a torpedo explosion in the nearby submarine *B-37*. The *S-350* was salvaged, repaired, and placed back in service in 1966.

In 1956 work began at SKB-112 on the more advanced Project 654, a medium-range submarine with a surface displacement of some 1,600 tons. This was the first Soviet submarine design based on the high-speed USS *Albacore* (AGSS 569). Construction of the lead submarine was begun at the Krasnoye Sormovo yard, but soon halted. This marked the end of the development and construction of medium-range submarines in the USSR.

## Specialized Submarines

Beyond combat submarines, the Soviet Navy constructed and converted submarines for a large number of specialized roles. In 1961 work began at SKB-112/Lazurit in Gor'kiy on the diesel-electric Project 690 (NATO Bravo), a target submarine. Developed under chief designer Ye. V. Krilov, the submarine had a special outer hull to absorb non-explosive strikes by ASW rockets and torpedoes. From 1967 to 1969 four ships of Project 690 were completed at the Komsomol'sk yard on the Amur River in the Far East.[21] These target submarines were employed for ASW training and weapons development. From the late 1970s a new target submarine was designed to replace these ships; however, work on this design, based on Project 877/Kilo and designated Project 690.2, was stopped in the mid-1980s.

Another specialized design produced at SKB-112 was for carrying submersibles that could be employed for salvage or to rescue survivors from submarines disabled on the ocean floor above their collapse depth. The submersible concept was tested on modified Whiskey-class submarines with the project numbers 613S and 666. In 1968 the Lazurit bureau initiated the design of Project 940 (NATO India), a large "mother" submarine with a surface displacement of 3,860 tons. The submarine carried two rescue submersibles (Project 1837) that could be launched while submerged to travel to and "mate" with the disabled submarine. After taking survivors aboard, the rescue submersibles would return to the submerged mother submarine to transfer the rescuees. The two ships of the India class were completed at Komsomol'sk in 1976 and 1979.

The U.S. Navy had earlier developed a similar but more flexible underwater rescue submersible, designated Deep Submergence Rescue Vehicle (DSRV). As part of an extensive deep submergence program initiated after the loss of the *Thresher* (SSN 593) in April

*Project 940/India was a specialized submarine that could "nest" two rescue submersibles. The submarine had sail-mounted diving planes. The markings on the after deck are underwater "landing aids" for the submersibles. This was the Pacific Fleet unit; the other was in the Northern Fleet.* (Japanese Maritime Self-Defense Force)

*The attack submarine* Pintado *(SSN 672) of the* Sturgeon *class modified to a DSRV "mother" configuration. The DSRV* Avalon *is "mated" to the submarine's after hatch. Temporary markings are evident on the* Pintado*'s sail. Submarines carrying the DSRVs retain their full combat capability.* (U.S. Navy)

1963, the U.S. Navy planned a force of 12 DSRVs, to be carried to sea in mother submarines as well as by specialized surface ships.[22] In the event, only two of these craft were completed by the Lockheed Corporation in 1971–1972. However, the DSRVs could be carried by and operated from standard attack submarines (SSN) that have minimal special fittings in contrast to the Soviet scheme, wherein only the two Project 940/India submarines could carry rescue submersibles. Eight U.S. SSNs were fitted as mother submarines without detracting from their combat capabilities, and two specialized DSRV surface support ships were built.[23] The DSRVs have a high degree of "strategic" mobility, being capable of being air transported to overseas ports by C-5 cargo aircraft. Each of the two rescue vehicles, the *Mystic* (DSRV-1) and *Avalon* (DSRV-2), is operated by a crew of three and can embark 24 survivors.[24] Their operating depth is 5,000 feet (1,525 m), that is, significantly beyond the collapse depth of U.S. submarines.

During their development the DSRV program helped to provide "cover" for deep-ocean search and recovery systems in highly covert or "black" programs, including equipment for the extensively modified *Seawolf* (SSN 575), *Halibut* (SSN 587), and *Parche* (SSN 683). Those submarines

*Closeup of the DSRV* Avalon, *while mated to a British submarine. One French and several British submarines as well as several U.S. submarines were configured to carry and support DSRVs. The openings in the bow are twin ducted thrusters for precise maneuvering; there are two more thrusters aft.* (U.S. Navy)

*The* Dolphin *was built specifically as a deep-diving research submarine. During her career she has undergone many modifications, with different sonars, hull materials, and sail configurations being fitted. This photo shows her configuration in 1983.* (U.S. Navy)

*Numerous U.S. World War II–era submarines were configured for various research roles and served into the mid-1970s. This is the* Grouper *(SS/AGSS 214), configured as a sonar test ship. The three large "fins" are antennas for the PUFFS passive ranging system.* (U.S. Navy)

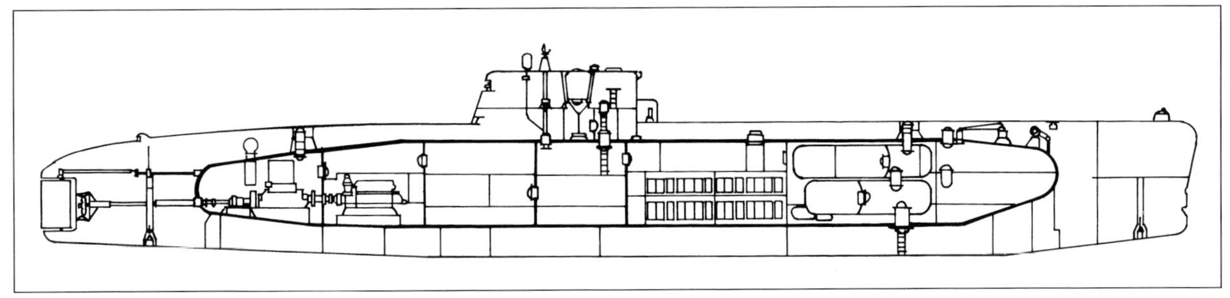

*Project 1840/Lima AGSS showing the two forward, inter-connected hyperbaric chambers. LOA 278 ft 10 in (85.0 m)* (©A.D. Baker, III)

were modified to carry out clandestine operations and to search for and recover objects dropped on the ocean floor (foreign weapons, re-entry vehicles, etc.).

*The enigmatic Project 1840/Lima, a research submarine. Forward she has two large hyperbaric chambers within her pressure hull; divers can easily transit from the chambers to the open sea.* (Malachite SPMBM)

The only special-purpose submarine built by the U.S. Navy in this period was the deep-diving research craft *Dolphin* (AGSS 555). Completed in 1968, the *Dolphin* is a small undersea craft with a length of 165 feet (50.3 m), a displacement of 860 tons surfaced, and 950 tons submerged. She was employed to test equipment and materials to depths of 3,000 feet (915 m). The *Dolphin* was unarmed, although when completed she was fitted with a single test torpedo tube (removed in 1970). She was the last diesel-electric submarine to be constructed by the U.S. Navy.

Into the mid-1970s the U.S. Navy also operated varying numbers of World War II–era submarines converted to various research configurations.

The Soviet Navy acquired several specialized submarines from the 1970s onward, most designed by the SDB-143/Malachite bureau and constructed at Sudomekh-Admiralty, that yard being known for building specialized submarines. The diesel-electric Project 1840 (NATO Lima) completed in 1979 was as an underwater laboratory for research activities, with two hyperbaric chambers. The diesel-electric Project 1710 (NATO Beluga) was a full-scale hydrodynamic test ship.

The Beluga had her beginnings in 1960 when Malachite engineers recommended creating a submarine to examine the means of reducing a submarine's water resistance or drag, the same concept as the *Albacore* in the U.S. Navy.[25] The proposal was well received by the Navy and by the Siberian Department of the USSR Academy of Sciences, which was engaged in research of underwater shapes. The project received preliminary approval and, as an early step, a large, buoyant-ascent model known as *Tunets* (Tuna) was constructed for research into various means of controlling the boundary layer, including the use of gas infusion and polymer ejection.

The 37¾-foot (11.5-m) model *Tunets* was designed by Malachite and built by the Siberian Department for testing at an acoustics range near Sukhumi on the Black Sea. Model tests with polymers demonstrated a decrease of 30 to 40 percent in total resistance to provide more than a 12 or 13 percent increase in speed.

The technical design for the Project 1710/Beluga submarine was developed in 1975, and the Admiralty-Sudomekh yard began construction in 1985. The lengthy interval was caused by having to overcome the opposition to the project by some scientists at the Krylov Research Institute who considered that all data required from the project could be obtained from laboratories. The diesel-electric Beluga was completed at Leningrad in February 1987 and transferred through inland waterways to the Black Sea for trials.

The Beluga's hull form is a body-of-revolution with a length-to-beam ratio of 7:1, similar to that of the USS *Albacore*. The hull—with a low-profile faired-in sail—reveals a very high degree of streamlining. A large sonar is provided in the bow section with fully retractable bow planes. The submarine has a system for delivering a polymer solution to the boundary layer flow over the hull, appendages, and propeller. The propeller and shaft are configured to permit changing the axial distance between the propeller and hull, providing researchers with the opportunity to investigate the effects of polymer on the thrust deduction factor. There is extensive interior space for research equipment.

After arriving at Balaklava on the Black Sea, in sea trials the Beluga demonstrated the performance promised by her designers and the *Tunets* tests. However, the Beluga's operations were curtailed because of the political conflict between the new Russian regime and newly independent Ukraine. Also halting research was the shortage of funds within the Russian Navy. During her relatively short period of trials, the Beluga did provide important insights into submarine hydrodynamics as well as scaling data to improve the accuracy of predictions based on model tank testing. Among the Beluga's notable achievements were a better understanding of submarine hull shaping and the effectiveness of polymers on both hull drag and propulsion efficiency.

Several Soviet diesel-electric as well as nuclear submarines were converted to specialized roles in this period, among them:

| Type | Project | NATO | Role |
|---|---|---|---|
| SS | 613RV | Whiskey | ASW missile trials |
| SS | 617 | Whiskey | ballistic missile trials |
| SS | 633V | Romeo | ASW missile-torpedo (2 units) |
| SSBN | 658 | Hotel | sensor trials |
| SSBN | 667A | Yankee | sonar trials (Proj. 09780) |
| SSGN | 675 | Echo II | support deep-sea operations |
| SSBN | 667A | Yankee | support deep-sea operations (Proj. 09774)[26] |
| SSBN | 667BDR | Delta III | support deep-sea operations[27] |

*Stern aspect of the Beluga, showing her body-of-revolution cross section. The doors atop the sail are visible. Her role was similar to that of the USS* Albacore, *completed almost 35 years earlier.* (Leo Van Ginderen collection)

Additional submarines of Projects 611/Zulu and 613/Whiskey classes were refitted for various research projects. (See Chapter 2.)

In the early 1990s the Rubin bureau proposed the conversion of a Project 667BDR/Delta III to an underwater Arctic research platform, capable of carrying one or two submersibles. That proposal was not pursued.

## The United States Perspective

The United States became the first nuclear power to cease the construction of diesel-electric submarines because of (1) the promised—and fulfilled—efficacy of nuclear submarines, (2) the long distances to the U.S. Navy's forward operating areas, and (3) the already high level of defense funding for much of the Cold War, especially for deterrence (e.g., Polaris, Poseidon, and Trident programs). Because the opposition by the U.S. submarine community to non-nuclear submarines was distinctive, an examination of this issue in the context of Cold War submarine requirements and developments is warranted.

As the U.S. conventional submarine force began to decline precipitously in the 1970s, several voices were heard advocating that the Navy procure new diesel-electric submarines. Foremost among the early advocates were retired Captain K. G. Schacht and Commander Arthur (Art) Van Saun, the latter a former commanding officer of the USS *Barbel*. Both men argued for a small number of non-nuclear submarines to supplement the

*Project 1710/Beluga was a hydrodynamics test platform. All of her masts and periscopes retracted into the sail, with doors covering the openings to reduce drag. The Beluga also carried out polymer ejection tests during her abbreviated sea trials.* (Malachite SPMBM)

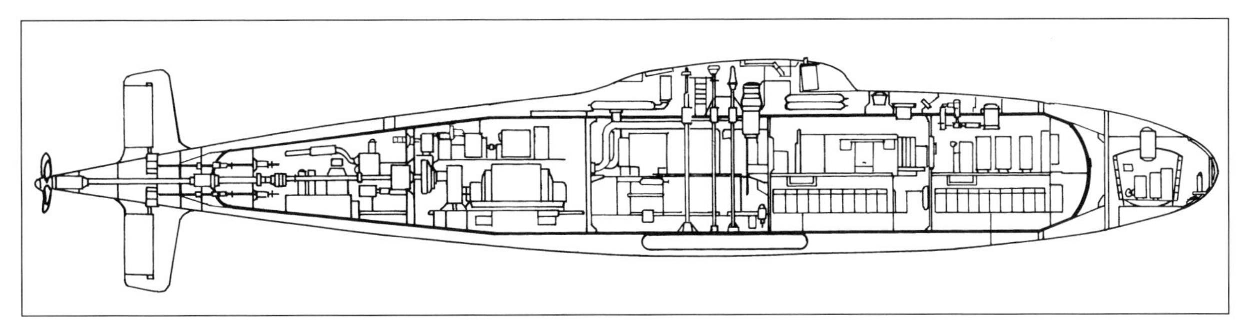

*Project 1710/Beluga AGSS. LOA 209 ft 11 in (64.0 m)* (©A.D. Baker, III)

Navy's growing and costly nuclear submarine fleet.[28]

The ranks of non-nuclear submarine advocates soon were joined by members of Congress, especially Representative William Whitehurst and Senator Gary Hart, as well as by several non-submarine flag officers, among them Admiral Elmo R. Zumwalt, former Chief of Naval Operations (1970 to 1974), and Vice Admiral John T. Hayward, a pioneer in the development of nuclear propulsion and nuclear weapons, who was the Navy's first Assistant CNO and then Deputy CNO for research and development (1957 to 1962).

The diesel-submarine issue was intensified by the Falklands War of 1982. The British nuclear submarine *Conqueror* sank an Argentine cruiser, forcing the Argentine Navy to return to its bases, except for submarines. Meanwhile, the inability of British nuclear submarines to effectively support commando operations forced the dispatch of the diesel-electric submarine *Onyx* from Britain to the South Atlantic, a distance of some 7,000 n.miles (12,970 km).[29]

During the war the Argentine submarine *San Luis,* a German-built Type 209, undertook a 36-day patrol, during which she located and operated in

Table 13-2

**U.S Submarine Force Levels, 1945–1980**

| Ship Type | 1945 | 1950 | 1953* | 1955 | 1960 | 1965 | 1970 | 1975 | 1980 |
|---|---|---|---|---|---|---|---|---|---|
| Submarines—*conventional* | | | | | | | | | |
| SS-SSK-SSR | 237 | 73 | 122 | 121 | 131 | 83 | 59 | 11 | 6 |
| SSG | — | — | 1 | 2 | 4 | — | — | — | — |
| AGSS auxiliary | — | — | 7 | 18 | 20 | 15 | 26 | 2 | 1 |
| Submarines—*nuclear* | | | | | | | | | |
| SSN-SSRN | — | — | — | 1 | 7 | 22 | 48 | 64 | 73 |
| SSGN | — | — | — | — | 1 | — | — | — | — |
| SSBN | — | — | — | — | 2 | 29 | 41 | 41 | 40 |
| **Total** | **237** | **73** | **130** | **142** | **165** | **149** | **174** | **118** | **120** |

Notes: * End of Korean War (June 1953).

the area of the British carrier force. Although the *San Luis* reported firing several torpedoes at British ships, she scored no hits because of faulty wiring of her fire control system (which had been overhauled during a recent yard period). British ASW forces prosecuted numerous suspected submarine contacts, expending a large number of weapons, but without success.[30]

U.S. Secretary of the Navy John Lehman observed,

> The ability of a modern diesel-electric submarine to engage a naval task force that is essentially stationary while operating in a specific area is not surprising. These submarines are extremely quiet when operated at low speeds and for this reason substantial helicopter, subsurface, and surface anti-submarine warfare defense is required whenever a naval task force is constrained to a limited area.[31]

After the war, Admiral Sandy Woodward, who had commanded the British forces in the Falklands campaign, said that the U.S. Navy probably should have a half dozen diesel submarines for high-risk submarine operations—inshore surveillance; operating in shallow, mined waters; landing agents; and other activities when one should risk a smaller, cheaper diesel submarine rather than a nuclear submarine. He also pointed out that a modern diesel costs one-third to one-fifth as much as an SSN.[32]

In the 1980s the Israeli Navy approached the U.S. government for funds and support for the construction of three modern diesel submarines to replace a trio of older boats. Almost simultaneously, U.S. shipyards were being approached by South Korean representatives who wished to build perhaps two submarines in the United States, to be followed by additional construction in Korea. Several U.S. yards that were *not* engaged in nuclear submarine construction expressed interest, and a tentative agreement was reached with the Todd Pacific Shipyards whereby Israel and South Korea would construct submarines of the same design, which had been developed for Israel by the Dutch firm RDM. The Australian Navy also expressed some interest in "buying into" the arrangement.[33]

Such a construction program, it was estimated, could have led to a production run of two or three submarines a year while providing employment for up to 7,000 shipyard workers and supporting-industry workers in the United States.[34] This was a critical factor in view of the numerous American shipyard closings during the 1970s and 1980s. Further, building those submarines could have actually facilitated *importing* technical data into the United States. This would have provided U.S. naval officers and members of Congress with an up-to-date view of this field while providing a basis for future understanding of the non-nuclear submarine threat to the U.S. Navy. The submarine community immediately opposed the program, citing nuclear submarine technology loss to other countries if diesel submarines were built at any U.S. shipyards. The vehemence of those objections led Representative William Whitehurst to write to Secretary of Defense Caspar Weinberger:

> I know that the Navy has been adamant in their opposition to a diesel submarine program of their own, but their going to *this* length [halting foreign construction in U.S. yards] absolutely confounds me. There is no possibility of the transfer of technology, so what on earth is their objection? The opportunity to create additional jobs for American workers and keep these shipbuilding companies viable ought to outweigh any suspicion the Navy might have that this represents the nose of the camel under the tent in forcing diesel submarines on them.[35]

The submarine community's objections prevailed. The Israeli submarines were built in Germany, the South Korean submarines were built to a German design in Germany and South Korea, and the Australian submarines were built in that country to a Swedish design.

The adamant objection to diesel submarines being built for the U.S. Navy or allied navies originated with Admiral H. G. Rickover, who had long claimed that "the Navy" had opposed nuclear submarines, preferring instead diesel submarines, which were less-capable (and less-costly) than "nukes." Although Admiral Rickover was forced out of the

Navy in January 1982, from mid-1982 until 1994, the U.S. Navy had a succession of nuclear submariners as Chief of Naval Operations, all Rickover protegés. Further, after Rickover's departure the head of naval nuclear propulsion continued to be a full admiral, appointed for an eight-year term (the CNO normally serves for only four years). With these and other nuclear submarine flag officers in key Navy positions in Washington, advocates of non-nuclear submarines were actively opposed by the senior submarine admirals, who were known as the "submarine mafia."[36]

But interest in non-nuclear submarines in the United States continued. In 1990 former Secretary of the Navy Paul H. Nitze and Admiral Zumwalt proposed acquisition of a small number of such craft to compensate for the expected decline in the number of nuclear submarines. The non-nuclear submarines would be employed for a variety of missions, both combat and support.[37] Increasingly, the principal role envisioned for non-nuclear submarines was anti-submarine training. With an all-nuclear U.S. submarine force, even the "smallest" attack submarines in service after the year 2001—the *Los Angeles* (SSN 688) class—could not effectively simulate diesel-electric submarine targets for ASW forces. The SSNs are too large, have very different signatures than the typical submarines operated by Third World navies, and are operated quite unlike the submarines the U.S. Navy would be countering. As one submarine officer wrote, "Substituting the artificially handicapped SSN for simulated enemy [diesel submarines] in any exercise borders upon the ridiculous."[38]

The post–Cold War cutback in U.S. SSNs made this situation critical, with few submarines being available for training air and surface ASW forces. This problem has been demonstrated periodically in several exercises with foreign navies. U.S. ASW forces have encountered unexpected difficulties in operations against some South American submarines during recent UNITAS operations, and Israeli submarines are said to always "sink" the high-value ships in exercises against the U.S. Sixth Fleet in the Mediterranean. The situation is further exacerbated in the relatively shallow waters of littoral regions, the expected primary "battleground" of future U.S. naval operations.

One officer in a *Los Angeles*-class submarine wrote,

> To overcome our lack of experience and perspective, we need a new solution. We must create an *aggressor* unit, the mission of which would be to portray, as accurately as possible, the capabilities of those diesel submarine forces about which we are most concerned. This aggressor unit must operate on one of those submarines of concern, preferably a [Soviet] Kilo or [German] Type 209. It's time for the U.S. Navy to build or buy one or more of these submarines.[39]

Another supporter of this concept, wrote,

> Buying a pair of diesel submarines would give us practice not only against Kilo-class submarines, but would help to return us to proficiency against all diesels. This is not to say that we aren't proficient, just that the lack of a real-diesel platform to continually train with has been a clear deficiency for all units that search for subs.[40]

The term "diesel" includes the Air-Independent Propulsion (AIP) submarines, which began entering service at the end of the century. These non-nuclear submarines, with several in operation and more under construction, have the ability to cruise at slow speeds for up to 30 days without snorkeling at very low noise levels. Vice Admiral John J. Grossenbacher, Commander Submarine Force, U.S. Atlantic Fleet, called the Swedish submarines and the German Type 212 submarines also fitted with AIP "A challenge to our undersea dominance."[41]

Beyond ASW training, another role appears highly suitable for non-nuclear submarines: special operations, landing and recovering SEAL teams and other personnel, and intelligence collection and surveillance operations in hostile areas. Today the U.S. Navy uses submarines of the *Los Angeles* class for this purpose. The U.S. Navy in 2002 was beginning the conversion of four Trident missile submarines, with a submerged displacement of 18,750 tons, to a combination cruise missile and special operations/transport role.[42] While these submarines would have great value in some scenarios, in most situations postulated for the future, a submarine will be required to deliver or remove a small number of people, perhaps a dozen or less, in a relatively shallow, restricted area. Should a large, nuclear-propelled submarine, with a crew of 130 or more be employed for such missions when a small, diesel-electric/AIP submarine with a crew of 30 to 55 could carry out many of those special missions?

The usual objection to employing non-nuclear submarines for this or any other combat mission is their limited transit speed. But a look at U.S. special operation submarines before the mid-1980s, when diesel-electric submarines were employed exclusively for this role, shows that they performed admirably. As noted, in the Falklands War of 1982 the Royal Navy had to dispatch a diesel-electric submarine, HMS *Onyx,* to the war zone for special operations because such craft were more suitable in that role than were nuclear submarines. Similarly, the *Onyx* and another British diesel-electric submarine, the *Otus,* carried out special operations in the Persian Gulf during the 1991 conflict, those operations being carried out thousands of miles from the nearest naval base.

By the early 1980s the U.S. submarine community's opposition to non-nuclear submarines had centered on three issues: (1) the perception that the Navy's leadership had long opposed nuclear submarines, preferring instead lower-cost diesel-electric craft; (2) the concern that nuclear submarine technology would be transferred to countries purchasing U.S.-produced diesel submarines; and (3) the fear that monies spent for non-nuclear submarines would take away funding from nuclear submarines. However, after more than two decades, these objections are no longer as convincing to objective consideration.

The first issue no longer has credibility, as from 1982 onward the Navy's leadership has included nuclear submariners in the most senior positions. With respect to the second issue, non-nuclear submarines now built in the United States would be based on foreign design as would their key components; U.S. sonars, pumps, quieting techniques, and other sensitive nuclear submarine components are not applicable to power-limited, non-nuclear submarines. Further, during the past two decades, several non-nuclear shipyards expressed interest in constructing such submarines, especially the Litton/Ingalls and Todd Pacific shipyards.

Only the third issue—fiscal—has any merit. On several occasions senior members of Congress have expressed interest in funding non-nuclear submarines for the U.S. Navy outside of the existing submarine budget.[43] However, the probable cost of an advanced non-nuclear/AIP submarine is about one-sixth the cost of a modern U.S. nuclear attack submarine. Coupled with the lower manning costs, and the availability of overseas bases for the foreseeable future, the nonnuclear submarine becomes especially attractive, especially in view of the continually stated shortfalls of nuclear submarines.[44] Non-nuclear submarines will be increasingly attractive as submarine requirements outstrip nuclear submarine availability, and nuclear submarine construction and manning costs continue to increase.

## Forever . . .

As then-Captain Brooks had observed, the Soviets consistently showed a major commitment to diesel-electric submarines to complement their growing fleet of nuclear-propelled submarines. In 1975, work began on the new generation submarine, Project 877 (NATO Kilo).[45] Designed at the Rubin bureau under chief designer Yuri N. Kormilitsin, the single-shaft ship has electric drive. The diesel engines (generators) are not connected to the propeller shaft, but serve to charge the batteries for the electric motors that drive the single propeller shaft. (In 1990 the Black Sea–based Kilo *B-871* began conversion to a pump-jet propulsor of 5,500 horsepower; however, progress was slow, and the submarine did not return to sea until 2000 because of the lack of batteries.[46] At the time she was the only fully operational Russian submarine in the Black Sea Fleet.)

The Kilo has six bow torpedo tubes and an advanced sonar installation. Two of the tubes are fitted for launching wire-guided torpedoes. The automated reloading system provides for rapid torpedo firing: The first salvo of six can be launched within two minutes and the second salvo five minutes later. The submarine can carry up to 24 tube-launched mines in place of torpedoes.[47]

Like previous Russian combat submarines, the Kilo retains the double-hull configuration and has bow-mounted diving planes. The 32 percent reserve buoyancy provides "surface survivability, the ability to remain afloat with at least one of the submarine's

*The Projects 636 and 877/Kilo were the last conventional submarine to be series produced in the USSR. A relatively simple design, the Kilo has been popular with foreign navies and licensed production may be undertaken in China.* (Japanese Maritime Self-Defense Force)

six compartments and adjacent ballast tanks flooded."

Construction began at three shipyards—Komsomol'sk in the Far East, Sudomekh/Admiralty in Leningrad, and the inland Krasnoye Sormovo yard at Gor'kiy—giving promise of a large building program. The lead ship, the *B-248,* was built at Komsomol'sk and entered Soviet service on 12 September 1980. Through 1991—and the breakup of the Soviet Union—13 submarines had been built at Komsomol'sk and 9 at Gor'kiy. Beginning with the 14th ship, these submarines were increased in length to accommodate improved machinery. (Two more Kilos were completed for the Russian Navy in 1992–1993.)

*The Kilo has an advanced hull form, but falls far short of most performance characteristics of the U.S.* Barbel *design, which went to sea two decades earlier. Except for SSBNs and a few specialized submarines, Soviet designers have avoided sail-mounted diving planes.*

The Project 877MK variant of the Kilo was a quieter submarine, and Project 636 was a greatly improved variant, initially intended only for Soviet service, but made available for export from 1993. The Project 636/Kilo has more powerful electric motors, increasing the underwater speed to 19 knots, while more effective machinery permits slower propeller rotation, significantly lowering noise levels. Also, Project 636 was fitted with the MGK-400EM digital sonar, which provided improved passive detection of submarine targets.

From the outset it was envisioned that Kilo variants would be supplied to Warsaw Pact and other navies, accounting for the non-standard project number and the Russian nickname *Varshavyanka* (woman from Warsaw). The Krasnoye Sormovo and Sudomekh/Admiralty yards worked together to construct ships for foreign use, most designated

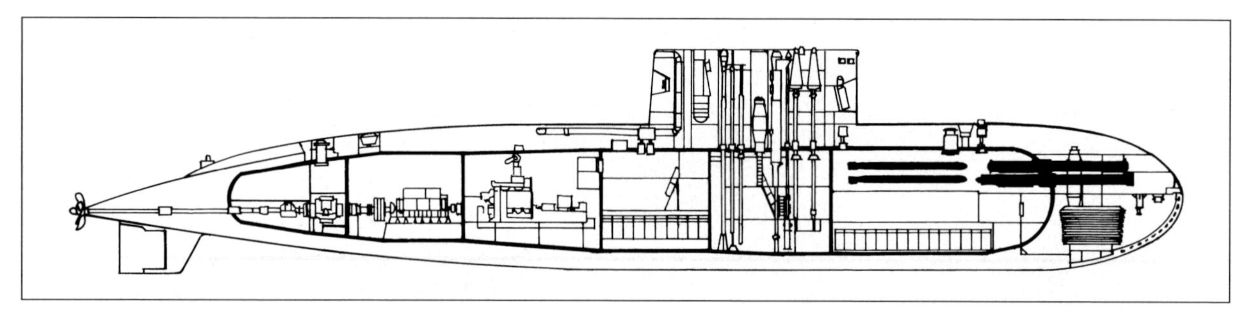

*Project 877/Kilo SS. LOA 238 ft 1½ in (72.6 m)* (©A.D. Baker, III)

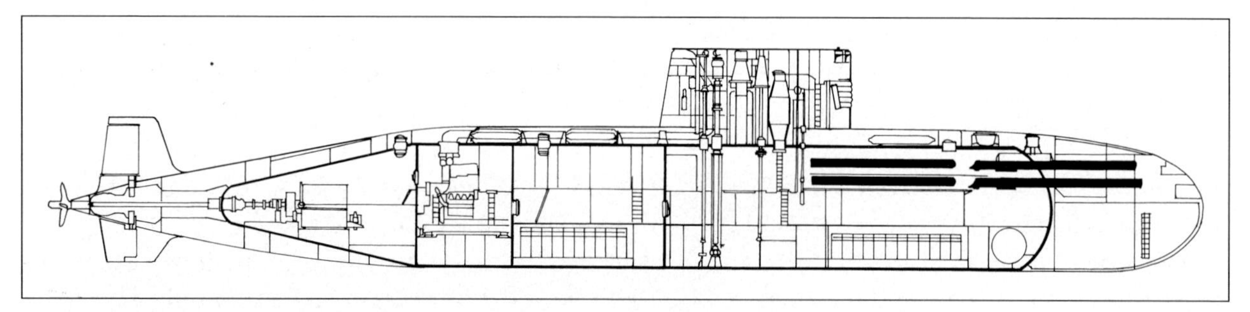

*Amur 1650/Project 677E SS. LOA 219 ft 10 in (67.0 m)* (©A.D. Baker, III)

877EKM. Since 1985 the yards have completed 21 submarines specifically for transfer: Algeria (2), China (4), India (10), Iran (3) Poland (1), and Romania (1). Some U.S. reports cited a large number of units being sought by China.[48] However, China's economic situation, coupled with its submarine construction capability, makes indigenous production—with Russian assistance—more likely. With some modifications, the foreign units are designated 877E and, with features for tropical operation, 877EKM.

In all, two dozen Kilos were produced for Russia, the final Kilo for Russian service being launched at Komsomol'sk on 6 October 1993. With another 21 built for foreign customers, the total production of Projects 636/877 was 45 submarines.

While the Russian submarines can launch both anti-ship and anti-submarine missiles, the foreign variants are armed only with torpedoes, except for the Indian units. Five or six of the Indian Kilos are fitted to fire the 3M-53E Klub-S anti-ship missile, which India also hoped to modify for land attack.

While Russian and many foreign sources are highly complimentary of the Kilo, a British submarine officer who had an opportunity to visit one was not as sanguine. Commander Jonathan Powess had just relinquished command of an *Upholder*-class SSK, Britain's last diesel-electric submarine design:

> The two classes had almost identical layouts. Talking to crew members, it was plain that the Russian boat had similar endurance and performance, but its noise isolation and shock resistance [were] suspect. Also, the larger crew—12 officers, 12 warrant officers, and 33 conscripts—was crammed in with little regard for comfort or ability to rest while off watch. The most obvious shortcoming of the Russian boat was the crude control-room equipment. It relied upon manual operation of all systems, using the remote controls as a backup. Interestingly, the Russian officers dismissed the *Upholder* class as being too delicate and likely to fail because of over-reliance on automated systems.[49]

Still, there may be some mitigation of the views of Commander Powess as the Project 877/Kilo was initially designed for Warsaw Pact navies to operate.

Kormilitsin subsequently designed a follow-on submarine, intended primarily for foreign transfer, to compete with the "low-end" diesel-electric submarines being marketed by several other nations. This was Project 677, given the Russian name Lada, with the export submarines designated as the

"Amur" series. Six variants were proposed, with the suffix number indicating the surface displacement; most Rubin marketing documents have emphasized three variants. The sixth and largest variant—1,950 tons—was intended as the follow-on to Project 636/Kilo in Russian service. (See table 13-3.)

Special emphasis in Project 677 was placed on quieting and versatility of weapons payload; like the Kilo, the larger Amur-type submarines would be able to launch anti-ship cruise missiles from torpedo tubes. Features include placing the forward diving planes on the sail (fairwater), the first Soviet submarines other than SSBNs to have this configuration. In addition, the Amur design provides for an optional AIP module to be inserted in the submarines to provide submerged endurance (without snorkeling) of 15 to 20 days.[50] The Kristal-27E fuel cell installation is some 33 to 39 feet (10–12 m) in length and employs hydrogen and oxygen. The single Russian unit that was under construction when this book went to press was being provided with fuel cells.

Construction of the first two submarines of the new design was begun at the Admiralty yard in St. Petersburg (formerly Leningrad) late in 1996 with the formal keel laying for both units on 26 December 1997—Project 677E for export and the larger Project 677 for the Russian Navy (named *Sankt Petersburg*/B-100). However, the anticipated purchase of the former submarine by India was not realized. Kormilitsin said that the two submarines were virtually identical, except that the export model would lack classified electronic and communications systems and would be modified for specific

TABLE 13-3
**Amur Series Conventional Submarines**

| | **Soviet Amur 550** | **Soviet Amur 950** | **Soviet Amur 1650 Project 677E** | **Soviet Amur 1950 Project 677** |
|---|---|---|---|---|
| Operational | — | — | — | (2003) |
| Displacement | | | | |
| surface | 550 tons | 950 tons | 1,765 tons | 1,950 tons |
| submerged | 700 tons | 1,300 tons | 2,300 tons | 2,700 tons |
| Length | 150 ft 11 in (46.0 m) | 196 ft 9 in (60.0 m) | 219 ft 10 in (67.0 m) | 236 ft 2 in (72.0 m) |
| Beam | 14 ft 5 in (4.4 m) | 18 ft 4 in (5.6 m) | 23 ft 3 in (7.1 m) | 23 ft 6 in (7.2 m) |
| Draft | | | | 14 ft 5 in (4.4 m) |
| Diesel engines* | 2 | 2 | 2 | 2 |
| horsepower | | | | |
| Electric motors | 1 | 1 | 1 | 1 |
| horsepower | | | 2,700 | 2,700 |
| Shafts | 1 | 1 | 1 | 1 |
| Speed | | | | |
| surface | | | 10 knots | 11 knots |
| submerged | 18 knots | 18 knots | 21 knots | 22 knots |
| Range (n.miles/kt) | | | | |
| snorkel | 1,500/ | 4,000/ | 6,000/ | 6,000/ |
| submerged | 250/3 | 350/3 | 650/3 | 500/3** |
| Test depth | 655 ft (200 m) | 820 ft (250 m) | 820 ft (250 m) | 820 ft (250 m) |
| Torpedo tubes*** | 4 400-mm B | 4 533-mm B | 6 533-mm B | 6 533-mm B |
| Torpedoes | 8 | 12 | 18 | 16 |
| Complement | 18 | 21 | 34 | 41 |

Notes: * Diesel generators.
** On battery; 15 to 20 days at 3 knots on AIP (fuel cells).
*** Bow torpedo tubes.

climatic conditions.[51] Series production of these submarines for the Russian Navy is doubtful because of the critical financial situation in Russia.[52]

**The U.S. Navy's decision in 1956 to cease the construction of conventional submarines as the first nuclear-propelled undersea craft was going to sea reflects the great confidence that the Navy's leadership—especially Admiral Arleigh Burke—had in nuclear technology. That decision, and subsequent efforts by senior Navy officers and civilians to build nuclear-propelled submarines, belies the subsequent myths that the Navy's leadership opposed nuclear submarine construction.**

**The other nations developing nuclear-propelled submarines initially continued the construction of diesel-electric submarines, especially the Soviet Union. While diesel submarine construction of the post-Stalin period did not approach the numbers of the immediate postwar programs, considered with the concurrent nuclear submarine programs, the Soviet conventional programs were also impressive, far exceeding the combined efforts of the Western navies.**

**Beyond providing large numbers of undersea craft for the Soviet Navy to complement nuclear-propelled submarines and for special purposes, this diesel-electric effort enabled a continuous flow of relatively modern submarines to Soviet allies and to Third World clients. This, in turn, provided additional orders for Soviet shipyards and contributed to Soviet influence on foreign military forces.**

*The sail of a Project 641/Foxtrot submarine operating in the Mediterranean in 1976. Her Quad Loop direction-finding antenna, Snoop Tray radar, and a radio aerial are raised from the fairwater, while the massive UHF antenna is raised against the after end of the sail. Several navies continue to operate Foxtrots. (James Bishop/U.S. Navy)*

**Soviet construction of diesel-electric submarines continued up to the end of the Cold War. New designs were under way when the Soviet Union fell, most significantly the Amur/Lada series, a "family" of undersea craft for both Russian use and export. Like many other post–Cold War military projects, however, the realization of this program is questionable.**

Table 13-4

**Second and Third Generation Conventional Submarines**

| | Soviet Project 633 NATO Romeo | Soviet Project 641 NATO Foxtrot | Soviet Project 641B NATO Tango | Soviet Project 877 NATO Kilo | Soviet Project 636 NATO Kilo | Soviet Project 1710 NATO Beluga | Soviet Project 1840 NATO Lima | U.S. *Dolphin* AGSS 555 |
|---|---|---|---|---|---|---|---|---|
| Operational | 1959 | 1958 | 1973 | 1980 | 1989 | 1987 | 1979 | 1969 |
| Displacement | | | | | | | | |
| surface | 1,330 tons | 1,957 tons | 2,640 tons | 2,300 tons | 2,350 tons | 1,407 tons | 1,870 tons | 861 tons |
| submerged | 1,730 tons | 2,475 tons | 3,560 tons | 3,036 tons | 3,126 tons | 2,500 tons | ~2,900 tons | 950 tons |
| Length | 251 ft 3 in | 299 ft 6 in | 295 ft 10 in | 238 ft 1½ in | 242 ft 1 in | 209 ft 11 in | 278 ft 10 in | 165 ft |
| | (76.6 m) | (91.3 m) | (90.2 m) | (72.6 m) | (73.8 m) | (64.0 m) | (85.0 m) | (50.29 m) |
| Beam | 21 ft 4 in | 24 ft 7 in | 28 ft 3 in | 32 ft 6 in | 32 ft 6 in | 29 ft 6 in | 42 ft 8 in | 19 ft 5 in |
| | (6.5 m) | (7.5 m) | (8.6 m) | (9.9 m) | (9.9 m) | (9.0 m) | (13.0 m) | (5.92 m) |
| Draft | 15 ft 1 in | 16 ft 9 in | 18 ft 8 in | 20 ft 4 in | 20 ft 4 in | 19 ft | 22 ft 4 in | 16 ft |
| | (4.6 m) | (5.1 m) | (5.7 m) | (6.2 m) | (6.2 m) | (5.8 m) | (6.81 m) | (4.9 m) |
| Diesel engines | 2 | 3 | 3 | 2* | 2* | 1 | 1 | 2 |
| horsepower | 2,000 | 6,000** | 5,570 | | | | | |
| Electric motors | 2 | 1 | 3 | 1 | 1 | 1 | | 1 |
| horsepower | 2,700 | 8,100 | 8,100 | 5,500 | 5,500 | 5,500 | | 1,650 |
| Shafts | 2 | 3 | 3 | 1 | 1 | 1 | 1 | 1 |
| Speed | | | | | | | | |
| surface | 15.5 knots | 16.8 knots | 13 knots | 10 knots | 11 knots | | 11 knots | 7.5 knots |
| submerged | 13 knots | 16 knots | 15 knots | 17 knots | 19 knots | 26 knots | 8 knots | 15 knots |
| Range (n.miles/kt) | | | | | | | | |
| snorkel | 9,000/9 | 17,900/8 | 14,000/7 | 6,000/7 | 7,500/7 | | | |
| submerged | 350/2 | 400/2 | 450/2.5 | 400/3 | 400/3 | | 245 n.miles | |
| Test depth | 985 ft | 920 ft | 985 ft | 820 ft | 985 ft | 985 ft | 1,230 ft | 3,000 ft |
| | (300 m) | (280 m) | (300 m) | (250 m) | (300 m) | (300 m) | (375 m) | (915 m) |
| Torpedo tubes*** | 6 533-mm B | 6 533-mm B | 6 533-mm B | 6 533-mm B | 6 533-mm B | none | none | removed |
| | 2 533-mm S | 4 533-mm S | 4 533-mm S | | | | | |
| Torpedoes | 14 | 22 | 24 | 18 | 18 | | none | none |
| Complement | 52 | 70 | 78 | 52 | 52 | 22 | 41 + 6# | 27 |

Notes: * The submarine has two 1,000-kW diesel generators (1,500-kW in Project 636). Also fitted are a 190-hp motor for "creeping" operations and two 102-hp emergency propulsion units.

** 5,475 horsepower in early units.

*** Bow + Stern.

# Divers in long-term decompression.

14

# Unbuilt Giants

*The U.S. Navy looked at several large and innovative submarine designs during the Cold War, but none was pursued except for the* Triton *(SSRN 586) and ballistic missile submarines. This painting, produced for* Mechanix Illustrated *magazine, shows a large transport submarine using "amphibious flying platforms" to land more than 2,000 marines on a hostile beach.* (U.S. Navy)

From the beginning of modern submarine development at the start of the 20th Century, submarine size increased continuously to accommodate more torpedoes and larger propulsion plants. However, the design and—on occasion—the construction of significantly larger submarines in any period came about because of the quest for undersea craft to carry:

- large-caliber guns
- cargo
- troops
- crude oil
- aircraft
- missiles

During the Cold War both Russian and American submarine designers sought large submarines for a number of reasons; although few of their proposals and designs were realized, to some degree they did provide the basis for the large Soviet and U.S. missile submarines that emerged as the largest undersea craft to be constructed.

The origins of many Cold War–era submarine projects can be traced to World War I. The largest submarines constructed by any nation in that period era were the British K-class submarines, intended to operate as scouts with the battle fleet. These submarines, led by the *K1* completed in 1917, were large mainly because of the steam turbines and boilers required to drive them at 24 knots on the

*The* K26 *was the last of the steam-propelled submarines built by the Royal Navy to operate with surface forces. Most of the K-boats were built during World War I; they were unsuccessful and were more dangerous to their own crews than to the Germans. Because of their numerous accidents, the crews of these submarines quipped that the "K" stood for "coffin."* (Imperial War Museum)

surface. When diving, the boilers were shut down and the trapped steam continued to turn the turbines as the ship submerged; electric motors were used for underwater propulsion. The engineering spaces reached temperatures in excess of 100°F (38°C) as they submerged.

Those submarines displaced 1,800 tons on the surface and were 338 feet (103.1 m) in length. They were armed with four bow and four beam-mounted 18-inch (457-mm) torpedo tubes, two 4-inch (102-mm) deck guns, and one 3-inch (76-mm) anti-aircraft gun. Seventeen of these ships were built. They suffered critical engineering problems, and five were operational losses.[1] An improved—and larger—series of K-class submarines was planned, but only one was completed, the *K26,* in 1923. (See table 14-1.) The problems with the steam plants of the K-boats led to abandoning that type of submarine propulsion until development of nuclear-propelled submarines a quarter century later.

At the start of World War I, submarines were fitted with small-caliber guns for attacking merchant ships and coastal craft. Deck guns were adequate for sinking the small, unarmed merchant vessels of the time. Further, the cost of a dozen rounds of small-caliber ammunition was far less than that of a single torpedo. And a submarine of that period could carry several hundred rounds of ammunition compared to perhaps a dozen torpedoes.

The largest guns ever mounted in submarines, however, were for land attack. The Royal Navy's three "submarine monitors" of the M class, led by the *M1* completed in 1918, were each armed with a 12-inch (305-mm) gun. Displacing 1,600 tons on the surface and 296 feet (90.24 m) long, these submarines were slightly smaller than the K class and had conventional diesel-electric propulsion.

The M-class submarines were developed specifically to bombard targets along the German-held coast of Belgium. The 12-inch weapon had sufficient elevation for the muzzle to break the surface while the submarine was submerged. There was a foresight on the muzzle and the periscope acted as a rear sight such that the gun could be fired while the submarine remained submerged. However, the craft had to surface to reload the gun, which could be accomplished in less than a minute. Still, surfacing and submerging took time. The 12-inch gun proved highly reliable and could be fired after having been loaded for a week, with the submarine periodically submerging. It fired an 850-pound (385.5-kg) shell.[2] Beyond their 12-inch Mk XI guns, taken from discarded battleships, the *M1* and *M2* had four 18-inch bow torpedo tubes; the *M3* had four 21-inch (533-mm) tubes.

The *M1,* the only one of the trio to see service before the war ended, was sent to the Mediterranean but had no contact with the enemy. After the war the 12-inch gun was removed from the *M2,* and she was converted in 1925–1927 with a hangar and catapult installed forward of her conning tower.[3] Thus rebuilt she carried a small, Parnall Peto floatplane.[4] The *M3* also had her big gun removed and was refitted to carry 100 mines. (A fourth M-class submarine was never finished and the hull was scrapped.)

*The largest guns ever mounted in an undersea craft—12-inch guns—were fitted in the three M-class submarines. The* M1 *was the only one to see service in World War I, but she never fired her gun in anger. The gun could be fired while submerged, but the submarines had to surface to reload.* (Imperial War Museum)

The largest guns mounted in submarines built in the 1920s and 1930s were 6-inch (152-mm) weapons, except for the French *Surcouf.* She was proceeded by another "submarine cruiser," the British experimental *X1,* completed in 1925. This concept evolved from the German submarine cruisers of World War I, being intended (like the *Surcouf*) for long-range operations against merchant convoys that had light warships as escorts. The *X1* carried four 5.2-inch (132-mm) guns in two cupola-type gun shields. The heavy gun armament and large fuel bunkers made the *X1* significantly larger than the previous K or M classes.[5] She also had six 21-inch torpedo tubes. But the submarine was not considered successful, suffered machinery problems, and was scrapped in 1931 after very brief service. The *X1* had the dubious distinction of being the only RN warship laid down after World War I to be broken up before the start of World War II.

More impressive was the French submarine cruiser *Surcouf,* completed in 1934, the world's largest undersea craft of the time. She carried twin 8-inch (203-mm) guns in a turret forward of the conning tower; a hangar aft could accommodate a Besson MB411 floatplane. Her torpedo armament consisted of eight 21.7-inch (550-mm) tubes and four 15.75-inch (400-mm) tubes.[6]

Again, this was a large submarine, displacing 3,304 tons on the surface with a length of 360⅚ feet (110.0 m). Her large fuel bunkers provided a long cruising range for hunting ocean convoys—10,000 n.miles (18,530 km) at ten knots on the surface.

## Underwater Cargo Ships

The development of cargo submarines began on the eve of World War I, when Germany produced a series of cargo-carrying undersea craft led by the *Deutschland.*[7] The origins of these submarines may have been a dinner in Berlin in 1909 at which American submarine designer Simon Lake had sat next to Alfred Lohmann, a director of the North German Lloyd shipping line. During a conversation of several hours Lake tried to convince Lohmann of the potential of cargo submarines.[8]

Soon after war began in Europe in August 1914, Lohmann proposed a class of specialized cargo-carrying submarines to penetrate the British blockade of Germany and bring home needed strategic metals and rubber. Rudolf Erback, a Krupp engineer, was named to design the cargo craft, based on the minelaying submarine *U-71.* The cargo submarine

*The world's first "commercial" cargo submarine was the German* Deutschland, *shown here returning to Bremen after her first of two successful voyages to the United States. She has the American flag at her foremast. Her two radio masts hinged down into recesses in the deck.* (U.S. Navy)

project was given the code designation "U-200" with two boats being initially ordered.

Launched at Flensburg on 28 March 1916 and fitted out at Kiel, Project U-200—named *Deutschland*—was the largest submarine yet built in Germany. She had a conventional design except for the enlarged space between her inner and outer hulls, as much as five feet (1.55 m) amidships. This space could be used to carry "wet" cargo in addition to the cargo stowed within her pressure hull. She had a 780-ton cargo capacity with a surface displacement of 1,440 tons. Of this cargo, some 250 tons of crude rubber were to be carried in floodable compartments external to the pressure hull. She was unarmed, being fitted with neither torpedo tubes nor deck guns. The *Deutschland* was credited with a surface range of 14,000 n.miles (25,940 km) at 9.5 knots and an underwater endurance of two hours at a maximum speed of 7.5 knots.

Her commanding officer was *Kapitaenleutnant* Paul L. König, a veteran merchant marine officer. He entered the Navy when the war began and was in action against the Russians. He and the two other officers—one Navy and one merchant service—were hurriedly given submarine training; the 30 sailors assigned to the submarine, however, were qualified submariners in the Imperial Germany Navy. The entire crew was

*The* Deutschland *at New London, Connecticut, on her second trans-Atlantic cargo voyage. The ship's broad beam and small conning tower are evident. In the foreground are her propeller guards, immediately aft of her retracted stern diving planes.* (U.S. Navy)

"discharged" from the Navy to serve in the *Deutschland* as merchant seamen.

After being placed in service in June 1916, the *Deutschland* made two voyages to the United States, on both occasions evading British warships blockading Germany and specifically hunting for the cargo submarine. In June–July 1916 she traveled to Baltimore, Maryland, carrying a cargo of dyes and precious stones. Simon Lake visited the *Deutschland* while she was in Baltimore. He reached an agreement with representatives of the North German Lloyd shipping firm to build additional cargo submarines in the United States. That project was stillborn.

On her return voyage to Germany in August 1916, the *Deutschland* carried 341 tons of nickel, 93 tons of tin, and 348 tons of crude rubber (of which 257 tons were carried external to the pressure hull). The cargo was valued at just over $17.5 million, a considerable sum for the time and several times the cost of the submarine. After safely returning to Bremen, the *Deutschland* made a second, equally successful trip to the United States. She arrived in New London, Connecticut, in December 1916 and, after loading cargo, she quickly sailed back to Germany.

The second submarine of this design, the *Bremen,* departed the city of that name en route to the United States in September 1916 under the command of *Kapitaenleutnant* Karl Schwartzkopf. The submarine disappeared at sea with all hands. An examination of British war records indicate no action against a U-boat that could have been the *Bremen.* Paradoxically, the *Bremen* may have been carrying financial credits for Simon Lake to begin building cargo submarines for Germany.

After the *Bremen*'s trials in the summer of 1916, the German government ordered six additional cargo U-boats of the same design. These submarines were launched from April to May 1917, after American entry into the war. The German decision to wage unrestricted submarine warfare, an act that led to U.S. entry into the war in April 1917, ended the voyages of cargo submarines. The German Navy acquired the *Deutschland* and the six similar submarines that were being fitted out. The *Deutschland* was fitted with two 21-inch (533-mm) torpedo tubes external to her pressure hull and two 5.9-inch (150-mm) and two 3.47-inch (88-mm) deck guns. Designated *U-155,* she proved a poor combat submarine, difficult to maneuver and taking a long time to dive. Still, she made three war patrols and was credited with sinking ten ships totaling 121,000 tons.[9]

The six similar cargo submarines—designated *U-151* through *U-154, U-156,* and *U-157*—were completed as U-boat "cruisers" with the same weapons as the rearmed *Deutschland*/*U-155.* Five of the class—including the *U-155*—survived the war. During the war there were proposals to further convert them to tankers to refuel other U-boats, or to use them as command submarines to coordinate U-boat attacks, a precursor of "wolf pack" operations of World War II.

The *Deutschland*'s two voyages to the United States had demonstrated the feasibility of a transocean merchant submarine. But her voyages were events never to be repeated. After the United States entered the war in April 1917, Lake held negotiations with the U.S. Shipping Board and Emergency Fleet Corporation. According to Lake, his proposal to construct a fleet of 11,500-ton submarines, each to carry 7,500 tons of cargo, was accepted by the government.[10] In the event, none was built. Lake continued to advocate cargo-carrying submarines until his death in June 1945.

(The German Navy designed, but never built, large submarine cruisers during World War I. The largest of several designs was Project P for *Panzerkreuzer* [armored cruiser]; the June 1916 characteristics called for a submarine with a surface displacement of 2,500 tons; a length of approximately 360 feet [110 m]; a surface speed of 21 knots; and an armament of four 105-mm deck guns and ten torpedo tubes of various sizes.[11] An even larger submarine cruiser—to have been propelled by four steam turbines—would have had a surface displacement of some 3,800 tons, a length of 410 feet [125 m], a surface speed of 25 knots, and an armament of three or four 150-mm guns plus six torpedo tubes.[12])

## World War II Cargo-Transport Submarines

During World War II several navies—and one army—developed cargo-carrying undersea craft. The U.S. Navy used submarines to carry ammunition and supplies to the beleaguered garrison on Corregidor Island in Manila Bay as the Japanese overran the Philippines early in 1942. Those submarines, running the Japanese naval blockade, brought out military nurses, communications specialists, and gold bullion. These were fleet-type submarines, with most of their torpedoes removed.

After this experience, President Franklin D. Roosevelt directed the Navy to develop cargo-carrying submarines. The three submarines of the *Barracuda* (SS 163) class, unsatisfactory as fleet boats, were converted to equally unsatisfactory cargo craft. They were never employed in the cargo role and saw secondary service in U.S. coastal waters during the war.[13]

American forces began their offensive in the Pacific in August 1942 in the Solomons. Soon Japanese-held islands were being bypassed by U.S. forces, and the Japanese Navy began using submarines to carry cargo and to tow cargo canisters to sustain cut-off garrisons. Subsequently, the Japanese Navy *and Army* developed and built several specialized cargo submarines, some of which saw extensive service.[14]

Germany used submarines to carry cargo to Norway during the assault on that nation in 1940; seven U-boats made nine voyages carrying munitions, bombs, lubricating oil, and aircraft fuel as well as spare parts for a damaged destroyer. Both Germany and Japan used submarines to carry high-priority cargoes between the two Axis nations, with Germany also employing Italian submarines in that role. Also, as Allied armies closed on the surviving German and Italian troops trapped in Tunisia in May 1943, the German naval high command began planning to employ U-boats to carry gasoline from Italian ports to North Africa.[15]

During the war the German Navy developed several specialized cargo submarine designs. The first of these were the so-called *milch cows*—specialized craft to provide fuel, torpedoes, and other supplies to U-boats at sea to enable them to operate at greater distances, especially off the U.S. Atlantic coast, in the Caribbean, and in the Indian Ocean.[16] The first *milch cow* was the *U-459*, a Type XIV submarine built specifically for the U-boat supply role. Completed in April 1942, she was relatively large, although she had no torpedo tubes; she did mount anti-aircraft guns for self-protection. The *U-459*'s cargo included almost 500 tons of fuel, sufficient to add an extra four weeks on station for 12 Type VII submarines or eight more weeks at sea for five Type IX submarines. Four torpedoes were carried outside of the pressure hull for transfer by rubber raft to other submarines.

The *U-459* carried out her first at-sea refueling on 22 April 1942 some 500 n.miles (925 km) north-east of Bermuda. In all, she successfully refueled 14 submarines before returning to her base in France. Subsequent refueling operations became highly vulnerable to Allied attacks, because they concentrated several U-boats in one location, and because the radio messages to set up their rendezvous often were intercepted and decoded by the Allies. Ten Type XIV submarines were built, all of which were sunk by Allied air attacks.[17] In addition, several other large submarines were employed in the *milch cow* role.

The first German U-boat designed specifically for cargo operations to and from the Far East was the Type XX. This U-boat was to have a surface displacement of 2,700 tons and was to carry 800 tons of cargo, all but 50 tons external to the pressure hull. The first Type XX was to be delivered in August 1944, with an initial production at the rate of three per month. No fewer than 200 of these submarines were planned, but construction was abandoned before the first delivery in favor of allocating available resources to the Type XXI program.[18]

In 1944, when it became evident that the Type XX program was suffering major delays, the Naval Staff considered the construction of Type XXI variants as supply and cargo submarines. No decision was made to pursue this project, although several designs were considered. They are significant because of the role of Type XXI submarines in the development of early U.S. and Soviet submarines of the Cold War era.

The Type XXID was to have been a *milch cow* and the Type XXIE a cargo submarine. The supply submarines were to carry out replenishment while both U-boats were submerged. The Type $XXID_1$ retained the Type XXI pressure hull and internal arrangement, but all but two torpedo tubes (and all reloads) were deleted, and the outer hull was enlarged (as in the earlier *Deutschland*). Light anti-aircraft guns were to be fitted, and later design changes included the addition of two bow torpedo tubes as a counter to ASW ships. This project was abandoned, however, in favor of the Type XXIV fuel supply boat, because construction changes would put too much of a demand on an already-critical situation in the shipbuilding industry. In the Type XXIV the outer hull configuration of the original Type XXI was retained with the exception of an enlarged upper deck. A large portion of the submarine's battery was eliminated to provide space for cargo. Still, the design could not come even close to the cargo capacity of the aborted Type XIV.

The Type $XXIE_1$, $E_2$, and T variants also were cargo ships. Their internal arrangements were to remain unaltered except for the torpedo armament; hence there was to be a major enlargement of the outer hull. There were variations in conning towers, gun armament, and torpedo tubes. To use as many standard sections as possible, the Type XXIT and (earlier) Type XXIV gained cargo space by reducing the size of the crew spaces, battery, and torpedo compartment. These changes would have provided for carrying 275 tons of cargo on a return voyage from East Asia—125 tons of rubber, 12 tons of concentrated molybdum, 67 tons of zinc in bars, 67 tons of concentrated wolfram, and 4 tons of miscellaneous cargo. None of these Type XXI variants was constructed before the collapse of the Nazi state.[19]

Details of these U-boat programs and plans were acquired and studied by the American, British, and Soviet navies after the war, and—coupled with their own wartime experiences—would influence submarine developments.

Early in World War II the Soviet Navy employed submarines to carry small numbers of people, usually saboteurs and "agents," and limited cargo. The situation changed when German forces began the siege of the Crimean port of Sevastopol. When Soviet defenses collapsed in the Crimea in the fall of 1941, about 110,000 troops, sailors, and marines remained in Sevastopol. Soviet ships and submarines, running a gauntlet of bombs and shells, brought men, munitions, and supplies into the city.

Heavy losses in surface ships led the commander of the Black Sea Fleet in April 1942 to order submarines to deliver munitions and food to Sevastopol and to evacuate wounded troops as well as the remaining women and children. The largest available Soviet submarines of the Series XIII (L class) could carry up to 95 tons of cargo, while the smaller units delivered far less. Not only was every available space within the submarines used for cargo (including containers of gasoline), but cargo also was loaded into torpedo tubes and mine chutes.

Some 80 runs were made into Sevastopol by 27 submarines. They delivered 4,000 tons of supplies and munitions to Sevastopol's defenders and evacuated more than 1,300 persons. (Sevastopol fell on 3 July 1942 after a siege of 250 days.)

Based on the Sevastopol experience, the Soviet Navy's high command initiated an urgent program to build transport submarines. First an effort was undertaken at TsKB-18 (later Rubin) to design submarine barges for transporting cargo—solid and liquid—that could be towed by standard submarines or a specialized submarine tug (Project 605). It was envisioned that these underwater barges could be rapidly built in large numbers. From the beginning of the project, the major complexity was not with the underwater barge, but with the towing operation by a submarine. The Navy was forced to cancel the project because of this problem.[20]

The Tactical-Technical Elements (TTE) requirement for a small cargo submarine was issued by the Navy in July 1942, becoming Project 607. This was to be a submarine with a capacity of 250 to 300 tons of solid cargo not larger than 21-inch torpedoes, and also 110 tons of gasoline in four ballast tanks. Two folding cargo cranes would be fitted.

The engineering plant was diesel-electric, with a single propeller shaft. No torpedo tubes would be provided, although two small guns were to be mounted. These cargo submarines were to use the same equipment and fittings in the small submarines of the earlier VI and VI-*bis* series (M class—202 tons submerged). This approach would simplify the design and construction of the submarines, which could be built at inland shipyards. The chief designer of Project 607 was Ya. Ye. Yevgrafov.

TsKB-18 began issuing blueprints by April 1943. But by that time the general military situation had changed in favor of the Soviets and the need for underwater transports disappeared; Project 607

TABLE 14-1
**Pre–Cold War Large and Cargo Submarine Designs**

| Country | Class/Ship | Comm. | Built | Displacement* | Length | Notes |
|---|---|---|---|---|---|---|
| Germany | *Deutschland* | 1916** | 8 | 1,440 tons / 1,820 | 213¼ ft (65.0 m) | 2 cargo submarines 6 cruisers |
| Britain | *K1* | 1917 | 17 | 1,800 tons / 2,600 | 338 ft (103.05 m) | steam powered |
| Britain | *M1* | 1918 | 3 | 1,600 tons / 1,950 | 296 ft (90.24 m) | 12-inch gun |
| Britain | *K26* | 1923 | 1 | 2,140 tons / 2,770 | 351¼ ft (107.09 m) | steam powered |
| Britain | *X1* | 1925 | 1 | 2,780 tons / 3,600 | 363½ ft (110.82 m) | submarine cruiser |
| USA | *Argonaut* | 1928 | 1 | 2,710 tons / 4,080 | 381 ft (116.16 m) | minelayer-transport |
| USA | *Narwhal* | 1930 | 2 | 2,730 tons / 3,960 | 371 ft (113.11 m) | fleet submarine |
| France | *Surcouf* | 1934 | 1 | 3,304 tons / 4,218 | 360⅝ ft (110.0 m) | submarine cruiser |
| Germany | Type XB | 1941 | 8 | 1,763 tons / 2,177 | 294¾ ft (89.86 m) | |
| Germany | Type XIV | 1942 | 10 | 1,688 tons / 1,932 | 220¼ ft (67.15 m) | *milch cow* |
| Germany | Type XX | — | — | 2,708 tons / 2,962 | 255 ft (77.74 m) | cargo submarine |
| USSR | Project 607 | — | — | 740 tons / 925 | 210 ft (64.0 m) | cargo submarine |
| USSR | Project 632 | — | — | 2,540 tons / (85.0 m) | 328 ft (100 m) | minelayer-transport-tanker |

*Notes:* * surface / submerged

** Completed as cargo submarine; transferred to the German Navy in 1917.

was canceled. No technical or operational problems were envisioned with the project. While the Soviet Union did not construct Project 607 submarines, the concept of cargo-transport submarines continued to occupy the thoughts of Soviet (as well as U.S.) submarine designers in the post–World War II era. The Soviets may also have considered ocean-going cargo submarines in this period. According to the memoirs of the U.S. ambassador to the USSR, Admiral William H. Standley, while discussing with Josef Stalin the problems of shipping war matériel to Russia, Stalin asked: "Why don't you build cargo submarines? Cargo submarines could cross the ocean without interference from Nazi submarines and could deliver their supplies directly to our own ports without danger of being sunk."[21]

Admiral Standley responded that he was "sure that the question of building cargo submarines has received consideration in my country." Stalin replied, "I'm having the question of cargo submarines investigated over here."

During World War II most U.S. submarines could carry mines in place of torpedoes, launched through standard 21-inch tubes. Mines were launched from special chutes in the Soviet Series XIII (L class) and XIV (K class), and the USS *Argonaut* (SM 1).[22] These larger submarines of both navies also were employed in commando raids and, in some respects, became the progenitors of specialized transport submarines.

## Soviet Submarine Transports

In 1948 TsKB-18 developed a draft design for Project 621—a landing ship-transport submarine to carry out landings behind enemy lines. This was to be a large submarine, with a surface displacement of some 5,950 tons. Designed under the direction of F. A. Kaverin, this underwater giant—with two vehicle decks—was to carry a full infantry battalion of 745 troops plus 10 T-34 tanks, 12 trucks, 14 towed cannon, and 3 La-5 fighter aircraft. The troops and vehicles would be unloaded over a bow ramp; the aircraft would be catapulted, the launching device being fitted into the deck forward of the aircraft hangar.[23] Both conventional diesel-electric and steam-gas turbine (closed-cycle) power plants for both surface and submerged operation were considered for Project 621. (The latter plant had been planned for a variant of the Type XXVI U-boat.)

TsKB-18 also developed the draft for Project 626, a smaller landing ship-transport ship intended for Arctic operations. The ship would have had a surface displacement of some 3,480 tons and was intended to carry 165 troops plus 330 tons of fuel for transfer ashore or four T-34 tanks.

But neither of these designs was constructed. About this time the SKB-112 bureau (later Lazurit) developed Project 613B, a variation of the Whiskey-class torpedo submarine fitted to refuel flying boats. This design also was not pursued.

Simultaneously, interest in specialized minelaying submarines was renewed.[24] In 1956 the Soviet Navy's leadership endorsed a TTE for a large minelayer capable of carrying up to 100 of the new PLT-6 mines plus transporting 160 tons of aviation fuel in fuel-ballast tanks. This was Project 632, with Yevgrafov of TsKB-18 assigned as chief designer.

Preliminary designs addressed carrying mines both "wet" and "dry" (i.e., within the pressure hull). Soon the heavy workload at TsKB-18 led to the design work on Project 632—estimated to be 33 percent complete at the time—being transferred to TsKB-16 (later Volna/Malachite), with Yevgrafov being reassigned to that bureau. The design was completed in two variants—wet storage for 90 mines or dry storage for 88 mines. A combined wet/dry configuration could carry 110 mines. A further variant of Project 632 showed a small increase in dimensions that would permit 100 troops to be carried in the mine spaces, with the minelaying gear being removable. The latter feature was a consequence of the wartime Sevastopol experience, which would require that these and other large submarines be able to transport aviation fuels and to be reconfigured at naval bases to transport combat troops or wounded (with medical attendants) in place of mines.

Project 632 was approved for construction in February 1958. Significantly, in October 1958 the design for a nuclear-propelled variant of the minelayer was approved as Project 632M, employing a small O-153 reactor plant.[25] This ship would be some 100 to 200 tons heavier than the basic 632 design. The nuclear variant would have had a submerged cruising range estimated at 20,000 n.miles, compared to 600 to 700 n.miles for the conventional propulsion plant. But when the Central Committee and Council of Ministers approved the seven-year shipbuilding program in December 1958, the Project 632 submarine was deleted.

In its place a replenishment submarine was proposed. Project 648, which was already under design, would have had a secondary minelaying capability.[26] N. A. Kiselev, new to TsKB-16, was named the chief designer of the project. This was the first submarine project of Kiselev, a state prize winner and the chief designer of the Project 68/*Chapayev*-class gun cruisers. The submarine's primary mission requirement would be to replenish and rearm submarines attacking Allied merchant shipping, that is, a *milch cow* role. Project 648 was to carry missiles (ten P-5/P-6 [NATO SS-N-3 Shaddock)] or torpedoes (40 21-inch and 20 15.75-inch) plus 34 tons of food,[27] 60 tons of potable water, and 1,000 tons of diesel fuel (or the equivalent in aviation fuels).[28]

The weapons and stores were to be transferred at sea to submarines, a considerable challenge, especially with respect to the cruise missiles. Diesel fuel was to be transferred to a submarine while both were submerged. Aviation fuel would be carried for transfer to seaplanes in remote operating areas. Again, the Sevastopol experience led to the TTE requiring the submarine to transport 120 troops and their weapons, or to evacuate 100 wounded personnel.

Preliminary at-sea replenishment tests were conducted in 1957 using a Project 611/Zulu and an older prewar L-class submarine. In late 1958 the Navy assigned the Project 611/Zulu submarine *B-82*, the Project 613/Whiskey submarine *S-346*, and a Project 30-*bis*/*Skoryy*-class destroyer to carry out

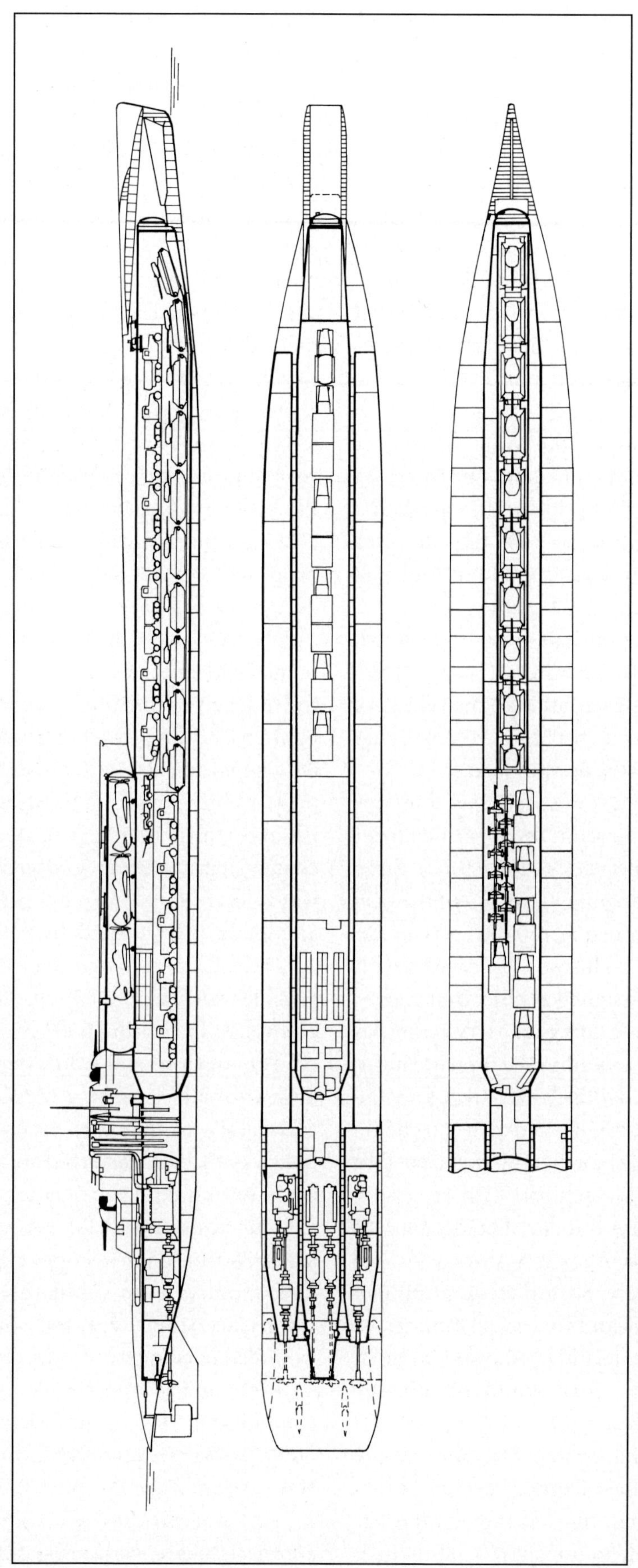

*Project 621 submarine LST.* (©A.D. Baker, III)

*An at-sea replenishment trial in 1957 with a Zulu (above) and a pre-war Series XIII L-class submarine. This operation occurred in the Sea of Okhotsk; the photo was taken by a U.S. Navy patrol plane. Deck guns have been removed from the L-class boat.* (U.S. Navy)

more complex submarine replenishment tests. The three ships were modified in 1960 at Severodvinsk, and at-sea testing began shortly thereafter.

Meanwhile, the design for Project 648 had been approved on 10 July 1958. Because of the termination of Project 632, the new submarine was to carry up to 98 mines in place of replenishment stores. Work on Project 648 began at the Severodvinsk shipyard (No. 402), with a section of the submarine's hull being fabricated with specialized replenishment equipment installed. The project was complex, and according to Russian historians, "As it was more profitable to construct the large-series orders for atomic submarines, the shipyard's director, Ye. P. Yegorov, tried in every possible way to shift construction of the transport-minelayer submarine to another yard or shut down the project overall."[29]

The difficulties in replenishing submarines at sea and interest in nuclear propulsion for a replenishment submarine led to cancellation of Project 648 in June 1961. There already was a preliminary design for Project 648M in which three of the ship's silver-zinc batteries and two diesel engines would be replaced by two small O-153 nuclear plants of 6,000 horsepower each. It was estimated that the nuclear capability would increase submerged endurance from the 600 hours with diesel-electric propulsion to 1,900 hours.

The design was presented to the Navy and shipbuilding committee, but this modification of the original Project 648 design was already being overtaken by the more ambitious Project 664 submarine.[30] Project 664 combined the characteristics of a "submarine LST" with a replenishment submarine having nuclear propulsion. Design work had began in 1960 at TsKB-16 under Kiselev. This would be a very large submarine, with a surface displacement of 10,150 tons, and would carry 20 cruise missiles or 80 21-inch torpedoes or 160 15¾-inch torpedoes for transfer to combat submarines. Liquid cargo would include 1,000 tons of diesel oil or aviation fuel, plus 60 tons of lubricating oil and 75 tons of potable water as well as 31 tons of food. In the LST role the submarine would carry 350 troops, although up to 500 could be carried for a five-day transit.

There obviously was interest in replenishment submarines at the highest levels of the Navy. A 1961 issue of *Voyennaya Mysl* (Military Thought), the senior (classified) Soviet military journal, contained an article by Admiral Yuri Panteleyev looking at future submarine operations. Among the technical problems he said must be resolved was: "to create a class of special submarine tankers and submarine transports for the shipment of combat supplies, equipment, and contingents of personnel."[31] (Panteleyev also called for "a system for all

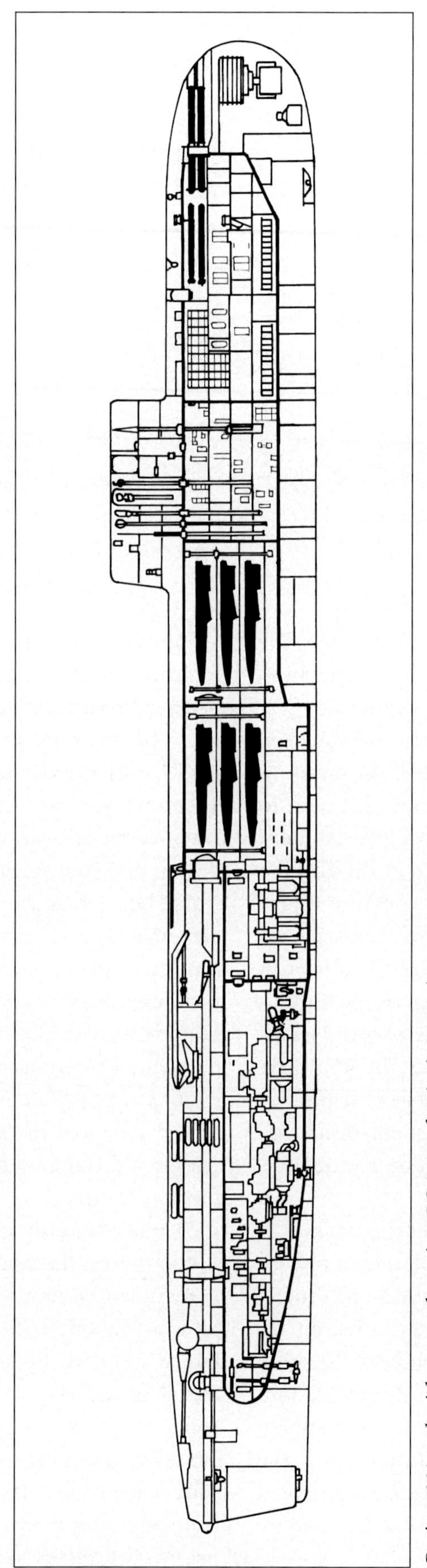

*Project 664 replenishment submarine. LOA 462 ft 3 in (140.9 m)* (©A.D. Baker, III)

types of underwater supply, for submarines lying on the bottom at points of dispersal and at definite depths and not moving."[32])

Construction of Project 664 began at Severodvinsk in 1964. But soon it was determined that combining three missions—replenishment, transport, and minelaying—in a single hull caused major complications, even in a nuclear-propelled submarine. Both range and operating depth were reduced. In May 1965 all work on the lead submarine was halted. The proposal was made to transfer the project to a Leningrad shipyard to make room at Severodvinsk for accelerated construction of Project 667A/Yankee SSBNs. But the project was halted completely.

Accordingly, in August 1965 TsKB-16 was directed to respond to the TTE for a large diesel-electric submarine LST—Project 748.[33] Kiselev again was assigned as chief designer. The design bureau, realizing the limitations of conventional propulsion for this submarine's missions, additionally initiated nuclear-propelled variants.

Six variants of Project 748 were developed, with surface displacements from 8,000 to 11,000 tons. Most variants had three separate, cylindrical pressure hulls side by side, encased in a single outer hull. The first variant met the basic TTE; the second variant carried a larger number of PT-76 amphibious tanks; the third variant had VAU-6 auxiliary nuclear power plants; the fourth variant had two OK-300 reactor plants generating 30,000 horsepower; the fifth variant had the VAU-6 system with a single pressure hull; and the sixth variant replaced the OK-300 plant with four VAU-6 units.[34]

This large submarine could carry up to 20 amphibious tanks and BTR-60P armored personnel carriers, and up to 470 troops. (See table 14-2.) In addition to a torpedo armament of four bow 21-inch torpedo tubes with 12 to 16 torpedoes, the submarine was to be fitted with two 57-mm anti-aircraft guns and surface-to-air missiles. And, of course, the submarine could serve as a minelayer.

TsKB-16 recommended proceeding with the fourth (nuclear-propelled) variant. Still, construction was not initiated, because the Navy, Ministry of Shipbuilding Industry, and General Staff of the Armed Forces ordered a review of the features of Projects 632, 648, 664, and 748 in an effort to develop a "ubiquitous" or multi-purpose nuclear submarine. TsKB-16 (now named Volna) was directed to develop a preliminary design for the submarine—Project 717—with Kiselev once more the chief designer.[35] The TTE called for the clandestine delivery of up to 800 marines and four personnel carriers; the transport of arms, munitions, fuel, and provisions; or 250 commandos and ten amphibious tanks and ten personnel carriers and the evacuation of troops and wounded, as well as minelaying. This was to be the world's largest submarine designed to that time, with a surface displacement of more than 17,600 tons and nuclear propulsion.

The preliminary design effort was completed early in 1969. In July the Navy and the Ministry of Shipbuilding Industry added to the TTE the requirement for "the rescue of the crews of sunken submarines with the aid of rescue apparatus." This change led to revised specifications, which were not formally approved until February 1970. Completion of the revised contract design for Project 717 was delayed until October 1971.

The Severodvinsk shipyard made preparations for constructing five submarines to this design. Full-scale mockups were made of the control room, cargo spaces, and other portions of the submarine. However, this project, too, was stillborn, because in the late 1970s the available building ways at Severodvinsk were needed for the construction of nuclear submarines, especially Project 941/Typhoon SSBNs that were being developed as a counter to the U.S. Trident program, the *Ohio* (SSBN 724) class.

Thus ended the design of large minelaying/transport/replenishment submarines in the Soviet Union. There still was some interest in submarine tankers. In the 1960s TsKB-57 undertook the design of a large submarine tanker, Project 681, intended primarily for commercial operation. With two VM-4 nuclear reactor plants, the submarine would have a cargo capacity of 30,000 tons. Subsequently, TsKB-16 began design of another nuclear-propelled submarine tanker in 1973 designated Project 927, but neither of these projects was pursued.

There again was interest in submarine tankers—and container submarines—in Russia in the 1990s. The Malachite bureau (former TsKB-16/SDB-143) put forward preliminary designs for a submarine

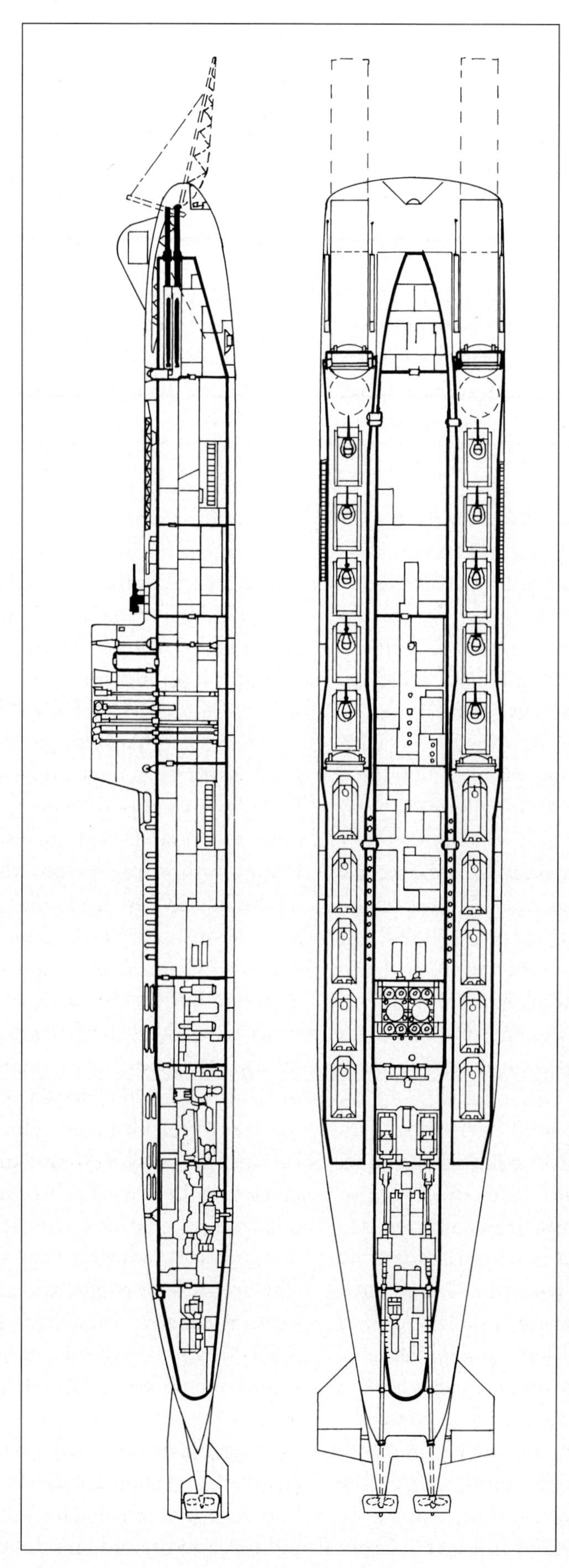

*Project 748 submarine LST. LOA 521 ft 6 in (159.0 m)* (©A.D. Baker, III)

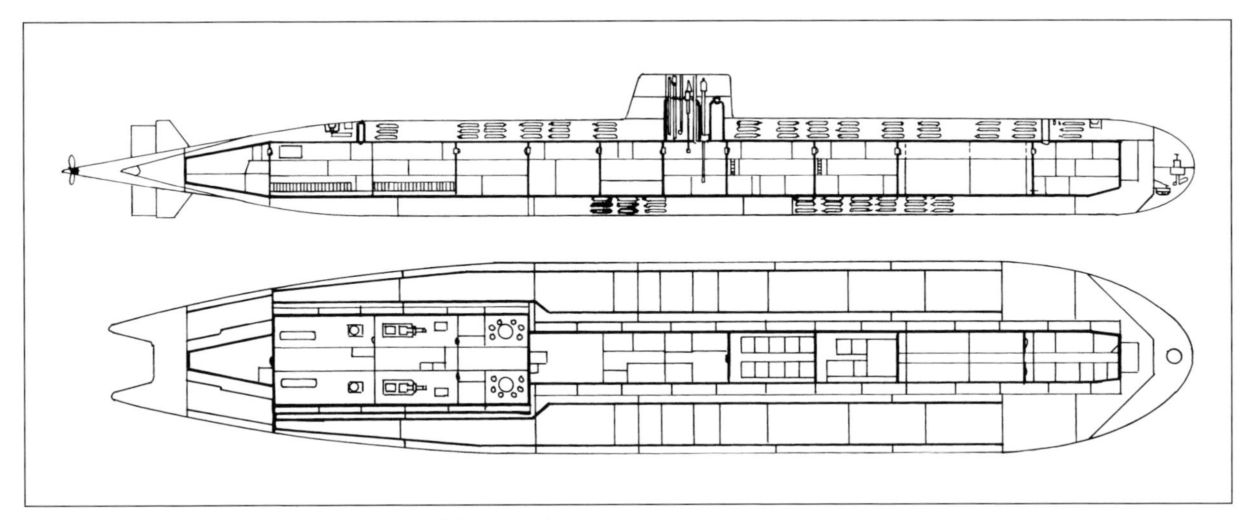

*Project 681 submarine tanker. LOA 515 ft (157.0 m)* (©A.D. Baker, III)

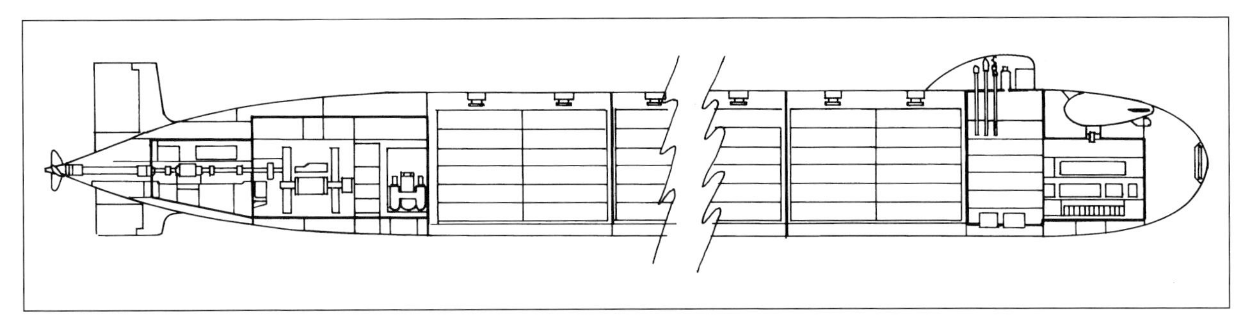

*Malachite design for submarine container ship.* (©A.D. Baker, III)

capable of transporting petroleum or freight containers, especially in the Arctic region. Envisioning under-ice navigation between European and Asian ports, and possibly northern Canada, according to Malachite designers, "Given equal cargo capacity, the efficiency of an underwater container ship is considerably higher, for example, than that of an icebreaker transport ship of the *Norilsk* type. The underwater tanker is competitive."[36]

Malachite proposed tanker and container variants of the same basic nuclear submarine design, employing an elliptical cross-section. The tanker variant would transport almost 30,000 tons of petroleum, which could be loaded and discharged from surface or underwater terminals. The underwater container carrier could transport 912 standard (20-foot/6.1-m) freight containers, loaded through a series of hatches. It was estimated to take 30 working hours to load or unload a full shipload. Large cargo hatches and an internal container moving scheme would facilitate those operations. (See table 14-3.)

A single-reactor, single-shaft propulsion plant was proposed with three diesel generators to provide for maneuvering in harbor and for ship electrical needs. Two of the diesel generators would be fitted to work as closed-cycle/Air-Independent Propulsion (AIP) systems for emergency under-ice operation. Thirty tons of oxygen was to be carried to provide an AIP endurance of 20 hours at a speed of seven or eight knots.

No detailed design or procurement followed, because Russia fell into financial extremis in the post-Soviet era.

## U.S. Transport Submarines

In the 1950s and 1960s, the U.S. Navy contemplated large submarines for amphibious landings. During World War II the Navy's three largest submarines, the *Argonaut,* built as a minelayer, and the slightly smaller *Narwhal* (SS 167) and *Nautilus* (SS 168) were used as amphibious transports.

In early August 1942 the *Argonaut* and *Nautilus* unloaded their spare torpedoes and took aboard

219 marine raiders. On 17 August the troops were landed by rubber boats on the Japanese-held Makin Island in the Marshalls.[37] After catching the Japanese garrison by surprise, the marines were soon confronted by a tenacious defense. The marines captured some documents and blew up a fuel dump before withdrawing. There were about 70 Japanese troops on the island; 46 were killed at the cost of 30 marine raiders lost.[38]

The second major U.S. submarine landing involved the *Nautilus* and *Narwhal;* their target was Attu, one of two Aleutian islands the Japanese had occupied in June 1942. The submarines conducted a rehearsal landing at Dutch Harbor in April and then took on board 200 Army scouts and set off for Attu as part of Operation Landcrab. The submarines arrived off Attu on 11 May 1943 and successfully landed their troops by rubber boat five hours before the main U.S. assault. The submarine landings were a useful complement to the main landings, but they were not a critical factor in the recapture of Attu.

Soon after the war the U.S. Navy converted several fleet boats to cargo and transport configurations.[39] The *Barbero* (SS 317) was converted to a cargo configuration at the Mare Island shipyard in 1948.[40] Her four stern 21-inch torpedo tubes were deleted, and she was provided with storage for just over 100 tons of aviation gasoline or 120 tons of dry cargo for landing ashore. From October 1948 she participated in several exercises, but her use was limited and she was inactivated in June 1950 (and subsequently converted to a Regulus missile submarine; see Chapter 6).

A second cargo submarine conversion was the *Guavina* (SS 362), converted to a tanker.[41] Her 1950 conversion at Mare Island provided "blister" tanks that could carry 585 tons of aviation fuel—160,000 gallons (605,650 liters) of transferrable fuel. Only two stern torpedo tubes were retained. Like the *Barbero,* she was to provide support for troops ashore during amphibious operations. And, like the *Barbero,* exercises revealed her limited effectiveness. However, in 1957 the *Guavina* was modified at the Charleston (South Carolina) naval shipyard to refuel seaplanes. She was fitted with a platform aft and refueled seaplanes astern and by floating fuel "bladders" to the aircraft.

The modified *Guavina* was intended to be one of several undersea craft supporting the Martin P6M Seamaster, a four-turbojet flying boat intended primarily for aerial minelaying and long-range reconnaissance. The P6M was to operate from the water for as long as six to eight months, being maintained, refueled, and rearmed by seaplane tenders and by submarines. Weapons were loaded through the top of the fuselage, enabling the aircraft to be rearmed and refueled while in the water. Twelve aircraft were completed before the program was cancelled in 1959, principally to help fund the Polaris SSBN program.

The U.S. Navy modified two fleet submarines to serve as troop transports, the *Perch* (SS 313) and *Sealion* (SS 315), both initially changed to SSP.[42] Modified at the Mare Island and San Francisco naval shipyards, respectively, they rejoined the fleet in 1948 with accommodations for 115 troops; a

*The minelayer* Argonaut *was the largest non-nuclear submarine built by the United States. She was built as the* V-4 *and designated as a fleet submarine (SF 7); renamed and changed to SM 1 in 1931, she also carried the designations SS 166 and APS 1. This 1931 photo clearly shows her two 6-inch guns, with practice torpedoes on deck near the forward 6-inch (152-mm) gun mount.* (U.S. Navy)

*A Martin P5M Marlin flying boat taxies up to the stern of the submarine tanker* Guavina *during a refueling exercise. The* Guavina's *fuel capacity has been increased by side blisters; a refueling station has been installed on her stern, with sailors standing by to handle lines and hoses. The* Guavina *had a streamlined fairwater.* (U.S. Navy)

hangar fitted aft of the conning tower could accommodate a tracked landing vehicle (LVT) carrying a jeep and towed 75-mm cannon; eight ten-man rubber boats also were carried.[43]

During the Korean War, on the night of 1–2 October 1950, the *Perch* used rubber boats to land 67 *British* marines behind communist lines to blow up a railroad tunnel that could not be struck by aircraft. (One marine was killed in the raid.) This was the only U.S. submarine combat action of the Korean War. Many similar raids were carried out by U.S. surface ships during the Korean War, questioning the need for the submarine in that operation. This was the largest submarine combat landing ever undertaken except for the U.S. submarine landings on Makin Island (1942) and Attu (1943).

Subsequently, the U.S. Navy employed a series of converted submarines in the transport role, with two normally being in commission into the 1990s. During the Vietnam War the transport submarine *Perch*, based at Subic Bay in the Philippines, conducted training operations on a regular basis with U.S. and British special forces. For example, as part of Jungle Drum III, a multination exercise in the

*The U.S. Navy converted several fleet submarines to troop transports during the Cold War. Here the well-worn* Sealion *unloads Marines into rubber boats in a very non-clandestine exercise in 1966. A Sikorsky HRS helicopter sits on the after deck; at times the* Sealion *carried a hangar aft of the conning tower for rubber boats or an amphibious tractor.* (U.S. Navy)

spring of 1965, the *Perch* landed 75 U.S. Marines on the Malay Peninsula.

Beginning in November 1965 the *Perch* landed marine "recce" troops and Navy swimmer personnel on the coast of South Vietnam, in areas held by or suspected to be under Viet Cong control. Often these operations were in preparation for landings by Marines from the Seventh Fleet's amphibious force.

In this period the *Perch* carried an armament of two 40-mm deck guns plus .50-caliber machine guns that could be mounted on her conning tower. During Operation Deckhouse III in August 1966, while Navy swimmers were conducting a beach reconnaissance, the Viet Cong began firing at the landing party. The *Perch* answered with cannon and machine gun fire. Similarly, when a South Vietnamese unit working on the beach in conjunction with Navy swimmers was threatened by Viet Cong, the *Perch* stood by 500 yards (457 m) offshore to provide gunfire support as the submarine took the 85 Vietnamese troops and several refugees off the beach and transported them out of the area.

The USS *Tunny* (SS 282), a former Regulus submarine, was converted to an APSS in 1966 at the Puget Sound (Washington) naval shipyard. She began operations off Vietnam in February 1967. Two operations proposed for the *Tunny* were stillborn: After the capture of the U.S. intelligence ship *Pueblo* by North Koreans in January 1968, it was proposed that the *Tunny* carry commandos into Wonsan Harbor to blow up the ship; that plan was vetoed by senior U.S. commanders. Another proposal called for the *Tunny* to carry commandos to a point off Haiphong Harbor to attack a major target near Hanoi, the North Vietnamese capital; that plan also was vetoed.

The *Grayback* (SSG/APSS 574), converted in 1968–1969 at Mare Island, had her two former Regulus missile hangars modified to carry self-propelled swimmer delivery vehicles and rubber boats. She participated in special operations in 1970–1972 off the coast of Vietnam.[44]

The *Grayback* was succeeded in the transport role by two ex-Polaris SSBNs, which had been withdrawn from the strategic missile role with their missile tubes deactivated, the *Sam Houston* (SSBN 609) and *John Marshall* (SSBN 611). They, in turn, were replaced by two later and larger ex-Polaris submarines, the *Kamehameha* (SSBN 642) and *James K. Polk* (SSBN 645); all being changed to SSN as transports, as the nuclear community eschewed specialized designations.

These ex-missile submarines could carry one or two dry-deck shelters, or hangars, fitted aft of the sail and used as lock-out chambers for swimmers or could carry a small swimmer delivery vehicle. The submarines were fitted to carry 65 Marines or swimmers and their gear. Like the *Grayback*, the ex-SSBNs retained their torpedo armament and—for political reasons—were designated as attack submarines (SSN) and not as transport submarines. (No Soviet submarines were converted to the transport role.)

In all, eight U.S. submarines were converted to the amphibious transport role during the Cold War and immediately afterward:[45]

| Number | Former | Name | Transport |
|---|---|---|---|
| APSS 282 | SSG | *Tunny* | 1966–1969 |
| SSP 313 | SS | *Perch* | 1948–1959 |
| SSP 315 | SS | *Sealion* | 1948–1969 |
| APSS 574 | SSG | *Grayback* | 1969–1984 |
| SSN 609 | SSBN | *Sam Houston* | 1986–1991 |
| SSN 611 | SSBN | *John Marshall* | 1984–1992 |
| SSN 642 | SSBN | *Kamehameha* | 1992–2002 |
| SSN 645 | SSBN | *James K. Polk* | 1993–1999 |

In addition, several attack submarines (SSN) were modified to carry one dry-deck shelter and a small number of swimmers. The retirement of the *Kamehameha* left the Navy with only SSNs for the transport role.[46] However, in 1999 the submarine community proposed conversion of four Trident missile submarines of the *Ohio* class, being retired under strategic arms agreements, to a combination cruise missile/transport role. Under that plan, the four ships would each be fitted to carry some 65 commandos or swimmers, their small boats, equipment, and an ASDS submersible. (See Chapter 16.) In the SSGN configuration two Trident missile tubes would be converted to large swimmer lock-out chambers (to be made with submersibles or deck chambers). The remaining 22 Trident tubes would be modified to carry commando/swimmer gear, unmanned vehicles, or seven vertical-launch Tomahawk land-attack missiles per tube.

The U.S. Navy, like the Soviet Navy, considered the construction of specialized, nuclear-propelled transport submarines, but the only design "sketch"

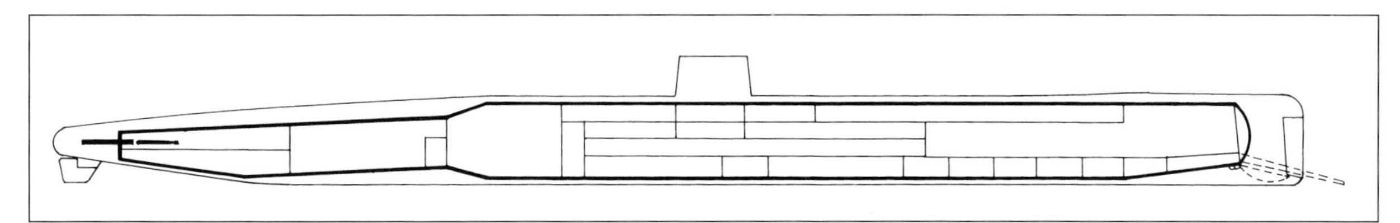

*U.S. submarine LST (preliminary design).* (©A.D. Baker, III)

to be developed was for a large submarine approximately 565 feet (172.26 m) long. Based on Navy information, artist Frank Tinsley drew a future transport submarine for *Mechanix Illustrated* magazine. It showed a 10,000-ton submarine with a length of 720 feet (219.5 m) and a beam of 124 feet (37.8 m) unloading Marines into high-speed, 100-m.p.h. "air rafts" for an amphibious assault. That submarine was to carry 2,240 troops and their equipment.

None of these proposals was pursued.

*The* Sturgeon-*class submarine* Silversides *(SSN 679)—a "straight" attack submarine—fitted with a dry deck shelter for special forces in a 1992 exercise. Converted Polaris submarines could each carry two shelters, which could be used to lock-out swimmers and to store rubber boats or small submersibles.* (Giorgio Arra)

## Commercial Tankers

More effort was expended in the United States to develop commercial submarine tanker designs. Shortly after the historic Arctic transit of the *Nautilus* (SSN 571) in August 1958, her commanding officer, Captain William R. Anderson, wrote:

> Man's last unknown sea, the ice-mantled Arctic Ocean, has yielded to the Atomic Age. Across the top of the world, blazed by thousands of instrument readings, lies a maritime highway of tomorrow—strategic, commercially promising, and, I am convinced, safe.
>
> ***
>
> This ice may forever bar the routine transit of surface ships. But it will not deny the Arctic Ocean to submarines—provided they are nuclear powered.[47]

Almost simultaneously, submarine designers in Britain, Japan, and the United States began detailed studies and preliminary designs of nuclear-propelled submarine tankers for Arctic operations. The Electric Boat yard, under contract to the U.S. Maritime Administration, in 1958 undertook a study of principal dimensions and power requirements for submarine tankers of 20,000, 30,000, and 40,000 deadweight tons with submerged speeds of 20, 30, and 40 knots.[48]

These tanker variants were examined in the most detailed study of the period in a paper entitled "Submarine Tankers," presented in November 1960.[49] That paper, prepared under the auspices of the U.S. Maritime Administration, reflected analysis and model tests to provide unprecedented detail of the variants. (The characteristics of the 20,000-ton and 40,000-ton variants are provided in table 14-3.) Significantly, to limit the tanker's draft to about 36 feet (11 m) to enable entry to a number of ports, the paper offered a rectangular outer hull form as well as the conventional circular hull cross-section. The former would have a central main pressure hull with four separate, cylindrical cargo tanks fitted parallel to the main pressure hull.

With respect to speed, the study addressed a range of speeds, noting:

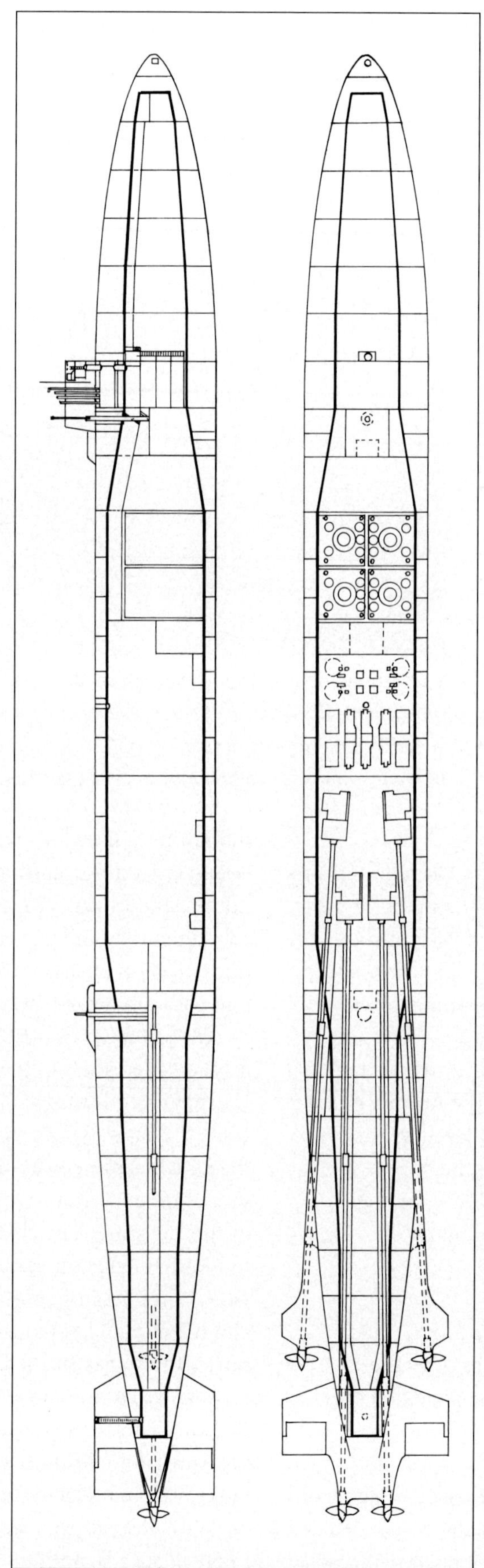

*U.S. commercial submarine tanker (40,000 deadweight tons; approx. 40 knots).* (©A.D. Baker, III)

> It is realized that there is no demand for 40 knots speed in the oil trade. It could even be granted that a submarine mode of transportation at such speeds would be uneconomical, unproven, difficult to man and manage, and so on. The fact remains, however, that based on the technology available today [1960], and the research study described in this paper, it could be possible, from an engineering viewpoint, to design an optimum submarine tanker which could carry 40,000 tons of oil products at about 37 knots with installed power of about 250,000 shp.[50]

The various tanker designs of the study required from one to four pressurized-water reactors, with turbines turning from one to four propeller shafts. Speeds of 38 knots (40,000-ton tanker) to 42 knots (20,000-ton tanker) could only be achieved with circular cross-section hulls; the largest and fastest tanker would require four reactors and four turbines generating 240,000 horsepower to four propeller shafts. That design provided a hull diameter of 102½ feet (31.25 m)!

Despite the apparent feasibility of such ships and the periodic reinforcement of these papers by other studies and articles, no action was forthcoming. In the early 1970s the American decision to exploit the oilfields at Prudhoe Bay, Alaska, led the U.S. National Oceanic and Atmospheric Administration as well as the Maritime Administration to again look into submarine tankers as an alternative to the pipeline across Alaska and the use of surface tankers to carry the petroleum from a southern Alaskan port to the continental United States. Their studies once more demonstrated the feasibility as well as certain advantages to using submarine tankers. Political factors, however, led to the construction of the trans-Alaskan pipeline and employment of surface tankers to move the petroleum.

There was still another effort to obtain backing for submarine cargo ships in the summer of 2000 when Russia's RAO Norilsk Nickel, a giant metals conglomerate, proposed the conversion of Project 941/Typhoon strategic missile submarines to cargo carriers.[51] Under this proposal the giant, 23,200-ton (surface) submarines would be converted to carry 12,000 tons of ores—nickel, palladium, platinum—from the firm's mining-and-smelter center in northwestern Siberia to the ice-free port of Murmansk. The submarines would be capable of under-ice transits up to 25 knots, according to a Norilsk spokesman. The cost of conversion was estimated at $80 million per submarine, much less than the replacement costs for the nuclear icebreakers and cargo ships now used in this role in the Arctic seas. A year later, Russian designers put forward a proposal to convert Typhoon SSBNs to Arctic container carriers. Their massive missile battery would be replaced by cargo bays and several internal elevators for handling hundreds of containers.[52]

However, the adverse publicity following the loss of the nuclear-propelled submarine *Kursk* in August 2000, coupled with the conversion costs and the concern for nuclear submarine operations in coastal waters, made such conversions unlikely. (At the time three of the original six Project 941/Typhoon submarines remained in naval service; the others were being decommissioned and dismantled.)

**Early in the 20th Century, relatively large undersea craft were proposed for a variety of specialized missions. In many respects the most significant "large" submarine of the pre–World War II era was the German merchant submarine *Deutschland.* Two wartime cargo voyages to the United States by the *Deutschland* were highly successful—but have never been repeated, in war or in peace.**

**Navies soon produced large undersea craft as submarine monitors (British M class), fleet-scouting submarines (British K class), long-range, big-gun cruisers (British *X1* and French *Surcouf*), minelayers, replenishment submarines, and aircraft-carrying submarines. (See Chapter 15.)**

**During World War II submarines were used extensively by the United States and Japan as troop carriers and to transport supplies to isolated garrisons. These experiences led the U.S. Navy, after the war, to convert fleet submarines to several specialized configurations—troop transport, cargo, tanker—to support amphibious operations. Subsequently, the U.S. Navy and, especially, the Soviet Navy undertook the design of large, specialized troop/equipment transports,**

primarily for amphibious operations. The Soviets also designed large supply submarines to replenish other submarines on the high seas. In the event, none of these submarines was constructed. However, this Soviet design experience would have a direct impact on the design of large nuclear-propelled combat submarines.

The U.S. government, the U.S. Electric Boat/General Dynamics firm, and the Malachite design bureau in Russia all designed large, nuclear-propelled submarine tankers for the bulk transport of petroleum products for the civil sector. Despite the feasibility and potential economic success of such craft, primarily in the Arctic region, none was pursued. But, as demonstrated by the proposal in 2000 to convert Project 941/Typhoon SSBNs to ore carriers, the idea of Arctic cargo-carrying submarines persists.

TABLE 14-2

**Large Military Submarine Designs**

| | Soviet Project 648 | Soviet Project 664 | Soviet Project 717 (Revised) | Soviet Project 748 Variant I | Soviet Project 748 Variant IV |
|---|---|---|---|---|---|
| Displacement | | | | | |
| surface | 5,940 tons | 10,150 tons | 18,300 tons | 9,800 tons | 11,000 tons |
| submerged | | 13,100 tons | ~27,450 tons | ~14,700 tons | ~17,000 tons |
| Length | 334 ft 8 in (102.0 m) | 462 ft 3 in (140.9 m) | 623 ft 6 in (190.1 m) | 521 ft 6 in (159.0 m) | 501 ft 10 in (153.0 m) |
| Beam | 42 ft (12.8 m) | $46\frac{7}{12}$ ft (14.2 m) | $75\frac{5}{12}$ ft (23.0 m) | 64 ft (19.5 m) | 69 ft 6 in (21.2 m) |
| Draft | 29 ft (8.85 m) | 32 ft 6 in (9.9 m) | 22 ft 4 in (6.8 m) | 16 ft 10 in (4.9 m) | 19 ft 8 in (6.0 m) |
| Reactors | — | 2 OK-300 | 2 | — | 2 |
| Turbines | — | 2 | 2 | — | 2 |
| horsepower | — | 30,000 | 39,000 | — | 30,000 |
| Diesel engines | 2 2D-43 | — | — | 2 1D43 | — |
| horsepower | 8,000 | — | — | 8,000 | — |
| Electric motors | 2 | — | — | 2 | — |
| horsepower | 8,000 | — | — | 5,200 | — |
| Shafts | 2 | 2 | 2 | 2 | 2 |
| Speed | | | | | |
| surface | 14 knots | | | | |
| submerged | 12.5 knots | 18 knots | 17–18 knots | 10 | 17 |
| Test depth | 985 ft (300 m) | 985 ft (300 m) | 985 ft (300 m) | 985 ft (300 m) | 985 ft (300 m) |
| Torpedo tubes* | 4 533-mm B<br>2 400-mm B<br>2 400-mm S | 6 533-mm B | 6 533-mm B | 4 533-mm B | 4 533-mm B |
| Torpedoes | 4 533-mm<br>20 400-mm | 12 | 18 | 16 | 14 |
| Mines | 98** | 162*** | 252# | — | — |
| Aircraft | — | — | — | — | — |
| Guns | — | — | 2 57-mm## | 2 57-mm## | 2 57-mm## |
| Complement | 77 | 79 | 117 | 80 | 80 |

*Notes:* * Bow; Stern.
** 98 AGM type or 96 APM Lira type or 92 PLT-6 type or 94 Serpei type.
*** 112 PM-1, RM-2, PM-3 types or 162 RM-1, APM Lira, Serpei, UDM types.
# 126 PMR-1, PMR-2, PMT-1 types or 252 PM-1, PM-2, PM-2G types.
## Surface-to-air missiles also would be fitted.

TABLE 14-3
**Large Submarine Tanker Designs**

| | **Soviet Project 681** | **Russian Malachite Tanker 30,000 DWT*** | **U.S. 20,000 DWT* 20 knots** | **U.S. 20,000 DWT* 30 knots** | **U.S. 40,000 DWT* 20 knots** | **U.S. 40,000 DWT* 30 knots** |
|---|---|---|---|---|---|---|
| Displacement | | | | | | |
| surface | 24,750 tons | 78,000 tons | 38,000 tons | 43,600 tons | 73,300 tons | 81,800 tons |
| submerged | | 87,500 tons | 41,800 tons | 47,900 tons | 92,000 tons | 90,000 tons |
| Length | 515 ft (157.0 m) | 780 ft 8 in (238.0 m) | 555 ft (169.2 m) | 570 ft (173.8 m) | 710 ft (216.46 m) | 780 ft (237.8 m) |
| Beam | 85 ft 3 in (26.0 m) | 87 ft 11 in (26.8 m) | 80 ft (24.4 m) | 90 ft (27.44 m) | 120 ft (36.58 m) | 120 ft (36.58 m) |
| Draft | 29 ft 6 in (9.0 m) | 54 ft 2 in (16.5 m) | 40 ft (12.2 m) | 40 ft (12.2 m) | 40 ft (12.2 m) | 40 ft (12.2 m) |
| Reactors | 2 VM-4 | 1 | 1 | 2 | 1 | 3 |
| Turbines | 2 | 1 | 2 | | | |
| horsepower | 30,000 | 50,000 | 35,000 | 105,000 | 53,300 | 168,000 |
| Shafts | 2 | 1 | 1 | 2 | 1 | 3 |
| Speed | | | | | | |
| surface | | | | | | |
| submerged | 17 knots | 20 knots | 20 knots | 30 knots | 20 knots | 30 knots |
| Test depth | 660 ft (200 m) | 330 ft (100 m) | | | | |
| Cargo | 15,000 tons | 30,000 tons** | 20,000 tons | 20,000 tons | 40,000 tons | 40,000 tons |
| Complement | 50 | 35 | | | | |

*Notes:* * DWT = Deadweight Tons.
** 29,400 tons as a container carrier.

**15**

# Aircraft-Carrying Submarines

*Submarine aircraft carriers—the Japanese* I-14 *(left) and* I-400 *and* I-401 *(order unknown) at Pearl Harbor in 1946. Japan was the only nation to operate submarine aircraft carriers in wartime. The* I-14*'s hangar door is completely open; the others have their aircraft cranes raised. Two U.S. fleet boats are in the photo.* (U.S. Navy)

Two revolutionary weapons were introduced early in the 20th Century: the aircraft and the submarine. Both were avidly pursued by the Russian and U.S. Navies. During World War I there were several precursory efforts to combine these two weapons. The British and German Navies used standard submarines to carry floatplanes to sea, submerging to float off the aircraft, which, after carrying out their mission, would return to a land base or put down at sea, to be scuttled after the crew was recovered.

Between the world wars, trials of operating floatplanes from submarines were carried out by Britain, France, Japan, and the United States. Britain converted the submarine monitor *M2* to carry a floatplane, while France built the submarine cruiser *Surcouf*, which operated a single floatplane. (See Chapter 14.)

*The British* M2 *launches a Parnall Petro floatplane from her catapult. The hangar door is open, and the ship's aircraft recovery crane is extended. The* M2 *was originally a submarine monitor, one of three such submarines that each mounted a 12-inch gun.* (Royal Navy Submarine Museum)

The U.S. Navy's effort was more modest, with the relatively small submarine *S-1* (SS 105), completed in 1920, being fitted with a hangar that could accommodate a small, collapsible floatplane. From 1923 to 1926 the *S-1* carried out trials with the Martin MS-1 and Cox-Klein XS-1 and XS-2 aircraft. There was no follow-up effort. Neither the German nor Soviet Navies experimented with aircraft aboard submarines between the world wars.

*The* S-1 *was the only U.S. submarine to operate a fixed-wing aircraft, shown here with a Cox-Klein XS-2 in 1926. Note the small hangar between the conning tower and the twin-float biplane. Later U.S. transport submarines would have helicopters land aboard, but were not hangared.* (U.S. Navy)

When World War II began in the late 1930s, only the French and Japanese fleets had aircraft-carrying submarines. During the war the *Surcouf*, operated by the Free French after June 1940, did not fly her aircraft before being sunk in a collision with a merchant ship in 1942. Japanese efforts in this field began in 1923, using a German Caspar-Heinkel U-1 biplane fitted with floats for trials aboard a submarine. A series of floatplanes was developed for submarine use beginning with the Watanabe Type 96 (E9W1), which entered service in 1938.[1] This biplane aircraft and the monoplane Yokosuka Type 0 (E14Y1), later given the Allied code name Glen, which entered service in 1941, had far-reaching service in the early years of World War II in the Pacific.

At the start of the war, the Japanese Navy had 12 large I-series submarines that could each carry a single floatplane. More aircraft-carrying submarines were under construction, several of which became operational during the war.[2] They had hangars for a single, disassembled floatplane, with a catapult built into the deck.

These planes flew missions throughout the southwest Pacific and Indian Ocean areas. In August 1942 the submarine *I-25* twice launched an E14Y1 Glen from a position off Cape Blanco,

Oregon, on incendiary bombing raids against the United States. The Japanese sought to ignite forest fires in the northwestern United States. These were the only aircraft attacks against the continental United States; they inflicted no significant damage and no casualties. (The Japanese also employed large submarines to refuel flying boats, including two seaplanes that bombed Pearl Harbor on the night of 3–4 March 1942.)

In 1942 the Japanese Navy initiated the *I-400*-class Special Submarines or *Sen-Toku* (STo).[3] The *I-400*s were the largest non-nuclear submarines ever constructed, and their unique design calls for special mention in this volume. These submarines were developed specifically to launch aircraft to bomb Washington, D.C., and New York City. While the first units were under construction, the changing course of the Pacific War caused the Navy's leadership to reassign the *I-400* submarines to strike the Panama Canal in an effort to slow the flow of U.S. reinforcements to the Pacific.

The original 1942 design of the *I-400* provided a hangar for two floatplanes; subsequently, the submarine was enlarged to handle three aircraft. The aircraft hangar, beneath the conning tower, opened to an 85⅓-foot (26-m) catapult track forward of the hangar. The aircraft were pre-warmed in the hangar through a system of circulating heavy lubricating oil while the submarine was submerged. The submarine then surfaced to launch aircraft.

The design had a vertical figure-eight hull configuration forward, evolving into a horizontal figure eight (with side-by-side pressure hulls) amidships. This arrangement permitted the submarine to have two forward torpedo rooms, one above the other, while housing four diesel engines, paired side-by-side amidships. The submarine's aircraft magazine could hold four aerial torpedoes, three 1,760-pound (800-kg) bombs, and 12 550-pound (250-kg) bombs. Beyond aircraft, each *I-400* was armed with eight 21-inch (533-mm) torpedo tubes forward. One 5.5-inch (140-mm) deck gun and ten 25-mm anti-aircraft guns were fitted.

Eighteen submarines were planned. The *I-400* was completed in December 1944, followed by the *I-401* and *I-402* in 1945. The *I-402* was modified before completion to a tanker configuration to carry fuel from the East Indies to Japan, but the war ended before she undertook a tanker mission.[4]

Air operations by the *I-400*s were delayed because of the lag in producing the aircraft designed specifically for these submarines, the high-performance Aichi M6A1 Seiran floatplane.[5] This was the world's only aircraft built specifically for bombing missions from submarines. The specifications initially called for no undercarriage; if feasible, after a bombing mission the aircraft would return to the submarine and come down at sea. As built, the large, single-engine aircraft had large, twin cantilever floats. It could carry a torpedo or 1,875-pound (850-kg) bomb or two 551-pound (250-kg) bombs.

During practice, the time to unfold an aircraft's wings and tail surfaces, and ready a plane for launching—in darkness—was less than seven minutes. Three aircraft could be readied for flight and launched within 30 minutes of a submarine coming to the surface. Although this was a long time for the submarine to be exposed, even at night, it was a remarkable achievement.

The two submarines of the AM type, the *I-13* and *I-14,* were slightly smaller than the *I-400* design and were configured to embark two M6A1 aircraft. They were intended to operate with the *I-400*s in long-range air strikes.[6]

On 26 July 1945 the *I-400* and *I-401* sortied from the Inland Sea to raid the U.S. naval anchorage at Ulithi with their attack aircraft, Operation Haikari. The *I-13* and *I-14* accompanied them, each with one aircraft to scout the lagoon at Ulithi. The war in the Pacific ended on 15 August, two days before the planned strike. All three completed submarines survived the war, never having fired a shot in anger. Future plans for the undersea craft—had the war continued—included replacing their aircraft with Baka rocket-propelled suicide aircraft,[7] and there were unconfirmed reports of proposals to use the submarines to launch aircraft against the United States carrying biological (germ) agents.

U.S. naval officers went through the submarines after the war. There was some consideration by the U.S. Navy of converting one or more of these giants to transport submarines. To meet U.S. Navy safety standards and rehabilitate the ships would take six months of yard work and would cost some $750,000 per submarine; this did not include later

*An* I-400-*class submarine comes alongside the U.S. submarine tender* Proteus *after Japan's surrender in August 1945. The base of the conning tower, offset to starboard, contains the three-aircraft hangar, which opens onto the catapult. There were proposals after the war to convert the surviving submarine carriers into cargo ships.* (U.S. Navy)

modifications that would be needed to employ U.S. batteries.[8] In the event, no work was undertaken. The *I-402* was scuttled off the coast of Japan in 1946; the *I-400* and *I-401* as well as the *I-14* were sailed to Pearl Harbor for further examination before being scuttled, also in 1946.

Several German U-boats also operated "aircraft" during World War II—towed autogiros. Unpowered, these devices were launched by the submarines on the surface to provide aerial surveillance around the U-boat. The aircraft—the Focke-Achgelis Fa 330—required a 17-mile-per-hour (27 km/h) airspeed, generated by submarine speed plus wind speed. Some 200 to 500 feet (61–152 m) of steel cable was usually winched out for operation, with a telephone line being imbedded in the cable to enable the airborne observer to communicate with the submarine. The free-wheeling rotor kept the craft aloft, with a normal altitude of about 330 feet (100 m). Although the Fa 330/U-boat operations were limited, the device was an interesting one, reflecting the highly innovative approach of the Germans to submarine development and operations.

*U.S. sailors examine the hangar of an* I-400-*class submarine. The massive hangar door is at left. The hangar could accommodate three M6A1 Seiran floatplanes, whose engines could be warmed up while the submarine was submerged. These underwater giants were conceived to attack U.S. cities.* (U.S. Navy)

The Germans also planned to employ real helicopters from submarines. The Flettner 282 was part of the massive German helicopter program of World War II. The diminutive aircraft—called *Kolibi* (Humming Bird)—was the world's first helicopter to become operational with a military service. After flight trials in 1940–1941, which included operations from the cruiser *Köln* during heavy seas in the Baltic, the Fl 282 was ordered into mass production for the German Navy.

The Fl 282 was unarmed and was used for anti-submarine reconnaissance. A production order for *one thousand* Fl 282s was placed in 1944 for both the Navy and Air Force, but less than 30 pre-production aircraft were finished before the war ended. Some of those were used as shipboard ASW aircraft in the Baltic, Aegean, and Mediterranean. The aircraft was designed with rapidly removable rotors, in part to facilitate possible stowage in canisters on submarines. (One Fl 282 was reported to have landed on the deck of a U-boat.)

After the war, in 1948, TsKB-18 (later Rubin) developed a draft design for Project 621—a large landing ship-transport submarine. In addition to a battalion of troops, tanks, and vehicles, the submarine was to carry three La-5 fighter aircraft in a hangar built into the conning tower. The aircraft would be launched by catapult. This was the only known Soviet aircraft-carrying submarine to reach this stage of design.

Although there was no serious Soviet consideration of aircraft-carrying submarines, U.S. intelligence in the early 1950s did give credence to a submarine-launched nuclear air attack against Strategic Air Command (SAC) bomber bases. A secret Project Rand study in 1953—sponsored by the U.S. Air Force—concluded that, "Using the submarine-launched or low-altitude Tu-4 surprise attack, *the*

*The large Aichi M6A1 Seiran was the world's only aircraft built as a submarine-launched bomber. With a crew of two, the aircraft had a maximum weight of 9,800 pounds. Its top speed was 295 miles per hour, and range was in excess of 700 statute miles. This aircraft—the only survivor of 28 produced—is in the collection of the National Air and Space Museum, Washington, D.C.* (Eric Long/NASM)

*A Focke-Achgelis Fa 330 is prepared for flight atop a U-boat's conning tower. The tethered, unpowered autogiro "kite" was intended to provide the submarine with long-range surveillance. Rarely used and with limited effect, the idea was yet another manifestation of German innovation in submarine warfare.*

*This Fa 330 has been fitted with wheels, the configuration used to train U-boat sailors how to "fly" these aircraft. They normally flew at about 330 feet. The steel cable (right) contained a telephone line for the flier to report ship sightings to the U-boat.* (U.S. Army)

*enemy can destroy a major part of SAC potential at relatively small cost in A-bombs and aircraft.* With no more than 50 aircraft and bombs, two-thirds or more of SAC bomber and reconnaissance aircraft could be destroyed."[9] (Emphasis in original)

The Rand study postulated that Soviet submarines each would carry one aircraft with performance similar to the North American F-86 Sabre, a Mach 1 fighter aircraft with the F-86H and F-86K variants being able to carry a nuclear weapon.[10] The submarine-launched attack, with each aircraft armed with a 40-kiloton bomb (i.e., twice the explosive power of the Hiroshima A-bomb), could strike all occupied SAC bomber bases in the United States and overseas within 700 n.miles (1,300 km) of the coast. Most bases in the continental United States and 15 overseas SAC bases would be targets of the proposed submarine attack. Only 8 of 39 U.S. strategic bomber bases were beyond that range. The Rand study estimated that only a slight increase in Soviet aircraft size would provide a range of about 1,200 n.miles (2,225 km), enabling attacks on the remaining eight U.S. bases.[11]

Without warning the submarine attack against the United States was estimated to be able to destroy all heavy bombers (B-36) and 76 percent of the medium bombers (B-47); with warning—defined as about one hour—the submarine-launched strike would still destroy 100 percent of the heavy bombers as well as 73 percent of the medium bombers.[12] Soviet submarine strikes were estimated to be less effective against overseas SAC bases because their larger size would make aircraft on them less vulnerable to 40-kiloton bombs.

But Soviet submarine-launched strike aircraft existed only in the deliberations of Rand study group.

Immediately after the war the U.S. Navy gave little thought to aircraft-carrying submarines (at the time designated SSV). A Submarine Officers Conference in 1946 noted, "No design studies should be made on this type of submarine at this time unless the Chief of Naval Operations believes that the need for such a type submarine may be required in the near future."[13]

The development of nuclear propulsion led to some interest in aircraft-carrying submarines by the Office of Naval Research (ONR). In response to an ONR solicitation, aircraft designer Ed Heinemann—he preferred to be called an innovator—developed a series of design sketches for a fighter aircraft that could be accommodated in the massive bow hangar of the Regulus missile submarine *Halibut* (SSGN 587), completed in 1960.[14] Heinemann's sketches for ONR indicated how a new-design aircraft or his versatile A4D Skyhawk could fit into the submarine's hangar with minimum modification. The basic *Halibut* hangar was 80 feet (24.4 m) long. The new-design aircraft was Douglas model 640, a turbo-jet attack aircraft that would be catapult launched from the surfaced submarine, would come down at sea on its flying boat hull, and would be recovered aboard the submarine by a telescoping crane. Depending on modifications to the hangar, the aircraft's wings, tail fin, or nose would fold for shipboard stowage.

The Navy did not pursue Heinemann's proposals.

## Project Flying Carpet

During this period there were several proposals for nuclear-propelled, aircraft-carrying submarines.

The most ambitious proposal was sponsored by the Navy's Bureau of Aeronautics, responsible for aircraft development. The extensive feasibility study of aircraft-carrying submarines—called Project Flying Carpet—was demonstrated by the Boeing Aircraft Company. The secret study employed the *Thresher* (SSN 593)-type S5W propulsion plant and, initially, hangar configuration and hull lines based on the *Halibut* design.[15]

The near-term submarine carrier configuration—designated AN-1—would carry eight high-performance aircraft in two large hangars built into the forward hull. The submarine would be some 500 feet (152.4 m) long and displace 9,260 tons on the surface, larger than any U.S. submarine then planned, including the Polaris missile submarines.

The starting point for AN-1 aircraft would be a modified Grumman F11F Tiger turbojet fighter.[16] The aircraft's standard folding wings (for carrier use) would be supplemented by a folding tail fin, and a large rocket booster would be used for launching from a "zero-length" catapult. The launchers would be elevated to the vertical (90°) to launch aircraft. The pilots would climb into the aircraft while they still were in the hangar, before being moved onto the launcher by an automated system.

The feasibility of stowing conventional aircraft in Regulus II missile hangars as well as submarine weight, stability, and equilibrium was conducted using the *Grayback* (SSG 574) with an F11F aircraft.[17]

An improved, Mach 3 aircraft eventually was to replace the F11F, a Mach 1+ fighter. The later aircraft would be recovered through the use of an innovative hook-and-cable arresting system. In an emergency, an aircraft set down at sea could be brought back aboard the submarine by crane.

Stowage would be provided for aircraft fuel, weapons, and other stores for ten missions per aircraft, that is, a total of 80 missions per submarine. During the preliminary design process, it appeared feasible to increase the number of missions to at least 160 with only minor changes in the submarine design. The pressure hull would have three "sections"—hangar I, hangar II, and the after section, which contained control, crew, reactor, machinery, and related spaces. The after section would have six compartments.

The AN-2 variant aircraft-carrying submarine had similar hull lines to the AN-1. However, this variant would operate Vertical Takeoff and Landing (VTOL) aircraft, carried in eight *vertical* hangars built into the hull forward of the sail structure. Thus the below-deck configuration of the forward hull (hangars I and II in the AN-1) would differ considerably from the AN-1.

Noting that "Flight deck operations in the conventional meaning of the word do not exist," the study indicated that four VTOL aircraft could be launched within 5½ minutes of surfacing and eight aircraft in just over nine minutes.[18] These times could be reduced substantially if the engine start and run-up time was accomplished by self-contained starters rather than using shipboard power. Under the most adverse operational launch sequence, the time to launch all eight aircraft was estimated to be 18 minutes. (Adverse sea conditions would be compensated for by moving the aircraft, via deck tracks, to the amidship launchers closest to the ship's center of buoyancy.)

The Boeing study calculated that the AN-1 submarine would cost about half again as much as a Polaris missile submarine (based on 1958 estimates):[19]

| | |
|---|---|
| *Nautilus* (SSN 571) | $ 75 million |
| *Halibut* (SSGN 597) | $ 85 million |
| Polaris SSBN | $100 million |
| AN-1 carrier | $140–150 million |

The aircraft-carrying submarine was not pursued. A number of reasons have been put forward: A questionable operational requirement for submarine-based aircraft, bureaucratic opposition from the Bureau of Ships to a ship concept developed by the Bureau of Aeronautics, and the shortage of submarine construction capability because the Navy was accelerating the construction of both torpedo-attack submarines and Polaris missile submarines.

Although the U.S. Navy did not pursue the design of an aircraft-carrying submarine, proposals continued to come forth from a variety of sources. The U.S. Patent Office has received numerous submarine aircraft carrier proposals: One dated 1930 shows a submarine with a hangar built into the superstructure, carrying two floatplanes that were to be launched on

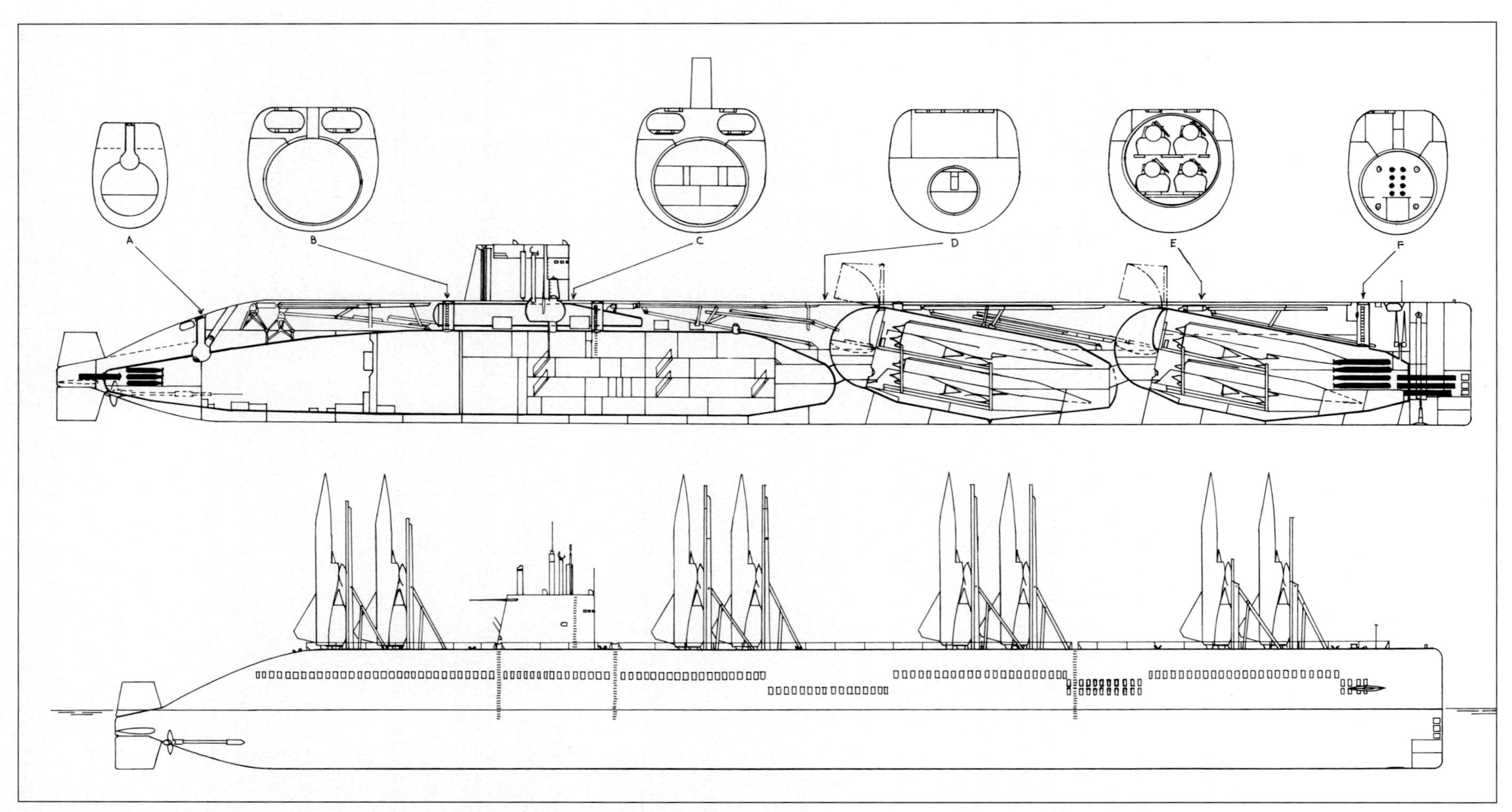


*U.S. submarine aircraft carrier AN-1 design. LOA 498 ft 6 in (152.0 m).* (©A.D. Baker, III)

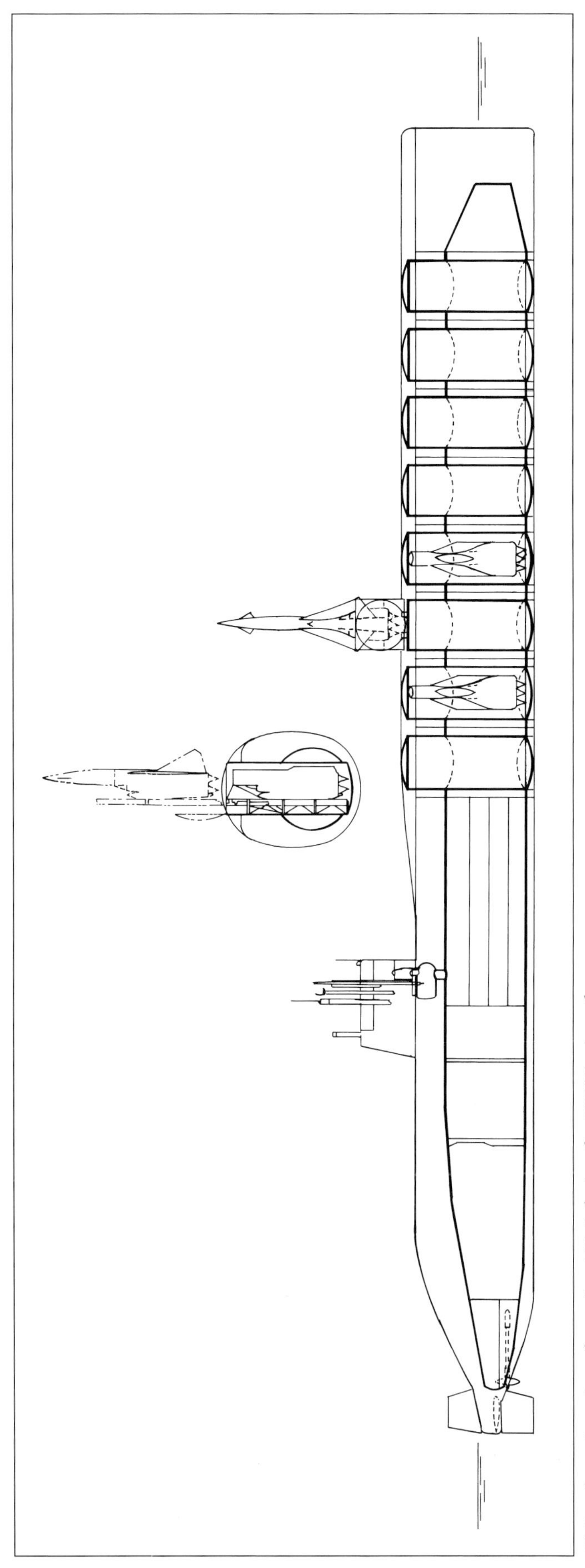

*U.S. submarine aircraft carrier AN-2 design for VTOL aircraft.* (©A.D. Baker, III rollers.[20] A post–World War II patent shows a conventional submarine with a large hangar within the pressure hull and an elevator to lift the floatplanes to the main deck, where they would take off. After alighting at sea, the floatplanes would come aboard the stern of the submarine.[21]

While few of these proposals were feasible from an engineering or operational viewpoint, they were interesting and help demonstrate the continued interest in this type of craft, combining the revolutionary weapons of the 20th Century.

**The only nation to make extensive use of aircraft-carrying submarines was Japan, during World War II. These submarines launched their aircraft on numerous reconnaissance missions during the war and, in 1942, on the only aircraft attacks against the United States.**

**With the advent of nuclear propulsion, the U.S. Navy sponsored the preliminary design of large, aircraft-carrying submarines. These designs, undertaken primarily by aircraft companies under the aegis of Navy agencies outside of the Bureau of Ships/Naval Sea Systems Command, were not fully developed, but they were an interesting conceptual excursion.**

**Rather, missiles, satellites, and other advanced-technology systems had overtaken the potential missions of the aircraft-carrying submarine.**

TABLE 15-1
**Aircraft-Carrying Submarines**

| | **Japanese *I-13* class** | **Japanese *I-400* class** | **U.S. Boeing AN-1** |
|---|---|---|---|
| Operational | 1944 | 1944 | — |
| Displacement | | | |
| surface | 3,603 tons | 5,223 tons | 9,260 tons |
| submerged | 4,762 tons | 6,400 tons | 14,700 tons |
| Length | 372 ft 9 in (113.7 m) | 400 ft 3 in (122.0 m) | 498 ft 6 in (152.0 m) |
| Beam | 38 ft 6 in (11.7 m) | 39 ft 4 in (12.0 m) | 44 ft 3 in (13.49 m) |
| Draft | 19 ft 4 in (5.9 m) | 23 ft (7.0 m) | 23 ft 7 in (7.19 m) |
| Reactors | — | — | 1 S5W |
| Turbines | — | — | 2 |
| horsepower | — | — | 15,000 |
| Diesel engines | 2 | 4 | — |
| horsepower | 4,400 | 7,700 | — |
| Electric motors | 2 | 4 | — |
| horsepower | 600 | 2,400 | — |
| Shafts | 2 | 2 | 2 |
| Speed | | | |
| surface | 16.75 knots | 19.7 knots | |
| submerged | 5.5 knots | 7 knots | 16 knots |
| Range (n.miles/kts) | | | |
| surface | 21,000/16 | 34,000/16 | |
| submerged | 60/3 | 90/3 | |
| Test depth | 330 feet (100 m) | 330 feet (100 m) | |
| Torpedo tubes* | 6 533-mm B | 8 533-mm B | 4 533-mm B<br>2 533-mm S |
| Torpedoes | 12 | 20 | |
| Guns | 1 140-mm<br>7 25-mm | 1 140-mm<br>10 25-mm | — |
| Aircraft | 2 | 3 | 8 |
| Complement | 108 | 144 | 163** |

Notes: * Bow + Stern.
** Includes 12 officer pilots and 2 flight officers.

# 16

# Midget, Small, and Flying Submarines

*The Project 865/Losos submarines—generally known by their Russian name* Piranya*—were the most advanced "midgets" developed during the Cold War. Although there was considerable interest in midget submarines at various times in the Soviet and U.S. Navies, few were produced.* (Malachite SPMBM)

Several navies developed and operated "midget" submarines during World War II. The most notable—and successful—efforts in this field were the Japanese midget submarines and British X-craft.[1] On the basis of their exploits, several nations expressed interest in this category of undersea craft during the Cold War.

The Japanese employed midget submarines—special attack units—at the outset of the war: Five Type A craft, each armed with two torpedoes, were released from five "mother" submarines to penetrate the U.S. naval base at Pearl Harbor (Hawaii) immediately before the surprise air attack on 7 December 1941. All five of the midgets were lost without scoring any torpedo hits, although one and possibly two did penetrate the harbor.[2] Interestingly, Vice Admiral Chuichi Nagumo, commander of the Pearl Harbor attack force, reported after the attack that "it is certain that extraordinary results were obtained by the most valiant attacks by the special attack units of the submarine force."[3] Subsequently, Japanese midgets were in action at Diego-Suarez, Madagascar, on 31 May 1942, when two midgets sank a British tanker and damaged the battleship *Ramillies.* Neither midget returned. At the same time, three midgets were launched into Sydney Harbor, Australia. Although several U.S. and Australian warships were in the harbor, the midgets inflicted no damage and none survived the attack.

Japanese midgets also operated off Guadalcanal, the Aleutians, and Philippines, but without significant success. When the war ended in August 1945, the Japanese had several hundred midget submarines being prepared to attack the U.S. assault forces in the invasion of the Japanese home islands, which was planned to begin in November 1945. About 440 midget submarines were completed in Japan from 1934 to 1945. All had two 18-inch (457-mm) torpedo tubes and were operated by crews of two to five men.

Britain's Royal Navy operated X-craft from 1942. These 50-foot (15.2-m), 27-ton submarines were towed to their operational areas by larger submarines.[4] The X-craft had a small, floodable chamber to "lock out" swimmers to place explosives under an enemy ship or to cut through barrier nets. The submarine's armament consisted of two curved, two-ton charges that were carried against the hull containing the explosive amatol. In combat, a four-man crew would pilot the craft beneath an anchored enemy warship, release the amatol charges under the hull, and slip away before they were detonated by a timing device. Fitted with a lockout chamber, the X-craft could also send out a swimmer to attach small limpet mines to an enemy ship.

The most famous X-craft operation of the war—code name Source—occurred in September 1943. Six large submarines each towed a midget underwater to a position off the Norwegian coast to attack the German battleship *Tirpitz* and other warships at anchor in Norwegian fjords. One midget was lost en route and another was scuttled after her explosive charges had to be jettisoned. Three of the six X-craft reached Altenfjord and two were able to lay their charges under the battleship *Tirpitz.* German sailors sighted the midgets, which were fired on, and both were scuttled, with six of the British submariners being captured. When the charges detonated, they lifted the 52,700-ton battleship about five feet (1.5 m), inflicting major damage on her propulsion machinery, guns, and rangefinders. The dreadnought was immobile for almost six months while repairs were made. The sixth X-craft, like most of the others, suffered mechanical problems, could not locate the nearby battleship *Scharnhorst,* and, after a rendezvous with her mother submarine, was scuttled. The raid had cost nine British submariners killed and six captured.[5]

The larger, 30-ton British XE-craft arrived in East Asia in mid-1945. Late on 30 July the two XE boats slipped into Singapore harbor, having been towed to the area by large submarines. Using limpet mines, swimmers from midgets attached six limpet mines to sink the Japanese heavy cruiser *Takao.* Both midgets successfully rendezvoused with their larger brethren waiting offshore.

In all, Britain constructed two prototype X-craft followed by five operational X-craft in 1942–1943, 25 larger XE-craft in 1944–1945 for use in the Pacific. Six similar XT-craft for training were delivered in 1944, for a total of 36 midgets produced for the Royal Navy during the war.

*A British X-craft, showing the midget's air induction mast and day and night periscopes. The later XE-craft had a flush deck. The U.S. Navy had great interest in these craft after World War II, with two being borrowed for demonstration purposes.* (Imperial War Museum)

Germany and Italy also developed midget submarines during World War II, but those craft accomplished little. The Type XXVII *Seehund* (sea dog) was the definitive variant of the German midgets, a 15-ton craft that could carry two 533-mm torpedoes. Their successes were few. By the end of the war in May 1945, 285 had been delivered of a planned 1,000 units. Under development was the Type 227 variant fitted with a Kreislauf closed-cycle diesel propulsion plant using liquid oxygen for underwater propulsion.[6]

In contrast to the limited operations of German and Italian midgets, the Italian "human torpedoes," or "pigs," were highly successful. These craft were piloted by two swimmers after being carried by a mother submarine. They would detach the warhead beneath an enemy ship and ride the vehicle back to the mother submarine. Several other nations also had such craft. Neither the Soviet nor U.S. Navies developed midget submarines during the war.

After the war Britain was the first country to resume the development of midgets. Initially two

XE-type were kept in operation, and in 1951 the Royal Navy ordered four new boats, the 36-ton *X51* through *X54;* a fifth unit was planned but not built.[7] But after their delivery in 1954 only two were kept in operation at any given time.

Their "normal" armament was a pair of three-ton charges of high explosives. *Nuclear mines* were proposed for these craft. In 1954 the Naval Staff proposed development of a nuclear mine—code named Cudgel—to be used by these craft to attack Soviet harbors, especially Kronshtadt in the Gulf of Finland, a major base of the Soviet Baltic Fleet. The nuclear mine was to be adopted from the Red Beard, an aerial bomb weighing 1,750 pounds (795 kg) with a 15-kiloton yield.[8]

Like their predecessors, these Cudgel-armed X-craft were to be towed to the proximity of their objective and then cast off to clandestinely plant the nuclear charge, which could be laid to depths of 300 feet (90 m). The time delay to detonate the nuclear weapon was up to seven days after being laid. Development of the nuclear charge was not pursued, and in 1958 the new X-craft were discarded.[9]

The Royal Navy also considered—very briefly—the possibility of a small nuclear-propelled submarine. In the late 1940s and 1950s, the U.S. Navy and U.S. Air Force both studied Aircraft Nuclear Propulsion (ANP) projects, with the Air Force actually flying a test reactor in a B-36 bomber.[10] The Navy looked into the feasibility of nuclear-propelled flying boats for cargo, ASW, and radar surveillance functions.[11] Among the candidates examined was the prototype Saunders-Roe S.R.45 Princess flying boat. The aircraft first flew in 1952 but never entered commercial service and, with two unfinished sisters, was soon mothballed. The Navy cancelled its ANP program in 1960, but interest was continued by some of the participating firms and the Royal Navy, the latter dubbing the effort the Empire Project.

The latter's interest ended when, in 1960, a formal briefing was made by one of the firms to a group of British officers. This was for a submarine of a few hundred tons to be manned by a crew of 12. "But we saw that the scheme was impractical. The provision of a large passive sonar, sufficient facilities, and weapons would have resulted in a full-size nuclear submarine."[12]

## The U.S. Navy's Midget Sub

The first—and only—midget submarine developed by the U.S. Navy during the Cold War was the *X-1* (SSX 1).[13] After World War II several U.S. naval officers proposed roles for midget submarines, especially attacking Soviet submarines at their home bases. More immediate, however, was interest in developing "anti-midget" tactics. A report to a Submarine Officers Conference in 1949 observed:

> The midget submarine promises to be as operationally desirable to the U.S. Navy as it would be to a probable enemy. . . . Inasmuch as the Russians have evinced definite interest in midget submarines it is particularly important that our defensive forces have the opportunity to develop the most effective defenses against such craft. This can only be done if we ourselves possess midget submarines with which to train our defense forces.[14]

By the late 1940s the leading advocate of midget submarines for the U.S. Navy was Commander Raymond H. Bass on the staff of Commander Submarine Force U.S. Atlantic Fleet. Bass envisioned several variants of midget submarines for a variety of roles and soon interested the Bureau of Ships in his quest. In 1950 the Navy considered the construction of two midget submarines, one of 25 tons with a 300-foot (91-m) operating depth and another of 70 tons that could operate to 250 feet (76 m). The larger craft would carry two "short" torpedoes with a crew of four to six men.

The fiscal year 1952 shipbuilding program, which was developed in 1950, provided for the smaller prototype craft, given the designation *X-1.* British X-craft were a major factor in the development of the *X-1:* In the summer of 1950 the British *XE7* was sent to the United States for three months to help the U.S. Navy evaluate the efficiency of such craft in penetrating harbor defenses; she was followed in 1952 by the *XE9,* and in 1958 by the *Sprat* (ex-*X54*).[15] Describing the *XE7* tests in penetrating anti-submarine nets in the Norfolk area during August 1950, Captain Lawrence R. (Dan) Daspit recounted:

> In those tests the X craft succeeded in going up to the net that had been laid, putting men [swimmers] outside the craft, cutting the net and getting through within 28 minutes under adverse tide conditions. There was no disturbance of the net to amount to anything at all.[16]

In discussing the requirement for the *X-1*, Daspit explained:

> This country has never operated any submarines of this size. . . . We feel that we need a small submarine to evaluate our harbor defenses. The Navy is responsible for the defense of harbors in this country and in advanced bases. A ship of this type, we feel, may succeed in going through the defenses we have today and in mining large units such as carriers or any other valuable ships.[17]

In the same meeting of the Navy's General Board, Commander Bass predicted, "After the Navy sees what these little craft can do, the demand will be created for their use in many places."[18]

The *X-1* was built by the Fairchild Engine and Airplane Corporation at Deer Park, Long Island, New York; she was launched at the nearby Jakobson's Shipyard in Oyster Bay, Long Island, on 7 September 1955, and was placed in service as the *X-1* on 7 October 1955 under Lieutenant Kevin Hanlon.[19]

The submarine was 49½ feet (15.1 m) long, with a surface displacement of 31.5 tons, making her slightly smaller than the *X51* series, which were 36 tons and 53¾ feet (16.39 m) long.

*The* X-1, *the U.S. Navy's only midget submarine built during the Cold War era, on trials in Long Island Sound on 6 October 1955. The submarine was built by an aircraft manufacturer and was the U.S. Navy's only submarine with closed-cycle propulsion.* (U.S. Navy)

*The* X-1 *"airborne" at the Philadelphia Naval Shipyard in 1960. At the time she was being taken out of "mothballs" and prepared for service as a research craft in the Severn River, near Annapolis, Maryland.* (U.S. Navy)

Manned by a crew of four, the *X-1* was developed ostensibly to test U.S. harbor defenses. But the craft was intended to have the capability—after being towed into a forward area by a larger submarine—of locking out swimmers for sabotage missions against enemy ships in port and of other combat tasks. Two divers could be carried for short durations, for a total of six men. Also, a 4-foot (1.2-m) section could be installed forward (between the craft's bow and main hull sections) to carry a single, 1,700-pound (771-kg) XT-20A mine. A JP-2 passive sonar was provided. However, on a realistic basis, the craft had no combat capability.

The most unusual aspect of the *X-1* was her propulsion plant. The BuShips design provided for a diesel-electric plant. Fairchild proposed—and the Navy approved—a diesel-hydrogen peroxide arrangement: While on the surface, the 34-horsepower

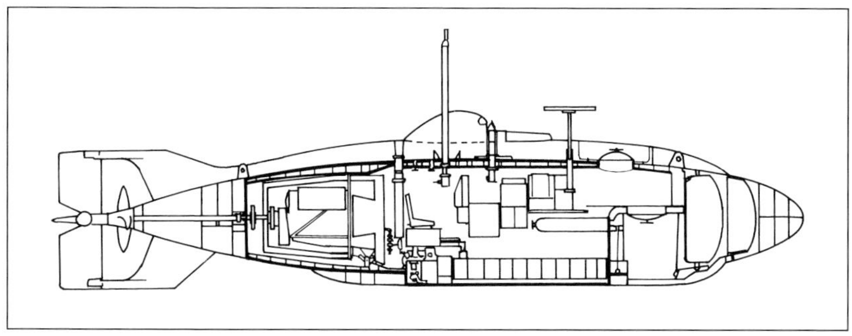

X-1 *(SSX 1). LOA 49 ft 7 in (15.1 m)* (©A.D. Baker, III)

three years later, in December 1960, she was towed to Annapolis, Maryland, and rehabilitated for use as a research craft in the upper Chesapeake Bay, where she was used for ASW-related studies of the chemical and physical properties of water. These activities continued through January 1973, with the craft again taken out of service and stricken the following month, on 16 February.[23] The U.S. Navy did not pursue other midget submarines—except for a "flying submarine."

diesel engine operated on air as did a conventional submarine; when submerged, the diesel engine ran on oxygen provided by the decomposition of hydrogen peroxide. A 60-cell electric battery and a small electric motor/generator were fitted for emergency submerged propulsion and electric power. The diesel-peroxide plant was intended to provide an underwater speed of eight knots, but six knots was the maximum achieved.[20]

Numerous difficulties were encountered with the propulsion plant. After initial trials in New London, yard work was planned at the Portsmouth Naval Shipyard. Three test stand runs on peroxide were undertaken in October–December 1956. On the initial peroxide test run, conducted on 26 October 1956, "The engine was very unstable and establishment of the fuel-to-peroxide ratio was almost impossible. During this test the engine stalled on the peroxide cycle at least once, according to an official record."[21] The damage with each test run was progressively more devastating to the piston and engine block. There also were problems with water contaminating the lubrication oil.

The problems continued into 1957. Even when there weren't problems, the maintenance requirements on the propulsion plant were horrendous. As one of her officers-in-charge, Richard Boyle, later recalled, "The power plant was such a nightmare that it is very difficult to visualize 'capability.' Can you imagine going up a fjord [on a mission] and being faced with changing the lube oil every 24 hours?"[22]

The *X-1* suffered a peroxide explosion at the Portsmouth Naval Shipyard on 20 May 1957 that blew off her bow section, which sank; the remainder of the craft remained intact and afloat. Rebuilt with a "straight" diesel-electric plant, she was taken out of service in December 1957 and laid up in reserve. But

## A Submersible Seaplane

The U.S. Navy began contemplating the merger of aviation and submarine technologies into a single vehicle as early as 1946.[24] By that time several Navy laboratories were looking into the required technologies. When asked by the press whether such a vehicle could be produced, Vice Admiral Arthur W. Radford, the Deputy Chief of Naval Operations for Air, replied: "Nothing is impossible."[25] It is significant that, despite various nomenclature, from the outset the concept provided for a submersible seaplane and not a flying submarine because of the differing structural approaches to a seaplane vis-à-vis a submarine.

A decade later, in 1955, studies were being conducted under contract from the Department of Defense by the All American Engineering Company while aviation pioneer John K. (Jack) Northrop was designing such craft. The All American vehicle was to alight on and take off from the water on "hydro-skis"; once on the water the craft could be "sealed" and could submerge.[26]

Although nothing resulted from these studies, by the early 1960s the U.S. Navy was ready to invest in such a vehicle. A Navy engineer working on the project, Eugene H. Handler, explained,

> There is . . . a tremendous amount of [Soviet] shipping in the Soviet-dominated Baltic Sea, the essentially land-locked Black Sea, the Sea of Azov, and the truly inland Caspian Sea. These waters are safe from the depredations of conventional surface ships and submarines.[27]

While noting that there were proposals to carry midget submarines in seaplanes as well as by a larger submarine for such operations, he explained, "The most needed improvements for the midget submarine appear to be increases in cruise speed and radius so that the submarine is able to return from missions without immediate assistance from other craft."[28]

In the early 1960s the Convair aircraft firm proposed a small submarine that could travel at high speeds on the ocean surface using hydro-skis or hydrofoils.[29] The firm had earlier developed the F2Y Sea Dart, a supersonic floatplane that employed water skis. This led to the Navy's Bureau of Naval Weapons—at the time responsible for aircraft development—awarding a contract to Convair in 1964 to examine the feasibility of a "submersible flying boat," which was being called the "sub-plane" by those involved with the project. The Convair study determined that such a craft was "feasible, practical and well within the state of the art."

The Bureau of Naval Weapons' specified a set of design goals:[30]

| | |
|---|---|
| air cruise speed | 150–225 mph (280–420 km/h) |
| air cruise altitude | 1,500–2,500 feet (460–760 m) |
| air cruise radius | 300–500 n.miles (555–925 km) |
| maximum gross takeoff | <30,000 lb (<13,600 kg) |
| submerged speed | 5–10 knots |
| submerged depth | 25–75 feet (7.6–23 m) |
| submerged range | 40–50 n.miles (75–95 km) |
| submerged endurance | 4–10 hours |
| payload | 500–1,500 lb (230–680 kg) |

takeoff and land in State 2 seas

ability to hover while submerged

Several firms responded to a Navy request, and a contract was awarded to Convair to develop the craft. The flying boat, which would alight and take off using retractable hydro-skis, would be propelled by three engines—two turbojets and one turbofan—the former for use in takeoff and the latter for long-endurance cruise flight. Among the more difficult challenges of the design was the necessity of removing air from the engines and the partially full fuel tank to reduce buoyancy for submerging. Convair engineers proposed opening the bottom of the fuel tank to the sea, using a rubber diaphragm to separate the fluids and using the engines to hold the displaced fuel.

To submerge, the pilot would cut off fuel to the engines, spin them with their starter motors for a moment or two to cool the metal, close butterfly valves at each end of the nacelles, and open the sea valve at the bottom of the fuel tank. As the seaplane submerged, water would rise up into the fuel tank beneath the rubber membrane, pushing the fuel up into the engine nacelles. Upon surfacing, the fuel would flow back down into the tank. The only impact on the engines would be a cloud of soot when the engines were started.

When the engines were started, their thrust would raise the plane up onto its skis, enabling the hull, wings, and tail surfaces to drain. The transition time from surfacing to takeoff was estimated to be two or three minutes, including extending the wings, which would fold or retract for submergence.

Only the cockpit and avionics systems were to be enclosed in pressure-resistant structures. The rest of the aircraft would be "free-flooding." In an emergency the crew capsule would be ejected from the aircraft to descend by parachute when in flight, or released and float to the surface when underwater. In either situation the buoyant, enclosed capsule would serve as a life raft. The sub-plane was to be constructed of a high-density material, probably stainless steel, with titanium also being considered. Titanium or fiberglass was considered for the pressure capsule containing the two-man crew. Cargo could consist of mines; torpedoes; or, under certain conditions, agents to be landed on or taken off enemy territory.

The Navy Department approved development of the craft, with models subsequently being tested

*An artist's concept of the planned U.S. Navy "submersible seaplane." The craft was to have three aircraft engines, two turbojet engines for takeoff and one turbofan for cruise flight. The concept appeared practical but was killed for mainly political reasons.* (General Dynamics Corp.)

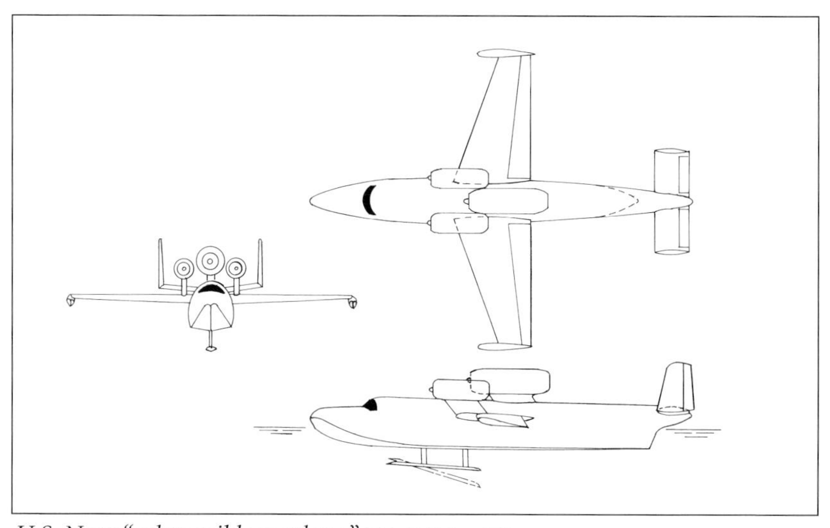

*U.S. Navy "submersible seaplane."* (©A.D. Baker, III)

in towing tanks and wind tunnels. But in 1966 Senator Allen Ellender, of the Senate's Committee on Armed Services, savagely attacked the project. His ridicule and sarcasm forced the Navy to cancel a project that held promise for a highly interesting "submarine." Although the utility of the craft was questioned, from a design viewpoint it was both challenging and highly innovative.

## Soviet Small Submarines

An unusual Soviet attempt at a small "submarine" occurred during the Khrushchev era. This concept was proposed by Nikita Khrushchev after visiting a naval base in Balaklava (Crimea), where he observed both high-speed surface craft and submarines. The Soviet leader expressed the opinion that in the nuclear era most warships should be "submersibles" and that a missile craft would be a good starting point.

At Khrushchev's direction, a submersible missile ship was designed between 1958 and 1964. The initial design effort was assigned to TsKB-19; in 1963 the bureau was joined with TsKB-5 to form the Almaz marine design bureau. The craft—designated Project 1231—was to have high speed (40 to 60 knots), carry four P-25 anti-ship missiles, and be capable of submerging to shallow depths. To achieve high surface speed, engineers experimented with various combinations of hull forms and hydrofoils, with models tested in towing tanks, on lakes, and in wind tunnels. After considering propulsion alternatives, the M507 diesel plant was selected, with a snorkel installation and electric motors for submerged operation.[31]

The effort proved technically difficult, in part because TsKB-19 engineers had no submarine experience, and also because of the problems of combining a high-speed craft and a submarine in a single platform. Project 1231 was terminated when Khrushchev was removed from office. "The design torment was ended," according to a later evaluation by submarine designer.[32]

Despite periodic reports of Soviet midget submarine operations, the Soviets did not develop such craft until well into the Cold War. In particular, there were continuous reports in the 1980s that Soviet midget submarines were invading Swedish territorial waters. According to a Swedish government statement,

> Thorough and expert analysis of the information that it has been possible to extract from the imprints documented on the sea floor has thus indicated, among other things, that we are dealing here with manned minisubmarines which are capable of proceeding both suspended in the water, driven by propeller machinery, and on the sea bottom, by caterpillar tracks, among other things. They can thus use channels and areas where the water is shallow and navigation conditions difficult, and they can be regarded as extremely difficult to locate and combat.[33]

However, as the Cold War came to an end, the Swedish government was forced to admit that the evidence was wrong: the caterpillar tracks were left by the trawling equipment of local fishermen, while submarine sounds detected by Swedish acoustic sensors were those of minks and other mammals searching for food in coastal waters.[34]

(Although Soviet midget submarines did not operate in Swedish coastal waters, larger undersea craft did so on a regular basis. The Project 613/Whiskey submarine *S-363* ran aground in October 1981 while far into Swedish territorial waters.)

The concept of a small—not "midget"—submarine was developed in Soviet research institutes for use in waters with broad, shallow shelf areas down to 650 feet (200 m), where operations of normal submarines either were limited because of the navigational conditions or were excluded altogether. In those areas small submarines were to counter enemy forces and conduct reconnaissance missions. These craft would differ from "midgets" in that they could travel to their operational area under their own power and not be carried by or towed behind a larger submarine.

Based on these concepts, in 1973 the Malachite design bureau began work on a small intelligence-collection submarine—Project 865.[35] Given the Russian name *Piranya* (NATO Losos), the design team initially was under chief designer L. V. Chernopiatov and, from 1984, Yu. K. Mineev. The Soviet shipbuilding industry lacked experience in the design and construction of such small submarines, and it was not possible to use the equipment developed for larger submarines, hence extensive research and development efforts were required.

The decision was made to construct the craft of titanium to save weight and provide strength. Also, the non-magnetic properties of titanium would help the craft evade magnetic mines. A conventional diesel-electric propulsion system was selected for the Piranya with heavy emphasis on machinery quieting. The craft would have a single propeller shaft and there would be conventional control surfaces, including bow diving planes.

The submarine would be manned by a crew of three (officers) and could carry up to six divers for ten-day missions. The small crew—one watch-stander normally on duty—was possible because of a very high degree of automation. The conditions of living and working in a small submarine for that period led to the fabrication of a full-scale simulator in a research center of the Ministry of Health.

The Piranya crews were to undergo training and psychological testing for ten-day periods in the simulator, which had inputs from navigation, sonar, and radar systems. The galley, sanitary, and berthing spaces as well as air-conditioning and air-cleaning systems were similar to standard submarine equipment, albeit on a smaller scale. Indeed, these features, coupled with the size of the craft—218 tons surface and 390 tons submerged displacement—could categorize them as coastal submarines. (The volume of the Piranya was six times that of the American *X-1*.)

A lockout chamber in the Piranya enabled access for divers. Their equipment—or mines or small torpedoes—were carried in two large containers mounted external to the pressure hull in the submarine's "humpback." The containers are 39⅓ feet (12 m) long and two feet (620 mm) in diameter and could be used down to the craft's operating depth (650 feet). They accommodate two PMT mines, or two 400-mm torpedoes, or two swimmer-delivery devices.

Construction of the lead submarine—the *MS-520*—was begun at Leningrad's Admiralty shipyard in 1981, with her keel being formally laid down in July 1984. The *MS-520* was completed in 1988, and the second unit, the *MS-521*, in 1990. Several sea

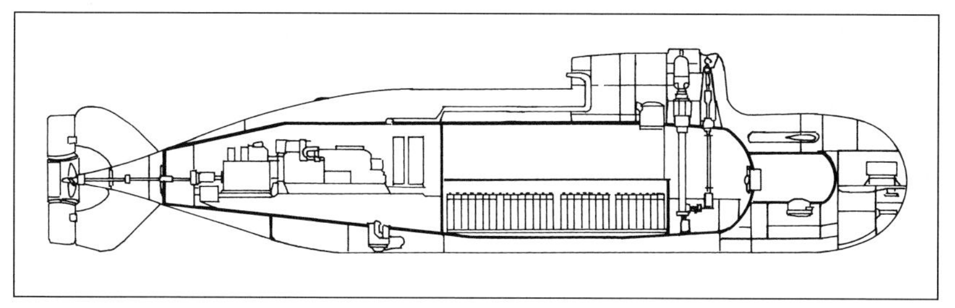

*Project 865/Piranya. LOA 92 ft 6 in (28.2 m)* (©A.D. Baker, III)

areas were considered for trials of the craft—the Baltic, Azov, Caspian, and Black Seas. Preference was given to the Baltic Fleet, specifically to the base at Paldiski (formerly Baltiski), which had enclosed waters and was located close to submarine test ranges. The last factor was significant because of the extensive trials and evaluations planned for the craft. At times the number of crewmen and technicians on board during this period reached 14.

Several variants of Project 865 were considered at the Malachite bureau, with surface displacements varying from 250 to 750 tons. In 1991 the Special Boiler Design Bureau in St. Petersburg completed development of the Kristall-20 Air-Independent Propulsion (AIP) system that would be applicable to the *Piranya* design. The system underwent extensive testing and was accepted by the Ministry of Defense for installation in future units. The demise of the Soviet Union and the severe cutbacks in naval construction halted these efforts. The two Russian craft were laid up in reserve in July 1992 and were stricken in 1997.

Although Piranya-type submarines had been offered for sale to other nations by the Russian government, none has been sold. There was one potential "buyer," however: in the 1990s three men were indicted in a Florida court for conspiring to purchase large quantities of cocaine and heroin to be transported from South America for distribution in the United States.[36] As part of the conspiracy, the defendants were alleged to have planned to use airplanes, helicopters, *and submarines* to transport the drugs. They had planned to procure the vehicles—including Piranya submarines—from Russia; reportedly, they already had acquired a half-dozen helicopters. But neither submarines nor even documentation were sold.

Whereas the Piranya-class small submarines were capable of operating independently in restricted areas, such as the Baltic Sea, a nuclear-propelled submarine was considered to provide long-range transport for the craft. Work on that project also halted with the end of the Cold War.

Separate from midget and small submarines, both the U.S. and Soviet Navies developed a variety of small submersibles to carry or propel swimmers for reconnaissance and to attack enemy shipping in port. These craft—some carried on the decks of submarines—exposed the pilots and swimmers to the sea, as in World War II craft. The U.S. Navy in the early 1990s began the construction of the so-called Advanced SEAL Delivery Systems (ASDS) in which the divers could travel in a dry environment and lock out upon reaching their objective, was in reality a midget submarine.[37] These craft are to be carried on the decks of several modified attack submarines of the *Los Angeles* (SSN 688) class as well as the *Seawolf* (SSN 21) class that have been provided with special fittings.

The ASDS, with a "dry" weight of 55 tons, carries a crew of two plus eight swimmers on missions lasting almost a day (i.e., 100 n.miles [185 km]). Propulsion and control is provided by a large ducted thruster plus four smaller, rotating thrusters. Among the submersible's other unusual features are a fixed-

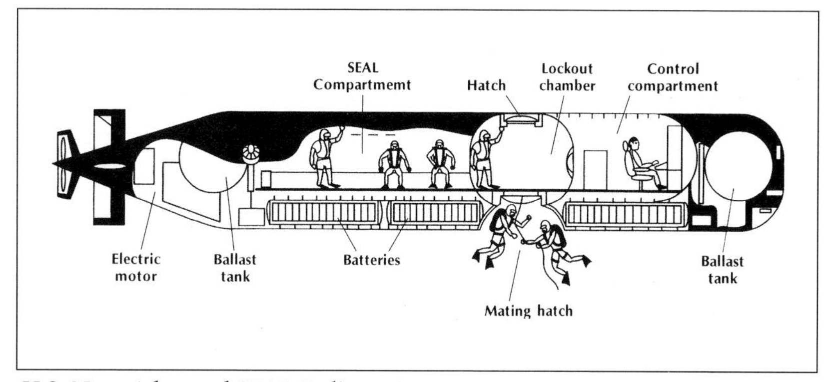


*U.S. Navy Advanced SEAL Delivery System.* (Chris Nazelrod/U.S. Naval Institute *Proceedings*)

length periscope and electronics masts, which fold down when not in use. This arrangement reduces hull penetrations and saves weight.

Six vehicles were planned when this volume went to press. The Northrop Grumman Corporation's ocean systems facility near Annapolis, Maryland, delivered the first craft for tests and training in 2000. This craft became operational in 2003 with a total of six planned for completion by about 2012—a remarkably long gestation period for a craft of great potential value for special operations (SEALs) and other clandestine activities.

*A drawing of a U.S. Navy Advanced SEAL Delivery System (ASDS) "in action." The craft is moored above the sea bed with twin anchors as swimmers emerge through the lower hatch to the craft's lockout chamber. The "patches" on the side of the hull indicated the retracted thrusters. The ASDS, long in development, is being procured at a very slow production rate. (U.S. Navy)*

*The first ASDS vehicle running trials off Honolulu, Hawaii. Submarines must have special modifications to carry the vehicle. The periscope (left) and electronics mast* fold *down and do not retract, saving internal space and providing more flexibility in internal arrangement. She has an eight-inch (203-mm) freeboard.* (U.S. Navy)

**Midget submarines were used extensively by several nations in World War II; however, they were not employed by the U.S. or Soviet Navies. In the Cold War era, both navies considered such craft, and prototypes were constructed, the United States doing so early in the period and the Soviet Union—producing much larger craft—late in the Cold War. Neither nation undertook large-scale production, nor did they provide their respective craft to foreign navies, although many nations operate such craft. The U.S. Navy's *X-1* was not successful; the Soviet era Piranyas were, but were not affordable when the Cold War ended.**

**The U.S. Navy also initiated a remarkable small submarine project in the "submersible seaplane" effort. That effort did not come to fruition because of political machinations.**

**While limited interest in such underwater craft continues, the only Soviet-U.S. effort to survive the Cold War has been the U.S. Navy's advanced swimmer delivery vehicle. This gives promise of providing considerable capabilities to the United States.**

TABLE 16-1
## Midget and Small Submarines

| | German Type 127 *Seehund* | British *X51* *Stickleback* | U.S. *X-1* SSX 1 | Soviet Project 685 *Piranya* (NATO Lasos) | U.S. Advanced SEAL Delivery Vehicle (ASDS) |
|---|---|---|---|---|---|
| Operational | 1944 | 1954 | 1955 | 1988 | 2001 |
| Displacement | | | | | |
| surface | 15.3 tons | 36 tons | 31.5 tons | 218 tons | 55 tons# |
| submerged | 17.4 tons | 41 tons | 36.3 tons | 390 tons | |
| Length | 39 ft (11.9 m) | 53 ft 10 in (16.4 m) | 49 ft 7 in (15.1 m) | 92 ft 6 in (28.2 m) | 65 ft (19.8 m) |
| Beam | 5 ft 7 in (1.7 m) | 6 ft 3 in (1.9 m) | 7 ft (2.1 m) | 15 ft 5 in (4.7 m) | 6 ft 9 in (2.06 m) |
| Draft | | 7 ft 6 in (2.29 m) | 6 ft 9 in (2.05 m) | 12 ft 9½ in (3.9 m) | 8 ft 3 in (2.5 m) |
| Diesel generators | 1 | 1 | 1 | 1 | nil |
| horsepower | 60 | 50 | | 220 | |
| Electric motors | 1 | 1 | 1 | 1 | 1 |
| horsepower | 25 | 44 | | 80 | |
| Shafts | 1 | 1 | 1 | 1 | 1 propulsor## |
| Speed | | | | | |
| surface | 7.7 knots | 6.5 knots | 5 knots | 6.4 knots | |
| submerged | 6 knots | 6 knots | 6 knots | 6.65 knots | 8 knots |
| Range (n.miles/kts) | | | | | |
| surface | 300/7 | | 185/5 | 1,450/6.4* | nil |
| submerged | 63/3 | | 185/6 | 260/4 | 125/8 |
| Test depth | 100 ft (30 m) | 300 ft (90 m) | 150 ft (45 m) | 650 ft (200 m) | 200 ft (60 m) |
| Torpedo tubes | nil** | nil | nil | 2 | nil |
| Torpedoes | 2 533-mm | nil | nil | 2 400-mm | nil |
| Complement*** | 2 | 4 + 1 | 4 + 2 | 3 + 6 | 2 + 8 |

Notes: * Maximum range with snorkel.
** Carried on brackets external to the pressure hull.
*** Crew + divers/swimmers.
# Dry weight in air.
## Electric motors power the main propulsor, a shrouded propeller; four small, trainable ducted thrusters provide precise maneuvering and hover capabilities.

17

# Third-Generation Nuclear Submarines

*The Project 971/Akula I shows an advanced hydrodynamic hull form, with the sail structure faired into the hull. The towed-array pod is prominent, mounted atop the upper rudder. Production of the Akula SSNs as well as other Soviet submarines was drastically cut back with the end of the Cold War.*

In 1967–1968 the U.S. Navy began planning for the follow-on to the *Permit* (SSN 594) and *Sturgeon* (SSN 637) classes, the *Sturgeon* herself having been completed in 1967. As the Office of the Chief of Naval Operations laid out requirements for the next-generation SSN, the Naval Sea Systems Command (NavSea) initiated a submarine design effort to respond to those requirements.[1] This effort came to be known as CONFORM, short for Concept Formulation.

Captain Myron Eckhart Jr., at the time head of preliminary ship design in NavSea, recalled that the Office of the Secretary of Defense "funded the initial design work at a level much higher than normal—allowing the Navy to perform exceptionally thorough research, design, and engineering right from the project's beginning."[2] And "The CONFORM project involved more of the nation's best design talents than had ever before been focused on the first stage of a submarine's design." CONFORM began with "a clean sheet of paper," according to Eckhart.

The CONFORM submarine was to use a derivative of the S5G Natural Circulation Reactor (NCR) plant that Vice Admiral H. G. Rickover was developing for the one-of-a-kind submarine *Narwhal* (SSN 671). A senior submarine designer called the S5G plant—with direct drive—"the best reactor ever built."[3] The NCR plant uses the natural convection of the heat exchange fluids (coolants) at low speeds rather than circulating pumps, a major noise source of propulsion machinery in nuclear submarines.

*A model of one of the CONFORM concepts for an advanced SSN. In this view the submarine's masts, periscopes, and small bridge structure are folded flush with the hull. This arrangement had several advantages over traditional telescoping masts and periscopes.*

The S5G plant was estimated to be capable of providing a submerged speed of more than 30 knots with a CONFORM submarine hull approximately the same size as the *Sturgeon*, that is, about 4,800 tons submerged. Innovative features being considered for the CONFORM submarine to enhance combat capabilities included hull shape, weapons, sensors, and even the periscopes. The last included a proposal for periscopes and masts that folded down onto the hull rather than retracting vertically into the hull. This feature would have reduced pressure hull penetrations, provided more flexibility in internal arrangement, and possibly alleviated the need for a sail.[4]

But Admiral Rickover scuttled the CONFORM project. Separate from the CONFORM effort, Rickover was working on what he foresaw as the next-generation submarine reactor plant that would have the "primary objective of high speed."[5] The Rickover plan adapted the D2G pressurized-water reactor plant used in surface warships, which developed about 30,000 horsepower per propeller shaft, to a submarine configuration. This evolved into the S6G reactor plant, which was to be evaluated in a single submarine, the SSN 688. Other one-of-a-kind submarine platforms for new propulsion plants had been the *Nautilus* (SSN 571), *Seawolf* (SSN 575), *Triton* (SSRN 586), *Tullibee* (SSN 597), *Narwhal*, and *Glenard P. Lipscomb* (SSN 685).[6]

This quest for a high-speed submarine was stimulated by two factors: First, the higher-than-expected speeds revealed by Soviet undersea craft—the Project 627A/November SSN's high-speed shadowing of the carrier *Enterprise* in January 1968, which had been a surprise to the U.S. Navy, and the 33-knot Project 671/Victor SSN.

*Detail of a CONFORM concept for an advanced SSN, showing the masts, periscopes, and small bridge structure in the raised position.*

Second, the *Thresher* (SSN 593) and *Sturgeon* classes employed the same 15,000-horsepower S5W reactor introduced in 1959 in the *Skipjack* (SSN 585); the later submarine designs were much larger, with more wetted surface, and hence were slower than the 33-knot *Skipjack*. The SSN 688 was intended to demonstrate "higher speeds than any U.S. submarine developed to date."[7]

The SSN 688 hull diameter was increased slightly, to 33 feet (10.1 m) compared to 31⅔ feet (9.65 m) for the *Thresher* and *Sturgeon*, but the new submarine was significantly longer—362 feet (110.3 m). The additional diameter was needed for the overhead cables and piping; the larger propulsion plant, with its quieting features, resulted in the longer and, hence, larger submarine, with essentially the same weapons and sensors. The larger size would sacrifice some of the speed increase that was expected from the greater horsepower. From the start, the Navy's leadership viewed the SSN 688 as a one-of-a-kind submarine, with the CONFORM project intended to produce the next SSN class.

In 1967 Rickover; Captain Donald Kern, the program manager for submarines in the Naval Sea Systems Command; and his boss, Rear Admiral Jamie Adair, had signed an agreement in which

> It was agreed that the fast submarine [SSN 688] would receive priority in new submarine design. It was further agreed that as soon as the fast submarine was authorized, Admiral Rickover would back to every possible extent the new attack submarine design as advocated by PMS-81 [the submarine branch under Kern].[8]

The agreement "lasted one day," according to Kern.[9] The Department of Defense, embroiled in reducing new military programs to help pay for the expanding war in Vietnam, sought to cut new ship programs. Why, the Navy was asked, were three different attack submarines being developed: the CONFORM, the SSN 688/S6G, and the *Lipscomb*. The *Lipscomb*—designated the Turbine Electric-Drive Submarine (TEDS)—was another attempt to reduce noise levels with turbo-electric drive, alleviating the need for reduction gears between the turbines and the propeller shaft, a concept similar to the electric-drive *Tullibee,* built a decade earlier. The *Lipscomb* was fitted with the S5Wa reactor plant which, because of the inefficiency of electric drive, delivered less power than the 15,000 horsepower of an S5W plant. The submarine was large, with a submerged displacement of 6,480 tons, some 35 percent greater than the *Sturgeon,* and a submerged speed of only 23 knots. Also, the *Lipscomb* suffered from heating problems, mostly because it had direct-current electric drive.

The CONFORM project—begun in the mid-1960s—examined a "multitude of alternative potential designs. . . ."[10] By November 1968 the initial CONFORM studies had produced 36 design concepts, examining such alternative features as twin reactors, single turbines, contra-rotating propellers, deep-depth capability, and larger-diameter torpedo tubes. According to Captain Eckhart, the contra-rotating arrangement for the CONFORM design gave promise of at least two knots higher speed than the SSN 688, while the design and the use of HY-100 steel could provide a test depth as great as 2,000 feet (610 m).[11]

Captain Kern recalled,

> The problem we had was [that] the CONFORM design became very competitive with [Rickover's] very conservative 30,000 [horsepower] 688. In other words, we started to show speeds that were competitive with his 688. We could produce a submarine with 20,000 shaft horsepower using counter-rotation [propellers] that could make about the same speed as his 688 with 30,000 shaft horsepower in a smaller submarine. So this got to be really difficult for him to handle. . . .[12]

The "front end" of the CONFORM submarine was to be greatly improved, with an increased number of weapons, advanced sonar, and a high degree of automation. And, with most CONFORM designs being smaller than the SSN 688 design, a CONFORM design probably would have been less expensive to build than the SSN 688.

But Rickover made the decision to series produce the SSN 688 regardless of the preferences of the Navy's leadership or the Department of Defense (DoD). Dr. John S. Foster Jr., the Under Secretary of Defense for Research and Engineering, was the DoD "point man" for the new SSN. He was the Pentagon's top scientist, who had spent most of his scientific career developing nuclear weapons.

Dr. Foster had several technical groups examining options for the new SSN, especially what was evolving from the CONFORM effort. As he reviewed Rickover's "high-speed" SSN 688 design, Foster expressed concern that the S6G reactor plant—nearly twice as large as the previous *Sturgeon*/S5W plant—would result in a larger submarine that would be only a couple of knots faster but would cost twice as much.[13] Also, not widely known at the time, the SSN 688 would not be able to dive as deep as previous submarine classes. (See below.)

Rickover would neither wait nor tolerate alternatives to the SSN 688. He used his now-familiar tactics of telling his supporters in Congress that Foster and other officials in DoD and the Navy were interfering with his efforts to provide an effective submarine force to counter the growing Soviet undersea threat. *Washington Post* investigative reporter Patrick Tyler concisely described the decision process that took place in March 1968:

> In the end, it was a political decision. Stories were beginning to circulate in Washington about the ominous nature of the Soviet submarine threat. Congress was intent on funding the SSN 688 whether the Defense Department put the submarine in the budget or not. Why should Foster needlessly antagonize men who had the power to act without him? Without any leadership in the Pentagon, the White House, or in the Navy to force Rickover to consider alternatives, Foster was powerless to stop the submarine's momentum.[14]

Although submarine force levels are beyond the scope of this book, Secretary of Defense Robert S. McNamara, in January 1968, announced that the attack submarine force requirement was for 60 first-line SSNs; the pre-*Thresher* SSNs were classified as second-line SSNs. Including all active ships and those under construction, only four additional SSNs would be required to attain 60 first-line submarines. The last two *Sturgeon*-class SSNs were authorized in fiscal year 1969. Two additional *Sturgeon* SSNs were scheduled for fiscal 1970, but this plan became meaningless when McNamara left the government in early 1968.

Rickover pushed the SSN 688 on DoD and the Navy, claiming that no other design was ready for production. In 1969 Secretary of Defense Clark Clifford told Congress that the first three "high-speed" SSN 688s would be requested in fiscal year 1970 and that four additional units would be requested in both the fiscal 1971 and 1972 programs.[15] "By that time, the new design (CONFORM) submarine should be ready for construction," he naïvely explained. While optimistic that CONFORM would go forward, Clifford was apprehensive about the *Lipscomb*. He noted that, while the TEDS project had risen considerably in cost, "we believe TEDS will be worth its cost since it will provide us unique and valuable operational and test experience with this new type of propulsion plant and other important quieting features. . . ."[16] The first three *Los Angeles* (SSN 688) submarines were estimated to cost an average of $178.7 million each; the *Lipscomb* was estimated to cost about $152 million but was expected by DoD officials to go as high as $200 million.

The SSN 688 and near simultaneous *Lipscomb* represented new propulsion plants, the sine qua non of Rickover's interests.[17] Both of those submarines would have the same weapons (four 21-inch torpedo tubes with 25 torpedoes) and sonar (updated AN/BQQ-2) as in the earlier *Thresher* and *Sturgeon* classes.

The CONFORM submarine proposals stressed innovation, with the reactor plant being a derivative of the *Narwhal*'s S5G NCR plant. Captain Eckhart would later write, "Rickover effectively terminated this affair when he made it known that, under his civilian authority [in the Department of Energy], he would not certify the reactor planned for the CONFORM sub—at least, not for that design application. That made the CONFORM design, as such, moot."[18]

The Department of Defense—at Rickover's urging and through his influence with Congress—was forced to accept the SSN 688 for series production.[19] The CONFORM project was gone, and Rickover ordered all copies of CONFORM documentation to be purged from official files. The funding of the single *Lipscomb*/turbo-electric drive submarine—long questioned by DoD and Navy officials—was forced by Congress on a reluctant administration.[20] The *Lipscomb* was built at Electric Boat and commissioned on 21 December 1974. Comparatively slow and difficult to support with a one-of-a-kind propulsion plant, she was in service for 15½ years, one-half of the design service life for an SSN.

## The *Los Angeles* Class

The Navy shifted to building the *Los Angeles* class, with its S6G pressurized-water plant (derived from the D2G) producing 30,000 horsepower; this was to provide a speed of 33 knots, some five knots faster than the previous *Sturgeon* class. Her 33-knot speed was the same as the *Skipjack* of 1959 (with her original propeller configuration) and also was about the original speed of the Soviet Project 671/Victor SSN, which had gone to sea in 1967.

The *Los Angeles* displaced 6,927 tons submerged—45 percent more than the *Sturgeon* while carrying essentially the same weapons and sensors albeit with quieter machinery. Comparative costs are difficult to determine; according to the Navy Department, in fiscal year 1978 dollars the last (37th) *Sturgeon* cost $186 million compared to the 32d *Los Angeles* with an estimated cost of $343 million in same-year dollars. The $186 million included upgrading the *Sturgeons* to the *Los Angeles* sonar and fire control systems at about $22 million each. Thus, more than three *Sturgeons* could be built for the cost of two *Los Angeles* SSNs.

Originally the *Los Angeles* was intended to be constructed of HY-100 steel, enabling the submarine to operate down to at least 1,275 feet (390 meters), the same as the previous SSN classes. However, difficulties were encountered with the HY-80 steel, and weight had to be cut because of the heavier reactor plant. This limited her test depth to an estimated 950 feet (290 meters).[21] Also, to save space and weight, the *Los Angeles* was not provided with under-ice and minelaying capabilities.

In 1971 the Navy awarded contracts for the 12 *Los Angeles*-class submarines authorized in fiscal years 1970–1972. The first submarine was ordered from Newport News Shipbuilding in Virginia, with that yard being awarded five hulls and Electric Boat the other seven. This was the first time that Newport News was the lead yard for a class of nuclear submarines, the yard having delivered its first submarine—for the Russian Navy—in 1905 and its first nuclear submarine, the USS *Robert E. Lee* (SSBN 601) in 1960.[22]

From 1974 Newport News and Electric Boat were the only U.S. shipyards constructing submarines. Both were building *Sturgeon* SSNs, while EB was preparing to build Trident SSBNs. Newport News laid the keel for the *Los Angeles* in 1972; the first of the class to be built at EB was the *Philadelphia* (SSN 690), also laid down in 1972. By that time both yards were encountering problems with their submarine programs and EB was facing a critical situation. The crisis was being caused by Newport News which—as lead yard for the SSN 688—was late in providing drawings to EB; the high costs for energy; design changes and interference by the Naval Sea Systems Command and Admiral Rickover; and inefficiencies at the EB yard, with hundreds of hull welds being found defective and, in some instances, missing.[23]

By the end of 1974 the EB yard had 18 SSN 688s under contract, their cost totaling $1.2 *billion*.[24] A "learning curve" had been expected wherein later submarines would cost less than the earlier units, hence the low amount. This projection proved false, and EB's internal audit was indicating a potential overrun of $800 million—67 percent—with most of the yard's submarines not yet started. Cost estimates for the *Los Angeles* class continued to increase. The *Los Angeles* (SSN 688) was launched on 6 April 1974 and commissioned on 13 November 1976. She would be the progenitor of the largest nuclear submarine series to be built by any nation.

In August 1977 the Navy announced a six-month delay in the completion of the first Trident missile submarine at Electric Boat, whose construction had begun at EB in mid-1974. In November 1977 a Navy spokesman announced that the first Trident SSBN was a year behind schedule and that her cost had increased by 50 per-

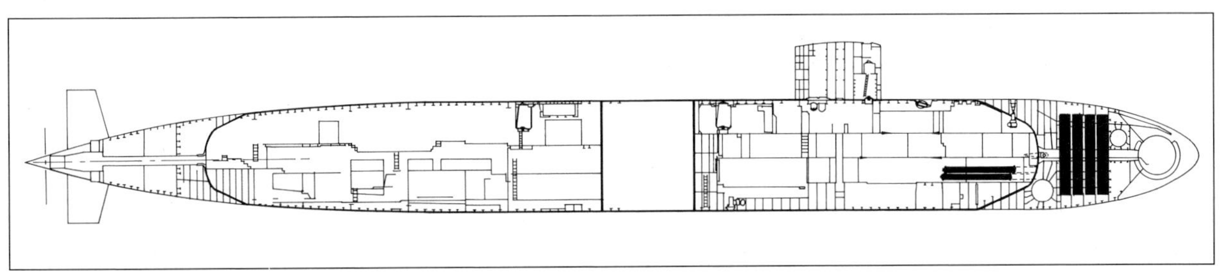

Toledo *(SSN 769) of Improved* Los Angeles *(SSN 688) class. LOA 362 ft (110.3 m)* (©A.D. Baker, III)

*The* Los Angeles *on her sea trials in September 1976. Most of the space aft of the sail is taken up with the reactor and machinery spaces. All of her masts and periscopes are fully retracted into the sail in this view.* (Newport News Shipbuilding)

cent from an estimated $800 million to about $1.2 *billion.*

Charges, countercharges, cover-ups, lies, secretly taped meetings, and congressional hearings followed. And the Trident's problems spilled over into the attack submarine program, with the head of the Naval Sea Systems Command telling a congressional committee that there was more than a half-billion-dollar overrun for the 16 attack submarines under construction at the EB yard. Vice Admiral Charles Bryan blamed management for the problems at Electric Boat.[25] Only the high priority of the Trident missile submarines stopped some Navy officials from attempting to terminate submarine construction at the EB yard. John Lehman, upon becoming Secretary of the Navy in February 1981, considered moving future Trident submarine construction to the Newport News shipyard.

*The* Los Angeles-*class submarine* Chicago *(SSN 721). The open hatch aft of the sail shows the position that a Dry Deck Shelter (DDS) would be mounted on the submarine for the special operations role. Unseen are the forward and after escape trunks with their hatches closed.* (U.S. Navy)

The entire U.S. nuclear submarine program was at risk. The Carter administration (1977–1981) had negotiated a settlement with General Dynamics, parent corporation of the EB yard, but the yard's myriad of problems still were not solved. When the Reagan administration took office, Lehman put together what he labeled a "draconian plan to clean up this disaster once and for all."[26] The final settlement between the Navy and General Dynamics was announced on 22 October 1981. The settlement came three weeks before the commissioning of the

*The* Annapolis *(SSN 760) was an Improved* Los Angeles*-class SSN with the forward planes fitted in the bow for under-ice operations. The submarine's AN/BPS-15 surface search radar is raised and her two periscopes are partially elevated. Just visible behind the scopes is the head of the snorkel intake mast.* (Giorgio Arra)

first Trident submarine—which was 2½ years behind schedule—and the firing of Admiral Rickover, which was a major factor in Lehman's plan.[27]

During the late 1970s the construction of the *Los Angeles* SSNs fell increasingly behind schedule and costs grew. Interestingly, there was a significant difference in costs between the two building yards. According to Rickover, although EB and Newport News were building identical submarines, the SSN 688 units being built at EB were costing about $50 million more per submarine than at Newport News: "The only possible explanation is that Electric Boat is less efficient than Newport News."[28]

At the request of the Senate Armed Services Committee, the Department of Defense examined the potential for producing lower-cost SSNs and SSBNs.[29] The growth in attack submarine size in successive classes had been accompanied by a closely related increase in costs. A major DoD "Submarine Alternatives Study" addressed the factors affecting submarine effectiveness:[30]

| | |
|---|---|
| Platform | Payload |
| Depth | Countermeasures |
| Endurance | Information Processing |
| Radiated/Self-Noise | Sensors |
| Reliability | Weapons |
| Speed | |
| Vulnerability | |

Two approaches were proposed to lowering SSN costs. First, the head of ship design in NavSea, Dr. Reuven Leopold, argued "There is a much greater potential for increasing combat effectiveness through combat systems performance gains than through platform performance gains."[31] This view called for giving up some platform features while increasing overall effectiveness through payload improvements. The second approach was to reduce submarine size through improvements in technology, with the DoD study noting that "power systems must be considered the number one candidate for investigation."[32]

Additional studies indicated that a smaller SSN, referred to as a Fast Attack Submarine (FAS) of perhaps 5,000 tons submerged displacement using advanced propulsion concepts, could carry approximately the same payload as a *Los Angeles*-class submarine. The "cost" would be the loss of perhaps five knots of speed, that is, to about that of the 27-knot *Sturgeon* class. Smaller than the *Los Angeles,* the new craft would be cheaper to build (after development costs). The head of research for the Secretary of the Navy, Dr. David E. Mann, estimated that billions of dollars could be saved if the Navy turned away from the larger submarines advocated by Vice Admiral Rickover and built the Fast Attack Submarine.[33]

Rickover savaged the FAS proposal—sometimes called "Fat Albert" because of its small length-to-beam ratio.[34] In a 49-page rebuttal, he reiterated the history of nuclear submarine development, stressing,

> I recognized that the potential for sustained high speed could be used to expand the role of the attack submarine beyond the traditional anti-shipping and anti-submarine missions. With virtually unlimited high speed

> endurance I felt they could provide a superior form of protection for our high value surface combatants [e.g., aircraft carriers].[35]

He continued with a discussion of the role of speed in naval warfare before returning to the importance of speed for nuclear submarines. Rickover concluded his paper urging that construction of the *Los Angeles* class continue, declaring:

> Recall, it was Congress that forced the Navy to build the *Los Angeles* in the first place, because of our continuing loss of speed capability in comparison to what was seen in Soviet nuclear submarines. The idea of intentionally building a slow submarine just to save money strikes at the core of absurdity.[36]

Yet, Rickover had previously advocated construction of the *Lipscomb* turbo-electric drive submarine—a comparatively slow and expensive submarine. In a congressional hearing he said, "and if the advantages over the other propulsion designs dictate, [electric drive] would then be incorporated into a later class design."[37]

"Fat Albert" died and, again, Rickover demanded the destruction of all documents proposing more innovative approaches to submarine propulsion plants. Mark Henry, a later head of the NavSea submarine preliminary design branch, said that the FAS "was not a good design. It was 'squeezed' to the point that it would have grown significantly during later design stages to get it 'to work.'"[38]

Rickover also defeated proposals to build a new SSBN that would be smaller and less costly than the Trident submarines of the *Ohio* (SSBN 726) class. There were several advocates for smaller SSBNs. As late as 1983—after Rickover's removal—a presidential panel observed, "The commission notes that—although it believes that the ballistic missile submarine force will have a high degree of survivability for a long time—a submarine force consisting solely of a relatively few large submarines at sea . . . presents a small number of valuable targets to the Soviets."[39]

Sending smaller Trident SSBNs to sea "would, as much as possible, reduce the value of each platform and also present radically different problems to a Soviet attacker than does the Trident submarine force," said the report. At one point Dr. Mann also suggested that a new SSBN—also carrying 24 Trident C-4 or D-5 missiles—with a submerged displacement slightly smaller than the *Ohio* class (some 15,000 tons compared to 18,750 tons) would cost about 30 percent less than the $1.5 *billion* per Trident submarine through use of a smaller reactor plant. There would be a loss of speed compared with the *Ohio* SSBN.

Mann suggested stopping the existing Trident program at 12 submarines and starting production of the new craft, referred to as SSBN-X. Most likely, the design costs of a new class of ships probably would have nullified expected savings. As with the Fast Attack Submarine, several proposals for smaller SSBNs sank without a trace. Rickover's successors as the head of the nuclear propulsion—all four-star admirals—adamantly opposed such proposals.

Despite his skill in stopping alternatives to his programs, by the 1970s Rickover's congressional power base had eroded to the extent that he was unable to initiate a new reactor program—his most ambitious to date—in the face of opposition by the Chief of Naval Operations. This nuclear plant was for a cruise missile submarine (SSGN), given the designation AHPNAS for Advanced High-Performance Nuclear Attack Submarine, and later called CMS for Cruise Missile Submarine.

In the late 1960s there was interest in providing U.S. submarines with both anti-ship and land-attack (nuclear) cruise missiles. The former was intended to counter the increasing number of large Soviet surface ships that were going to sea; the latter was seen as an additional weapon for attacking the Soviet homeland as both super powers continued to increase their nuclear arsenals. Precision guidance was becoming available for cruise missiles that would enable small nuclear warheads to destroy discrete targets.

This attempt at an anti-ship weapon was called the Submarine Tactical Missile (STAM).[40] The weapon, with a 30-inch (760-mm) diameter, would require special launch tubes. Admiral Rickover immediately embraced the STAM submarine as a platform for his proposed 60,000-horsepower AHPNAS reactor plant. Carrying 20 STAM weapons with a range of 30 n.miles (56 km), the proposed SSGN would have about twice the displacement of a *Los Angeles* SSN. (See table 17-2.) The AHPNAS reactor plant would drive the submarine at 30-plus knots. Design of the submarine began in 1970 and the preliminary design was completed by 1972.

Admiral Rickover told Congress that the AHPNAS program was the "single most important tactical development effort the Navy must undertake."[41] He also explained:

> The cruise missile would provide a totally new dimension in submarine offensive capability. In essence, the U.S. submarine would have the ability to react quickly, to engage the enemy on the submarine's own terms, and to press this initiative until each unit had been successfully attacked, regardless of the speed and tactics of the enemy. The very existence of this advanced high-performance nuclear attack submarine would constitute a threat to both the naval and merchant arms of any maritime force.
>
> * * *
>
> The submarine could act as an escort operating well ahead of a high-speed carrier task force, clearing an ocean area of enemy missile ships. . . . Employed in the role of an escort for combatants or merchant ships, the advanced high-performance submarine could operate independently or in conjunction with other escort vessels.
>
> * * *
>
> In addition to having an antisurface ship capability far superior to any the United States presently has at sea or plans to develop, the advanced high-performance submarine would also possess as a minimum the same substantial antisubmarine capabilities of the SSN 688 class submarine; it would be capable of detecting and attacking submerged submarines. . . . This new submarine will be fully capable of surviving at sea against the most sophisticated Soviet ASW forces.[42]

Admiral Rickover's proposed AHPNAS would truly be a "submarine for all seasons."

In July 1970 Admiral Elmo R. Zumwalt became Chief of Naval Operations. Zumwalt and his staff opposed the AHPNAS concept. First, Zumwalt feared that the large SSGN would be produced in significant numbers, ravaging the shipbuilding budget. Second, as a surface warfare specialist (the first to serve as CNO since Arleigh Burke in 1955–1961), Zumwalt wanted to provide offensive strike capability in cruisers and destroyers as well as submarines. At that time, the Navy's conventional strike force consisted of the aircraft on 15 large aircraft carriers.

Admiral Zumwalt promoted placing the Harpoon anti-ship missile and the Tomahawk multipurpose missile in surface ships. He wanted the missile's size constrained to permit launching from standard submarine torpedo tubes. (See Chapter 18.) The decision to provide Harpoons to submarines was made in 1972, following a Zumwalt-convened panel that recommended the encapsulated Harpoon as the fastest way to put cruise missiles on board U.S. submarines. The first flights of the Harpoon occurred that December and the missile became operational in surface ships and submarines in 1977.

The *Los Angeles* SSNs had four 21-inch torpedo tubes plus space for 21 reloads. Carrying, for example, eight Tomahawk cruise missiles would leave 17 spaces for torpedoes, SUBROC anti-submarine (nuclear) rockets, and Harpoon short-range missiles. Could the payload of the *Los Angeles* be increased without a major redesign? Further, with only four launch tubes, probably two tubes would be kept loaded with torpedoes leaving at most two tubes for launching missiles. Could these limitations be overcome?

In one of the most innovative submarine design schemes of the era, engineers at Electric Boat proposed fitting 12 vertical launch tubes for Tomahawk missiles into the forward ballast tanks of the *Los Angeles*, providing an almost 50 percent increase in weapons. Admiral Rickover rejected the proposal, preferring instead his AHPNAS/CMS cruise missile submarine. But that cruise missile submarine program ended in 1974 and with it died the large reactor project.

With Tomahawk planned for large-scale production and the cruise missile submarine program dead, the decision was made in the late 1970s—shortly before Rickover's removal—to pursue the vertical-launch configuration for submarines. Beginning with the USS *Providence* (SSN 719), commissioned in 1985, 12 vertical-launch Tomahawk tubes were installed in each submarine.[43]

*This view looking forward from the sail of the* Oklahoma City *(SSN 723) shows the 12 open hatches for the submarine's 12 vertical-launch Tomahawk missiles.* (U.S. Navy)

The first vertical-launch of a Tomahawk was made from the USS *Pittsburgh* (SSN 720) on 26 November 1986 on the Atlantic test range. Thirty-one submarines of the 62 ships of the *Los Angeles* design had vertical-launch tubes. The early submarines (SSN 688–718) could carry eight torpedo-tube-launched Tomahawks; the later submarines could carry up to 20 in addition to 12 vertical-launch missiles.

Beginning with the *San Juan* (SSN 751), the last 23 SSN 688s were designated as the "improved" or 688I class. In addition to vertical-launch Tomahawks, these submarines had minelaying and under-ice capabilities and improved machinery quieting.

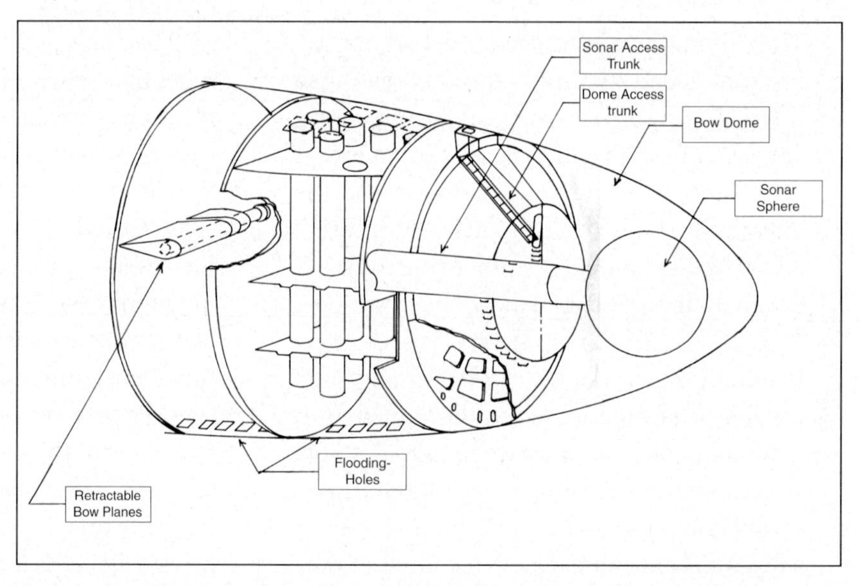


*Bow section of VLS-configured* Los Angeles-*class submarine.* (©A.D. Baker, III)

A U.S. nuclear research submarine was also built in the 1960s, the *NR-1,* initiated as a test platform for a small submarine nuclear power plant. In service the craft has been employed as a deep-ocean research and engineering vehicle. In explaining the craft's significance, Admiral Rickover told a congressional committee, "You will be looking at a development that I believe will be as significant for the United States as was the *Nautilus.*"[44] Rickover envisioned a series of these craft, hence his designation *NR-1.*

The veil of secrecy imposed on the *NR-1* by Rickover was partially lifted on 18 April 1965, when President Lyndon Johnson publicly revealed development of the craft. A White House press release, citing the severe endurance and space limitations of existing research submersibles, stated: "The development of a nuclear propulsion plant for

*The* NR-1 *at her launching. She has a short sail structure with sail-mounted diving planes and a fixed TV periscope (flying the General Dynamics banner in this photograph). Admiral Rickover envisioned the* NR-1 *as the first of a fleet of nuclear-propelled submersibles.* (General Dynamics)

*The* NR-1 *underway. The gear mounted forward of the sail is for surface or submerged towing. She must be towed to her operating areas; submerged towing facilitates clandestine operations although she is very noisy. The* NR-1*'s upper rudder and sail structure are painted international orange.* (U.S. Navy)

a deep submergence research vehicle will give greater freedom of movement and much greater endurance of propulsion and auxiliary power. This capability will contribute greatly to accelerate man's exploration and exploitation of the vast resources of the ocean."[45]

Beyond her nuclear plant, which gives her a theoretical unlimited underwater endurance, the *NR-1*'s most remarkable feature is her operating depth of 3,000 feet (915 m). While this was not as deep as the U.S. Navy's other manned research submersibles could dive, their underwater endurance was only a few hours and their horizonal mobility were severely limited.

The *NR-1* surface displacement is 365.5 tons, and she is 393 tons submerged; length overall is 145¾ feet (44.44 m). The Knolls Atomic Power Laboratory in Schenectady, New York, designed the reactor, while the submersible was designed and subsequently built by the Electric Boat yard in Groton. She was launched on 25 January 1969, underwent initial sea trials that summer, and was placed in service on 27 October of that year.

The craft is fabricated of HY-80 steel with the living/working spaces forward and the after portion of the craft containing the sealed reactor and engineering spaces. She is fitted with two large wheels that permit her to roll and rest on the ocean floor. The retractable tires are normal truck tires, with inner tubes filled with alcohol. In addition to her twin screws, which provide a submerged speed of 3.5 knots, the craft has paired ducted thrusters fitted in the hull, forward and aft, to provide a high degree of maneuverability. Extensive external lights, viewing ports, close-range sonars, a remote-controlled mechanical arm, and a recovery cage provide considerable capabilities. Three bunks are provided for the crew of five Navy operators plus two scientists; crew endurance is limited to a maximum of 30-day missions, mainly because of food storage.

The *NR-1's* principal operational limitation is her deployment mobility. Because of her slow speeds, she must be towed to her operating area, either by a surface ship or (underwater) by a nuclear submarine. Her surface towing speed is up to six knots; submerged it is just under four knots. She has been engaged in numerous research, ocean-engineering, and deep recovery missions, in addition to supporting the SOSUS program.

About 1971 Admiral Rickover initiated an *NR-2* program.[46] This would be a deeper-diving craft, reportedly capable of reaching more than 3,000 feet (915+ m). The craft was to have a reactor plant similar to that of the *NR-1.* However, neither the Navy nor Congress would support an *NR-2* at that time. About 1978 the Navy's leadership decided to proceed with the *NR-2* as a Hull Test Vehicle (HTV) to test the suitability of HY-130 steel for submarines. In 1983 the Navy decided that sections of HY-130 would be added to the conventional deep-diving research submarine *Dolphin* (AGSS 555) instead of constructing the nuclear-propelled HTV.

Despite Rickover's prediction that the *NR-1* would be the first of a fleet of nuclear-propelled submersibles for research, ocean engineering, and other tasks, no additional nuclear vehicles were built by the United States.

## Soviet Third-Generation Submarines

The Soviet Union's third-generation submarine program was far broader than that of the U.S. Navy. In the United States, extensive controversy, discussions, and debates led to the development and production of two classes, the *Los Angeles* SSN (62 ships) and *Ohio* Trident SSBN (18 ships) plus the one-of-a-kind *Lipscomb.* By comparison, the Soviet "second-plus" and third-generation nuclear submarines consisted of five production designs plus a one-of-a-kind torpedo-attack submarine as well as several special-purpose nuclear designs. But the production of most of these classes would be truncated by the end of the Cold War.

TABLE 17-1
**Soviet Third-Generation Nuclear Submarines***

| Number** | Type | Project | NATO | Comm. |
|---|---|---|---|---|
| 8 (4) | SSGN | 949 | Oscar | 1980 |
| 6 | SSBN | 941 | Typhoon | 1981 |
| 1 | SSN | 685 | Mike | 1983 |
| 3 (1) | SSN | 945 | Sierra | 1984 |
| 14 (6) | SSN | 971 | Akula | 1984 |
| 7 | SSBN | 667BDRM | Delta IV | 1984 |

Notes: * Technically the Project 667BDRM and Project 685 are an "intermediate" generation, sometimes referred to as the 2.5 or 2+ generation.
** Numbers in parentheses are additional submarines completed after 1991.

The first third-generation nuclear "attack" submarine was Project 949 (NATO Oscar), the ultimate cruise missile submarine and—after Project 941/Typhoon—the world's largest undersea craft.[47] Development of the P-700 Granit anti-ship missile system (NATO SS-N-19 Shipwreck) was begun in 1967 to replace the P-6 missile (NATO SS-N-3 Shaddock) as a long-range, anti-carrier weapon. The Granit would be a much larger, faster, more-capable weapon, and it would have the invaluable characteristic of submerged launch. The missile would be targeted against Western aircraft carriers detected by satellites.

A short time later design was initiated of an associated cruise missile submarine at the Rubin bureau under Pavel P. Pustintsev, who had designed earlier SSG/SSGNs.[48] The possibility of placing the Granit on Project 675/Echo II SSGNs also was considered, but the size and capabilities of the missile required a new submarine.

The Granit would be underwater launched and have a greater range than the previous P-6/SS-N-3 Shaddock or P-70/SS-N-7Amethyst missiles. Being supersonic, the missile would not require mid-course guidance upgrades, as did the earlier Shaddock. The decision to arm the Project 949/Oscar SSGN with 24 missiles—three times as many as the Echo II or Charlie—meant that the new submarine would be very large. The missiles were placed in angled launch canisters between the pressure hull and outer hull, 12 per side. As a consequence, the submarine would have a broad beam—giving rise to the nickname *baton* (loaf). The 59⅔-foot (18.2-m) beam would require a

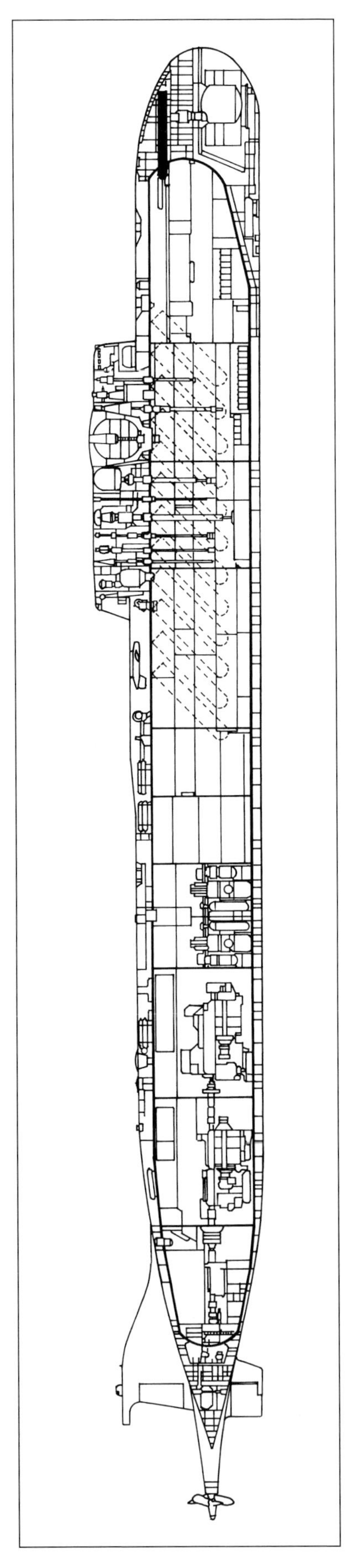

*Project 949A/Oscar II. LOA 508 ft 4 in (155.0 m)* (©A.D. Baker, III)

*A Project 949A/Oscar II in choppy seas, revealing the massive bulk of the submarine—called* baton *or "loaf" (of bread) in Russian. The six missile hatches on the starboard side, forward of the limber holes, each cover two tubes for large, Shipwreck anti-ship missiles.* (U.S. Navy)

length of 472⅓ feet (144 m), providing for nine compartments. Although the submarine was not as long as U.S. Trident SSBNs, the Oscar's greater beam meant that the submarine would have a submerged displacement of 22,500 tons, some 20 percent larger than the U.S. Trident SSBNs.

Beyond 24 large anti-ship missiles, the Oscar SSGN has four 533-mm and four 650-mm torpedo tubes capable of launching a variety of torpedoes and tube-launched missiles.[49] The submarine has the MGK-500 low-frequency sonar (NATO Shark Gill), as fitted in the Typhoon SSBN, and the MG-519 (NATO Mouse Roar) sonar.

The large size of the Oscar and the need for a speed in excess of 30 knots to counter U.S. aircraft carriers required a propulsion plant based on two OK-650b reactors powering turbines and twin screws. The twin-reactor plant produces 100,000 horsepower, according to published reports. This plant and the similar Typhoon SSBN plant are the most powerful ever installed in submarines.

The keel for the lead ship of this design was laid down on 25 June 1978 at the Severodvinsk shipyard. The submarine was commissioned on 30 December 1980 as the *Minskiy Komsomolets* (K-525).[50] Series production followed. Beginning with the third unit—Project 949A/Oscar II—the design was lengthened to 508⅓ feet (155 m). The additional space was primarily for acoustic quieting and the improved MGK-540 sonar, with a tenth compartment added to improve internal arrangements. These SSGNs—as well as other Soviet third-generation submarines—are considered to be extremely quiet in comparison with previous Soviet undersea craft. The changes increased the surface displacement by some 1,300 tons.

The large size and cost of these submarines led to some debate within the Soviet Navy over the means of countering U.S. aircraft carriers. As many as 20 submarines of Project 949 were considered, although according to some sources, the cost of each was about one-half that of an aircraft carrier of the *Admiral Kuznetsov* class.[51] Two naval officer-historians wrote: "It is obvious . . . the ideological development of the [SSGN], namely Project 949, overstepped the limits of sensible thought and logic."[52]

Project 949/Oscar SSGNs posed a major threat to U.S. aircraft carriers and other surface forces because of their missile armament and stealth. The massive investment in these submarines demonstrated the continuing Soviet concern for U.S. aircraft carriers as well as the willingness to make massive investments in submarine construction.

Through 1996 the Severodvinsk yard completed 12 Oscar-class SSGNs—two Project 949 and ten improved Project 949A submarines. The 11th Oscar II was launched in September 1999, apparently to clear the building hall, and is not expected to be completed.

These submarines periodically undertook long-range operations, a concern to the U.S. Navy because their quieting made them difficult to detect and track. On 12 August 2000, as Soviet naval forces held exercises in the Barents Sea, reportedly in preparation for a deployment to the Mediterranean, the *Kursk* (K-141) of this class suffered two violent explosions. They tore open her bow and sent the submarine plunging to the sea floor. All 118 men on board died, with 23 in after compartments surviving for some hours and perhaps a day or two before they succumbed to cold, pressure, and rising water.[53] The cause of the disaster was a Type 65-76 torpedo fuel (hydrogen-peroxide) explosion within the forward torpedo room, followed 2 minutes, 15 seconds later by the massive detonation of 2 torpedo warheads.

*An Oscar, showing the rounded bow and fully retracted bow diving planes (just forward of the missile hatches). The large sail structure has an escape chamber between the open bridge and the array of masts and scopes. One of the raised masts is the Quad Loop RDF antenna.*

The *Kursk,* ripped apart by the two explosions, sank quickly. The four previous Soviet nuclear-propelled submarines that had sunk at sea had also suffered casualties while submerged, but were able to reach the surface, where many members of their crews had been able to survive. They were able to do this, in part, because of their compartmentation and high reserve buoyancy; the explosions within the *Kursk* were too sudden and too catastrophic to enable the giant craft to reach the surface.

Design of the Soviet Navy's third-generation nuclear torpedo-attack submarines (SSN) began in 1971 at TsKB Lazurit. Titanium was specified for the design, continuing the development of a key technology of Projects 661/Papa and 705/Alfa.

Beyond reducing submarine weight, titanium would provide a greater test depth in third-generation SSNs—1,970 feet (600 m)—which would be coupled with other SSN advances, including improved quieting, weapons, and sensors. Progress in working titanium—alloy 48-OT3V—allowed it to be welded in air, albeit in special, clean areas; smaller parts were welded in an argon atmosphere.[54]

Developed under chief designer Nikolay Kvasha, the new attack submarine—Project 945 (NATO Sierra)—would be a single-screw design with a single nuclear reactor. This was the first Soviet combat submarine to have a single-reactor plant except for the Project 670/Charlie SSGN, also designed by Lazurit. The Sierra's OK-650a reactor plant would generate 50,000 horsepower, considerably more than that of the Charlie. As part of the quieting effort, the reactor employed natural convection for slow speeds (up to five or six knots) to alleviate the use of pumps. A single reactor was accepted on the basis of highly redundant components throughout the propulsion plant. The Sierra's plant would produce a maximum submerged speed of 35 knots.

The Sierra's weapons system is impressive: The Project 945/Sierra I has two 650-mm and four 533-mm torpedo tubes. The definitive Project 945A/Sierra II has four 650-mm and four 533-mm torpedo tubes, that is, more and larger torpedo tubes than in the contemporary U.S *Los Angeles* class. Further, the Sierra has internal stowage for 40 torpedoes and missiles (including those in tubes) compared to 25 weapons in the early *L.A.* class and 25 plus 12 vertical-launch missiles in the later ships of the class. The Sierra has a fully automated reload system.

The advanced MGK-500 sonar and fire control systems are provided; the large pod atop the vertical tail surface is for the towed passive sonar array. Built with a double-hull configuration, the Sierra had a large, broad sail structure, described as a compromise between the "wing" or "fin" sail of Rubin designs and the low "limousine" sail of Malachite designs. Like all third-generation com-

*The* Kursk *resting in the Russian Navy's largest floating dry dock at Roslyanko, near Murmansk, after her remarkable salvage in October 2001. The Oscar II's massive bulk is evident from the size of the worker at far left. The Oscar was the world's second largest submarine.* (Russian Navy)

*The Project 945/Sierra SSN was intended for large-scale production. However, difficulties in meeting titanium production schedules led to the program being reduced and, with the end of the Cold War, cancelled entirely after four ships were completed.*

bat submarines, the Sierra would have an escape chamber that could accommodate the entire crew.

Also, third-generation Soviet submarines would have smaller crews than their predecessors. The Sierra SSN initially was manned by 31 officers and warrant officers plus 28 enlisted men, compared to some 75 men in the Project 671/Victor SSNs. (Similarly, the Oscar SSGN's 107 personnel compare to more than 160 in the smaller U.S. *Ohio* SSBN, also a 24-missile undersea craft.)

The overall size of the Sierra was constrained by the requirement to produce the submarines at inland shipyards, that is, Gor'kiy and Leningrad and, to a lesser degree, Komsomol'sk-on-Amur. The Shipbuilding Ministry, responding to political influence, brought pressure to have the new class produced at Gor'kiy, reflecting the ministry's role in naval shipbuilding programs.[55] Krasnoye Sormovo thus became the third shipyard to produce titanium submarines, following Sudomekh/Admiralty in Leningrad and the Severodvinsk yard.

The lead Project 945/Sierra SSN was laid down at the Krasnoye Sormovo yard, adjacent to the Lazurit design bureau in Gor'kiy. This was the *K-239*, whose "keel" was placed on the building ways on 8 May 1982. Construction progressed at a rapid pace. After launching the following year, the submarine was moved by transporter dock through inland waterways to Severodvinsk for completion, trials, and commissioning on 21 September 1984.

Subsequent construction at the Krasnoye Sormovo yard progressed far slower than planned. Following the first two Sierra SSNs a modified design was initiated (Project 945A/Sierra II), providing a slightly larger submarine, the additional space being for quieting features and habitability, with the number of compartments being increased from six to seven. The self-noise levels were lower than the Sierra I.

During the early planning stages, some Navy officials envisioned a Sierra SSN program of some

*A Project 945A/Sierra II with her bow diving planes extended. The Sierra II has an enlarged, ungainly sail structure. Her masts and periscopes are offset to starboard. There are dual safety tracks running along her deck.* (U.K. Ministry of Defence)

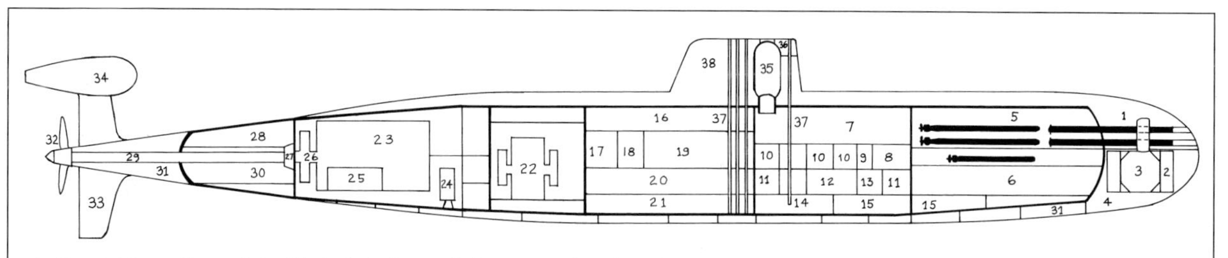

1. Floodable portions of double hull. 2. Sonar. 3. Pressure chamber of antenna. 4. Ballast tank. 5. Torpedo room. 6. Sonar room and crew berthing. 7. Control room. 8. Wardroom. 9. Pantry. 10. Officer cabins. 11. Crew berthing. 12. Medical room. 13. Heads. 14. Food storage. 15. Storage battery. 16. Electrical equipment room. 17. Propulsion plant control room. 18. Gyro room. 19. Communications and radar room. 20. Electrical switchboards. 21. Reserve propulsion system. 22. Reactor. 23. Main turbine. 24. Main circulation pump. 25. Turbogenerator. 26. Reduction gear. 27. Thrust bearing. 28. Electrical switchboards. 29. Propeller shaft. 30. Diesel generator and evaporators. 31. Ballast tank. 32. Propeller. 33. Vertical stabilizer/rudder. 34. Towed sonar-array pod. 35. Escape chamber. 36. Bridge. 37. Masts and periscopes. 38. Sail.

*Project 945/Sierra SSN (not to scale). LOA 351 ft (107.0 m)* (©A.D. Baker, III)

40 submarines to be constructed at both Krasnoye Sormovo and Severodvinsk.[56] Difficulties in producing titanium plates and difficulties at the Krasnoye Sormovo yard led to delays in producing these third-generation SSNs. In the event, only two Sierra I and two Sierra II SSNs were completed, all at Krasnoye Sormovo, the last in 1993, that is, four submarines delivered over ten years.[57]

When the program was halted, there were two additional, unfinished hulls at the Gor'kiy yard. Other than two Project 865 midget submarines, there would be no additional construction of titanium submarines by any nation. The Sierras were the last nuclear submarines to be built at Krasnoye Sormovo; a diesel-electric Kilo/Project 636 submarine for China was the last combat submarine built at the yard.

Anticipated delays in building the titanium Sierra led to the decision to pursue a "steel-hull Sierra"—essentially the same submarine built of steel, responding to the same TTE requirements. This approach was proposed by the Malachite design bureau and, after consideration by the Navy leadership and Shipbuilding Ministry, contract design was undertaken in 1976 at Malachite by a team led by Georgi N. Chernishev. He had been the chief designer of the Project 671/Victor SSN.

Beyond using steel in place of titanium, the new design differed in having the displacement limitation removed, which provided the possibility of employing new weapons and electronics on a larger displacement.[58] The new submarine—Project 971 (NATO Akula)—would be larger than the Sierra, with the use of AK-32 steel adding more than 1,000 tons to the surface displacement while retaining the same test depth of 1,970 feet (600 m).[59] The changes reduced the reserve buoyancy from the Sierra's 29 percent to 26 percent, still almost twice that of comparable U.S. SSNs. The submarine has six compartments.[60] Significantly, the later USS *Seawolf* (SSN 21) was highly criticized for being "too large"; the Akula SSN was still larger.

With the same OK-650 propulsion plant as the Sierra, the larger size of the Akula reduced her maximum submerged speed to about 33 knots. Major innovations appeared in the Akula, both to reduce noise and to save weight (e.g., ten tons were saved by using plastics in place of metal). The former include modular isolated decks and active noise cancellation. The anechoic coating on the later submarines was reported to have a thickness of 2½ inches (64 mm).

The Akula is fitted with the more advanced MGK-540 sonar and has four 533-mm and four 650-mm torpedo tubes. In the later Akula SSNs there are also six large tubes forward, flanking the torpedo loading hatch, each for launching two MG-74 "Impostor" large acoustic decoys. These are in addition to smaller decoys and noisemakers launched by Soviet (as well as U.S.) submarines to counter enemy sonars and torpedoes.[61]

The lead Project 971/Akula—the *Karp* (K-284)—was laid down at Komsomol'sk in 1980. This was the first lead nuclear submarine to be built at the Far Eastern yard since the Project 659/Echo I was laid down in 1958. After launching on 16 July 1984, the *Karp* was moved in a transporter dock to Bolshoi Kamen (near Vladivostok) for completion at the Vostok shipyard. Sea trials followed, and she was commissioned on 30 December 1984. This was only three months after the first Project 945/Sierra SSN was placed in service. The *Karp* spent several years engaged in submarine and weapon trials.

Series production of the Akula continued at Komsomol'sk and at Severodvinsk, with the latter yard completing its first submarine, the K-480, named *Bars* (panther), in 1989. The Akula became the standard third-generation SSN of the Soviet Navy. Design improvements were initiated early in the Akula program; the ninth Akula introduced still more improved quieting features, with four units being produced by the two shipyards to that configuration. Subsequently, Severodvinsk began producing an improved variant, called Akula II in the West. This Project 971 variation featured a modified pressure hull and was significantly quieter. The *Vepr'* (wild boar), the first ship of this type, was laid down at the Severodvinsk shipyard on 21 December 1993, the yard having been renamed Sevmash. Despite the slowdown in naval shipbuilding after the demise of the USSR, the *Vepr'* was placed in commission on 8 January 1996. However, the second Akula II, the *Gepard* (cheetah), was not commissioned until 4 December 2001.[62]

## Quieting Submarines

The Akula II is almost 230 tons greater in surface and submerged displacement and eight feet (2.5 m) longer than the earlier Akulas, the added space being for acoustic quieting with an "active" noise-reduction system. The *Vepr'* became the first Soviet submarine that was quieter at slower speeds than the latest U.S. attack submarines, the Improved *Los Angeles* class (SSN 751 and later units).[63]

This Soviet achievement was of major concern to Western navies, for acoustics was long considered the most important advantage of U.S. undersea craft over Soviet submarines. The issue of why the Soviets—belatedly—had undertaken the effort and cost to quiet their submarines, and how they did it were immediately discussed and debated in Western admiralties. Soviet designers have stated that the decision to undertake a major quieting effort came from major developments in acoustic technologies during the 1980s.

Western officials believed differently: U.S. Navy Chief Warrant Officer John A. Walker sold secrets to the Soviets from about 1967 until arrested in 1985.[64] While he principally sold access to American encryption machines, Walker undoubtedly sold secret material related to U.S. submarines—he had served in two SSBNs and was in sensitive submarine assignments during the early period of his espionage. His merchandise included SOSUS and operational submarine information, and this intelligence may well have accelerated Soviet interest in quieting third-generation submarines.

Also, in 1983–1984 the Japanese firm Toshiba sold sophisticated, nine-axis milling

*A Project 971/Improved Akula. The hatches for her large, MG-74 "Imposter" decoys are visible in her bow, three to either side of the torpedo loading hatch. The sail structure is faired into her hull.* (U.K. Ministry of Defence)

The Vepr', *the first of two Project 971/Akula II SSNs, underway. The submarine's masts are dominated by the two-antenna Snoop Pair radar fitted atop the Rim Hat electronic intercept antenna. Only SSNs mount the large towed-array sonar pod atop the vertical rudder.* (Courtesy Greenpeace)

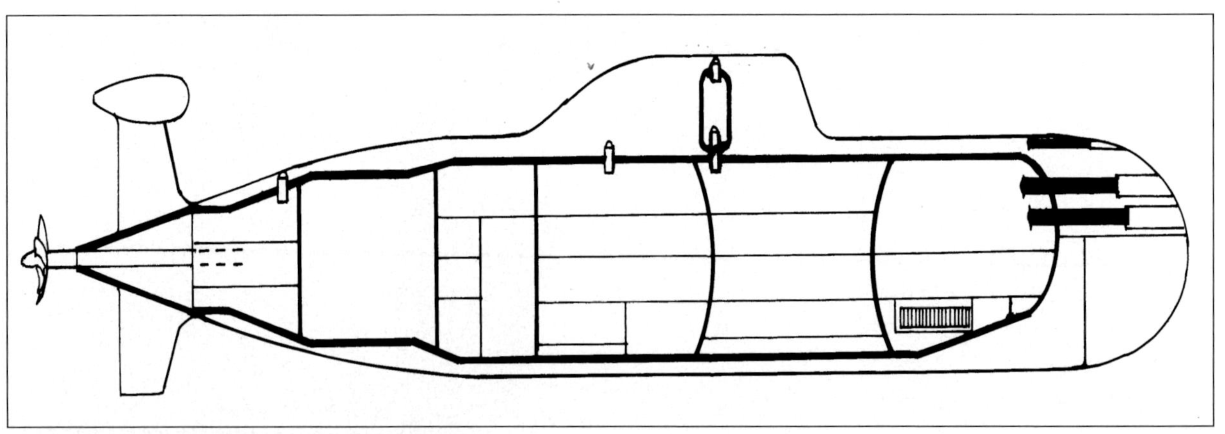

*Project 971/Akula SSN (not to scale). LOA 361 ft 9 in (110.3 m)* (©A.D. Baker, III)

*The sail of an Akula, with several of her periscopes and masts raised; doors cover them when they retract. The sensors on the pedestal between the masts, on the front of the sail, and in pods adjacent to the diving planes measure temperature, radioactivity, turbulence, and other phenomena.* (U.K. Ministry of Defence)

equipment to the USSR, while Köngsberg Vaapenfabrikk, a state-owned Norwegian arms firm, sold advanced computers to the USSR for the milling machines. U.S. congressmen and Navy officials charged that this technology made it possible for the Soviets to manufacture "quieter" propellers.[65] This interpretation was oversimplified because the Sierra and Akula designs and their propellers were developed much earlier. At the most, the Japanese and Norwegian sales led to more efficient production of Soviet-designed submarine propellers.

Improvements in automation also were made in the successive Akula variants. Submarine command and control as well as weapons control are concentrated into a single center, as in the Project 705/Alfa SSN—the ship's main "command post" (GKP), which promoted the reduction in manning.[66]

But Akula production slowed soon after the breakup of the Soviet Union in 1991. Komsomol'sk delivered its seventh and last Akula SSN in 1995; Severodvinsk's last Akula SSN, also its seventh, was the *Gepard* (K-335), completed late in 2001. No additional SSNs of this design are expected to be completed, although two more were on the building ways at Sevmash/Severodvinsk and another at Komsomol'sk. Those unfinished Akula hulls periodically fueled speculation that one or two Akula SSNs would be completed for China or India.[67]

(India, which had earlier leased a Project 670/Charlie SSGN, has had a nuclear submarine program under way since 1974. The first submarine produced under this effort is scheduled for delivery about 2010.)

An additional—and remarkable—SSN was built in this period of the "2.5" or "2+" generation, the Project 685 (NATO Mike). Although by the 1960s the Lazurit and Malachite design bureaus were responsible for Soviet SSN design, the Rubin bureau was

*The ill-fated* Komsomolets, *the only Project 685/Mike SSN to be constructed. She was configured to evaluate new technologies and systems for future attack submarines. There were proposals to construct additional submarines of this type, but all such efforts died with the submarine.*

able to convince the Navy of the need for an advanced combat submarine to evaluate a dozen new technologies.[68]

Design of the Project 685/Mike SSN was led by a design team under N. A. Klimov and, after his death in 1977, Yu. N. Kormilitsin.[69] The design effort went slowly. The large submarine would have titanium hulls, the material being used to attain weight reduction and, especially, great test depth. The submarine would reach a depth of 3,280 feet (1,000 m)—a greater depth than any other combat submarine in the world.[70] A solid-fuel, gas-generation system was installed for emergency surfacing. Propulsion was provided by a single OK-650b-3 reactor powering a steam turbine turning a single shaft to provide a speed in excess of 30 knots. The shaft was fitted with two, fixed four-blade propellers (an arrangement similar to those on some other Soviet SSNs). The double-hull Mike SSN had seven compartments.

Although primarily a development ship, this would be a fully capable combat submarine, being armed with six 533-mm torpedo tubes. Advanced sonar was fitted, although details of her 12 special/experimental systems have not been published. Like contemporary Soviet submarines, Project 685 was highly automated. The manning table approved by the Ministry of Defense in 1982 provided for a crew of 57 men. The Navy later increased the crew to 64—30 officers, 22 warrant officers, and 12 petty officers and seamen. While a small crew by Western standards, the change involved a qualitative shift, according to the ship's deputy chief designer, with enlisted conscripts filling positions originally intended for warrant officers, and more junior officers being assigned than were envisioned by the submarine designers.

Construction at Severodvinsk was prolonged, in part because of the experimental features of the submarine. The *K-278* was laid down in 1978, but she was not launched until 9 May 1983, and she entered service late in 1984. On 5 August 1985 she successfully reached her test depth of 3,280 feet. Under her first commanding officer, Captain 1st Rank Yu. A. Zelenski, the *K-278*—named *Komsomolets* in 1988—carried out extensive trials and evaluations.[71]

The U.S. intelligence community estimated that the *Komsomolets*—because of her large size and the impending availability of the RPK-55 Granat (NATO SS-N-21 Sampson) cruise missile—was the lead ship in a new class of submarine intended to attack land targets in Eurasia and possibly the United States.[72] The Central Intelligence Agency assessment also noted, "The apparent absence, however, of some features which have been noted on other new Soviet SSN classes raises the alternative, but less likely, possibility that the [Mike]-class may serve as a research platform for developmental testing of submarine-related technology."[73]

Several times the management of the Rubin design bureau proposed building additional Project 685 submarines. "The proposals were not accepted [although] Northern Fleet command was in favor of additional units," said deputy chief designer D. A. Romanov.[74]

Meanwhile, a second crew, under Captain 1st Rank Ye. A. Vanin, had begun training to man the submarine. On 28 February 1989 the second crew took the *Komsomolets* to sea on an operational patrol into the North Atlantic. This crew was poorly organized; for example, "the crew lacked a damage control division as a full-fledged combat unit."[75] And the crew lacked the experience of having been aboard the *Komsomolets* for four years as had the first crew.

The *Komsomolets* was at sea for 39 days. In her forward torpedo room were conventional weapons and two nuclear torpedoes. On the morning of 7 April 1989, while in the Norwegian Sea en route to her base on the Kola Peninsula, fire erupted in the seventh compartment. Flames burned out the valve of the high-pressure air supply in the compartment and the additional air—under pressure—fed the conflagration. Attempts to flood the compartment with chemicals identified as LOKh, which were intended to act as a halon-like fire suppressant, failed, and temperatures soon reached between 1,920 and 2,160°F (800–900°C).

The submarine, cruising at about 1,265 feet (385 m), was brought to the surface. The fire spread. The crew fought to save the ship for six hours. With the submarine beginning to flood aft, Captain Vanin ordered most crewmen onto the sea-swept deck casing and sail. On deck the men saw sections of the anechoic coatings sliding off of the hull because of the intense heat within the after portion of the submarine.

Vanin was forced to order the submarine abandoned. One of two rafts housed in the sail casing was launched after some delays because of poor crew training, into the frigid sea—the water temperature being near freezing. After ordering his crew off, Vanin went back into the sinking *Komsomolets*. Apparently some men had not heard the order and were still below. With four other officers and warrant officers, Vanin entered the escape chamber in the sail of the submarine as the *Komsomolets* slipped beneath the waves, stern first, at an angle of some 80 degrees.

Water entered the escape chamber and the men had difficulty securing the lower hatch and preparing it for release. Toxic gases also entered the chamber (probably carbon monoxide). Efforts were made to release the chamber, but failed. Then, as the craft plunged toward the ocean floor 5,250 feet (1,600 m) below, the chamber broke free and shot to the surface. As the chamber was being opened by a survivor, the internal pressure blew off the upper hatch. The gases had caused three men, including Vanin, to lose consciousness. Two warrant officers were thrown out of the chamber as it was battered by rough seas, flooding it and sending it plunging into the depths, carrying with it the unconscious Vanin and two others.

Aircraft overflew the sinking submarine and dropped a life raft, and ships were en route to the scene, some 100 n.miles (185 km) south of Bear Island. An hour after the *Komsomolets* disappeared a fishing craft pulled 30 men from the sea. Of 69 men on board the submarine that morning, 39 already were lost, including her commanding officer and one man killed in the fire. The effects of freezing water and smoke eventually would take the lives of three more, to bring the death toll to 42. Thus perished one of the major test beds for the Soviet Navy's next-generation submarines.

In this period of third-generation nuclear submarines, both the United States and USSR produced specialized nuclear undersea craft. Several deep-diving nuclear submarines were developed and built by the Malachite-Admiralty shipyard team, all believed to have titanium hulls. Project 1851 (NATO X-ray) was a relatively small submarine, somewhat analogous to the U.S. Navy's smaller *NR-1*. The Soviet craft, completed in 1986, had a surface displacement of approximately 300 tons and a length of some 98½ feet (30 m). Her operating depth was in excess of 3,800 feet (1,000 m).[76] Two modified follow-on craft were completed through 1995. (The NATO code name Paltus was sometimes applied to the two later craft.)

*A Project 1910/Uniform deep-diving submarine. The craft's specific missions are not publicly known. There are several unusual features evident; the bridge windshield is raised as is the high-frequency radio mast, which folds down into a recess aft of the small sail.*

Less is known about the succeeding Soviet submarines of this type, the deep-diving Project 1910 Kashalot (NATO Uniform) design. From 1986 the Admiralty yard produced three submarines of the Uniform design, the last completed in 1993, with a fourth hull reported to be left unfinished when the Cold War ended. These craft displaced 2,500 tons submerged and were just over 239 feet (73 m) long. The specific role of these submarines, other than deep-sea research, has not been publicly identified. Their test depth—achieved through the use of titanium hulls—is estimated to be 3,000 to 4,000 feet (915 to 1,220 m).

**After extensive controversy, discussions, and debates, U.S. third-generation nuclear submarines consisted of two designs, the *Los Angeles* SSN (62 ships) and *Ohio* Trident SSBN (18 ships), plus the one-of-a-kind *Glenard P. Lipscomb,* another attempt to produce a very quiet nuclear submarine. Advocates within the Navy ship design community and Navy headquarters had argued for the potentially more-capable CONFORM approach but were routed by Admiral Rickover who, when the third generation was initiated, was at the apex of his career.**

**By comparison, the Soviet "second-plus" and third-generation nuclear submarines consisted of five production designs, plus the one-of-a-kind advanced-technology *Komsomolets* and several special-purpose submarines. In addition there were several special-mission nuclear as well as several diesel-electric designs. However, the production of most of these classes was truncated by the end of the Cold War, with only 45 nuclear-propelled combat submarines being completed by the end of 1991 (compared to the 81 U.S. units).**

**Significantly, in the first two nuclear submarine generations, the USSR surpassed the United States in numbers of submarines produced as well as in numbers of non-nuclear undersea craft. In the third generation, Soviet submarines displayed significant increases in quality and continued to exhibit extensive innovation in design, materials, weapons, and automation. Also significant were the development of specialized submarines by the Soviet Navy in the form of new construction as well as by conversion and the modernization of the Soviet diesel-electric submarine fleet. All of these efforts contributed to the increasing effectiveness of the Soviet undersea force.**

**U.S. submarines, however, appear to have retained an advantage in quality of personnel and training and, until late in the Cold War, in acoustic sensors and quieting. Although progress was made in the selection and training of Soviet**

*A third-generation U.S. attack submarine, the* Greeneville *(SSN 772), maneuvers off of Guam in November 2001. Note the fittings on her after deck for the ASDS SEAL delivery vehicle. When embarked the vehicle will sit atop her after access hatch. The sheath for her towed-array sonar is fitted to starboard along most of her upper hull. (Marjorie McNamee/U.S. Navy)*

submarine personnel, there still were severe shortcomings, as the loss of the *Komsomolets* and other accidents demonstrated.

Most ominous had been the trend in Soviet acoustic quieting. Admiral James Watkins observed:

> When my first command, USS *Snook* [SSN 592], was launched [1960], we were infinitely ahead of the Soviets in nuclear submarine technology. Later, about ten years ago, we were about ten years ahead of them. Today [1983] we are only about five to eight years ahead of them and they are *still* in relentless pursuit.[77]

Three years later—in 1986—senior U.S. submarine designer Captain Harry A. Jackson wrote, "There will come a time in the not too distant future when Soviet submarine silencing will have been improved to the extent that detection by passive sonar is possible only at short ranges, if at all."[78]

By the late 1980s—with the appearance of the improved Project 971/Akula SSN—Soviet quieting technology caught up with U.S. technology. A 1988 appraisal of ASW by Ronald O'Rourke, the perceptive naval analyst at the Congressional Research Service of the U.S. Library of Congress, concluded:

> The advent of the new generation of quieter Soviet submarines has compelled the US Navy to institute an expensive and high-priority program for maintaining its traditional edge in ASW. By raising the possibility, if not the probability, of an eventual reduction in passive-sonar detection ranges, these quieter Soviet submarines may change the face of US ASW, and particularly the role of US SSNs in ASW operations.[79]

But within a few years the Soviet Union imploded, and the remarkable momentum developed in submarine design and construction quickly stalled.

While U.S. submarines apparently retained the lead in acoustic sensors and processing, by the 1980s Soviet submarines were displaying an array of non-acoustic sensors. These devices, coupled with Soviet efforts to develop an ocean surveillance satellite system that could detect submerged submarines, had ominous (albeit unproven) implications for the future effectiveness of Soviet ASW surveillance/targeting capabilities.

Table 17-2
**Third-Generation Submarines**

| | **U.S. *Los Angeles* SSN 688** | **AHPNAS SSGN*** | **Soviet Project 949 NATO Oscar I** | **Soviet Project 945 NATO Sierra I** | **Soviet Project 971 NATO Akula I** | **Soviet Project 685 NATO Mike** |
|---|---|---|---|---|---|---|
| Operational | 1976 | — | 1980 | 1984 | 1985 | 1983 |
| Displacement | | | | | | |
| surface | 6,080 tons | 12,075 tons | 12,500 tons | 6,300 tons | 8,140 tons | 5,680 tons |
| submerged | 6,927 tons | 13,649 tons | 22,500 tons | 8,300 tons | 10,700 tons | 8,500 tons |
| Length | 362 ft (110.3 m) | 472 ft (143.9 m) | 472 ft 4 in (144.0 m) | 351 ft (107.0 m) | 361 ft 9 in (110.3 m) | 388 ft 6 in (118.4 m) |
| Beam | 33 ft (10.06 m) | 40 ft (12.2 m) | 59 ft 8 in (18.2 m) | 40 ft (12.2 m) | 44 ft 7 in (13.6 m) | 35 ft 5 in (11.1 m) |
| Draft | 32 ft (9.75 m) | 32 ft 10 in (10.0 m) | 30 ft 2 in (9.2 m) | 31 ft 2 in (9.5 m) | 31 ft 9 in (9.68 m) | 24 ft 3 in (7.4 m) |
| Reactors | 1 | 1 | 2 OK-650b | 1 OK-650a | 1 OK-650b | 1 OK-650b-3 |
| Turbines | 2 | 2 | 2 | 1 | 1 | 2 |
| horsepower | 30,000 | 60,000 | 100,000 | 50,000 | 50,000 | 50,000 |
| Shafts | 1 | 1 | 2 | 1 | 1 | 1 |
| Speed | | | | | | |
| surface | | | 15 knots | | 13 knots | 14 knots |
| submerged | 33 knots | 30+ knots | 30+ knots | 35 knots | 33 knots | 30+ knots |
| Test depth | 950 ft (290 m) | 1,300 ft (395 m) | 2,000 ft (600 m) | 1,970 ft (600 m) | 1,970 ft (600 m) | 3,280 ft (1,000 m) |
| Missiles | 12 Tomahawk** | 20 STAM*** | 24 SS-N-19 | SS-N-16 | SS-N-16 SS-N-21 | nil |
| Torpedo tubes# | 4 533-mm A | 4 533-mm A | 4 533-mm B 4 650-mm B | 4 533-mm B 2 650-mm B## | 4 533-mm B 4 650-mm B | 6 533-mm B |
| Torpedoes | 25 | | 28 | 28 533-mm 12 650-mm | 28 533-mm 12 650-mm | |
| Complement | 141 | 111 | 107 | 59 | 73 | 64 |

Notes: * AHPNAS = Advanced High-Performance Nuclear Attack Submarine; baseline design data.
** 12 Tomahawk vertical-launch tubes in 31 submarines of class (beginning with the SSN 719).
*** STAM = Submarine Tactical Missile (anti-ship).
# Angled+ Bow.
## Increased to four 650-mm tubes in later units.

# 18

# Submarine Weapons

*A Tomahawk Land-Attack Missile/Conventional (TLAM/C) in flight, with its wings, fins, and air scoop extended. Although initially developed for the strategic (nuclear) strike role, the Tomahawk has provided an important conventional, long-range strike capability to U.S. and British submarines as well as to U.S. surface warships.* (U.S. Navy)

During the Cold War the traditional attack submarine weapons—torpedoes and mines laid from torpedo tubes—became increasingly complex as navies sought more capabilities for their submarines. And cruise missiles became standard weapons for attack submarines (SS/SSN).[1]

In the U.S. Navy the Mk 48 torpedo was introduced in 1972 as the universal submarine torpedo, replacing the Mk 37 torpedo and the Mk 45 ASTOR (Anti-Submarine Torpedo).[2] The former, a 19-inch (482-mm) diameter acoustic homing torpedo, had been in service since 1956.[3] The ASTOR was the West's only nuclear torpedo, in service from 1958 to 1977, carrying the W34 11-kiloton warhead. The Mk 48's range of some ten n.miles (18.5 km) and its accuracy were considered to make it more effective than a nuclear torpedo.

The Mk 48 is 21 inches (533 mm) in diameter, has a length of just over 19 feet (5.8 m), and carries a warhead of approximately 650 pounds (295 kg) of high explosives. The weapon is estimated to have a speed of 55 knots and a range of 35,000 yards (32 km). As with the Mk 37 torpedo, upon launching the Mk 48 is initially controlled from the submarine through a guidance wire that spins out simultaneously from the submarine and the torpedo. This enables the submarine to control the "fish" using the larger and more-capable passive sonar of the submarine. Subsequently the wire is cut and the torpedo's homing sonar seeks out the target.[4]

The Mk 48 Mod 3 torpedo introduced several improvements, including TELECOM (Telecommunications) for two-way data transmissions between submarine and torpedo, enabling the torpedo to transmit acoustic data back to the submarine. Later upgrades have attempted to overcome the challenges presented by high-performance Soviet submarines; mutual interference in firing two-torpedo salvos; launching the Mk 48 under ice, where there could be problems from acoustic reverberations; and, especially, using the Mk 48 against quiet, diesel-electric submarines in shallow waters. The ultimate ADCAP (Advanced Capability) series has been in service since 1988. This is a considerably updated weapon, although problems have persisted.[5]

*An Mk 48 torpedo "in flight" during tests at the U.S. Navy's torpedo station in Keyport, Washington. The later, ADCAP variant of the Mk 48 is the U.S. Navy's only submarine-launched torpedo. There is a shroud around the torpedo's twin propellers.* (U.S. Navy)

The effectiveness of U.S. torpedoes has been complicated by Soviet submarine evolution, such as increasing "stand-off" distances between their pressure and outer hulls, multiple compartmentation, strong hull materials, high speed, and other features, as well as the extensive Soviet use of anechoic coatings, decoys, and jamming to deter torpedo effectiveness. A 1983 review of U.S. torpedoes by the Secretary of the Navy's Research Advisory Committee (NRAC) saw a minority report that raised major questions about the Mk 48's efficacy and proposed consideration of small nuclear torpedoes to complement conventional torpedoes.[6] The "nukes" could overcome most, if not all, of the submarine countermeasures against conventional weapons. Further, "insertable" technology would permit specially built conventional torpedoes to be converted to low-yield nuclear weapons on board the submarine.

There was immediate opposition to this proposal by the U.S. submarine community. Charges were brought that open discussion of the torpedo issue would reveal classified information.[7] In fact, the submarine community was reacting to U.S. submarine vulnerability to Soviet ASW weapons. Despite support from several quarters for a broad discussion of torpedo issues, senior U.S. submarine officers eschewed all discussion of nuclear or conventional torpedoes to complement the Mk 48.

As late as the 1980s, however, a nuclear torpedo-tube-launched weapon called Sea Lance was considered as a successor to the nuclear SUBROC, which was in the fleet aboard the *Permit* (SSN 594) and later attack submarines from 1964 to 1989. (See Chapter 10.) The Sea Lance, initially called the ASW Stand-Off Weapon (SOW), was conceived as a common ship/submarine-launched weapon and later as a submarine-only weapon.[8] Sea Lance was intended for attacks out to the third sonar Convergence Zone (CZ), that is, approximately 90 to 100 n.miles (167 to 185 km).

The weapon was to mate a solid-propellant rocket booster with a Mk 50 12.75-inch (324-mm) lightweight, anti-submarine torpedo. Although often labeled a successor to SUBROC, the Sea Lance was to have only a conventional (torpedo) warhead, whereas the SUBROC carried only a nuclear depth bomb. The Mk 50 payload would limit the Sea Lance's effective range to only the first CZ, that is, some 30 to 35 n.miles (55.6 to 65 km). The Sea Lance was to have been stowed and launched from a standard 21-inch torpedo tube in a canister, like the earlier Harpoon anti-ship missile. When the capsule reached the surface, the missile booster was to ignite, launching the missile on a ballistic trajectory toward the target area. At a designated point the torpedo would separate from the booster, slow to re-enter the water, and seek out the hostile submarine.

The technical and program difficulties proved too great for a dual surface/submarine-launched weapon. The surface-launched weapon evolved into the Vertical-Launch ASROC (VLA). In 1991 additional technical problems led to complete cancellation of the Sea Lance project, leaving the Mk 48 as the only U.S. submarine-launched ASW weapon.

In U.S. submarines the 21-inch-diameter torpedo tube has been standard since the *S-11* (SS 116) of 1923.[9] This size restricted the development of submarine-launched missiles as well as large torpedoes. The three nuclear submarines of the *Seawolf* (SSN 21) class, the first completed in 1997, have 26.5-inch (670-mm) torpedo tubes. However, no large-diameter torpedoes or missiles were proposed for those launch tubes, although had all 29 planned *Seawolf* submarines been built, such development may have been justified.[10]

## Anti-Ship Missiles

The STAM anti-ship weapon proposed for Admiral H. G. Rickover's APHNAS program also was short lived. (See Chapter 17.) Instead, two more-capable and more-flexible missiles went aboard U.S. torpedo-attack submarines: Harpoon and Tomahawk. The Harpoon was developed for use by ASW aircraft against surfaced cruise missile submarines—Project 651/Juliett SSGs and Project 675/Echo II SSGNs. Surfaced submarines were immune to the U.S. Navy's aircraft- and surface ship-launched ASW torpedoes, which had a feature to prevent them from attacking surface ships, that is, the launching ship.

A subsonic missile, the Harpoon is 17¼ feet (5.2 m) long, initially had a range of some 60 n.miles (111 km) and delivered a 510-pound (230-kg) warhead.[11] For submarine use the Harpoon was "encapsulated" in a protective canister that was ejected from a torpedo tube. The canister rose to the surface, and the missile's solid-propellant rocket motor ignited, leaving the canister and streaking toward its target. Active radar in the missile provided terminal guidance. Harpoons were first placed on board SSNs in 1977.[12]

The larger, more-versatile Tomahawk is the size of a torpedo and is launched from torpedo tubes or vertical-launch tubes.[13] Development of Tomahawk began in 1972 as a nuclear land-attack missile for launching from submarines as a means of enhancing U.S. strategic attack forces. Secretary of Defense Melvin R. Laird looked upon the SLCM—Submarine-Launched Cruise Missile—as both a hedge against a breakdown of détente with the Soviets and as a "bargaining chip" for future arms control discussions with the Soviets. Some observers believe that Laird used the SLCM program as a means to win support or at least acceptance of the Joint Chiefs of Staff and of conservative politicians for the Strategic Arms Limitations (SALT) agreement with the Soviet Union.

Early in the SLCM/Tomahawk development process Admiral Elmo R. Zumwalt, the CNO, directed that the missile be sized to permit launching from submarine torpedo tubes. After being launched from a submarine torpedo tube, the Tomahawk reached the surface, where a solid-propellant booster rocket ignited, and stub-wings and control fins extended as did the air-intake for the turbofan jet engine. Flight was subsonic. The range of the nuclear Tomahawk carrying a W80 warhead with a variable yield of from 5 to 150 kilotons was in excess of 1,000 n.miles (1,850 km).

The USS *Barb* (SSN 596) conducted the first submarine launch of a Tomahawk on 1 February 1978. The deployment of the nuclear-armed Tomahawk Land-Attack Missile (TLAM/N) began in 1984 with attack submarines carrying "several" missiles in place of torpedoes on their forward-area deployments. The later *Los Angeles* (SSN 688) submarines were fitted with 12 vertical-launch tubes for Tomahawk missiles in addition to those carried internally.

Development of the TLAM/N was followed by a series of conventional missiles, the TLAM/C carrying a 1,000-pound (454-kg) high-explosive warhead, and TLAM/D carrying small "bomblets" as well as a warhead carrying carbon-fiber spools for attacking electric power facilities. Also produced were Tomahawk Anti-Ship Missiles (TASM) for use against Soviet surface ships.

The accuracy of the early TLAM missiles using a ground-mapping system was on the order of 33 feet (ten meters); later missiles—employing the global positioning system—have increased accuracy. The TASM missiles employed a homing radar.

Various types of Tomahawks were deployed on U.S. surface warships as well as submarines (and in British submarines, beginning with HMS *Splendid* in 1999). The first operational use of the Tomahawk was in the 1991 Gulf War (Operation Desert Storm); U.S. Navy surface warships fired

276 missiles and submarines launched 12 missiles—the *Pittsburgh* (SSN 720) fired four and the *Louisville* (SSN 724) fired eight. These submarine TLAMs represented 4 percent of the 288 missiles fired in the Gulf War. Subsequently, TLAM/C missiles have been fired by submarines against targets in the Afghanistan, the Balkans, Iraq, and Sudan.

*As seen through the periscope of the USS* Pittsburgh *(SSN 720), a TLAM/C missile streaks from the water, transitions to flight, and streaks toward a target in Iraq during Operation Desert Storm (1991). The debris near the missile is from the protective cover fitted over the air scoop before it deploys.* (U.S. Navy)

The TLAM/N weapons were removed from Navy surface ships as well as submarines following President George Bush's announcement on 27 September 1991 that tactical/theater nuclear weapons would be cut back and that all would be taken off of warships. The approximately 100 nuclear Tomahawks carried in surface ships and SSNs were taken ashore.[14] Only conventional TLAMs are now carried in U.S. surface ships and submarines.

## Soviet Submarine Weapons

The second- and third-generation Soviet submarines were armed with an array of advanced weapons developed from the 1960s onward. Like the U.S. Navy, the Soviet Navy rapidly developed torpedo-tube-launched missiles. The first of these were anti-submarine weapons.[15]

A 1960 decree of the Soviet government directed the development of a new generation of submarine-launched ASW weapons. The decree assigned several of the country's top research activities to the program, with the initial effort placed under a most unusual chief designer, Lieutenant General Fyodor Petrov, an artillery and rocket specialist. Two variants of the new weapon would be produced as the RPK-2 V'yuga (blizzard), with both given the NATO designation SS-N-15 Starfish. The V'yuga-53 variant had a diameter of 533-mm and—like the U.S. SUBROC—would carry a nuclear depth bomb; the Soviet weapon had a 20-kiloton warhead. After launching from a standard torpedo tube, the weapon would streak to the surface, where its solid-propellant rocket would ignite; the missile would travel out to some 21½ n.miles (40 km) with the nuclear depth bomb entering the water to detonate at a preset depth. It was developed specifically for launching from the Project 705/Alfa SSN.

The larger V'yuga-65 variant was a 25.5-inch (650-mm) weapon, launched from the large-

*The Project 633RV/Romeo (top) and Project 613/RV/Whiskey were submarines converted to test advanced submarine-launched weapons. Both ships were fitted with two 650-mm torpedo tubes in a deck housing forward; they both retained two internal 533-mm tubes. The conversions were designed by a Malachite team led by R. A. Shmakov.* (Malachite SPMBM)

*A torpedo is lowered through the loading hatch of an Akula-class SSN. The Soviet Navy employed a number of different torpedoes—including nuclear weapons—at any given time during the Cold War. Some were highly innovative in concept and could provide flexibility in combat.* (Russia's Arms Catalog)

diameter tubes fitted in the Project 671RTM/ Victor III and later submarines. This weapon had a solid-propellant rocket to deliver a small, 400-mm diameter ASW homing torpedo. The range was to be similar to the 533-mm weapon.

After initial underwater trials with a fixed launcher, a T-43 minesweeper was fitted with a 650-mm launch tube for sea trials of the V'yuga-65. Then the Whiskey-class diesel submarine *S-65* was converted to a test platform (Project 613RV), being fitted with a fairwater over the bow that housed two 650-mm tubes. The *S-65* carried out sea trials of both weapons in 1964–1968, using internal torpedo tubes for the 533-mm variant. The missiles could be launched down to depths of 200 feet (60 m).

Both weapons became operational in 1969.

The V'yuga/SS-N-15 evolved into the follow-on RPK-6 Vodopad (waterfall) and RPK-7 Veter (wind) missiles. Again, both were given a single NATO designation, SS-N-16 Stallion. The Vodopad was a 533-mm weapon and the Veter a 650-mm weapon. They could carry either a 400-mm ASW homing torpedo or a nuclear depth bomb. Also, the weapons would be torpedo-tube launched; streak to the surface, where their solid-propellant rocket engines would ignite; and fly toward the target. The 533-mm missile had a range of some 20 n.miles (37 km), while the larger missile could reach 55 n.miles (100 km) and carried a larger payload. The missiles could be launched from depths down to 655 feet (200 m).

Two Project 633/Romeo-class diesel submarines were reconfigured as test craft (Project 633RV), with two 650-mm torpedo tubes fitted in a fairwater built over their bows. The modified *S-49* was delivered in 1973 and the similar *S-11* in 1982. After successful trials, the 533-mm Vodopad became operational in Soviet submarines in 1981, followed by the 650-mm Veter in 1984. They were placed aboard Project 671RT/ Victor II and Project 705/Alfa and later SSNs as well as the Project 949/Oscar SSGN.

Still unknown outside of official circles is the method of detection envisioned when using ASW weapons of the range of the SS-N-16. The 55 n.mile range was beyond the sonar detection capability of Soviet submarines; external (off-board) sensors possibly were planned for use with this weapon.

The next Soviet submarine-launched missile to enter service was the P-700 Granit (NATO SS-N-19 Shipwreck), a large anti-ship missile that would be the successor to the earlier submarine-launched anti-ship missiles.[16] Work on the new missile began in 1969. Developed specifically for attacking Western aircraft carriers, the ramjet-powered SS-N-19 has the advantages of greater range and speed as well as underwater launch compared to earlier weapons. Maximum range of the SS-N-19 is about 300 n.miles (555 km). Being supersonic—capable of Mach 2.5 speeds—the missile does not require mid-course guidance, as did the subsonic SS-N-3; guidance is inertial with terminal radar homing. If multiple missiles are launched, the missiles can exchange data upon detecting surface ships to concentrate on striking the optimum target. The SS-N-19 can carry a conventional or nuclear warhead.

The launching submarine—the Project 949/Oscar SSGN—need not surface to carry out an attack. It has retracting antennas that can be raised above the surface as well as a "surfacing antenna buoy" to enable the submarine to receive radio communications, target designation, and navigation signals while remaining submerged. The 24 weapons fitted in the submarine significantly exceeded the number in previous SSG/SSGNs, providing increased firepower as well as greater performance. The SS-N-19 entered service on board the Oscar SSGN in 1983.[17]

*The Project 667AT/Yankee Notch was a conversion of outdated Soviet SSBNs to carry Tomahawk-like land-attack missiles. In this view of a Yankee Notch, the new amidships section is visible; she has four horizontal launch tubes on each side.* (U.K. Ministry of Defence)

The next Soviet submarine-launched cruise missile was a land-attack missile, the RK-55 Granat (NATO SS-N-21 Sampson). It is similar to the U.S. Tomahawk Land-Attack Missile (TLAM), being a 510-mm weapon launched in a canister from the 533-mm torpedo tube. The SS-N-21 is longer than the U.S. Navy's 533-mm TLAM. Both missiles have cruise speeds of about Mach 0.7.

Development began in 1976. The missile's configuration, folding fins and wings, turbojet engine, terrain-following guidance, and other features generally resemble the U.S. Tomahawk, giving rise to the Western nickname "Tomahawkskii." It has a subsonic speed (Mach 0.7), with a maximum range of some 1,620 n.miles (3,000 km) while carrying a 100-kiloton nuclear warhead. The missile entered service in 1987, being carried by several SSN classes —Project 671RT/Victor II, 671RTM/Victor III, 945A/Sierra II, 971 Akula, and possibly 705/Alfa.

The SS-N-21 also was put aboard three Project 667A/Yankee SSBNs rebuilt to carry the missile.[18] Their ballistic missile compartments (No. 4 and 5) were removed, and a new section was fitted that contained eight 533-mm horizontal launch tubes, angled outboard four per side, with 32 missiles being carried. Improved navigation and other equipment was fitted. The submarines retained their four 533-mm bow torpedo tubes. These submarines—the *K-253*, *K-395*, and *K-423*—were redesignated Project 667AT Grusha (NATO Yankee Notch). Their conversions were completed between 1986 and 1991.

The SS-N-21 was a nuclear land-attack missile and, under the bilateral U.S.-Soviet initiatives of 1991, all were removed from the fleet. Two of the Yankee Notch SSGNs were quickly retired with the third, the *Orenburg* (K-395), being retained in service into the 21st Century albeit without the SS-N-21 missile.

A fourth rearmed Yankee SSBN, the *K-420*, was modified in 1981–1982 to carry the larger P-750 Grom (thunder) land-attack cruise missile, given the NATO designation SS-NX-24 Scorpion.[19] This

*An amidships view of a Project 667AT/Yankee Notch conversion to an SSGN configuration to carry land-attack missiles. The later U.S. program to convert Trident missile submarines to an SSGN conversion was similar in concept, with the U.S. conversions to also provide capabilities for carrying special operations forces—seemingly incompatible roles. (U.K. Ministry of Defence)*

was a Mach 2 weapon carrying a nuclear warhead. The SSBN *K-420* was designated Project 667M and nicknamed Andromeda, for the Greek goddess. Missile tests were completed in 1989 but were not successful, and the SS-NX-24 program was cancelled a short time later.[20]

Beginning in 1991 the Soviet and U.S. Navies removed all tactical nuclear weapons from their fleets. Plans for additional Yankee SSGN conversions were cancelled.

Two advanced tactical cruise missiles were initiated late in the Cold War, the P-800 Oniks (NATO SS-N-26 Yakhont) and the P-10 Klub-S (NATO SS-N-27). The P-800 is a high-speed, sea-skimming anti-ship missile, launched in a container from 533-mm torpedo tubes (as well as from surface ships). Carrying a relatively small conventional warhead of 440 pounds (200 kg), the solid-propellant SS-N-26 has a terminal speed of Mach 2.5; in low-level flight the missile has a range of some 65 n.miles (120 km), while at higher altitudes the missile can reach 160 n.miles (300 km).

The Klub-S is one of a family of anti-ship missiles for surface ship and submarine use. Launched from 533-mm torpedo tubes, it has a range of some 135 n.miles (250 km); after launching, the booster drops off and a sustainer engine propels the missile at subsonic speed until it approaches its target. It has inertial and active-radar guidance. Upon selecting a ship target, the rocket-propelled warhead separates and accelerates to Mach 2.9. A conventional warhead of about 440 pounds is fitted. There are several variants of the missile, including land attack.

Beyond being fitted in Russian ships, the Klub-S has been procured by the Indian Navy for Kilo-class submarines as well as for surface ships.

TABLE 18-1

**Submarine-Launched Cruise Missiles**

| | RPK-2 V'yuga NATO SS-N-15 Starfish | RPK-6 Vodopad NATO SS-N-16 Stallion | RPK-7 Veter NATO SS-N-16 Stallion | P-700 Granit NATO SS-N-19 Shipwreck | RK-55 Granat NATO SS-N-21 Sampson | P-800 Oniks NATO SS-N-26 Yakhont | P-10 Bifyuza NATO SS-N-27 Klub-S | UGM-84A Harpoon | BGM-109 Tomahawk |
|---|---|---|---|---|---|---|---|---|---|
| Operational | 1969 | 1981 | 1984 | 1981 | 1987 | 1999 | | 1977 | 1983*** |
| Weight | 3,970 lb (1,800 kg) | 5,390 lb (2,445 kg) | (7,000 kg) | 15,430 lb | 3,750 lb (1,700 kg) | 8,600 lb (3,900 kg) | 5,510 lb (2,500 kg) | 1,757 lb (798.6 kg) | 3,200 lb (1,450 kg) |
| Length | 26 ft 11 in (8.2 m) | 26 ft 11 in (8.2 m) | 36 ft 1 in (11.0 m) | 32 ft 10 in (10.0 m) | 26 ft 6 in (8.09 m) | 29 ft 2 in (8.9 m) | 27 ft (8.22 m) | 17 ft 2 in (5.2 m) | 20 ft 3 in (6.17 m) |
| Diameter | 21 in (533 mm) | 21 in (533 mm) | 25½ in (650 mm) | 33½in (850 mm) | 20 in (510 mm) | 21 in* (533 mm) | 21 in (533 mm) | 13½ in (343 mm) | 20½ in# (520 mm) |
| Propulsion | solid-fuel rocket | solid-fuel rocket | solid-fuel rocket | turbojet | turbojet + solid-fuel booster | solid-fuel rocket | turbojet + solid-fuel booster | turbojet + solid-fuel rocket | turbofan + solid-fuel rocket |
| Range | 19+ n.miles (35+ km) | 20 n.miles (37 km) | ~38 n.miles (~70 km) | 300 n.miles (555 km) | 1,620 n.miles (3,000 km) | 65 n.miles** (120 km) | 135 n.miles (250 km) | 60+ n.miles (111 km) | 750+ n.miles (1,390+ km)## |
| Warhead | nuclear depth bomb | UMGT-1 torpedo *or* nuclear depth bomb | UMGT-1 torpedo *or* nuclear depth bomb | nuclear *or* conventional | nuclear 100 KT | conventional 440 lb (200 kg) | conventional ~440 lb (200 kg) | conventional 510 lb (231 kg) | conventional 1,000 lb (454 kg) *or* nuclear (W80 warhead)### |

Notes: * Launched with 650-mm canister.

** Low-altitude range; 160 n.miles (300 km) high-altitude range.

*** TASM variant in submarines; TLAM/N in 1987.

# Launched with 533-mm canister from torpedo tubes.

## TASM variant; TLAM/N was 1,400+ n.miles (2,600+ km).

### W80 warhead credited with a variable yield of from 5 to 150 kilotons.

Also, several Soviet submarines were "armed" with surface-to-air missiles. On Project 949/Typhoon SSBNs and Project 636/877 Kilo-class diesel submarines, there is space provided in the after portion of the sail structure for a man to use a shoulder-held launcher for short-range, infrared-homing missiles. Each submarine has stowage for eight missiles, to be used when the submarine is on the surface.

These weapons are not mounted on the submarines, as were similar missiles contemplated by the U.S. and British navies. In the 1980s the U.S. Defense Advanced Research Projects Agency undertook development of the SIAM (Self-Initiating Anti-aircraft Missile), which was to be launched from submerged submarines against low-flying aircraft.[21] The missiles, radar- and infrared-homing, would be launched upon the aircraft being detected by towed-array sonar or by periscope or radar. Earlier the U.S. Navy had considered a submarine-launched variant of the Sidewinder, dubbed "Subwinder." But neither SIAM nor Subwinder reached the submarine test stage.

The British system, known as SLAM for Submarine-Launched Anti-aircraft Missile, used a four-tube launcher for the Blowpipe missile. The submarine's sail was to broach the water, and the SLAM launcher would be extended to engage the aircraft. SLAM was evaluated in the submarine *Aeneas* in 1972, but it did not become operational, in part because of the missile's short range. Rather, a submarine's ability to submerge and its inherent stealth were considered the best defense against air attack.

## New Torpedoes

Simultaneously with their cruise missile programs of the 1960s, the Soviet Navy took delivery of several new torpedoes. Both 533-mm and 650-mm diameter torpedoes were produced as well as 400-mm "short" anti-submarine torpedoes for use with the SS-N-15 and SS-N-16 ASW missiles (and from ASW aircraft).

Whereas the U.S. Navy successively relied on the Mk 37, Mk 48, and now Mk 48 ADCAP (Advanced Capability) torpedoes, albeit with several modifications of each, the Soviet Navy developed several different submarine torpedoes in this period. The Type 53-61 torpedo that entered the Soviet fleet in 1962 had introduced active-passive acoustic homing to Soviet submarine torpedoes. (The SAET-50 of 1950 had been the first Soviet torpedo with passive homing guidance.[22]) The 53-65 was the first torpedo to be propelled with a gas turbine employing hydrogen-peroxide as a fuel; this was the world's fastest "conventionally propelled" torpedo, being capable of 68 knots for 6.5 n.miles (12 km) and 44 knots for 12 n. miles (22 km).[23] The later 53-65M variant replaced hydrogen-peroxide with oxygen.

In 1972 the appearance of the Project 671RT/Victor II SSN introduced the 650-mm torpedoes to submarines. These weapons were intended specifically for attacking aircraft carriers and other large warships. Beginning with the Type 65-73, these weapons had a speed of 50 knots and a maximum range of some 27 n.miles (50 km); reportedly, at 30 knots the 650-mm torpedoes could travel up to 55 n.miles (100 km). As one Russian text noted, "With a nuclear warhead such a torpedo could be effectively used also against shore marine constructions and beach objects."[24]

In general Soviet torpedoes had speeds similar to U.S. torpedoes, up to some 50 knots. However, Soviet torpedoes often had significantly larger warheads; for example, the 53-series have 640- to 660-pound (200-300 kg) warheads; the 650-mm DST-90 has a warhead of about 1,235 pounds (560 kg); and the DT torpedo carries 990 pounds (450 kg) of high explosive.

Wire guidance was fitted in torpedoes from 1969 onward. The appearance of wake-homing torpedoes in the 1980s was especially threatening to Western warships. U.S. officials expressed considerable concern over this weapon; one admiral, when asked, in 1987, the most effective way to counter the weapon, replied: "The only way we can stop the wake-homer now is to put a frigate in the wake of each carrier!"[25] A subsequent U.S. Navy appraisal of Soviet submarine torpedoes stated:

> The Soviets have been emphasizing torpedo development. . . . They have an imposing torpedo arsenal when analyzed from any aspect: variety, capability, warhead size, or quantity. According to some estimates, *a hit by one* of the Soviets' largest diameter

torpedoes could put a CV [aircraft carrier] out of action.[26] (Emphasis added)

The U.S. Navy report also speculated:

> the advantages from the Soviet viewpoint, of a torpedo attack are beginning to outweigh the advantages of a cruise missile attack. All of the above suggests that torpedo attacks will likely compete with cruise missiles as the primary method of surface warship attack in the future. A change of emphasis in Soviet ASUW [anti-surface warfare] attack tactics—from cruise missiles to torpedoes—although not guaranteed, should not be unexpected.[27]

The largest Soviet torpedo is the DT, with a weight of 9,920 pounds (4,500 kg) and a length of 36 feet (11 m). This weapon is intended for attacking large surface ships—battleships, aircraft carriers, and super tankers. The newest torpedo in service is the UGST, entering the fleet in the 1990s as a replacement for the TEST-71 and 53-65 weapons. The UGST is a 50-knot-plus torpedo, with a wire guidance capability of 13.5 miles (25 km), about one-half of the torpedo's range. This torpedo weighs 4,850 pounds (2,200 kg), with a relatively small warhead of 440 pounds (200 kg); it is 23 feet, 7 inches (7.2 m) long.

The Soviet nuclear torpedoes had warheads of about 20 kilotons. They, too, were taken off submarines following the 1991 U.S.-Soviet tactical/theater nuclear weapons agreement. Previously most or all submarines on combat patrols carried two or more nuclear torpedoes in their weapons loadout.[28]

The large number of torpedo-type weapons created support and loadout problems for Soviet submarine force commanders. These were compensated for, in part, by the larger number of weapons found in later Soviet attack submarines. At the same time, the variety of weapons caused substantial problems for U.S. naval officials developing torpedo countermeasures.

Unquestionably the most difficult Soviet "torpedo" to counter is the rocket-propelled VA-111 Shkval (squall). This is a 200-knot torpedo that is launched from standard 533-mm torpedo tubes. It is an ASW weapon that may also have an anti-torpedo capability.

During World War II some rocket-propelled torpedo development was conducted in Germany and Italy. After the war interest was shown in this type of weapon in the USSR and the United States. In the United States test vehicles reached an underwater speed of 155 knots, but such weapons were not developed. Only the Soviet Union developed rocket-propelled torpedoes. The aircraft-launched RAT-52 torpedo of 1952, for use against surface ships, had an underwater speed of 68 knots; the air-launched ASW torpedoes APR-1 and APR-2 used solid-rocket propulsion to attain speeds just over 60 knots.

In 1960 the NII-24 rocket research institute began development of a submarine-launched rocket torpedo, initially

TABLE 18-2
**Soviet Submarine-Launched Torpedoes Since 1962**

| | Diameter | Type* | Operational | Guidance** | Warhead*** |
|---|---|---|---|---|---|
| SAET-60 | 533 mm | ASUW | 1961 | AH | C |
| SET-40 | 400 mm | ASW | 1962 | AH | C |
| 53-61 | 533 mm | ASW | 1962 | AH | C |
| SET-65 | 533 mm | ASW | 1965 | AH | C |
| 53-65 | 533 mm | ASUW | 1965 | AH, WH | C |
| 53-68 | 533 mm | ASUW | 1968 | AH | N |
| STEST-68 | 533 mm | ASW | 1969 | AH, WG | C |
| TEST-71 | 533 mm | ASW | 1971 | AH, WG | C |
| 65-73 | 650 mm | ASUW | 1973 | AH | C, N |
| 65-76 | 650 mm | ASUW | 1976 | AH | C, N |
| USET-80 | 533 mm | ASW | 1980 | AH, WG | N |
| DST-90 | 650 mm | ASUW | | AH, WH | C |
| DT | 650 mm | ASUW | | AH | C |
| UGST | 533 mm | ASW, ASUW | | AH, WG | C |

Notes: * ASUW = Anti-Surface Warfare; ASW = Anti-Submarine Warfare
** AH = Acoustic Homing; WG= Wire-Guided; WH= Wake-Homing.
*** C= Conventional; N= Nuclear

*The Shkval (squall) is a rocket-propelled torpedo that can reportedly achieve a speed of 200 knots—perhaps four times that of conventional torpedoes. Originally fitted with a nuclear warhead, the Shkval also carries a high-explosive warhead, with some versions of the conventional weapon being available for foreign sale.*

under the direction of chief designer I. L. Merkulov.[29] The requirement for a very high-speed torpedo led to the use of a solid-propellant rocket motor coupled with moving the weapon through the water surrounded by a gaseous envelope or "bubble" that is created by the torpedo's shape and sustained by the rocket's exhaust. This technique is known as "artificial cavitation." A flat plate at the front of the torpedo creates a local cavity around the weapon's nose; rocket exhaust gases are then fed into that region to create the cavity that encases the projectile as it streaks through the water. Stabilization of the torpedo is achieved through the use of four fins that extend from the midbody after launch.

The preliminary design of the Shkval was completed as early as 1963, and the experimental launches began the following year at Lake Issyk-Kul in west-central Asia. The test submarine *S-65* (Project 613RV) was further modified to launch Shkval test vehicles, with the first launches in May 1966 near Feodosiya on the Black Sea. Tests were halted in 1972 because of problems identified in the Lake Issyk-Kul work and the weapon underwent further development.

Seven launches were made from the *S-65* in June–December 1976. The M-5 variant of the VA-111 system was declared ready for operational use in November 1977—17 years after the project was initiated. The 21-inch torpedo is 27 feet (8.2 m) long and weighs 5,950 pounds (2,700 kg). Estimated range is more than 10,000 yards (9.15 km) at approximately 200 knots. Initially fitted with a nuclear warhead, later variants are reported to have terminal guidance and a conventional, high-explosive warhead of 460 pounds (210 kg).[30]

The Shkval had been a part of the Soviet counterattack tactics, realizing that their submarines could have been detected before they knew a U.S. submarine was present. When alerted by the noise of the American submarine launching torpedoes, the Shkval would have been fired "down the bearing" toward the enemy while the Soviet submarine took evasive maneuvers and launched decoys and countermeasures.

Regarding the Shkval, Rear Admiral Valeri Aleksin, a former submarine commander, said "This was a complete new-stage breakthrough in the development of underwater weapons. And as far as I know, it is impossible to protect yourself against this kind of torpedo, and the Americans are behind in the development of this kind of technology."[31] When this book went to press, there were reports that a 300-knot version of the Shkval was in development.[32]

**Soviet and U.S. submarine torpedoes in the early Cold War period were severely limited in capability, especially against other submarines. Both nations quickly sought more effective weapons, initially improved torpedoes and then introduced various rocket-propelled ASW weapons, the latter to extend weapons ranges and speed of flight to take advantage of longer-range sonar detections. Further, there was early interest in nuclear weapons because of the limitations of ASW torpedoes and of sonar accuracy; both the Soviet and U.S. navies deployed submarine-launched ASW weapons with nuclear warheads.**

With the advent of the Mk 48 torpedo, the U.S. Navy abandoned its Mk 45 ASTOR nuclear torpedo as well as the SUBROC nuclear ASW missile. The Mk 48—currently the U.S. Navy's only submarine-launched torpedo—is intended for the anti-surface as well as ASW role. In contrast, the Soviet Union developed and retained in service an array of submarine-launched weapons, that is, several different torpedoes and several ASW missiles.

Although the Soviet Union led the world in the development of underwater-launched cruise missiles, the U.S. Navy led in the development of cruise missiles launched from a torpedo tube. This effort began with modification of the air-launched Harpoon anti-ship missile. This launch technique was necessitated because of the late 1950s decision to abandon specialized cruise missile submarines. The Soviet Navy, with a large force of specialized SSG/SSGN submarines for the anti-ship role, followed the U.S. lead in this field.

By the end of the Cold War the U.S. had but three tube-launched weapons—the Mk 48 ADCAP torpedo, the Harpoon anti-ship missile, and the multi-role Tomahawk anti-ship/land-attack missiles. The two missiles, however, were limited by the 533-mm diameter of U.S. torpedo tubes whereas the Soviets had put to sea submarines with 533-mm and 650-mm tubes. Coupled with the highly innovative approach to torpedo development (e.g., wake-homing, nuclear, and very-high-speed), the Soviet Navy put to sea considerably more innovative submarine tactical weapons during the Cold War.

(Both navies also developed submarine-laid mines during the Cold War.)

19

# Fourth-Generation Nuclear Submarines

*The USS* Seawolf *on her initial sea trials in July 1996. One of the most controversial submarines in U.S. naval history, the administration of President George H.W. Bush sought to limit the* Seawolf *program to one ship, but three have been built. The "bands" around her hull are test gauges. On trials she exceeded her 35-knot design speed.* (U.S. Navy)

The USS *Seawolf* (SSN 21)—the last U.S. submarine design of the Cold War—is probably the most controversial submarine in American history.[1] The origins of the *Seawolf* are somewhat clouded. The decision to construct a new SSN class was reached in July 1982, only one year after a Navy decision *not* to develop a new SSN. In 1981 the head of the Naval Sea Systems Command told Congress:

> We are not designing [an advanced SSN]. We reviewed this back in early 1981, all the design efforts that we had going on in Electric Boat, and decided to stop the design efforts on all the new classes of submarines and to proceed on a course of upgrading the [SSN] 688 class as the attack submarine [for future production].

> No conspicuously cost-effective candidate to follow the SSN 688 class emerged from the studies of alternative attack submarines. The CNO [Chief of Naval Operations Admiral Thomas B. Hayward] directed that efforts toward the design of the Fast Attack Submarine, a smaller, less capable submarine, and the SSNX, a larger, more capable submarine, should be discontinued. The CNO also directed that highest program priority be placed on improving the capability of the SSN 688 class and existing ASW weapons.[2]

Interest in a follow-on to the *Los Angeles* (SSN 688) class began in June 1982, when Admiral James D. Watkins became the Chief of Naval Operations. He was the first nuclear submariner to attain the highest Navy's position.[3] When Watkins became CNO, the astute and outspoken John Lehman had been Secretary of the Navy for more than a year. Lehman had strong support from President Ronald Reagan to "rebuild" the U.S. Navy, and he created an ambitious program to attain a fleet of 600 ships, with an emphasis on aircraft carriers, battleships, and strategic missile submarines; he also increased the planned attack submarine force level from 90 to 100 SSNs. His "take charge" attitude left little freedom of action for Watkins. Indeed, describing how he administered the various components of the Navy Department, Lehman said that he had "left Watkins with the submarine programs to run."[4]

At almost the same time, in January 1982, Admiral H. G. Rickover left office after more than 30 years as head of the Navy's nuclear propulsion directorate. Lehman, with the full backing of the White House staff, had managed to outmaneuver Rickover in the biennial ritual used to allow the admiral to remain on active duty beyond the statutory retirement age. Many members of Congress who had long supported Rickover had died or retired; the newer members lacked the anti-Soviet fervor that Rickover had exploited in congressional committees to gain him unprecedented support for nuclear ships and for himself.

Although a highly qualified, four-star submarine officer succeeded Rickover, it was the Chief of Naval Operations, Admiral Watkins, who became the leading advocate of a new attack submarine.[5] To propel a new SSN the Naval Reactors Branch had proposed a pressurized-water reactor plant in the range of 45,000 horsepower, that is, a 15,000 horsepower increase over the *Los Angeles* class.

The Naval Sea Systems Command (NavSea) undertook a large number of conceptual studies for a new SSN; preliminary designs were then undertaken by a team composed of NavSea, Electric Boat (EB), and Newport News. The Naval Reactors Branch contracted with Electric Boat for a detailed design of the engineering spaces; Newport News—after winning a competition with EB—was awarded a contract for the detail design of the "front end" of the submarine and for integrating the "total ship."

The primary emphasis in designing what came to be called the SSN 21 was quieting, the longtime quest of the U.S. submarine community. The SSN 21 also would have a new combat system—weapon launchers and sensors—and thus would be "the first major effort to improve the 'front end' [forward of the reactor plant] of attack submarines in more than two decades," according to a senior submarine officer.[6] While the SSN 21 was in the preliminary design stage, the Deputy CNO for Submarine Warfare, Vice Admiral Nils R. (Ron) Thunman, convened a group of submarine officers and engineers to help determine the characteristics for the SSN 21. Known as Group Tango (the phonetic word for the *T* in Thunman), this all-Navy, classified study group addressed seven characteristics: (1) speed, (2) depth, (3) torpedo tubes, (4) weapons load, (5) Arctic capability, (6) radiated noise, and (7) sonar effectiveness.[7] Group Tango established "goals" for the SSN 21, which were believed to be realistic and attainable. According to Navy sources, the SSN 21's characteristics would meet only three of the seven goals: Arctic capability, radiated noise, and sonar effectiveness; two SSN 21 characteristics would fall short of the "minimum" goals set by Group Tango with respect to operating depth and number of torpedo tubes.

The SSN 21 would be a large submarine, exceeding 9,000 tons in submerged displacement, that is, more than 30 percent larger than the improved *Los Angeles*-class submarines. The S6W reactor plant was intended to provide a speed of 35 knots.[8] The "tactical speed" of the *Seawolf*—the speed at which the submarine can detect an enemy submarine before being detected herself—would be in excess of 20 knots compared with Soviet SSN tactical speeds estimated at between "only six to eight knots maximum."[9] Also, the SSN 21's radiated noise levels would be lower than any previous U.S. nuclear-propelled submarine, according to official sources. A Navy official told Congress that if the SSN 21 met its acoustic goals, "the Soviets . . . are still going to be some five to ten years behind us" in quieting.[10]

Contributing to quieting, in place of a conventional propeller the SSN 21 has a pump-jet or ducted-thruster propulsion system. It is similar to the propulsor in the earlier British *Trafalgar*-class SSNs.[11] The thruster ingests water and, through the use of several stages of fixed (stator) and rotating (rotor) blades, expels the water aft to propel the submarine. The device reduces cavitation and other noise sources associated with conventional propellers.[12] However, the additional surface area and weight introduced by the duct and stators can present drag and longitudinal stability problems. Before being fitted in the SSN 21, a similar propulsor was evaluated in the USS *Cheyenne* (SSN 773) of the *Los Angeles* class. (The lower propulsive efficiency of the pump-jet in comparison with advanced open propeller designs was perceived by Soviet designers as a significant limitation to this system.[13])

The use of polymers apparently was considered for the *Seawolf.* After the *Albacore* (AGSS 569) research into polymer ejection in 1970–1972, the USS *Jack* (SSN 605) had been used in polymer work and, later, the USS *William H. Bates* (SSN 680). The *Jack* trials had looked into noise reduction, while the *Bates* trials evaluated the value of polymers in drag reduction. All of these efforts encountered problems with chemical mixtures and limited polymer storage capacity. An effective system at the time of those trials would have required large amounts of polymer, probably affecting the size of the ship.

Subsequently, the Navy and Pennsylvania State University conducted small-scale tests and a laboratory-scale experiment with a 20-foot (6.1-m) model to demonstrate that polymers could be distributed effectively and that they would reduce drag. This polymer research was nearly complete, and a full-scale demonstration was planned on board a *Los Angeles*-class submarine, but the project was cancelled in the early 1990s. A Navy statement asserted, "The new advanced polymers and injection system is projected to achieve submerged speed increases of 20%. Quieter operations at tactical speeds are also expected from its use."[14]

The SSN 21's pressure hull would be fabricated from HY-100 steel, which had been proposed for but not used for the *Los Angeles* class. Using HY-100 would return the SSN 21 to the operating depth of some 1,300 feet (395 m) of pre-*Los Angeles* SSNs. And it was planned that starting with the fourth SSN 21, the Navy would employ HY-130 steel, providing an even greater depth. Early difficulties with the improved steel led to the decision to build all SSN 21s with HY-100. Also, the SSN 21 has bow-mounted diving planes (vice sail mounted), which can retract into the bow for under-ice operations, as in the later SSN 688s.

For the SSN 21's primary mission of ASW the submarine was provided with eight 26.5-inch (670-mm) torpedo tubes. This was the first change in tube diameter since the 21-inch (533-mm) tube was introduced in U.S. submarines with the S class of 1920. While no torpedo beyond improved versions of the 21-inch Mk 48 was envisioned, the larger tubes provide "growth potential" for future weapons. No vertical-launch tubes for Tomahawk missiles were fitted; rather Tomahawks as well as Harpoon anti-ship missiles and mines would be launched from torpedo tubes.

The mission of penetrating Soviet defenses to operate against undersea craft in the shallow Barents Sea or Sea of Okhotsk was an arduous one, hence a large torpedo battery was desired. The massive torpedo room would accommodate more than 40 long weapons; with eight weapons in tubes the SSN 21 could carry a total of some 50 weapons, that is, one-third more than in a modified *Los Angeles*-class SSN (37 weapons). This large torpedo loadout was intended to enable the SSN 21, after penetrating into Arctic waters, to assail Soviet missile and attack submarines at a prodigious rate, remaining on station for long periods. Tomahawk attacks—against land or surface ship targets—were considered a lesser requirement, hence the exclusion of VLS tubes.

Coupled with the weapons, a new sonar/fire control system was to be provided, the SUBACS (Submarine Advanced Combat System). When initiated in 1980, SUBACS was planned for backfit into the later *Los Angeles* submarines as well as installation in the SSN 21. SUBACS would integrate the bow-mounted passive and active sonars, wide-aperture passive array panels fitted on the sides of the SSN 21, and advanced towed-array sonars. SUBACS was considered a major factor in providing U.S. superiority over new Soviet submarines.

From the start SUBACS encountered development, cost, and management problems. The planned optical data bus—using fiber-optic technology to transmit data—encountered difficulties, causing a redesign to employ more-conventional electronic technology. Next there were problems in producing the multilayer computer circuit boards. And there were management problems on the part of the prime contractor (the IBM Corporation) and the Navy. These contributed to unprecedented cost increases—in a single year SUBACS costs increased $1 *billion* over budget.

Several redesigns as well as changes in management were made, with key managers replaced in the Navy and at IBM, and SUBACS was renamed AN/BSY-1. It would go in later *Los Angeles* submarines and the more advanced and complex AN/BSY-2 in the SSN 21.[15] Cost "caps" were placed on the BSY systems, and were continually exceeded. At the request of Congress, the General Accounting Office (GAO) monitored the BSY-series development.

A series of GAO reports addressed BSY-series shortfalls. In 1985: "Under the latest Navy plan . . . SUBACS will provide less performance than originally intended, require additional funds, and may delay the delivery of the first two SUBACS-equipped SSN 688 submarines."[16] In 1987: "the time required for software development and integration was underestimated."[17] In 1991: "Most of the risks identified by IDA [Institute for Defense Analyses] . . . still exist. According to IDA, if left unresolved, those risks could significantly impair system development through increased costs, schedule delays, and degraded system performance. IDA recommended specific actions to mitigate these risks. However, the Navy did not implement 5 of 6 IDA recommendations,"[18] Also in 1991: "The risks that the Navy has allowed in the development of its BSY-2 combat system are serious. . . ." And, "the Navy is not following some sound management principals and practices. . . . the Navy could find itself with combat systems that fall short of their promised capability and could cost millions to enhance."[19] In 1992—as the BSY-1 was being evaluated in the USS *San Juan* (SSN 751)—the GAO reported:

> Two AN/BSY-1 critical operational issues (reliability and maintainability) were unsatisfactory. System failures decreased AN/BSY-1 reliability. The failures were not corrected in the required time. Changes are being made that are expected to correct these deficiencies.
>
> Limitations resulted in incomplete demonstration of two critical operational issues (weapons employment and navigation) and unrealistic operational testing.[20]

Certainly the BSY-series problems could be corrected in time, but only at tremendous cost, and cost would become an increasing concern in the post–Cold War era. Problems continued with the BSY-series, serving as a lightning rod for the increasing criticism of the *Seawolf*/SSN 21 program. As early as 1984, a classified study undertaken for the Under Secretary of Defense (Policy), Dr. Fred Iklé, criticized the *Seawolf*, mainly on the basis of cost and the Navy's modeling of the submarine's effectiveness being based on a narrow set of ASW engagements. The Navy planned to build three *Seawolf* SSNs per year; the analysis predicted that for the same funds the Navy could procure *seven* of the *Los Angeles*-class submarines.[21] And following Secretary Lehman's expected departure, in 1986–1987 Navy funding was severely reduced, making it impossible to construct three *Seawolfs* per year to reach the force objective of 100 SSNs.[22]

While a *Seawolf* SSN would individually be superior to a *Los Angeles* submarine in virtually all characteristics, obviously the 3:7 ratio meant that more submarines could be at sea and forward deployed if the less-expensive submarine was procured. Further, three *Seawolfs* would carry 150 weapons compared to 259 in seven *Los Angeles*-class SSNs. Still, the arduous mission of ASW operations in Soviet regional waters could require a submarine more capable than the *Los Angeles*, especially in view of the reduced noise levels of the Soviet Akula-class SSNs.

## The Battle Is Joined

Admiral Watkins quickly attacked the study as well as the integrity of the analyst.[23] But the battle over the *Seawolf* was already joined in Congress, the Reagan administration, and the press. When the program was initiated the Navy planned to ask for funding for one SSN 21 in fiscal year 1989 (beginning 1 October 1988), two in fiscal 1991, and then three per year thereafter with a program goal of 29 SSN 21s.

The *Seawolf* was duly authorized in fiscal 1989. But within a year the technical and management problems with the project as well as increasing costs led the powerful Senate Armed Services Committee to delete all construction funds for *Seawolf* submarines in fiscal 1991, instead substituting two more submarines of the *Los Angeles* class; the House Armed Services Committee approved funding for one *Seawolf* after an urgent appeal by the Navy's senior civilians and admirals. The public hearing into *Seawolf* problems by that House committee on 24 July 1990 led a journalist to write that the one *Seawolf* was approved "Probably out of pity."[24] Thus the second *Seawolf* was funded.

President Bush's Secretary of Defense, Dick Cheney, halved future construction rates for the *Seawolf*. Following an intensive review of defense programs, on 13 August 1990, Cheney announced that only 1½ submarines would be bought per year instead of the three wanted by the Navy. With an anticipated 30-year service life, this eventually would provide a submarine force of only 45 SSNs, fewer than half the 100 SSNs approved by the Department of Defense.

By 1991—with the Soviet Union undergoing political upheaval—the Trident strategic missile submarine program had been curtailed at 18 SSBNs. With the reduction in the SSN building rate, there was concern that only a single shipyard would be retained to construct nuclear submarines. At the start of the *Seawolf* program, the Electric Boat (EB) yard in Connecticut was constructing both *Los Angeles* and Trident submarines; Newport News Shipbuilding in Virginia was building *Los Angeles*-class submarines, but also nuclear aircraft carriers and commercial cargo ships, while seeking other surface ship work. With more control over the EB yard because of its single-product work, the head of the Navy's nuclear propulsion directorate, Admiral Bruce DeMars, sought to give the construction contract for the entire *Seawolf* program to the EB yard.

Newport News brought a lawsuit against the Navy for awarding the contract to build the *Seawolf* at Electric Boat. Newport News costs were less, but the Navy prevailed in the courts, and after some delays, EB began building the SSN 21. It meant that in a few years Newport News would be out of submarine work; the Navy would have gone from seven yards constructing nuclear submarines in the 1960s to only one.

Troubles continued to plague the *Seawolf*'s BSY-2 system, and on 1 August 1991, the Navy revealed that massive weld failures had been discovered in the *Seawolf*'s hull. The problem—first identified in June 1991—resulted from a procedure that allowed the welds to cool at an unacceptably fast rate for the high carbon weld wire being used. This resulted in the welds being too brittle. Reportedly, all welds had to be replaced. At the time the submarine was 17 percent complete. As a result, the *Seawolf*'s cost increased and she was further delayed.

Another factor in the rising costs was the low production rate of attack submarines. Also, EB had begun the last Trident SSBN in 1991, and the yard's last SSN 688 would be started in 1993; the end of those programs would increase the overhead costs being carried by the *Seawolf* program, which would be the yard's only construction effort.

In January 1992—with the Cold War over—Secretary Cheney announced that the entire *Seawolf* program would be canceled; only the SSN 21 would be completed. The funds previously voted by Congress for the SSN 22 and SSN 23 were to be rescinded. No additional submarines would be constructed until the fiscal 1997 program, when a new, "lower-cost" attack submarine would be initiated. The Clinton administration, entering office in January 1993, affirmed the one-ship *Seawolf* program.

*The* Seawolf *under construction, showing her massive bulk. There are three sonar panels on each side; her ducted propulsor is visible at the far left. The* Seawolf *was long on the building ways and was several years in trials and undergoing yard work before she made her first overseas deployment in mid-2001.* (U.S. Navy)

But pressure from the congressional delegations of Connecticut and neighboring states, promoted by Admiral DeMars, forced the construction of the SSN 22, which had been funded in fiscal 1991. Similarly, the SSN 23, which had been partially funded in fiscal 1992, was forced onto the administration by Congress. Admiral DeMars claimed that $1.5 *billion* had already been committed for building the third submarine, hence it could be completed for only an additional $1.5 *billion.* Although many of these commitments of money still could have been stopped, Congress approved the third unit—at an estimated cost of $3 *billion,* or more than one-half the cost of a 100,000-ton, nuclear-propelled aircraft carrier.

In the style of Admiral Rickover, Admiral DeMars sought to squelch all dissent against the *Seawolf* program, both within and outside of the Navy. At DeMars's direction, the Naval Investigative Service investigated analysts who opposed the *Seawolf.* He attacked the editor of *The Submarine Review,* the journal of the Naval Submarine League, a private, professional organization that supports submarine programs.[25] The DeMars letter caused the firing of the editor of *The Submarine Review,* a distinguished author, submarine officer, and combat veteran, who had founded the journal in 1983. A subsequent investigation of this situation by the Inspector General of the Department of Defense determined that the DeMars letter "had a chilling effect on public debate of unclassified submarine issues."[26]

The *Seawolf* went to sea on her initial trials on 3–5 July 1996. Several of the submarine's wide-aperture sonar arrays (three per side) were damaged during the trials, in part because of their poor design and because the ship had not yet had her anechoic coatings installed, exposing the arrays to more turbulence and flow forces than would occur when the submarine was in service. Admiral DeMars, on board for the trials, claimed that the *Seawolf* had gone faster than any previous American submarine.[27] Press reports said that the submarine had exceeded 40 knots, although that number seems highly unlikely. Those sea trials were conducted before installation of the anechoic coating, without the submarine being instrumented, and not on a measured range, that is, not an official "standardization trial."

The *Seawolf* was commissioned on 19 May 1997—after a construction period of more than eight years.[28] Problems persisted. The submarine's protracted tests and trials were halted in the fall of 2000 because of welding problems related to her high-pressure air system. There also were problems with the craft's Tomahawk launch capability, pump-jet propulsor, and sonar. The *Seawolf* did not undertake her first operational deployment until June 2001, having spent four years undergoing shipyard work and additional trials.[29]

The second submarine of the class, the *Connecticut* (SSN 22), was commissioned on 11 December 1998. She also required extensive ship-

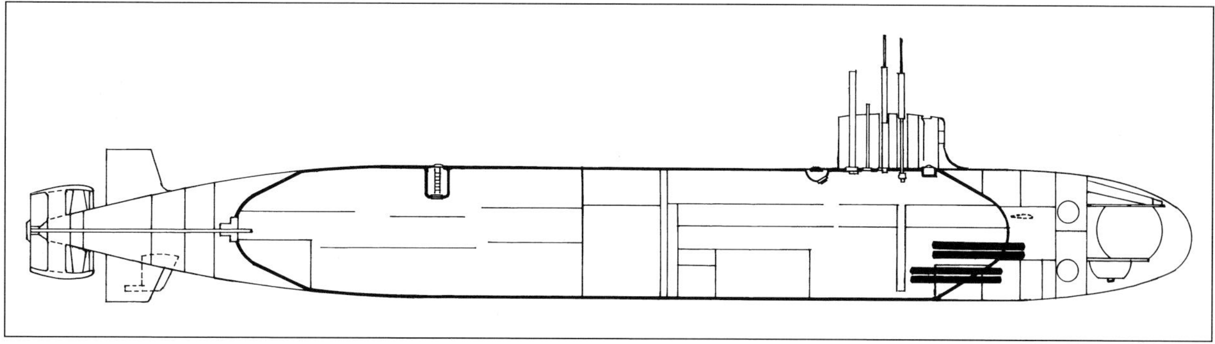

Seawolf *(SSN 21). LOA 353 ft (107.6 m)* (©A.D. Baker, III)

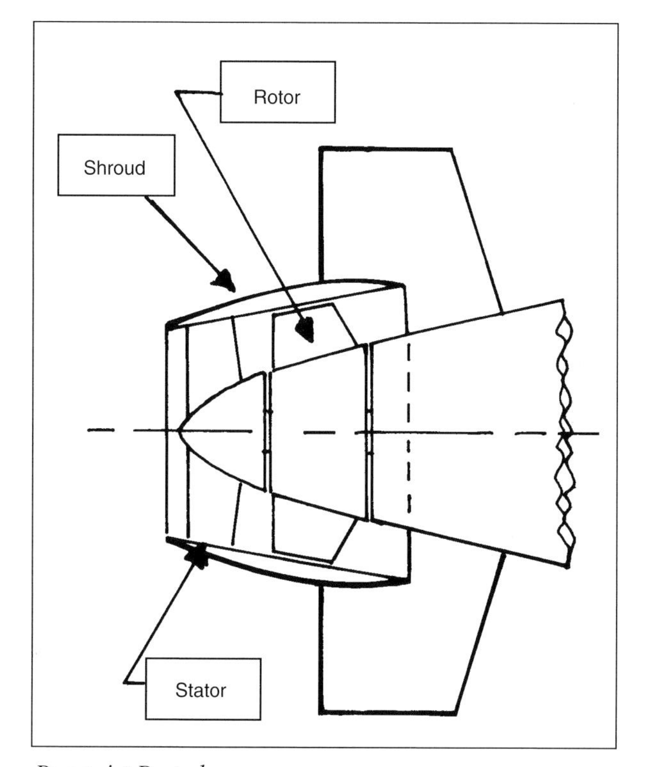


*Pump-jet Propulsor.*

yard modifications and "tuning up" before she was considered to be operational.[30] However, in mid-2001 the *Connecticut* did undertake a brief Arctic cruise. The third unit, named *Jimmy Carter* (SSN 23), was started in December 1995.[31] The submarine community gained approval in 1999 to complete that submarine as a specialized submarine to replace the *Parche* (SSN 683) in deep-ocean search, research, and recovery operations. There also were reports that the *Jimmy Carter* would have a special operations (SEAL) capability.[32] These changes were expected to delay completion of the submarine for more than two years, at least until mid-2004 and would cost about $1 *billion.*

Questions have been asked about why the Navy was taking one-third of the *Seawolf*-class—reputed to be its best submarines—to convert the *Carter* to essentially a non-combatant role. Although the *Carter* will retain weapons and sensors, a review of the operations of her predecessor, the *Parche,* indicates that she is likely to be employed almost exclusively in search/recovery operations.

In the mid-1980s the official cost estimate for a 29-ship *Seawolf* program was $38 *billion* (then-year dollars). In 1999 a submarine officer, with full access to Navy and shipyard sources, estimated the cost of the three-ship *Seawolf* program—including direct research and development costs—at almost $16 *billion.*[33] At the time of that estimate, none of the submarines was fully operational. On an adjusted cost basis, the three *Seawolf*-class submarines are probably the most expensive warships yet constructed except for nuclear-propelled aircraft carriers.

As the high cost of *Seawolf* construction and congressional opposition became manifest, in 1990 the Chief of Naval Operations, Admiral Frank B. Kelso II, proposed a lower-cost SSN, initially given the project name "Centurion"—a submarine for the new century.[34] The Navy's goal for the program was a multimission SSN that was (1) substantially less expensive than the *Seawolf,* (2) capable of maintaining U.S. undersea superiority against a greatly reduced but continuing Russian submarine effort, (3) more capable than the *Seawolf* or improved *Los Angeles* classes for operations in littoral areas, and (4) better able than the *Seawolf* or improved *Los Angeles* designs to incorporate major new submarine technologies as they became available. The last—generally called "technology insertion"—was considered one of the most important features of the new SSN design; it was to consist of both modules and systems/features that would be improved with successive submarines of the class.[35]

Through mid-1991 the Navy had maintained that the Centurion/NSSN would be a complement to the *Seawolf* rather than a successor. In late June 1991 there were reports that the *Seawolf* procurement would cease about the year 2000 to permit acceleration of the new, lower-cost attack submarine. On 28 August 1992 the Department of Defense approved concept definition studies for the new attack submarine and directed that the Navy keep the cost of the Centurion SSN program at $1 *billion* or less per submarine and examine a variety of attack submarine alternatives for the Centurion (including conventional submarines). At the time the *Seawolf* SSNs were estimated to cost about $2 *billion* each in series production. The Navy and the Electric Boat and Newport News shipyards studied a large number of alternative SSNs.

(The term Centurion was dropped in 1993 as the New Attack Submarine program was initiated, designated NSSN, then briefly NAS, and changed back to NSSN. The lead submarine subsequently was named *Virginia*/SSN 774; that is used for clarity in the remainder of this volume.)

Construction of the *Virginia* was authorized for fiscal 1998, and she was laid down on 3 October 1997.

## Congressional Intervention

By the late 1980s there was increasing concern within the U.S. intelligence community over the quieting levels of Soviet third-generation nuclear submarines. As stated in a subsequent report by a congressional advisory panel,

> Our current anti-submarine (ASW) capability rests almost entirely on listening for the sounds generated by Soviet submarines. That approach has been successful and our anti-submarine forces have grown potent because the Soviets have traditionally built relatively noisy submarines. But the future of that approach is now very much in doubt because the Soviet Union has begun to produce quiet submarines.[36]

In response to this concern of Soviet submarine quieting, Representative Les Aspin, chairman of the House Armed Services Committee, convened an advisory panel in 1988–1989 to look into U.S. submarine and anti-submarine warfare.[37] The panel consisted of ten of the nation's most qualified technologists and submarine specialists.[38]

Congress already had provided some funds for anti-submarine research by the Central Intelligence Agency, a novel undertaking outside of the Navy, which has prime responsibility for ASW. The panel recommended that Congress provide continuing funding for separate advanced-technology submarine/ASW programs by the Navy as well as by the Defense Advanced Research Projects Agency (DARPA).[39] The panel's report also was critical of the Navy's efforts, stating that its "in-house" technical community—laboratories and shipyards—had become too oriented toward maintenance of the existing fleet at the expense of research and development for the future. With respect to countering quieter Soviet submarines, the panel noted that virtually every aspect of U.S. ASW capabilities was based on passive sonar and that "better passive sonars are not the answer."[40] The panel called for increased funding and a new architecture for the nation's ASW research and development program. But increased funding for research and development, even for ASW, which periodically enjoyed high-level Navy support, was rarely forthcoming.

In response, in 1988 Congress established the SUBTECH (Submarine Technology) program under DARPA to explore new submarine technologies. SUBTECH, funded for several years, was strongly opposed by the Navy's submarine leadership and, subsequently, was transferred to the Navy, where the program quickly dissipated into the Navy's labyrinth of traditional submarine/ASW programs.

At the same time, DARPA's earlier sponsorship of the advanced reactor development by the Westinghouse Electric Corporation was "folded in" to a joint DARPA-Navy program. This effort included a helium-cooled plant that promised a weight of about 25 pounds (11.3 kg) per horsepower and a liquid-metal plant of about 35 pounds (15.9 kg) per horsepower compared to existing U.S. submarine plants of some 200 pounds (90 kg) per horsepower. Although this program was initiated to be independent of the nuclear propulsion directorate, it soon died because of the directorate's intimidation of Westinghouse.[41]

By the early 1990s, as the independent SUBTECH effort ended and problems with the *Seawolf* program were increasingly evident, Congress became more involved in the Navy submarine programs. At the urging of Speaker of the House Newt Gingrich, in the fall of 1995 Representative Duncan Hunter, chairman of the subcommittee on military procurement of the House National Security Committee, asked four civilians to serve as a panel to examine contemporary submarine issues and report back to him and Mr. Gingrich.[42] Such congressional involvement in submarine matters was unprecedented.

The panel voiced strong exception to the Navy's new attack submarine program (i.e., *Virginia* class) in private discussions with the House leadership and in formal hearings of the National Security Committee from 1995 to 1997.[43] In his conclusions, panel member Anthony (Tony) Battista, a former Navy engineer and who had been provided access to relevant intelligence sources, both U.S. and British, stated:

- Russian submarine and anti-submarine warfare technology is even more advanced than the assessment provided by U.S. intelligence sources.
- The quietest—both acoustically and non-acoustically—operational attack submarine in the world today is owned by the Russians—the Akula.
- The U.S. is not spending enough on submarine and ASW technology.
- The U.S. is paying too much for its submarines and in terms of the threat is getting a very poor return on investment.
- The entire U.S. submarine and ASW program should be restructured—in part as proposed by the House National Security Committee—and should be a high priority national program.

Further, a Navy-sponsored panel, headed by retired Vice Admiral Albert J. (Al) Baciocco, also criticized the *Virginia* program:

> The strategy for incorporation of future improvements is not clear. While the [*Virginia*] program [manager] noted there were plans for mission specific hull sections, preplanned product improvements, and technology insertion, the panel was unable to determine that these plans existed . . . the panel noted that the baseline design lacked certain desirable features which would probably be needed in the future and could still be incorporated into an early hull with vigorous action. These desired features could include an improved sail, a hybrid propulsor, and fiber optic towed arrays.[44]

That panel identified a key difference between U.S. and Soviet investment in new submarine designs:

> The panel noted that core submarine technology investments are generally too small to investigate important and/or revolutionary options in a timely manner. As a result future modifications or new designs will be limited to evolutionary improvements over their predecessors.[45]

## The *Virginia* Class

Meanwhile, responding to the troubled submarine situation, the Congress passed the fiscal 1996 defense authorization bill that included: (1) $700 million toward construction of the third *Seawolf* (SSN 23), (2) $704.5 million for long-lead-time and advance construction costs for an attack submarine to be authorized in fiscal 1998 (i.e., the *Virginia*) and to be built at Electric Boat, and (3) $100 million in advanced construction funds for an SSN in fiscal 1999 to be built by Newport News Shipbuilding.

Further, the 1996 defense authorization bill specified that the Secretary of Defense should prepare "a detailed plan for development of a program that will lead to production of a more capable, less expensive submarine program than the submarine previously designated as the New Attack

Submarine." Although such an objective was an age-old quest, it was believed that contemporary technology could achieve such a goal. As part of this plan, the Secretary was to provide for the construction of four nuclear attack submarines that would be authorized and funded in fiscal years 1998 through 2001, "the purpose of which shall be to develop and demonstrate new technologies that will result in each successive submarine of those four being a more capable and more affordable submarine than the submarine that preceded it."

During this period the U.S. House Committee on Armed Services held extensive hearings on the future of the submarine force. Newt Gingrich, the Speaker of the House at the time, personally participated in meetings with senior Navy officials and with defense experts, both within and outside of the Department of Defense. The outcome of these efforts was congressional direction for the Navy to develop competitive prototypes for the next generation of attack submarines, with two submarines to be built at Newport News Shipbuilding and two at the Electric Boat yard. The yards were to have predominant influence in their designs, and the one judged to be the most effective would be adapted for production. It was believed that the end of the Cold War and the rapid decline of Soviet military forces would permit the delay in initiating new construction programs.

In the event, the Navy submarine community ignored this direction and developed a program in which Newport News Shipbuilding and Electric Boat would share the construction of each submarine of the new *Virginia* class. This concept, while innovative, caused higher unit costs while destroying potential competition in U.S. submarine design and construction.[46]

To counter congressional criticism over the lack of innovation in U.S. submarine design, the submarine community proposed that new systems and features be "inserted" in successive ships of the *Virginia* class rather than developing new designs that, admittedly, would have cost considerably more.

Also, the submarine community joined with the Defense Advanced Research Projects Agency (DARPA) to sponsor an extensive and objective study of future submarine concepts. This effort evolved from the findings of a Defense Science Board task force "Submarines of the Future," chaired by John Stenbit. The task force report, published in July 1998, stressed the future importance of nuclear attack submarines.[47] Among its recommendations, the task force noted:

> The next generation SSN must be a highly capable warship with rapid response capability.
>
> It should have flexible payload interfaces with the water, not torpedo tubes, VLS [Vertical Launching System] and other special purpose interfaces
>
> It should not constrain the ship and size of weapons, auxiliary vehicles, and other payloads when they are used.

The Stenbit panel was particularly concerned that existing SSN 21-inch diameter torpedo tubes, 21-inch VLS cells, and smaller-diameter countermeasures ejectors inhibited the development of submarine combat capabilities.[48] The report of the Stenbit panel led to a Navy-DARPA effort initiated in December 1998 for "A clean-slate, technically aggressive" look at future submarines, stressing (1) connectivity—the ability to communicate effectively, (2) sensors, (3) payloads, and (4) supporting platforms.[49] The Navy and DARPA provided a list of eight examples of revolutionary capabilities as a guide for contractor teams. These included: submarines towing and supporting manned or unmanned submersibles; redesigning submarines to accommodate "bomb-bay" type weapon and submersible carriage for more flexible launching; and increased use of off-board sensors. Such features, it was hoped, would permit more flexible weapon and sensor payloads, enabling future submarines to play a greater role in ballistic missile defense, task force protection, naval fire support, special operations, and even "urban warfare."

More than two years later, the two contractor teams selected to develop such concepts responded to the Navy-DARPA proposals with rather mundane solutions to the challenges. Indications were that there would be no forthcoming innovations. Meanwhile, in response to congressional pressure and professional criticism in this area, the third *Seawolf* submarine, the *Jimmy Carter*, was redesigned to carry and operate "larger payloads," primarily unmanned underwater vehicles for reconnaissance, mine detection, deep-ocean search, and other activities, as discussed above.

The Stenbit panel and the DARPA-Navy efforts addressed only the "front end" of the *Virginia* class and future submarines—"payloads and sensors." The panel felt that existing nuclear propulsion plants were "excellent and don't need improvement."[50]

Meanwhile, the submarine community ignored earlier congressional guidance to have the shipyards develop competitive prototypes. Instead the *Virginia* design was accelerated, creating the problems identified by the Baciocco panel in 1996. And, initially, there was another attempt to freeze Newport News Shipbuilding out of submarine construction. As with the *Seawolf* class, the Navy proposed building all future submarines at one shipyard. But the strong congressional delegation from the state of Virginia blocked that attempt. Subsequently, at the Navy's urgings, the two yards proposed a joint construction program, with each yard building portions of all submarines. Newport News Shipbuilding builds the bow, stern, sail, and habitability sections; auxiliary machinery; and weapons handling spaces of all units. Electric Boat builds the command and control spaces, engine room, and main propulsion unit raft for all units. Each yard builds the reactor plant module and performs final outfitting, testing, and delivery for alternate submarines.

While the Navy stated that this teaming approach would be less expensive than building all of the first four SSNs at Electric Boat—$10.4 *billion* versus $13.6 *billion*—the fifth and subsequent submarines would be more expensive under the teaming arrangement. At the time it was estimated that an SSN built at EB would cost $1.55 *billion* compared with $1.65 *billion* for a team-built submarine. In 2001—when the Navy announced a major cost increase in the *Virginia* program—the estimated costs were in excess of $2 *billion* per unit, and expected to rise further.

The design requirements for the *Virginia* called for a submarine less expensive than the *Seawolf* with the same level of quieting. The resulting design was for a submarine of 7,700 tons submerged displacement, with a length of 377 feet (114.94 m); the smaller hull is fitted with four 21-inch torpedo tubes, a regression to the *Los Angeles* torpedo armament in tube size and number. In addition, 12 vertical launch tubes for Tomahawk TLAM missiles are fitted forward.

Quieting is a more difficult characteristic to ascertain before a submarine's completion and sea trials. Propulsion is provided by an S9G pressurized-water reactor with a steam turbine providing an estimated 25,000 horsepower to a ducted propulsor (as in the *Seawolf*).

Several innovative features are provided in the *Virginia,* such as two AN/BVS-1 photonics masts, which are non-penetrating. Because they are non-penetrating, these masts provide more flexibility on sail placement, that is, they do not have to be directly above the control room. The AN/BQQ-10 Acoustic Rapid COTS Insertion (ARCI) is provided, which is an upgrade to the BQQ-5/BQQ-6/BSY-1 "legacy" sonars found in earlier SSNs.[51] Another innovation in the *Virginia* is the torpedo room that can be rapidly reconfigured, with most weapons and stowage trays removed, to provide "Tokyo hotel" style bunks for up to 50 special forces personnel.

There is little change in automation compared with previous SSNs. The *Virginia* has a planned manning requirement of 134 (14 officers plus 120 enlisted)—the same as the *Seawolf* and only a few men less than the *Los Angeles.*

While the *Seawolf* and *Virginia* introduced several new "front end" features, their reactor plants are similar to the basic design of earlier pressurized-water plants. Natural convection, which reduced the need for pumps at low power, had been developed with the S5G propulsion plant in the *Narwhal* (SSN 671). Significantly, the Naval Reactors Branch has continually extended the life of nuclear fuel cores; the *Virginia* SSN is expected to have a 30-year core life, meaning that the submarines will not be refueled during their service life.

Virtually all discussion of submarine propulsion plants within the Navy or industry had long been stifled by the U.S. nuclear submarine community. Possibly the last major effort by a U.S. Navy agency to address the issue came in 1978 when the Office of Naval Research (ONR) compiled a report on "Proposed Future Submarine Alternatives." ONR had

*The* Seawolf *at high speed on the surface. The* Seawolf *design was optimized for ASW, having eight torpedo tubes and carrying 50 torpedo-size weapons (including Tomahawk missiles). However, the lack of vertical-launch tubes limits her effectiveness in "littoral operations," the principal post-Cold War role of the U.S. Navy.* (Jim Brennan/Electric Boat Corp.)

sought ideas from industry and from qualified consultants on technologies and systems that could advance submarine capabilities. There were many responses, with ONR compiling a comprehensive report. Beyond "front-end" technologies and systems, the report included suggestions for new types of reactor plants that could lead to smaller or more-efficient propulsion systems, including reactors employing gas and liquid-metal as heat transfer media, gaseous-fueled reactors, and—possibly—fusion reactors. Also addressed were more efficient means of converting nuclear fission to propulsive power, such as gas turbines and magnetohydrodynamic systems. The report only sought to describe submarine alternatives "that should be seriously explored if the United States is to maintain submarine tactical and strategic superiority over the Soviets."

When Admiral Rickover saw a copy of the draft report, he immediately demanded that it be withdrawn from circulation. It was. A short time later he used his influence to have appointed the first nuclear submarine officer to head ONR.

## The Russian Fourth-Generation

> A nuclear submarine fleet is the future of the Armed Forces. The number of tanks and guns will be reduced, as well as infantry, but a modern navy is a totally different thing. The governments of all developed countries understand this very well.[52]

These words of Marshal Pavel Grachev, the Russian Minister of Defense, were spoken in June 1993, as the former USSR disintegrated. Within the context of the massive cutbacks of Russian military programs, three weapon areas appeared to be receiving emphasis: (1) space, (2) tactical aircraft, and (3) submarines. In all three areas there seemed to be a continuing, high tempo of research and development, at a higher rate than observed in other weapon-system areas.

At that time construction was continuing, albeit at a slow rate, on several third-generation submarines of the Project 949A/ Oscar II SSGN, Project 971/Akula I/II SSN, and Project 877/Kilo SS classes. Work had halted on two unfinished Project 945/Sierra II SSNs at Gor'kiy, while two Akula SSNs at Komsomol'sk also would not be completed.

However, design work was under way as were preparations for the construction of at least two new submarines, SSN Project 885, the *Severodvinsk* (NATO Grany), and SSBN Project 955, the *Yuri Dolgorukiy.* The decision was made to construct all future nuclear submarines at the Severodvinsk shipyard, now known as *Sevmash* (Northern Machine Building Enterprise).[53] However, Komsomol'sk in the Far East, with unfinished Project 971/Akula submarines, retained a nuclear construction capability as did the Sudomekh-Admiralty yard in St. Petersburg (formerly Leningrad).

Sevmash is the world's largest submarine construction facility, with an employment at the end of the Cold War of some 40,000 men and women.[54] Non-nuclear submarines would continue to be constructed at the Admiralty-Sudomekh complex in St. Petersburg (Leningrad).

On 21 December 1993, the keel for Project 885/*Severodvinsk* was laid down at Sevmash. Designed at the Malachite bureau by a team under the leadership of V. N. Pyalov, she was to be the lead ship for a class of SSNs. She was to carry torpedoes and tube-launched missiles, launched from probably eight 533-mm bow torpedo tubes. Reportedly the tubes were to be angled about ten degrees from the centerline to accommodate the bow sonar dome. Also fitted would be eight vertical-launch tubes, each accommodating probably four 650-mm P-800 Oniks (NATO SS-NX-26 Yakhont) anti-ship missiles.[55] The vertical tubes would be fitted aft of the sail (similar to an SSBN configuration, but probably angled forward).

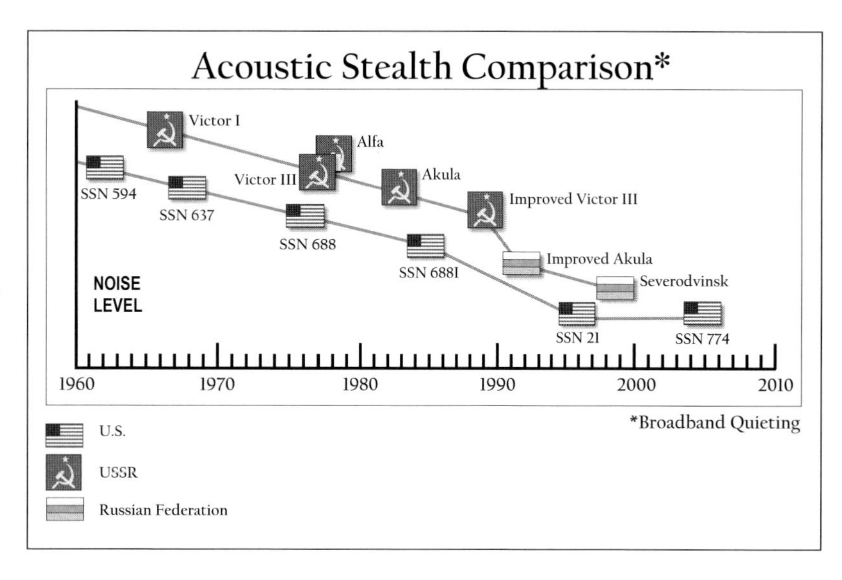


The *Severodvinsk*'s design provided for a large, spherical-array bow sonar—the first fitted in a Russian combat submarine, a feature found in U.S. submarines from the early 1960s. Improved quieting and advanced hull coatings, both anechoic and noise-attenuating, were also provided in the Soviet SSN.

The U.S. Director of Naval Intelligence, Rear Admiral Edward D. Sheafer, stated that he expected that the *Severodvinsk* "will be quieter than the improved *Los Angeles*-class submarines that we are building today." He told a journalist that this evaluation "represents a significant challenge for us. I don't think you would find any disagreement in the intelligence community over that statement."[56] A U.S. intelligence analyst added "There is some modeling data that shows in some parts of the acoustic regime the Russian submarine will be as quiet as *Seawolf*," although, he added, "by no means as quiet across the board."[57]

About that time the U.S. Navy released a chart showing the broad-band (machinery) noise levels of U.S. and Soviet-Russian submarines. The chart showed the broadband noise levels of both nations' SSNs declining up to the fourth-generation submarines. The U.S. *Seawolf* and *Virginia* were credited with the same noise level. Similarly, the Russian Project 971/Akula and *Severodvinsk* were shown with the same noise level—higher than in the U.S. *Seawolf* and *Virginia* SSNs. When shown the chart, a senior designer at the Malachite bureau asked, "Why do you assume that we would make no progress in this field in the decade between their designs?"[58]

The *Severodvinsk* was to have been followed by series production with the second submarine of the class to have been laid down at Sevmash in 1996. However, by that time work had halted on the lead submarine, although when this book went to press, some work on the hull was being reported.

A new SSBN, designed by the Kovalev team at the Rubin bureau, was laid down at Sevmash on 2 November 1996. The submarine was given the name *Yuri Dolgorukiy* (Soviet class name Borey).[59] While few details of this ship are available, indications were that it would carry 12 to 16 ballistic missiles, with a submerged displacement of some 19,400 tons. Although much smaller than the Project 941/Typhoon SSBN, the *Yuri Dolgorukiy* was to be larger than other Russian submarines except for the Oscar SSGN. The new ballistic missile submarine was to have a twin-reactor plant and a single propeller shaft, the latter feature a first for Russian SSBNs.

The *Yuri Dolgorukiy* was scheduled to be completed in 2002, with a construction of one per year planned to maintain a force level of 14 to 18 modern SSBNs through 2010, that is, these submarines and the Project 667BDRM/Delta IV submarines.[60]

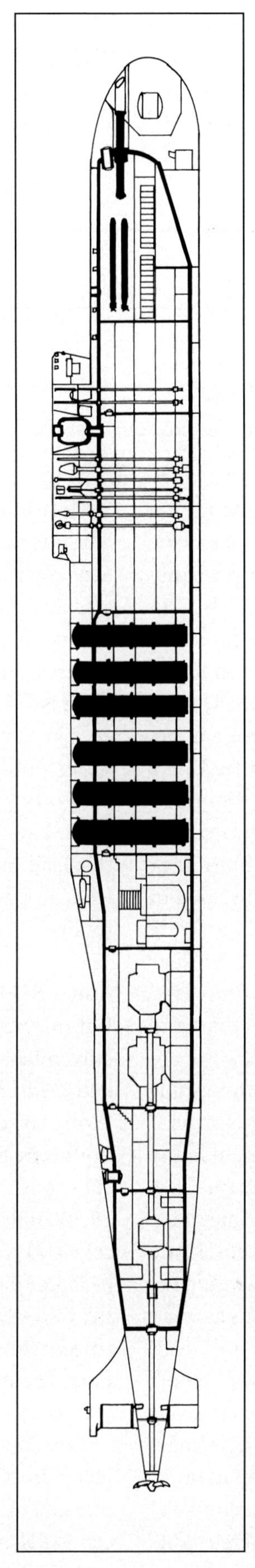

*Project 955/Yuri Dolgoruki SSBN (early design). LOA approx. 557 ft 6 in (170.0 m)* (©A.D. Baker, III)

(The Rubin bureau also began design of a fourth-generation SSGN, to be fitted with vertical-launch missile tubes, possibly for 24 anti-ship missiles. She also was expected to have a single-shaft nuclear propulsion plant.)

Construction of the *Severodvinsk* SSN halted soon after being started for lack of funds. The *Yuri Dolgoruki* SSBN was similarly halted in mid-1998 when less than 5 percent complete because of financial problems and difficulties with the planned R-39UTTH (NATO SS-N-28 Grom) submarine-launched ballistic missile. That weapon was cancelled in 1998, possibly in favor of a navalized variant of the RS-12M Topol-M ICBM (NATO SS-27).[61] The Topol-M is a three-stage, solid-propellant missile credited with a range of almost 5,650 n.miles (10,470 km), carrying a single reentry vehicle of 550 kilotons. The use of a single warhead indicates a high degree of accuracy. This is a large missile, weighing just over 104,000 pounds (47,200 kg) and is 74½ feet (22.7 m) long, with a diameter of six feet (1.86 m). The land-based SS-27 was accepted into Soviet service in December 1997, having been delayed by irregular and limited funding. The new missile required a redesign of the submarine. In 2001 the commander-in-chief of the Russian Navy, Admiral Vladimir Kuroyedov, said that the redesigned *Yuri Dolgoruki* would be completed in 2005 although even that schedule seemed unlikely when this volume went to press.

As the Rubin bureau redesigned the next-generation SSBN, the Malachite bureau began work on the post-*Severodvinsk* SSN. In the spring of 1997 the commander-in-chief of the Russian Navy, Admiral of the Fleet Feliks Gromov, visited Malachite's offices to initiate design of the new SSN, referred to as a "true" fourth-generation ship.[62]

(Lazurit, the third submarine design bureau, located in Nizhny Novgorod, formerly Gor'kiy, ceased work on military submarine designs. It now develops tourist-carrying undersea craft, offshore oil facilities, submarine rescue systems, and other noncombat projects.)

Beyond the halt in submarine construction caused mainly by economic conditions, there also were problems at Sevmash. For example, at times in the 1990s there was no money available to pay local power-plant workers at Severodvinsk, and fuel reserves were extremely low, a factor even in summer because of power and water requirements for the yard complex and the adjacent city of some 300,000.

Not only was most new construction halted, but submarine force levels also dropped precipitously, because older submarines were retired and, at a slower rate, defueled and scrapped. Many "hot" submarines were laid up with minimal maintenance, awaiting space at disposal facilities. The active submarine force was cut back to only the newest units. Long-range, out-of-area operations also were reduced, with submarine deployments being the main long-range force of the Russian Navy. The effort to maintain *a single* SSBN on deployment has been difficult, with a three-month "gap" in even this austere schedule when an SSBN suffered a fire. Ship and submarine crews—officers and enlisted alike—were having difficult times, because pay, food, and accommodations were in short supply. A 1994 statement by the chief of the Financial and Economic Directorate of the Pacific Fleet noted, "The Fleet cannot even pay for bread for the table of compulsory-service seamen, not to mention paying plants for ship repairs. In the near future complete disconnection of naval forces from energy supplies is expected."[63]

Thus ended the massive Soviet submarine programs of the Cold War.

**Neither the United States nor the Soviet Union were able to carry out its program for fourth-generation nuclear submarines. The Cold War ended with little warning (and, fortunately, with no violence between the super powers).**

**For the United States the events of 1990–1991 led to immediate cancellation of the *Seawolf* program and a slowdown in the procurement of the successor SSNs of the *Virginia* class. But even before the historic events of those years, the *Seawolf* program was in trouble. Problems with welding the HY-100 steel (not the HY-130 planned for later ships), the near-catastrophic problems with the SUBACS/BSY system, and massive cost increases contributed to the scrutiny given to the program. Once the Cold War was over and the Soviet "threat" began to dissipate, the *Seawolf* program was cut to a single submarine. Still, despite**

the demise of Admiral Rickover, the "submarine mafia" was able to coerce sufficient congressional support to construct two additional *Seawolf*-class SSNs. Subsequent events vindicated critics of the program, because costs continued to increase beyond all estimates, construction time of the *Seawolf* exceeded that of any previous U.S. submarine, delays in readying the submarine for operational service have been unprecedented, and problems have persisted with several major systems. The *Virginia* SSN has emerged from this legacy. This program, too, has been behind schedule and significantly over cost estimates, in part because of the very low production rate. With respect to combat potential, the *Virginia* will be less capable than the *Seawolf* except in quieting, the price paid in the initiation of the *Virginia* program as a lower-cost complement to the *Seawolf.*

However, the *Virginia* class was in production in the 1990s, while Soviet submarine construction was almost halted. The end of the Cold War marked the demise of the USSR as a super power. And in that environment were lost the economic infrastructure and military requirements that had sustained the world's largest and—in several respects—most advanced submarine force. Ongoing third-generation submarine programs were truncated and the fourth-generation submarines were stillborn. Research and development continued on some fourth-generation weapons and other systems, but no new submarine designs emerged from the building halls of St. Petersburg, Severodvinsk, Gor'kiy, or Komsomol'sk for more than a decade after the end of the Cold War.

It should be noted that in candid and private discussions with Western intelligence officers, submarine commanders, and technical analysts as well as with Soviet submarine designers, there has been a strong indication that Soviet fourth-generation undersea craft would have achieved performance equal or superior to most aspects of their U.S. counterparts.

TABLE 19-1
**Fourth-Generation Nuclear Submarines**

| | U.S. *Seawolf* SSN 21 | U.S. *Virginia* SSN 774 | Russian Project 885 *Severodvinsk#* | Russian Project 955 *Yuri Dolgoruki#* |
|---|---|---|---|---|
| Operational | 1997* | (2005) | — | — |
| Displacement | | | | |
| surface | 7,467 tons | | 9,500 tons | 14,720 tons |
| submerged | 9,150 tons | 7,835 tons | 11,800 tons | 19,400 tons |
| Length | 353 ft (107.6 m) | 377 ft (111.94 m) | 393 ft 7 in (120.0 m) | 557 ft 7 in (170.0 m) |
| Beam | 40 ft (12.2 m) | 34 ft (10.37 m) | 49 ft 2 in (15.0 m) | 44 ft 3 in (13.5 m) |
| Draft | 35 ft (10.67 m) | 30 ft 6 in (9.3 m) | 32 ft 10 in (10.0 m) | 29 ft 6 in (9.0 m) |
| Reactors** | 1 S6W | 1 S9G | 1 OK-650KPM | 2 OK-650b |
| Turbines | 2 steam | 2 steam | 1 steam | 2 steam |
| horsepower | approx. 40,000 | approx. 25,000 | 43,000 | ~90,000 |
| Shafts | 1 (pump-jet) | 1 (pump-jet) | 1 (pump-jet) | 1 |
| Speed | | | | |
| surface | | | 16 knots | 15 knots |
| submerged | 35 knots | 25+ knots | 31 knots | 29 knots |
| Test depth | 1,300 ft (396 m) | | 1,970 ft (600 m) | 1,475 ft (450 m) |
| Torpedo tubes*** | 8 670-mm A | 4 533-mm A | 4 to 8 533-mm B | 4 533-mm B |
| Torpedoes/missiles | 50 | 37 | 24 | 12 |
| Vertical-launch tubes | nil | 12 | 8 (32) SS-NX-26 | 12 SS-N-28 |
| Complement | 134 | 134 | ~50 | ~100 |

Notes: * Commissioned in 1997; first operational deployment in 2001.
** See Appendix C for U.S. nuclear plant designations.
# Unofficial estimates.
*** Angled + Bow.

20

# Soviet Versus U.S. Submarines

*Soviet and U.S. submarines had a major role in the Cold War. These Project 641/Foxtrot submarines and their crews ranged far and wide in support of Soviet political and military interests. Like many of their American counterparts, key features of these submarines can be traced to the German Type XXI.*

Submarines were an important component of the military and political strategies of both the Soviet Union and the United States during the Cold War. At the start of this period both nations saw their submarines as traditional naval weapons, for use against enemy warships and merchantmen, and both rebuilt their submarine forces on the basis of German technology, primarily the superlative Type XXI U-boat. The Type XXI design included hull streamlining, array sonar, minimum conning tower fairwater, large electric batteries, deletion of deck guns, and other advanced features. While the Soviet Union expanded its prewar development of closed-cycle propulsion systems to include German technology, the United States gave little attention to this technology, instead pushing forward with nuclear propulsion, an effort that had its beginnings in the U.S. Navy as early as 1939.[1]

The USS *Nautilus* (SSN 571), the world's first nuclear-propelled vehicle, went to sea in January 1955. Three and a half years later the Soviet *K-3* was underway on nuclear power. The interval had been less than the four years between detonation of the first U.S. atomic bomb and the first Soviet atomic explosion. These intervals are significant because the USSR received extensive information from spies on the U.S. atomic bomb project, while there is no evidence that significant classified information on U.S. nuclear submarine development was obtained by the Soviets.

Both the Soviet Union and United States were highly innovative in their early Cold War submarine designs, those with diesel-electric propulsion as well as nuclear (and closed-cycle in the Soviet Navy). This thesis is easily demonstrated by looking at the numbers and multiplicity of early submarine designs, power plants, and weapons. However, following the loss of the USS *Thresher* (SSN 593) in 1963, the U.S. Navy became very conservative in submarine design, construction, and, to some degree, in operations. This conservative approach in design can be seen, for example, in the extended dependence on HY-80 steel, first employed in the *Skipjack* (SSN 585) completed in 1959 and retained through the *Los Angeles* (SSN 688) class, including the Polaris SSBNs.[2] Not only was there no increase in U.S. submarine operating depth, but there also was a reduction in depth for the 62 ships of the *Los Angeles* (SSN 688) series.[3] The decision to retain HY-80 steel rather than higher-strength HY-100 or HY-130 in U.S. submarines was dictated by their becoming "weight critical" because of increasing propulsion plant size, and difficulties in working the higher-strength steels. The penalties for this course of action have not only been restrictions in operating depth, but also a possible reduction in pressure-hull shock resistance, reduction in the number of watertight compartments, a reduction in reserve buoyancy, and a minimal weight margin to accommodate future growth. In this same period Soviet submarine operating depths increased significantly.

The *Seawolf* (SSN 21) introduced HY-100 steel, with some difficulties. The use of that steel in the *Seawolf* and *Virginia* (SSN 774) classes probably return those submarines to a 1,300-foot (400-m) test depth.

With respect to conservatism in operations, following the loss of the *Thresher*, no U.S. submarines undertook under-ice operations for five years. At the same time there were restrictions on operating depths until the submarines had undergone the various SUBSAFE modifications. Significantly, intelligence collection operations by U.S. submarines did become more aggressive during this period; those activities were directed principally by the National Security Agency, the Office of Naval Intelligence, the Forty Committee of the National Security Council, and executed by the submarine force commanders.[4]

The conservatism in U.S. submarine programs and policies was primarily caused by Admiral H. G. Rickover. After the *Thresher* loss he became more conservative in his approach to submarine design and construction, while at the same time his influence and authority in those areas increased significantly. Rickover told a congressional committee:

> My program is unique in the military service in this respect: You know the expression "from womb to the tomb"? My organization is responsible for initiating the idea for a project; for doing the research, and the development; designing and building the equipment that goes into the ships; for the operation of the ship; for the selection of the officers and men who man the ship; for their education and training. In short, I am responsible for the ship throughout its life—from the very beginning to the very end.[5]

Under Rickover's supervision the quality control in U.S. submarine construction was improved significantly. And, although under his direction the design and construction of nuclear reactor plants continued to be highly conservative, they incorporated a much higher degree of safety than that of their Soviet counterparts, and reactor plant safety procedures and engineer personnel training were far more demanding than those of the Soviet Navy.

Early in the Cold War there were largely independent submarine design groups at the Electric Boat yard, the Mare Island Naval Shipyard, and the

Portsmouth Naval Shipyard. In the nuclear era the yards' design staffs soon were stripped of their quasi-independent status, with the Bureau of Ships and its successor organizations in Washington assuming total direction of submarine design. In the 1960s Admiral Rickover gained de facto control of this centralized effort.

With his increasing control of submarine design, Rickover and his subordinates could also influence U.S. naval intelligence evaluations of Soviet submarine developments. Most participants in these evaluations have declined to discuss publicly the debates between the Naval Reactors Branch and the intelligence community. One submarine analyst, however, has publicly addressed Rickover's reluctance to accept that the Soviets were building high-speed, titanium-hull submarines. Gerhardt B. Thamm wrote: "Only after the CIA graciously signaled that it would publish this analysis no matter what ONI's [Office of Naval Intelligence] position, did NISC [Naval Intelligence Support Center] rush to publish its Alfa-class SSN study."[6]

Addressing the desire of Admiral Rickover's office to exaggerate the Soviet progress in quieting, Thamm recalled:

> I was once pressured by a very senior nuclear submariner to state at a conference that the Victor III-class SSN was engineered to be quieter than the Victor I-class SSN. I refused, because the consensus of the NISC acoustic analysts, the NISC photo analysts, and the NISC submarine analysts, as well as CIA and Defense Intelligence Agency analysts, was that the Victor III's engineering spaces were essentially the same as those of the Victor I's. The parting shot from the admiral was, "Well, I don't believe what you spooks say anyway!"[7]

Within the Soviet Union, several submarine design bureaus had the responsibility for the design of submarines during the Cold War. (See Appendix D.) The size of the design bureaus was massive by Western standards; at the end of the Cold War the Rubin bureau had some 3,000 employees, the Malachite bureau some 2,500, and the Lazurit bureau some 1,500 employees.

Each bureau was headed by a chief designer, a title changed in the mid-1980s to general designer (*glavni konstruktor*), reflecting the increased complexity of submarines and the increase in authority of the chief designers. These men had a high degree of independence in the management of their bureaus. Although they reported to the Ministry of Shipbuilding and the Defense-Industry Commission, the Navy Ministry, especially the long-serving Admiral S. G. Gorshkov, specified submarine and system characteristics and had a great influence on the bureaus.

Although the submarine design bureaus were largely specialized with regard to submarine types, they were in direct and, at times, intensive competition. This competition and the large number of related technical and scientific research institutes (somewhat akin to the U.S. Navy's laboratory complex) were a major factor in the advances in Soviet submarine development that led to superiority over U.S. undersea craft in several key performance criteria.

The effective expenditure of money was not a major criterion in the Soviet system. Funding was more of a consideration in U.S. programs because of the multicommittee, two-party Congress providing funds; the monitoring of programs by the Congressional Budget Office, Congressional Research Service, and General Accounting Office; and press coverage of defense programs. All of these activities provided checks and monitoring of expenditures and programs.

There is only limited information available to the public about the political machinations of Soviet submarine design and construction. Obviously, Admiral Gorshkov, holding the dual positions of deputy minister of defense and Commander-in-Chief of the Navy for almost 30 years—from January 1956 until December 1985—had enormous influence on Soviet naval programs.[8] He supported and garnered political and fiscal support for large submarine programs and for large submarines. But unlike Rickover, Gorshkov was engaged in developing and directing an entire navy, not just a single branch.

With respect to Soviet submarine design and construction, one man has had a very strong voice in the decision-making process, Academician Igor D. Spassky, the head of the Rubin design bureau

*Admiral of the Fleet of the Soviet Union S. G. Gorshkov* (U.S. Navy)

*Igor D. Spassky* (Rubin CDB ME)

since 1974.[9] His influence and his ability to outmaneuver his counterparts at other design bureaus (which have had several leadership changes during Spassky's tenure), as well as authorities at the First Central Research Institute and the Ministry of Shipbuilding, are reflected in his success in obtaining approval for series production of Project 941/Typhoon SSBNs and Project 949/Oscar SSGNs. Production of those massive submarines was opposed by some senior naval officers, ministry officials, and industrial competitors. Their objections were based primarily on cost and resource availability, and there was the continuing question of the relationship between the size and intensity of various physical fields (signatures) generated by large submarines. There is evidence that the First Central Research Institute had recommended smaller submarines specifically to reduce their physical fields.[10] Papers by other Soviet submarine designers have supported this position.[11]

But Spassky's views have prevailed as he demonstrated a Rickover-like political acumen during his long career as head of the Rubin design bureau.

The early limitations of Soviet technology and quality control led to considerable redundancy in the propulsion plants of the first two generations of Soviet nuclear submarines.[12] This included retaining twin reactor plants and twin propeller shafts. The double-hull configuration, with major internal compartmentation, was retained in Soviet submarines. The double-hull configuration provided additional protection for Soviet submarines against weapon impact or collisions with ice or with other submarines. Their double hulls, with ballast tanks distributed along the length of the submarine, as well as major internal compartmentation (significantly reduced in U.S. submarines), have enabled Soviet submarines that suffered major casualties to reach the surface and, often, to survive. This feature was part of the Soviet concept of "surfaced unsinkability"—the ability of a damaged submarine, with at least one compartment and adjacent ballast tanks flooded, to remain afloat after reaching the surface and retain stability and trim.[13] Double-hull construction also provided multiple hull surfaces (up to four) for anechoic and internal noise-attenuating coatings.

There were also "cultural" rationales for these characteristics. The Soviet Navy in World War II had largely operated in coastal waters, where there was considerable danger from mines. Similarly, U.S. sonar advantages intimated that U.S. submarines would fire the first shot. Thus Soviet submarine designers had more "cultural" concerns about survivability than their American counterparts.

(Double-hull submarine construction does have disadvantages, such as increased corrosion and maintenance problems, increased weight and hydrodynamic surface area, and possibly less crew habitability within smaller pressure hulls.)

Beyond submarine designs, the USSR pursued a much broader approach to submarine weapons development. Cruise missiles—originally intended for land attack—were adapted for the anti-carrier role. Torpedo development was also broader than U.S. efforts (e.g., wake-homing torpedoes, rocket-torpedoes, deep-depth launch).

## Differences

The asymmetry of naval forces, available resources, and national military policies saw a much greater Soviet emphasis on submarines for most of the Cold War. Although major surface warship programs were initiated in the USSR in the late 1940s (under Josef

Stalin) and again from the late 1960s onward (under Leonid Brezhnev), the most important—and threatening—component of the Soviet Navy was the submarine. Under the shipbuilding program of the late 1940s, submarines were envisioned as a component of a "balanced fleet," to be comprised of large and small surface warships, submarines, and a large land-based aviation component, with aircraft carriers in the planning stage. When that shipbuilding program was aborted upon the death of Stalin in March 1953, submarines and, subsequently, small, missile-armed surface ships and land-based aircraft were promoted by Nikita Khrushchev, who considered them a more effective and affordable alternative to large surface warships.

*Commander Thomas A. Jewell guides his command, the USS* John C. Calhoun *(SSBN 630), alongside a submarine tender. U.S. submarine personnel were drawn from a well-educated population base, and were given excellent education and training for submarine service.* (U.S. Navy)

Should a conventional war have been fought between NATO and the Soviet Union early in the Cold War, the large number of torpedo-attack submarines and anti-ship cruise missile submarines would have had the anti-shipping role, to fight a new Battle of the Atlantic against Anglo-American merchant shipping. Counting some older boats of wartime design, the massive Soviet submarine program provided a peak strength of about 475 submarines by 1958. (This number was often compared to Germany starting the Battle of the Atlantic in September 1939 with 47 oceangoing submarines.)

This principal role of Soviet submarines soon changed to the anti-carrier role as, from the early 1950s, U.S. aircraft carriers were fitted to launch nuclear strike aircraft against the Soviet homeland. This defensive role expanded in the 1960s to the anti-SSBN role, while the debut of the Project 667A/Yankee strategic missile submarine in 1967 added the strategic offensive mission (including protection of Soviet SSBNs) to the submarine force.

The considerable momentum in submarine design and production built up during the Khrushchev regime (1953–1964) continued under the Brezhnev regime (1964–1982), but with the submarine program sharing naval resources with the large surface fleet being constructed. This emphasis was reflected in the retention of a relatively large Soviet undersea force, more innovative submarines (designs

*Outdated Soviet submarines were converted for a variety of roles. This Project 658M/Hotel II SSBN has been reconfigured as a sensor test and evaluation ship, part of the Soviet Navy's continuing ASW effort.* (U.K. Ministry of Defence)

both built and not built), and, in several respects, performance superior to their U.S. counterparts.

Those submarine roles generated in the 1950s and 1960s continued through the end of the Cold War. And, large numbers of submarines were designed and constructed to carry out those roles. In addressing submarine design and construction during the Cold War several related aspects of submarine activities must be considered, among them: defense industry, manpower, and operations. All had a direct impact on submarine design and construction.[14]

*Industry.* Soviet submarine designers proved to be competent and innovative. However, in the early stages of the Cold War the Soviet defense industry suffered greatly from the need to rebuild from the devastation inflicted by the German invasion of the USSR as well as the need to rebuild other parts of Soviet industry and society.[15]

Even after the massive reconstruction, the USSR suffered from a second-rate industrial infrastructure. Sergei Khrushchev, a missile guidance engineer, told of the early Soviet failure to effectively reproduce and manufacture their version of the U.S. Sidewinder air-to-air missile. Nikita Khrushchev, his father, had asked, "Since they've organized mass production in America, they work without producing defective parts. What's keeping us from doing the same?"[16]

A committee was formed to answer Khrushchev's question:

> Electronics experts blamed machine builders for the lack of the equipment they required. When the machine builders were asked for their explanation, they complained about the poor quality of metal. They couldn't do any better with the metal they were given. Metallurgists simply threw up their hands. The quality of ore they were given was so bad that the machine builders should be thankful for what they got. The mining industry in turn referred to the poor quality of equipment supplied by the machine-building industry for processing ores. That closed the cycle.
>
> Everyone gradually got used to the unbelievable percentage of defective parts, which slowly decreased as people mastered production techniques.[17]

The issue of poor material quality plagued every aspect of the Soviet armed forces. Subsequent improvements, which in some respects produced submarine hull materials (steels and titanium) superior to those of the United States came about during the Cold War. But there were other problems. Equipment was rushed into production without thorough testing, the infrastructure to support a large submarine fleet—with adequate spares, maintenance personnel, and facilities—was not developed, and the large number of different submarine designs and weapons exacerbated the other problems while increasing costs. Coupled with the personnel limitations in the submarine force, these problems resulted in reduced readiness and major operational problems. These material and personnel situations combined to cause the loss of several diesel submarines as well as nuclear-propelled submarines.

Another, related, factor was the intense specialization of Soviet engineers. Sergei Khrushchev described this as the "stove pipe" system wherein Soviet engineers gained great depth of knowledge as they became more senior, but were relatively narrow in scope as they concentrated in a single field of endeavor. American engineers were more broadly educated and had much greater professional/job mobility than their Soviet counterparts.[18] Both systems had advantages and disadvantages; but the differences were significant and affected the products of their respective societies. In undersea warfare, Soviet submarines tended to be more special purpose craft; U.S. submarines were more multimission craft.

The American defense industry benefited during the Cold War from a relatively high quality of component products and access to a large, high-technology industrial base, in part related to high consumer demands (e.g., electronics, appliances, and, later, computers). Further, the relative openness of the U.S. society permitted a greater "cross-fertilization" of industries while private/corporate ownership could permit more rapid responses to military development and procurement needs.[19]

Despite greater quality control, the U.S. submarine industry was not immune to material failures, to a large degree because of personnel shortcomings. These failures, and especially poor quality control, were a constant target of Admiral Rickover and his demands for the highest possible quality in submarine construction. Material failures were responsible for the loss of U.S. and Soviet submarines. The USS *Thresher* suffered material casualties that caused her loss; she also had a design flaw in the context of her inability to blow to the surface from test depth. The five Soviet nuclear submarine losses mostly appear to have been caused by a combination of material and personnel failures.

*Manpower.* The large size of the Soviet submarine force placed heavy demands on personnel training and assignment within a conscript system that was always hard pressed to provide an efficient military force. Working with short-term, conscripted enlisted men and mostly career officers, the development of a coherent, efficient manpower base was difficult. Those problems plagued the Soviet submarine force until the demise of the USSR.

Senior Soviet officers operated within a system that valued quantity over quality. They touted statistics and stressed appearances over reality. This in turn encouraged "gun-decking"—faking data or falsifying reports—and coverups rather than initiative and the exposure and correction of shortfalls. The extent to which this situation was understood by senior officers and submarine designers contributed to the large percentage of officers in submarine crews and to adoption of a high degree of automation for Soviet submarines. Senior designers believed that removing the sailor from the "control loop" was a key to ship safety; the U.S. Navy held the opposite view.[20]

For the first 15 years of the Cold War the U.S. submarine force was manned by a relatively small, elite, all-volunteer group.[21] Subsequently, the Polaris program brought about a crisis in submarine manning. In a six-year period 82 large (136-man), highly trained crews were needed for the Polaris force, plus crews for some 20 new attack submarines of approximately 100 men each. These requirements placed stress on all aspects of personnel recruitment, training, and assignment for the U.S. submarine force. Admiral Rickover's personal and dogmatic interview and selection policies created a bottleneck, exacerbating the manning difficulties; in reality officers and some senior enlisted men were "drafted" into the submarine force in the 1960s. It can be argued that the U.S. submarine force never recovered from the officer shortfalls and policies of that period until the massive submarine force reductions at the end of the Cold War.

The U.S. Navy has not sought to solve the manpower problems by adopting large-scale automation in submarines. Recent U.S. submarine classes have had a crew reduction of only about 5 percent over more than two and a half decades. This situation is apparently satisfactory; the head of the naval nuclear propulsion program in 2001, Admiral Frank L. Bowman, stated:

> if you could walk on board tomorrow's *Virginia,* you'd wonder where you are. There are fewer components—fewer *primary* components. Systems you worked long and hard to qualify on, are not even installed. There are fewer watchstanders.[22]

In the same period, Soviet submarine manning declined significantly. The Soviet Project 945/Sierra SSN was manned by some 60 men—less than one-half the number in contemporary U.S. attack submarines.

Perhaps the more interesting comparison of manning levels is between U.S. Trident missile submarines of the *Ohio* (SSBN 726) class and the Soviet cruise missile submarines of Project 949/Oscar design. Note that the Soviet undersea craft has a two-reactor plant compared to the single-reactor plant in the *Ohio.*

While Soviet submarine manning was lower than in their U.S. counterparts, the former submarines required many more maintenance and support personnel ashore. Also, Soviet submarines required a much higher percentage of officers and warrants than did U.S. submarines, primarily because of the severe limitations in the training and experience of short-term conscript enlisted men.[23] Still, men aboard ship are more expensive to pay, train, and provide for than are those at shore bases.

*Operational.* A comparison of Soviet and U.S. submarine programs also must address operating policies. U.S. torpedo-attack submarines as well as ballistic missile submarines spent a much higher percentage of their time at sea than did their Soviet counterparts. Beginning in the early 1960s, the United States endeavored to keep more than one-half of its SSBN force at sea at any time. And, almost one-third of the SS/SSNs were kept at sea during much of the Cold War.

In comparison, some 15 percent of the Soviet Union's combat submarines were normally at sea during much of the Cold War era. Much of the remainder of the force was kept in a high state of readiness so that, with little strategic warning, a large force of submarines could be sent to sea. This was demonstrated in several large-scale exercises. The policy reflected the Soviet view that their intelligence activities could provide adequate warning of possible conflict with the West and that rapidly generating a large, fully armed and provisioned force was more effective than keeping a higher percentage of the submarines at sea on a continuous basis. The relatively low amount of time at sea by Soviet submarines reduced the at-sea training for personnel and was

TABLE 20-1
**Submarine Manning**

| Class | Launched | Displacement* | Weapons** | Manning |
|---|---|---|---|---|
| SSN 688 *Los Angeles* | 1974 | 6,927 tons | 25 | 141 |
| SSN 688 *Los Angeles* (VLS***) | 1986 | 7,102 tons | 37 | 141 |
| SSN 21 *Seawolf* | 1995 | 9,137 tons | 50 | 134 |
| SSN 774 *Virginia* | 2005# | 7,835 tons | 37 | 134 |
| SSN Project 945 Sierra | 1983 | 8,300 tons | 40 | 60 |
| SSN Project 971 Akula | 1982 | 10,700 tons | 40 | 73 |

Notes: * Submerged displacement.
** Torpedoes and cruise missiles.
*** Vertical-Launching System.
# Projected.

*Soviet submariners observe their observers. The Soviet Navy was plagued by a system of conscript enlisted men, integrated with career officers and* michmen, *or warrant officers. Coupled with poor shore support and facilities—especially housing —it is a tribute to these men that they performed so well.* (U.K. Ministry of Defence)

reflected in their reduced operational experience in comparison to their American counterparts.

Related to operational issues, Soviet ship repair facilities never kept pace with submarine construction. In some cases, new designs with unique maintenance and support requirements were in the fleet for a decade or more before required facilities became available. "The shortage of ship repair facilities drove the utilization of Soviet submarines," according to a U.S. intelligence official. "They had no place to put them for overhauls . . . it was a driving factor in their deployment schedules."[24]

Further, the multiplicity of Soviet submarine classes greatly complicated the production as well as support of the submarine force. Several Soviet naval officers and designers criticized this approach to submarine design and construction; one officer, then-Captain 3d Rank Georgi I. Sviatov, went so far as to meet in 1965 with Dimitri F. Ustinov, at the time first deputy chairman of the influential Council of Ministers and the longtime head of armaments production, to criticize the multi-ship programs.[25] Using the U.S. Navy as an example, Sviatov calculated that the USSR could have a submarine force of twice its size for the same cost if only two nuclear submarine classes were produced, and if both used the same reactor plant.

## The End of the Cold War

The end of the Cold War brought a major cutback in U.S. submarine building and a virtual halt to Soviet submarine construction. No U.S. submarines were authorized for construction from fiscal year 1992 (*Jimmy Carter*/SSN 23) through fiscal 1998 (*Virginia*); that six-year hiatus was the longest period in U.S. submarine history in which no new submarines were authorized.

In the Soviet Union there was a massive cutback in naval operations as well as all military and naval acquisition programs as government policies began to shift in the late 1980s. The Navy announced in 1989 that the surviving first-generation (HEN) nuclear-propelled submarines would soon be retired as would some second-generation craft. That year nine submarines were launched, and there were eight submarine classes in series production.[26] Despite the political upheavals in the Soviet Union under Mikhail Gorbachev (e.g., perestroika), at the time there was no indication that there would be any near-term impact on the submarine programs.[27]

TABLE 20-2
**Missile Submarine Manning**

| Class | Launched | Displacement* | Weapons** | Manning |
|---|---|---|---|---|
| SSBN 726 *Ohio* | 1976 | 18,750 tons | 4+24 | 163 |
| SSGN Project 949 Oscar | 1983 | 22,500 tons | 8+24 | 107 |

Notes: * Submerged displacement.
** Torpedo tubes + ballistic/cruise missiles.

*Another example of submarine "recycling" was the use of retired Hotel SSBNs and Project 629/Golf SSBs in the communications relay role (SSQN/SSQ). This is an extensively converted Project 629R/Golf SSQ.*

But following the collapse of the Soviet Union in 1991, there were massive layups of submarines until, by the mid-1990s, there were, at times, difficulties in keeping a single SSBN on patrol. From the end of the Cold War through 2002 two new-design Soviet-Russian nuclear submarines were laid down, the SSN *Severodvinsk,* in 1993, and the SSBN *Yuri Dolgorukiy,* in 1996. When this book went to press, neither submarine had been launched, and it was likely that they would be replaced in the building halls at Sevmash (Severodvinsk) by more-advanced designs. There was some progress in non-nuclear submarine development, in part because of the promise of foreign sales that could help subsidize those programs. (See Chapter 13.)

Little has been said about the potential successors to those "3+" or "3.5"-generation submarines. A book published in 1988 offered some tantalizing glimpses into the thinking of the Soviet Navy's leadership. *The Navy: Its Role, Prospects for Development and Employment* was written by three naval officers with a foreword by Admiral Gorshkov (who died the year the book was published).[28] The book appeared as Gorbachev was restructuring the Soviet bureaucracy and shifting economic priorities. The 1985–1991 five-year plan already had been decided, and *The Navy* sought to make a strong case to Gorbachev's civilian advisors and other important sectors of Soviet society for funding subsequent naval programs and force levels.

In discussing prospects for submarine development, the authors predicted continuing improvement in all aspects of undersea craft (albeit often referring to "the foreign press" as a data source, a common ploy when addressing classified or sensitive subjects). *The Navy,* perhaps optimistically, predicted "in the near future":

- operating depths of 6,560 feet (2,000 m) or greater
- interconnected spherical pressure hulls for deeper operating depths[29]
- major reductions in "physical fields" (i.e., detectable characteristics of the submarine)
- 50 to 60 knot speeds (and over 100 knots in the longer term)
- increased reactor plant output with reduction in specific weight (with gas-cooled and single-loop reactors being cited)
- unified engine-propulsor units operating on the principal of hydrojet engines with a steam plant
- decreasing water resistance through improved hydrodynamic properties, including the use of polymers[30]
- torpedo speeds of up to 300 knots
- improved ballistic and cruise missiles for various roles
- consideration of "arming" submarines with *manned aircraft* [emphasis in original]
- increased automation

The conclusion of this section of *The Navy* included this summary:

> The traditional name "submarine" hardly will be applicable to them. These will be

*The beginning of the end: The Project 941/Typhoon SSBN* TK-202 *being dismantled at the Zvezdochka shipyard at Severodvinsk. This April 2000 photo shows that the anechoic tiles having been partially stripped. Air conditioning and dehumidification piping cover the deck. These submarines will be remembered as the world's largest undersea craft.* (U.S. Navy)

> formidable nuclear powered ships invulnerable to the action of other forces and capable of monitoring enormous ocean expanses, deploying covertly and rapidly to necessary sectors, and delivering surprise, powerful strikes from the ocean depth against a maritime and continental enemy. Their unlimited operating range in combination with high speed makes them an almost ideal means for military marine transportation and for accomplishing other missions.

While the strategic or tactical value of some of these characteristics could be questioned, all would have had some military value. Could they have been achieved? If one looks at the start dates of such advanced submarines as the Project 705/Alfa and such weapons as the Shkval torpedo in comparison with contemporary Western undersea projects, the answer is certainly yes *for some features.* Work proceeds on advanced undersea craft at the surviving Russian submarine design bureaus and related institutes and laboratories. Although their staffs have been severely reduced, and even the remaining employees are not always paid on time, the momentum of some of these programs and the enthusiasm of many of these men and women continue.

Significantly, there were several omissions in submarine design goals presented in *The Navy.* While the Soviets emphasized quieting, no major breakthroughs were envisioned by Western intelligence.[31] But the Soviets had been pursuing non-acoustic detection means as well as "non-acoustic quieting" of their submarines and making improvements to virtually all submarine signature characteristics.

The trends were ominous. And one is reminded of then-Secretary of Defense James R. Schlesinger's statement of 30 October 1975 about the Soviet military buildup:

> If we are to maintain a position of power, the public must be informed about the trends. Some years from now, somebody will raise the question, why were we not warned, and I want to be able to say indeed you were.

**By the end of the Cold War, the United States claimed superiority in only two areas of subma-**

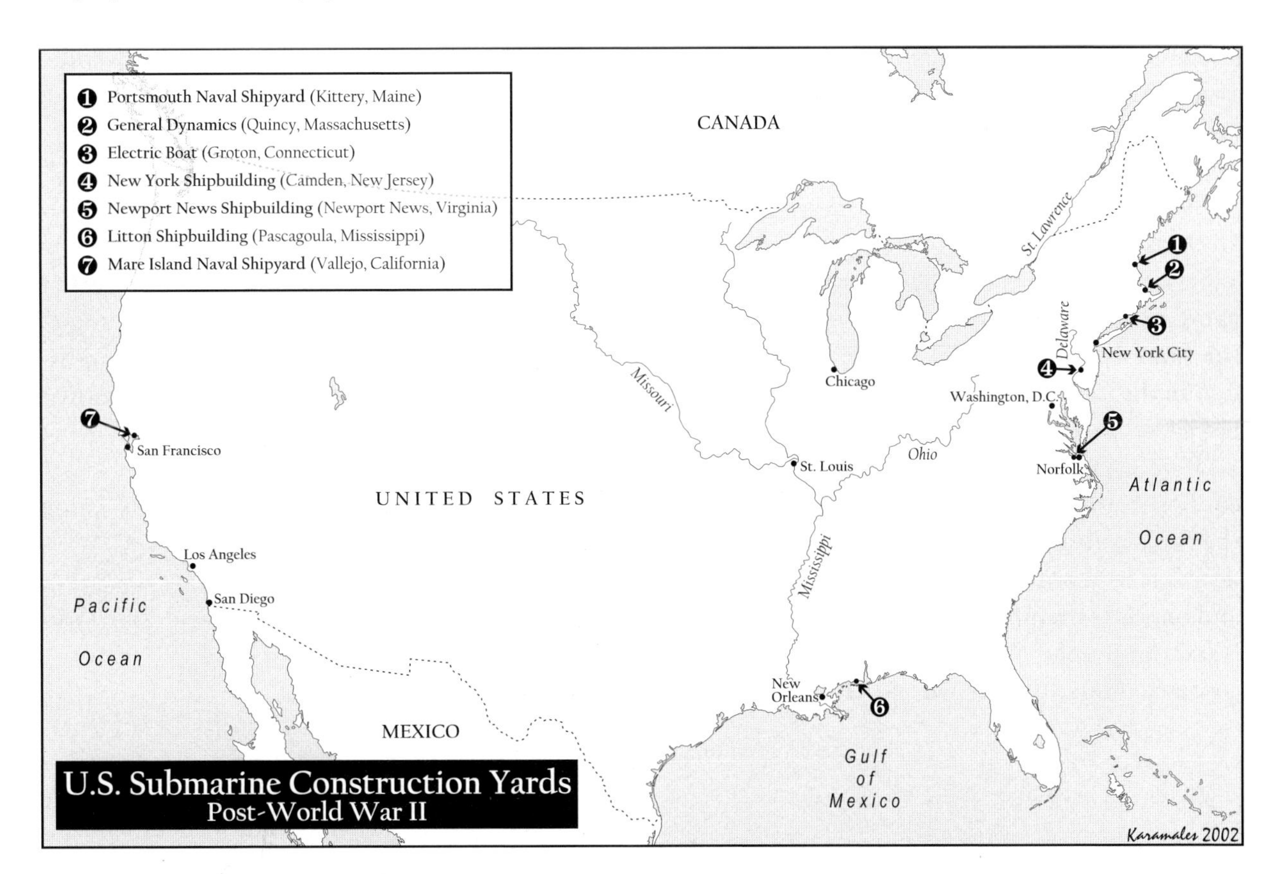


U.S. Submarine Construction Yards
Post-World War II

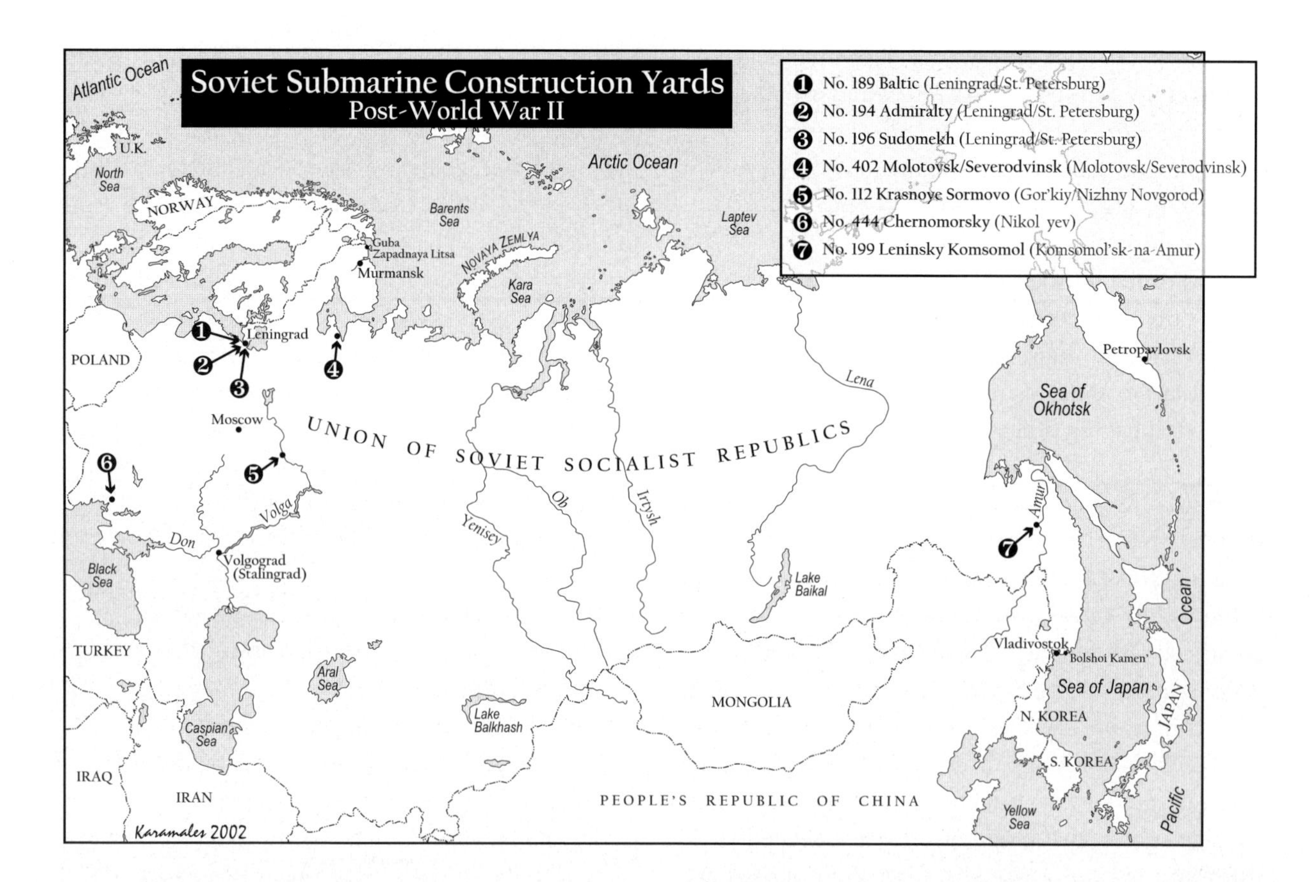


rine technology: passive sonar performance and acoustic quieting. When the Soviet Union collapsed, the latest Soviet submarines had achieved the U.S. level of acoustic quieting at low speeds and were ahead of U.S. undersea craft in "non-acoustic quieting" (e.g., wake generation, magnetic signature, hull flow noise).

From an operational perspective, it is apparent that the training and extensive operational experience of U.S. submariners has produced a cadre of submarine personnel without peer. At the same time, personnel failures—while existent in the U.S. force—were far more prevalent in Soviet submarines.

Since it appears unlikely that there will be a conflict between the United States and Russia in the foreseeable future, especially in view of the decrease in Russian submarine production rates and operating tempos, it is unlikely that U.S. and Russian submarines will ever be measured in combat. Thus, the answer to the question of which approaches to submarine research and development, design, construction, manning, training, support, and operations are superior may never be answered, or at least not until some time in the future when currently classified information is revealed.

Still, when considering the industrial, manpower, and operational limitations of the Soviet state, the Soviet achievements in submarine design and construction appear even more impressive. In discussing those achievements at the Malachite submarine design bureau in St. Petersburg, one of the Russian engineers leaned across the conference table and asked one of the authors of this book:

"Do you know how this situation came about?"

In response to our puzzlement he declared: "We had competition in submarine design. You [in Rickover] had Stalinism!"

# APPENDIX A: U.S. Submarine Construction, 1945–1991

*The USS* Nautilus *(SSN 571): The world's first nuclear-propelled submarine, shown in early 1955, shortly after being placed in commission.* (U.S. Navy)

U.S. submarine programs reached hull number SS 562 during World War II, with hulls 526–562 being cancelled late in the war. Subsequently, five of these numbers were assigned to postwar submarines, including two built in foreign shipyards:

| | |
|---|---|
| SS 551 | changed to SSK 2 |
| SS 552 | changed to SSK 3 |
| SS 553 | Norwegian *Kinn* (built with U.S. funds) |
| SS 554 | Danish *Springeren* (built with U.S. funds) |
| AGSS 555 | *Dolphin* (research submarine) |

All U.S. submarines built from 1900 until 1996 were numbered in a single series with the exception of six "small," initially unnamed submarines (SSK and SST series and the midget *X-1*).[1] Later the submarine community adopted the designation SSN 21 for the design that evolved into the *Seawolf,* the "21" having indicated an SSN for the 21st Century. Thus, the submarine hull numbers 21 to 23 were repeated. In addition, the U.S. Navy constructed one nuclear-propelled submersible, the *NR-1,* which is in the category of a "service craft" and not a ship.

The designations of numerous SS-type submarines were changed after modification for specialized missions, always retaining the same hull number; other specialized submarines were built with modified hull numbers within the same numbering system (e.g., AGSS, SSBN, SSG, SSR, SSRN). Several SSBN-type submarines had their missile systems deactivated and were employed as attack submarines and special operations transport submarines, both types reclassified as SSN.

The following table lists U.S. submarines placed in commission from 15 August 1945 to 31 December 1991. Submarines built in the United States specifically for foreign transfer are not included. The Ship Characteristic Board numbers were a means of tracking post–World War II ship designs.

| Built | Hull No. | Name/Class | SCB[2] | Commission | Reactors[3] |
|---|---|---|---|---|---|
| **Non-Nuclear Submarines** | | | | | |
| 13 SS | 285 | *Balao* class | — | 1945–1948 | — |
| 10 SS | 417 | *Tench* class | — | 1946–1951 | — |
| 1 AGSS | 555 | *Dolphin* | 207 | 1968 | — |
| 6 SS | 563 | *Tang* class | 2 | 1951–1952 | — |
| 1 AGSS | 569 | *Albacore* | 56 | 1953 | — |
| 2 SSR | 572 | *Sailfish* class | 84 | 1956 | — |
| 2 SSG | 574 | *Grayback* class | 161 | 1958 | — |
| 1 SS | 576 | *Darter* | 116 | 1956 | — |
| 3 SS | 580 | *Barbel* class | 150 | 1959 | — |
| 3 SSK | 1 | *K1* class | 58A | 1951–1952 | — |
| 2 SST | 1 | *T1* class | 68 | 1953 | — |
| 1 Midget[4] | | *X-1* | — | 1955 | (closed-cycle) |
| **Nuclear-Propelled Submarines** | | | | | |
| 1 SSN | 571 | *Nautilus* | 64 | 1954[5] | STR/S2W |
| 1 SSN | 575 | *Seawolf* | 64A | 1957 | SIR/S2G[6] |
| 4 SSN | 578 | *Skate* class | 121 | 1957–1959 | S3W or S4W[7] |
| 6 SSN | 585 | *Skipjack* class | 154 | 1959–1961 | S5W |
| 1 SSRN | 586 | *Triton* | 132 | 1959 | (2) S4G |
| 1 SSGN | 587 | *Halibut* | 137A | 1960 | S3W |
| 14 SSN | 593 | *Thresher* | 188 | 1961–1968 | S5W |
| 1 SSN | 597 | *Tullibee* | 178 | 1960 | S2C |
| 5 SSBN | 598 | *George Washington* class | 180A | 1959–1961 | S5W |
| 5 SSBN | 608 | *Ethan Allen* class | 180 | 1961–1963 | S5W |
| 31 SSBN | 616 | *Lafayette* class | 216 | 1963–1967 | S5W |
| 37 SSN | 637 | *Sturgeon* class | 188A[8] 300.65 | 1967–1975 | S5W |
| 1 SSN | 671 | *Narwhal* | 245 | 1969 | S5G |
| 1 SSN | 685 | *Glenard P. Lipscomb* | 300.68 | 1974 | S5Wa |
| 47 SSN | 688 | *Los Angeles* class | 303.70 | 1976–(1996[9]) | S6G |
| 12 SSBN | 726 | *Ohio* class | 304.74 | 1981–(1997[10]) | S8G |
| — | SSN 21 | *Seawolf* class | — | (1997–2005[11]) | S6W |
| 1 Submersible | — | *NR-1* | — | 1969 | NR-1 |

# Appendix B: Soviet Submarine Construction, 1945–1991

*The Project 651/Juliett SSG at high speed off the coast of Spain. These attractive submarines were considered a major threat to U.S. aircraft carriers.* (U.S. Navy)

All naval ships built during the Soviet era were assigned design numbers (to some extent similar to the U.S. Navy's Ship Characteristics Board numbers). In addition, submarines built through World War II were assigned series numbers (Roman numerals).

The following table lists Soviet submarines completed from 15 August 1945 to 31 December 1991. Submarines built in the Soviet Union specifically for foreign transfer are not included.

| Built | Type | Project/ | Series/Name | NATO Name | Commission | Reactors |
|---|---|---|---|---|---|---|
| **Non-Nuclear Submarines** | | | | | | |
| 7 | SS | IX-*bis* | S class | — | 1946–1948 | — |
| 1 | SS | X-*bis* | Pike | — | 1946 | — |
| 1 | SSX | 95 | | — | 1946 | (closed-cycle) |
| 53 | SS | XV/96 | | — | 1947–1953 | — |
| 21 | SS | 611 | | Zulu | 1953–1958 | — |
| 4 | SSB | AV611 | | Zulu | 1957 | — |
| 215 | SS | 613 | | Whiskey | 1951–1958 | — |
| 1 | SSX | 615 | | Quebec | 1953 | (closed-cycle) |
| 29 | SSX | A615 | | Quebec | 1955–1958 | (closed-cycle) |
| 1 | SSX | 617 | | Whale[1] | 1956 | (closed-cycle) |
| 22 | SSB | 629 | | Golf | 1959–1962 | — |
| 1 | SSB | 629B | | Golf | 1961 | — |
| 20 | SS | 633 | | Romeo | 1959–1962 | — |
| 58 | SS | 641 | | Foxtrot | 1958–1971 | — |
| 18 | SS | 641B | | Tango | 1972–1982 | — |
| 16 | SSG | 651 | | Juliett | 1963–1968 | — |
| 4 | SST | 690 | | Bravo | 1967–1970 | — |
| 2 | Midget | 865 | Piranya | Losos | 1988–1990 | — |
| 22 | SS | 877 | Varshavyanka | Kilo | 1980–1991 | — |
| 2 | AGSS | 940 | Lenok | India | 1976, 1979 | — |
| 1 | AGSS | 1710 | Markel' | Beluga | 1987 | — |
| 1 | AGSS | 1840 | | Lima | 1979 | — |
| **Nuclear-Propelled Submarines** | | | | | | |
| 1 | SSN | 627 | | | November 1958 | 2 VM-4 |
| 12 | SSN | 627A | | | November 1959–1964 | 2 VM-4 |
| 1 | SSN | 645 | | | November 1963 | 2 VT |
| 8 | SSBN | 658 | | Hotel | 1960–1962 | 2 VM-A |
| 5 | SSGN | 659 | | Echo I | 1961–1963 | 2 VM-A |
| 1 | SSGN | 661 | Anchar | Papa | 1969 | 2 VM-5 |
| 34 | SSBN | 667A | | Yankee | 1967–1972 | 2 VM-2-4 |
| 18 | SSBN | 667B | Murena | Delta I | 1972–1977 | 2 VM-4B |
| 4 | SSBN | 667BD | Murena-M | Delta II | 1975 | 2 VM-4B |
| 14 | SSBN | 667BDR | Kal'mar | Delta III | 1976–1982 | 2 VM-4S |
| 7 | SSBN | 667BDRM | Delfin | Delta IV | 1984–(1990[2]) | 2 VM-4SG |
| 11 | SSGN | 670 | Skat | Charlie I | 1967–1972 | 1 VM-4 |
| 6 | SSGN | 670M | Skat-M | Charlie II | 1973–1980 | 1 VM-4 |
| 15 | SSN | 671 | Ersh | Victor I | 1967–1974 | 2 VM-4P |
| 7 | SSN | 671RT | Semga | Victor II | 1972–1978 | 2 VM-4P |
| 25 | SSN | 671RTM | Shchuka | Victor III | 1977–1990 | 2 VM-4P |
| 29 | SSGN | 675 | | Echo II | 1963–1968 | 2 VM-A |
| 1 | SSN | 685 | Plavnik[3] | Mike | 1983 | 1 OK-650b-3 |
| 4 | SSN | 705 | Lira | Alfa | 1971–1981 | 1 OK-550 |
| 3 | SSN | 705K | Lira | Alfa | 1977–1981 | 1 BM-40A |
| 6 | SSBN | 941 | Akula | Typhoon | 1981–1989 | 2 OK-650 |
| 2 | SSN | 945 | Barracuda | Sierra I | 1984–1987 | 1 OK-650a |
| 1 | SSN | 945A | Kondor | Sierra II | 1990–(1993[4]) | 1 OK-650a |
| 2 | SSGN | 949 | Granit | Oscar I | 1980, 1983 | 2 OK-650b |
| 6 | SSGN | 949A | Anteiy | Oscar II | 1986–(1996[5]) | 2 OK-650b |
| 8 | SSN | 971 | Shchuka-B | Akula I/II | 1984–(2001[6]) | 1 OK-650b |
| 3 | AGSSN | 1851 | | X-ray | 1986–(1995[7]) | 1 |
| 3 | AGSSN | 1910 | Kashalot | Uniform | 1983–(1993[8]) | 1 |

# Appendix C: U.S. Submarine Reactor Plants

| Project | Symbol[1] | Design[2] | Startup | Installation/Notes |
|---|---|---|---|---|
| Submersible Reactor | | | | |
| | NR-1 | Knolls | 1969 | submersible *NR-1.* |
| Submarine Reactor Small (SRS) | | | | |
| | S1C | ([3]) | 1959 | land-prototype, Windsor, Conn. |
| | S2C | | 1960 | USS *Tullibee.* |
| Submarine Intermediate Reactor (SIR) | | | | |
| | S1G | Knolls | 1955 | land-prototype, West Milton, N.Y. |
| | S2G | Knolls | 1957 | USS *Seawolf* (SSN 575); replaced by S2Wa 1958–1960. |
| Submarine Advanced Reactor (SAR) | | | | |
| | S3G | Knolls | 1958 | land-prototype, West Milton, N.Y. |
| | S4G | Knolls | 1959 | USS *Triton;* only U.S. two-reactor submarine plant. |
| Natural Circulation Reactor (NCR) | | | | |
| | S5G | Knolls | 1965 | land-prototype, Arco, Idaho; also fitted in USS *Narwhal.* |
| | S6G | Knolls | 1976 | adapted from D2G reactor plant; USS *Los Angeles.* |
| Modifications and Additions to Reactor Facility (MARF)[4] | | | | |
| | S7G | Knolls | 1980 | land-prototype only; West Milton, N.Y. |
| | S8G | Knolls | 1981 | land-prototype, West Milton, N.Y. also fitted in USS *Ohio.* |
| Next Generation Reactor | | | | |
| | S9G | Knolls | . . . . | USS *Virginia.* |
| Submarine Thermal Reactor (STR) | | | | |
| | S1W | Bettis | 1953 | land-prototype, Arco, Idaho. |
| | S2W | Bettis | 1954 | USS *Nautilus;* refitted in *Seawolf* (SSN 575). |
| Submarine Fleet Reactor (SFR) | | | | |
| | S3W | Bettis | 1957 | USS *Skate.* |
| | S4W | Bettis | 1958 | USS *Swordfish.* |
| Submarine High Speed Reactor/Fast Attack Submarine Reactor | | | | |
| | S5W | Bettis | 1959 | USS *Skipjack.*[5] |
| Advanced Fleet Reactor | | | | |
| | S6W | Bettis | 1994 | land-prototype, West Milton, N.Y.; also fitted in USS *Seawolf* (SSN 21). |

# Appendix D: Soviet Submarine Design Bureaus

The design bureau in the Soviet-Russian defense industry is largely an unfamiliar concept in the West. In Western societies, weapons, aircraft, tanks, and warships are designed primarily by government agencies or by the producing industrial corporations. Within the naval sphere there are a few specialized ship design firms in the private sector (e.g., J. J. McMullen in the United States).

Within the Soviet-Russian state there have long been specialized design bureaus, to a significant degree separate from the government agencies sponsoring the weapons and from the industrial entities that produce the weapons. Some of these design bureaus have been well known in the West, especially those in the aviation field—Antonov (An), Tupolev (Tu), Mikoyan and Gurevich (MiG), Sukhoi (Su); later, missile design bureaus also came to popular attention, such as those of Chelomei and Korolev.

The first design bureau established specifically for submarines was the Technical Bureau No. 4, created in 1926 with a small group of engineers and designers from the Baltic (Baltiysk) Shipyard in Leningrad.[1] The first chief of the bureau was Boris M. Malinin, who would lead the Soviet submarine design effort for the next two decades.

By 1928 the bureau had completed the design of the first Soviet-era submarines, the Series I, or *Dekabrist* class, with six ships completed from 1930 to 1931. Technical Bureau No. 4 subsequently was changed to the Technical Bureau for Special Shipbuilding (TsKBS) No. 2, remaining subordinate to the Baltic Shipyard.

## TsKB-18 (Rubin)

In January 1937 bureau TsKBS-2 was changed to TsKB-18 and the following year, on 1 April 1938, the bureau was removed from subordination to the Baltic yard and, as a separate agency, was placed directly under the Peoples Commissariat of

*A Project 671/Victor I SSN with her periscopes and an antenna mast raised. Only Project 667-series SSBNs and a few special-purpose submarines had sail-mounted diving planes.*

Defense Industry. This bureau, in its various forms, designed most Soviet submarines constructed through World War II with four exceptions: the Type IV (*Pravda* class) and Type VI (M class) were designed by A. N. Asafov, the former while incarcerated in the Special Technical Bureau (OTB) operated by the state security agency OGPU. That bureau was transformed into OKTB-2 in April 1931 and was responsible for design of the Project V (*Shch* [Pike]) class. Also, the Type IX (S class) was designed by the Dutch-German firm Deshimag (based on the German Type IA), and Project 95 (*M-401*) was designed by Abram S. Kassatsier while in OKB-196, also a "prison bureau," located in the Sudomekh shipyard. During World War II the OKB was evacuated from Leningrad to the city of Zelenodolsk. In November 1946, Kassatsier, released from confinement, was transferred to TsKB-18 to work on Project 615 (Quebec SSC). In 1948 OKB-196 became a branch of TsKB-18, and in 1953 the OKB was abolished.

During the war TsKB-18—the principal submarine design agency—was evacuated to the inland city of Gor'kiy (now Nizhny Novgorod).

In September 1956 TsKB-18 began the design of nuclear-propelled cruise missile and ballistic missile submarines. Subsequently, it has specialized in nuclear-propelled cruise and ballistic missile submarines as well as diesel-electric attack submarines and deep-diving research vehicles. In the post–Cold War era, the yard has continued to design military submarines—nuclear and non-nuclear—as well as ocean-engineering equipment.

The bureau was renamed "Rubin" in 1966.

## Antipin Bureau

After World War II a design bureau was established in East Germany to coordinate work on adopting the Walter steam-gas turbine propulsion plant for Soviet submarines. (See Chapter 3.) The organization, formally established in 1947, was headed by Engineer-Captain 1st Rank A. A. Antipin, head of TsKB-18.

Plans, drawings, and documentation related to the German Type XXVI submarine were collected by the bureau and formed the basis for Project 617 (Whale), with the initial design for that submarine undertaken at Central Research Shipbuilding Institute No. 45 (now the Krylov Central Research Institute) under the supervision of B. M. Malinin. The Antipin bureau was abolished in 1948.

## SKB-143 (Malachite)

A new submarine design bureau was established in Leningrad on 30 March 1948 for the further development of Project 617 and other high-speed undersea craft. This was SKB-143, initially staffed by specialists from TsKB-18 and the Antipin Bureau in Germany (including ten German engineers), as well as members of the special power plant department of the Central Research Shipbuilding Institute No. 45. Antipin was appointed to head the SKB-143 bureau and named the chief designer of Project 617.

SKB-143 had two Leningrad locations, in the Shuvalovo suburb and at the Sudomekh Shipyard, where the research departments were located, that developed new power plants and the related test facilities.

Development of the first Soviet nuclear-propelled submarine, Project 627 (November SSN), was begun at SKB-143 in March 1953. Subsequently, the bureau has specialized in nuclear-propelled attack submarines.

In 1966 the bureau was renamed "Malachite." After the Cold War the bureau continued to develop nuclear submarine designs as well as commercial underwater systems.

Following the death of Stalin in March 1953, among other changes in the Soviet government, there was a major reorganization of Soviet design bureaus in several military fields.

## TsKB-16 (Volna)

Design bureau TsKB-16, which previously designed surface warships, was transferred to submarine projects in May 1953. The first chief designer of the bureau was N. N. Isanin. Initially the bureau took over Project A615 (Quebec) from TsKB-18. In 1955 that project was returned to TsKB-18, and TsKB-16 specialized in diesel-electric submarines armed with ballistic missiles, both conversions of Project 611 (Zulu SSB) and the design of Project 629 (Golf SSB).

Subsequently, the bureau initiated the design of the advanced nuclear-propelled Project 661 (Papa SSGN). In 1970 the bureau was assigned deep-diving vehicles previously under development at TsKB-18.

In 1966 TsKB-16 was renamed "Volna." The bureau was combined with SKB-143/Malachite in 1974.

*N. N. Isanin*
Rubin CDB ME

## SKB-112 (Lazurit)

A submarine design group was established in 1930 at the Krasnoye

Sormovo shipyard in Gor'kiy as construction began on the *Shch* Series III submarines at the yard. In June 1953 the design group was established as design bureau SKB-112, still subordinate to the shipyard. At that time Project 613 (Whiskey SS), entering large-scale production, was transferred from TsKB-18 to SKB-112.

In February 1956 this organization became the independent design bureau TsKB-112 under chief designer V. P. Vorobyev. Beyond Project 613 and numerous variations, the bureau subsequently designed the second-generation nuclear submarine Project 670 as well as diesel-electric undersea craft and specialized vehicles.

In 1968 the bureau was named "Sudoproyekt" and, in 1974, its name was changed to "Lazurit." With the end of the Cold War the bureau ceased work on combat submarines, instead specializing on research, rescue, and commercial undersea craft and structures.

# Appendix E: Soviet Ballistic Missile Systems

*A decommissioned Project 667-series SSBN at the Zvezda repair yard at Bolshoi Kamen, near Vladivostok, being defueled and made ready for scrapping in 1994.* (Courtesy Greenpeace)

The weapon system in Soviet ballistic missile submarines includes the missile, fire control apparatus, etc.

| System | Missile | | Submarines | |
|---|---|---|---|---|
| | Soviet | NATO | Soviet Project | NATO |
| D-1 | R-11FM | SS-1c Scud-A | V611, AV611 | Zulu IV, V |
| D-2 | R-13 | SS-N-4 Sark | 629 | Golf I |
| | | | 658 | Hotel I |
| D-3[1] | R-15 | | | 639 |
| D-4 | R-21 | SS-N-5 Serb[2] | PV611 | Zulu IV |
| | | | 613D4 | Whiskey |
| | | | 629A, 629B | Golf II |
| | | | 658M | Hotel II |
| D-5 | RSM-25/R-27 | SS-N-6 | 613D5 | Whiskey |
| | | | 667A | Yankee I |
| | | | 679 | |
| | R-27K/SS-NX-13 | | 605 | Golf IV |
| | | | 679 | |
| D-5M[1] | | | 687 | |
| D-5U | R-27U | | 667AU | Yankee I |
| D-6[1] | | | 613D6 | Whiskey |
| D-7[1] | RT-15M | | 613D7 | Whiskey |
| D-8[1] | | | 702 | |

| System | Missile | | Submarines | |
|---|---|---|---|---|
| | Soviet | NATO | Soviet Project | NATO |
| D-9 | RSM-40/R-29 | SS-N-8 Sawfly | 601 | Golf III |
| | | | 667B | Delta I |
| | | | 667BD | Delta II |
| | | | 701 | Hotel III |
| D-9R | RSM-50/R-29R | SS-N-18 Stingray | 667BDR | Delta III |
| D-9RM | RSM-54/R-29RM | SS-N-23 Skiff | 667BDRM | Delta IV |
| D-11 | RSM-45/R-31 | SS-N-17 Snipe | 667AM | Yankee II |
| | | | 999 | |
| D-19 | RSM-52/R-39 | SS-N-20 Sturgeon | 619 | Golf V |
| | | | 941 | Typhoon |
| D-19UTTKH[1] | RSM-52/R-39M | SS-N-28 Grom | 941U | Typhoon |

# Notes

## Perspective

1. GUPPY = Greater Underwater Propulsive Power; the letter *y* was added for pronunciation.
2. These are NATO code names; see Appendix B.
3. Melvyn R. Paisley, Assistant Secretary of the Navy (Research, Engineering, and Systems), testimony before the Committee on Appropriations, House of Representatives, 2 April 1985.
4. This position is "double-hatted," i.e., the admiral who held this position is both an executive within the Navy's shipbuilding organization (Bureau of Ships, subsequently Naval Ship Systems Command, and now Naval Sea Systems Command) and within the Atomic Energy Commission (subsequently Department of Energy). From August 1948 until January 1982, this position was held by Adm. H. G. Rickover.
5. The 60,000-shaft-horsepower plant was never developed by the United States. The Soviet Union developed 50,000-shp single-reactor plants (Akula, Sierra) and approximately 100,000-shp twin-reactor plants (Oscar, Typhoon).

## Chapter 1: Genesis

1. Winston S. Churchill, *Their Finest Hour* (Boston, Mass.: Houghton Mifflin, 1949), p. 598.
2. Winston S. Churchill, *The Hinge of Fate* (Boston, Mass.: Houghton Mifflin, 1950), p. 125.
3. Capt. S. W. Roskill, RN (Ret), *The War at Sea,* vol. II (London: Her Majesty's Stationery Office, 1957), pp. 367–68.
4. All displacements are surface unless otherwise indicated.
5. The best English-language reference on German submarines is Eberhard Rössler, *The U-boat: The Evolution and Technical History of German submarines* (Annapolis, Md.: Naval Institute Press, 1981).
6. Dönitz, who had commanded a U-boat in World War I, was head of the German submarine force from 1936 to January 1943, when he succeeded Grand Adm. Eric Raeder as Commander-in-Chief of the German Navy. However, in the latter role Dönitz retained direct operational control of U-boats.
7. Under the terms of the Treaty of Versailles (1919) that ended World War I, Germany was not allowed to design or possess U-boats. German engineers, however, clandestinely designed submarines in the 1920s for several other countries.
8. Adm. Karl Doenitz, *Memoirs: Ten Years and Twenty Days* (Cleveland, Ohio: World Publishing, 1959; reprinted in 1990), p. 353.
9. The Germans adapted the snorkel from a Dutch invention of Kapitainlieutenant J. J. Wichers, who had devised it to enable Dutch submarines operating in the Far East to operate their diesels underwater as a means of escaping from the tropical heat.
10. Supreme Command of the Navy, 2d Division Naval Staff B.d.U., *Ueberlegungen zum Einsatz des Typ XXI* (Considerations on the Use of Type XXI), 10 July 1944, p. 1.
11. This hull design was adapted from Walter's closed-cycle design, the lower bay originally intended as a storage tank for hydrogen-peroxide.
12. These massive shelters were built at U-boat bases in Norway, the Atlantic coast of France, and Germany.
13. GHG = *Gruppenhorchgerät* (group listening apparatus). The German Navy had experimented with the antecedents of the GHG as early as 1927. Sonar = Sound Navigation And Ranging. The term was invented in 1942 by F. V. (Ted) Hunt, director of the Harvard Underwater Sound Laboratory, as a phonetic analog to the acronym *radar* (Radio Detection And Ranging). According to Hunt, only later was the acronym developed, originally defined as Sound*ing*, Navigation, And Ranging).
14. U.S. Navy, Portsmouth Naval Shipyard, "Surrendered German Submarine Report, Type XXI," (July) 1946, Living and Berthing Section, p. 2. For a less-favorable view by a U.S. submariner, see Lt. Comdr. M. T. Graham, USN (Ret), "*U-2513* Remembered," *The Submarine Review* (July 2001), pp. 83–92.
15. Ibid., Sanitation section, p. 1.
16. Doenitz, *Memoirs: Ten Years and Twenty Days,* p. 355.
17. U.S. Naval Technical Mission in Europe, "Technical Report No. 312–45, German Submarine Design, 1935–1945" (July 1945), p. 3.
18. Schnee had previously commanded three U-boats, from 1940 to 1942; he was assigned to the *U-2511* in 1944 while the submarine was under construction.
19. Doenitz, *Memoirs: Ten Years and Twenty Days,* p. 429.
20. Dr. Gary E. Weir, Review of *Hitler's U-Boat War: The Hunters, 1939–1942* by Clay Blair, in *The Journal of Military History* (July 1997), pp. 635–36.

## Chapter 2: Advanced Diesel Submarines

1. U.S. shipyards thus completed an average of 4½ submarines per month for the 45 months of the Pacific War; in comparison, German yards completed 1,096 U-boats during the 68 months of World War II in Europe or more than 16 per month.
2. The so-called collapse depth for the pressure hull of a 400-foot submarine was considered nebulous, but in "neighborhood" of 800 feet (240 m) or a "little more"; the internal bulkhead strength for the eight compartments of these submarines was 600 feet (185 m), except that the end (torpedo room) compartments were rated at 1,100 feet (335 m) to facilitate the use of the McCann rescue chamber that could be lowered from the surface should the submarine become disabled on the ocean floor. Depths were measured from the submarine's keel.
3. An excellent account of the development of the fleet boat is found in Comdr. John D. Alden, USN (Ret), *The Fleet Submarine in the U.S. Navy* (Annapolis, Md.: Naval Institute Press, 1979).
4. U.S. submarines were effective against Japanese merchant shipping and warships from the spring of 1943 when the major defects with their torpedoes had been remedied. By that time the decisive battles of Coral Sea, Midway, and Guadalcanal had been won by U.S. naval surface and air forces.
5. Rear Adm. T.B. Inglis, USN, Director of Naval Intelligence, statement to the General Board of the Navy, 18 September 1947.
6. U.S. Navy, Office of the Chief of Naval Operations (OPNAV), Memorandum for the Director of the Operational Research Group, "Future Trends in Submarine Warfare," 17 October 1945. The memorandum was signed by two civilian scientists in OPNAV, Dr. W. J. Horvath and Dr. W. E. Albertson.
7. The Submarine Conference or Submarine Officers Conference, initiated in 1926, was a meeting of submarine officers in the Washington, D.C., area to discuss submarine policy, design, weapons, tactics, and personnel matters. During 1946, for example, the meetings were attended by some 70 to 85 officers (mostly lieutenant commanders to rear admirals), plus a few civilian scientists. However, the conference was reconstituted in late 1950, becoming smaller with representatives from specific Navy offices; it continued in that (less useful) form into the mid-1970s.
8. Details of Anglo-American intercepts of Abe's accounts of U-boat programs are discussed in F. H. Hinsley, et al., *British Intelligence in the Second World War,* vol. 3, part 1 (London: Her Majesty's Stationery Office, 1984), and part 2 (1988). These communications were sent in what the Americans called the "Purple" code, which could be read by U.S. cryptologists from 1940.
9. Subsequently transferred by Britain to France.
10. The letter *Y* was added to GUPPY for phonetic purposes.
11. Rear Adm. C. B. Momsen, USN, statement to the General Board of the Navy, 8 November 1948.
12. V. P. Semyonov, Central Design Bureau Rubin, letter to N. Polmar, 25 August 1997.
13. Comdr. W. S. Post, USN, statement to the General Board of the Navy, 19 November 1946.
14. U.S. submarines were credited with sinking 23 Japanese undersea craft during the war; British submarines sank 15 German, 18 Italian, and 2 Japanese boats.
15. Prairie Masker generated bubbles into the surrounding water to reflect the acoustic energy from active sonars seeking to detect the submarine. Prairie was derived from Propeller Air Ingestion Emission; Masker was a word.
16. The *Cochino* suffered a hydrogen (battery) explosion and fire and sank on 26 August 1949 while operating in the Norwegian Sea. At the time the *Cochino* was on an intelligence collection mission.
17. In 2002 former U.S. GUPPY submarines still were in service with the navies of Taiwan and Turkey.
18. Details of the GUPPY and conversion programs are found in Alden, op. cit.; Norman Friedman, *U.S. Submarines since 1945: An Illustrated Design History* (Annapolis, Md.: Naval Institute Press, 1994); and James C. Fahey, *The Ships and Aircraft of the U.S. Fleet,* 6th edition (1954) and 7th edition (1958). The Fahey volumes have been reprinted by the Naval Institute Press.
19. The initial (1945) plan called for building at least two and preferably four of these interim submarines, the program divided between the Electric Boat Co. yard and the Portsmouth Naval Shipyard; minutes of Submarine Officers Conference, 6 November 1945 (Report of 14 November 1945, p. 2).
20. German research had demonstrated that about 40 percent of the drag of a Type XXI was caused by its streamlined fairwater; *source*: Rössler, *The U-Boat,* p. 237.
21. BQR-4 was a U.S. military AN-series designation; the *B* indicates underwater (submarine) installation, *Q* sonar, and *R* receiving (passive detection). The letter *S* in the third position indicates an active (pinging) sonar. The BQS-4 was a submarine version of the surface ship SQS-4 sonar; it retained the designation SQS-4 in early nuclear submarines.
22. The added length reduced congestion in the officers' quarters, gave the commanding officer a private stateroom, and increased storeroom capacity.
23. The U.S. Navy's post–World War II construction program is tabulated in N. Polmar, *Ships and Air-*

*craft of the U.S. Fleet,* 17th edition (Annapolis, Md.: Naval Institute Press, 2001), pp. 630–33.

24. Statement by Rear Adm. R. F. Good, USN, at Submarine Conference, 46/09, p. 5.
25. The "R" class boats were an excellent, highly maneuverable design. Eight were scrapped soon after the war with two retained to help train ASW forces, the *R-10* until 1928 and the *R-4* until 1932.
26. Minutes of Submarine Officers Conference, 20 April 1949 (report of 13 May 1949).
27. Minutes of Submarine Officers Conference, 20 April 1949 (report of 13 May 1949).
28. By comparison, the *Tang* pressure hull diameter was 18 feet (5.5 m).
29. Rear Adm. C. B. Momsen, USN, statement to the General Board of the Navy, 20 December 1948.
30. Operations Evaluation Group (OEG), Office of the Chief of Naval Operations, "Force Requirements for Surface Craft and Submarines in Anti-Submarine Barriers," 19 February 1951, OEG Study No. 434, (LO)321–51, p. 1.
31. Capt. L. R. Daspit, USN, statement to the General Board of the Navy, 8 November 1948.
32. All U.S. submarines were assigned letter names from 1911 to 1924. The practice was reinstated for the SSKs.
33. Capt. L. R. Daspit, USN, statement to the General Board of the Navy, 8 November 1948.
34. One fleet submarine was converted to a tanker, the *Guavina* (SS 362). Converted from March 1949 to February 1950 to an SSO (later AOSS), she could refuel submarines on the surface or while submerged, reportedly down to 100 feet (30 m). Employed mostly to support seaplanes, she could carry 160,000 gallons (605,650 liters) of transferrable fuel.
35. The *K3* (SSK 3) was renamed *Bonita* and reclassified SS 552 on 15 December 1955. She was used as a submerged target for two underwater nuclear bomb tests at Eniwetok atoll in 1958; she was manned for the test on 16 May 1958 (test Wahoo of nine kilotons) and unmanned for the test on 8 June 1958 (test Umbrella of eight kilotons).
36. Capt. Ned Kellogg, USN (Ret), "Operation Hardtack as a Submerged Target," *The Submarine Review* (October 2002), p. 136.
37. Ibid., p. 149.
38. All seven SSK conversions from fleet boats were submarines of the *Gato* (SS 212) class, not the deeper-diving *Balao* (SS 285) class.
39. The great advance in plotting/tracking was developed by Lt. Comdr. J. E. Ekelund, USN; see Capt. Frank A. Andrews, USN (Ret), "Submarine Against Submarine" in *Naval Review, 1966* (Annapolis, Md.: U.S. Naval Institute, 1965), p. 50.
40. The U.S. weapon was the small, air-launched Fido acoustic homing torpedo, designated Mk 24 *mine,* to disguise its true purpose. During 1943–1945 some 340 Fido torpedoes were used by U.S. and British aircraft against U-boats, sinking 68 submarines and damaging another 33, a very high success rate for an anti-submarine weapon.

    The principal German acoustic torpedo of the war was the *Zaunkönig* (Wren), designated T5. This torpedo was specifically intended to attack convoy escort ships. (The Allies referred to the T5 torpedo as GNAT, for German Naval Acoustic Torpedo.) German naval historian Jürgen Rohwer credits the T5 with a 14.4 percent hit ratio, which is higher than Allied estimates.
41. The German Navy was developing wire-guided torpedoes at the end of World War II.
42. John Merrill and Lionel D. Wyld, *Meeting the Submarine Challenge: A Short History of the Naval Underwater Systems Center* (Washington, D.C.: Government Printing Office, 1997), p. 145. See Chapter 10 of this volume.
43. The term SOSUS, believed to have been coined in 1952, was itself classified until about 1967; the unclassified code name Caesar was used as a cover name for production and installation of the arrays.
44. Six hydrophones initially were installed at Eleuthera—three at a depth of 40 feet (12 m), two at 960 feet (290 m), and one at 1,000 feet (305 m). Subsequently, the first deep-water array was installed at Eleuthera; this was a 40-hydrophone, 1,000-foot long array at a depth of 1,440 feet (440 m). The 1,000-foot apertures apparently became the standard.
45. The Hudson Bay arrays terminated at the U.S. Naval Facility at Argentia, Newfoundland.
46. The installation of SOSUS off North Cape—Operation Bridge—is described in Olav Riste, *The Norwegian Intelligence Service 1945–1970* (London: Frank Cass, 1999), pp. 164–90.
47. Antisubmarine Plans and Policies Group, 15 December 1954, Op-312 File, Serial 00511P31, p. 2.
48. Alec Durbo, "Sounding Out Soviet Subs," *Science Digest* (October 1983), p. 16.
49. Chief of Naval Operations, subj: Projected Operational Environment (POE) and Required Operational Capabilities (ROC) Statements for the Integrated Undersea Surveillance Systems (IUSS), 14 March 1995, OPNAV Instruction C3501.204A, ser N874/5C660536, Enclosure (6) "Required Operational Capability for Naval Ocean Processing Facilities," 14 March 1995, p. 1.
50. The U.S. Navy boarding party consisted of five men—unarmed; none could speak Russian and conversations were conducted in French.

51. I. D. Spassky, "The Role and Missions of Soviet Navy Submarines in the Cold War," presentation to Naval History Symposium, U.S. Naval Academy, Annapolis, Md., 23 October 1993; Spassky has been director of Central Design Bureau Rubin (formerly TsKB-18) since 1974.

52. This was the *U-250,* sunk on 30 July 1944; six members of the crew, including the commanding officer, were captured by the Soviets. Codes, instructions, an Enigma cipher machine, and T-5 acoustic torpedoes were taken from the U-boat.

53. U.S., Joint Intelligence Committee, Intelligence Section, "Review of Certain Significant Assumptions, Summaries, and Conclusions as Presented in Joint Intelligence Committee (Intelligence Section, JIG), Estimates of the Capabilities of the USSR," (January 1948), p. 10. For an unclassified analysis of the disposition of German U-boats see "Mystery of Missing Submarines," *U.S. News & World Report* (11 December 1953), pp. 41–43; also, various editions of the annual *Jane's Fighting Ships.*

54. Eberhard Rössler, *U-Boottyp XXI* (Munich: Bernard & Graefe Verlag, 1980), pp. 58, 60, 159. The two U-boats that Rössler reported to have been towed to Kronshtadt were from the series *U-3538* through *U-3542.*

55. Wilhelm Hadeler, "The Ships of the Soviet Navy," in *The Soviet Navy* (New York: Frederick A. Praeger, 1958), p. 159. Only one Type XXI remains, the *U-2540,* scuttled in May 1945, salvaged by West Germany in 1957, and employed as a research craft named *Bauer;* in 1984 she became a museum ship at Bremerhaven.

56. Soviet engineers state that no Type XXIs were placed in operational service; for example, V. P. Semyonov discussion with N. Polmar, St. Petersburg, 5 May 1997, and Academician S. N. Kovalev discussion with N. Polmar in St. Petersburg, 6 May 1997. Kovalev called particular attention to the superiority of the Type XXI's overall design, battery capacity, pressure hull, and radio equipment in comparison with contemporary undersea craft.

57. A fourth unit, manned by a Russian crew, was accidently sunk by a British aircraft while the submarine was en route to the Soviet Union.

58. V. P. Semyonov discussion with N. Polmar, St. Petersburg, 5 May 1997.

59. *Alberich* consisted of a four-millimeter layer of rubber to muffle active sonar reflections (echoes) in the 10 to 18 kHz range; its effectiveness was estimated at up to 15 percent. The term ASDIC was coined by the Admiralty in 1918 to designate a submarine detection system; the term was *not* an acronym for Allied Submarine Detection Investigation Committee, as stated by the Admiralty in 1939 in response to an inquiry from the Oxford University Press. *Source:* Willem Hackmann, *Seek & Strike: Sonar, anti-submarine warfare and the Royal Navy, 1914–54* (London: Her Majesty's Stationery Office, 1984), p. xxv.

60. In the postwar period NATO assigned Soviet submarines letter names, with the military phonetic being used. Project 613 was designated "W" and was invariably known as Whiskey.

61. Peregudov was succeeded as chief designer of Project 613 by Yakov Yevgrafov and, soon after the keel laying of the first Project 613 submarine, by Zosima Derebin. Peregudov was assigned to more advanced submarine concepts. (See Chapter 5.)

62. See I. D. Spassky (ed.), *Istoriya Otechestvennogo Sudostroeniya* (The History of Indigenous Shipbuilding), vol. V *1946–1991* (St. Petersburg: Sudostroenie, 1996), pp. 81–83; Capt. 1/R B. Tyurin, "Medium Diesel-Electric Submarines of Project 613," *Morskoy Sbornik* (no. 8, 1995), pp. 74–79; Siegfried Breyer, "The Soviet Submarines of the W-Class as a Family of Classes," *Soldat und Technik* (no. 1, 1971), pp. 10–15; and Alexander Mozgovoi, "Finest Whiskey," *Military Parade* [Moscow] (November 2001), web site www.milparade.com/2001/48/0701.shtml.

63. V. P. Semyonov discussion with N. Polmar, St. Petersburg, 29 October 1997. The Soviets had attempted to develop similar creeping motors before World War II, but were not successful.

64. In 1955 the *S-70,* one of the first Project 613 submarines to be fitted with a snorkel, carried out a 30-day submerged cruise—with major difficulties. See Mozgovoi, "Finest Whiskey."

65. The world's first specialized minelaying submarine was the *Krab,* built in Russia in 1909–1915.

66. The SKhL-4 had a yield strength of approximately 57,000 pounds per square inch (4,000 kg/cm$^2$).

67. Gor'kiy was Nizhny Novgorod before 1932 and after 1990.

68. In 1960 a Soviet submarine tender and up to eight submarines had been based in Albania at Saseno (Sazan) Island in the Gulf of Valona.

69. The NATO code name Boat Sail referred to the similarity of the large air-search radar antenna to a sail; Canvas Bag referred to the canvas covering often fitted over the radar when it was not in use.

70. According to some sources, the *S-148*'s Navy crew was retained in the civilian research role. There were published reports in the West that a second Project 613 conversion took place, that submarine being named *Slavyanka;* however, there was no second conversion.

71. Nuclear torpedoes are described in V. P. Kuzin and V. I. Nikol'skiy, *Voyenno-morskoy Flot SSSR, 1945–1991* (The Navy of the USSR, 1945–1991) (St. Petersburg, Historical Oceanic Society, 1996), and

A. B. Shirokorad, *Sovetskiye Podvodnyye Lodki Poslevoyennov Postryki* (Soviet Submarines of Postwar Construction) (Moscow: Arsenal Press, 1997).

72. This was the 22d nuclear detonation by the Soviet Union. The RDS-9 warhead was suspended to a depth of 10½ feet (35 m) under a barge. The first U.S. underwater nuclear detonation was Test Baker at the Bikini atoll on 25 July 1946—the fifth U.S. nuclear explosion. It was an Mk 3 atomic bomb of approximately 20 kilotons.
73. Later Soviet nuclear torpedoes were estimated in the West to have warheads as large as 200 kilotons; see Rear Adm. John Hervey, RN (Ret), *Submarines* (London: Brassey's, 1994), p. 127.
74. The Mk 13 torpedo became available for U.S. naval aircraft in 1938; the later variants had a range of 6,000 yards (5,488 m) at 33 knots.
75. J. R. Oppenheimer, letter to Capt. W. S. Parsons, USN, "Design Schedule for Overall Assemblies," 27 December 1943.
76. Memorandum from Chairman, Submarine [Officers] Conference, to Chief of Naval Operations, "Submarine Conference on 6 November 1946—Report of," 18 November 1946, Op-31B:ch (SC) A3-2, p. 5.
77. I. D. Spassky (ed.), *The History of Indigenous Shipbuilding,* vol. V, pp. 84–85; V. A. Kucher, Yu. V. Manuylov, and V. P. Semyonov, "Project 611 Submarines," *Sudostroenie* (no. 5-6, 1995), pp. 58–62; and E. V. Zarkov.
78. The previous three-shaft Russian/Soviet submarine was the *Akula,* completed in 1911.
79. SAET = *Samonavodyashayasya Akustitcheskaya Torpeda* (homing acoustic torpedo).

## Chapter 3: Closed-Cycle Submarines

1. Two useful descriptions of the Walter submarines are Erwin Sieche, "The Walter Submarine-1," *Warship* (no. 20, 1981), pp. 235–46, and Michael Wilson, "The Walter Submarine—2," *Warship* (no. 20, 1981), p. 247–53; also published in book form as *Warship,* vol. V, by Conway Maritime Press and the Naval Institute Press. Also see Rössler, *The U-boat.*
2. Royal Navy, *German Naval History: The U-Boat War in the Atlantic, 1939–1945,* vol. III (London: Her Majesty's Stationery Office, 1989), p. 87. This originally was published as a classified series, written by Hessler under the auspices of the British Admiralty and U.S. Navy Department. Hessler was a U-boat commander and later the staff officer to the Flag Officer Commanding U-boats as well as being Dönitz's son-in-law.
3. Ibid.
4. The German Air Force required hydrogen-peroxide to fuel guided missiles.
5. Minutes of Submarine Officers Conference, 16 November 1945 (report of 14 November 1945, p. 3).
6. Minutes of Submarine Officers Conference, 23 October 1946 (report of 31 October 1946, pp. 5–8). Listed were: (1) external combustion condensing steam cycle, (2) internal combustion condensing steam cycle, (3) semi-closed gas turbine cycle, (4) Walter system, (5) recycle diesel system (Krieslauf), and (6) free-piston gas generator-gas turbine.
7. Capt. L. R. Daspit, USN, statement before the General Board of the Navy, 8 November 1948.
8. Comdr. T. W. Hagen, USN, Chairman, Ad Hoc Committee to the Submarine Conference, to the Chairman, Submarine Conference, enclosing "Study of Advantages of a Nuclear Propelled Submarine Over a Closed Cycle Submarine," 27 April 1949.
9. "Study of Advantages of a Nuclear Propelled Submarine Over a Closed Cycle Submarine," report of the Ad Hoc Committee to the Submarine Conference (27 April 1949), p. 5.
10. See Wilson, op. cit., pp. 251–53, and various editions of *Jane's Fighting Ships* in the 1950s for published material on the British turbine submarines.
11. Michael Wilson letter to N. Polmar, 22 September 1997.
12. Michael Wilson letter to N. Polmar, 8 October 1997.
13. Robert Gardiner (ed.), *Conway's All the World's Fighting Ships, 1947–1982,* part I: *The Western Powers* (London: Conway Maritime Press, 1983), p. 168. Another reason for abandoning high-test peroxide was the loss of the British submarine *Sidon* at Portland on 16 June 1955 after a peroxide-fueled torpedo blew up.
14. The NKVD also operated aircraft design bureaus; among the more noted engineers incarcerated in such facilities were A. N. Tupolev, the dean of Soviet aircraft designers; his protégé Vladimir Petlaykov; Tupolev's son-in-law and aircraft designer Vladimir Myasishchev; chief rocket/spacecraft designer Sergei Korolev; V. N. Peregudov, later chief designer of nuclear submarine Project 627 and chief of SKB-143 from 1953 to 1958; and M. G. Rusanov, later chief designer of submarine Projects 653 and 705.
15. The German engineers worked at the Antipin Bureau under contract and not by order.
16. The code name "Whale" was not part of the recently instituted U.S.-NATO designation scheme; the *W*-name in that scheme (i.e., Whiskey) was assigned to Project 613.
17. A detailed technical description of Project 617 appears in I. D. Spassky and V. P. Semyonov, "Project 617: The Soviet 'Whale'," *Naval History* [U.S. Naval Institute] (November/December 1994), pp. 40–45.
18. V. P. Semyonov discussion with N. Polmar, St. Petersburg, 10 May 1994.
19. The suffix *bis* indicated modification.

20. See, for example, Hackmann, *Seek & Strike,* p. 339.
21. The M-designations for these boats indicated that they were considered small (*malyutka*) submarines.
22. There were minor changes between the prototype Project 615 and the production A615, the principal one being the substitution of one oxygen tank for two with the same volume.
23. The German code name for the hull coatings was Alberich.
24. A. A. Postnov, "Experimental Submarine Design 613Eh with Electrochemical Generators," *Sudostroenie* (no. 2, 1998), pp. 26–28.
25. In the same period West Germany carried out trials with the submarine *U-205,* fitted with a 100-kilowatt electrochemical generator. There was interest in such a propulsion system in the United States in the 1980s by the Office of Naval Research and the Los Alamos National Laboratory for a submarine for Arctic operation. A four megawatt test-bed submarine conversion and a subsequent eight megawatt new construction were proposed. The effort was halted after opposition from the nuclear propulsion community. See, for example, Arctic Energies Ltd., *Evaluation of Fuel Cell Propelled Arctic Submarines* (8 June 1988), and Los Alamos National Laboratory technical staff discussions with N. Polmar during several meetings in 1982–1985.

## Chapter 4: U.S. Nuclear-Propelled Submarines

1. The Naval Research Laboratory was established in 1923, the result of an earlier recommendation by inventor Thomas Edison that the government establish a "great research laboratory."
2. From Norman Polmar and Thomas B. Allen, *Rickover: Controversy and Genius* (New York: Random House, 1982), p. 118; based on Polmar interviews with Gunn in 1961; also see Richard G. Hewlett and Francis Duncan, *Nuclear Navy, 1946–1962* (Chicago: University of Chicago Press, 1974), pp. 15–22.
3. Memorandum to the Director, Naval Research Laboratory, "Submarine Submerged Propulsion," 1 June 1939.
4. Two days after Japan surrendered in 1945, Gunn and Abelson were both presented with the Distinguished Civilian Service Award for their uranium separation process.
5. The Bureau of Steam Engineering and Bureau of Construction and Repair were merged on 20 June 1940, creating the Bureau of Ships. BuShips was responsible for the design, construction, and maintenance of all Navy surface ships and submarines.
6. Memorandum from Brig. Gen. Leslie Groves, USA, in Atomic Energy File, Navy Operational Archives, Washington, D.C. (undated).
7. BTU = British Thermal Unit. The report, "Proposed Submarine Powered by Nuclear Transformations," dated 19 November 1945, is appended to Minutes of Submarine Officers Conference, 6 November 1945 (Report of 14 November 1945).
8. Capt. Laning interview with N. Polmar, Orlando, Fla., 23 December 1979. Laning would become the first commanding officer of the second nuclear submarine, the *Seawolf* (SSN 575).
9. "Atomic Energy Submarine," 28 March 1946.
10. Vice Adm. Charles A. Lockwood, USN (Ret), *Down to the Sea in Subs: My Life in the U.S. Navy* (New York: W. W. Norton, 1967), p. 348.
11. Memorandum from Lt. Comdr. C. N. G. Hendrix, USN, Subject: "Submarine Future Developments," 5 February 1946.
12. Henry DeWolf Smyth, *Atomic Energy for Military Purposes* (Princeton, N.J.: Princeton University Press, 1945). The book was written at the direction of Gen. Groves.
13. Several naval officers were assigned to the Manhattan Project; the weapon officer/mission commander on the B-29s that dropped the atomic bombs on both Hiroshima and Nagasaki were naval officers.
14. Rickover's only command was the USS *Finch,* a former minesweeper in the Far East, for three months while the *Finch* was on the Yangtze River.
15. Mumma had been head of the U.S. Navy mission to Europe at the end of World War II to recover U-boat technology. He served as chief of the Bureau of Ships from April 1955 to April 1959.
16. Nimitz, who had commanded all U.S. forces in the Pacific Oceans theater during World War II, was a submarine officer and had specialized in the development of diesel engines for the U.S. Navy.
17. See Appendix C for U.S. reactor designations; in 1955 the Submarine Thermal Reactor was redesignated S1W for the land prototype and S2W for the *Nautilus* plant.
18. Office of the Chief of Naval Operations, Ship Characteristics Board, "First Preliminary Characteristics for Submarine (Nuclear Power), Shipbuilding Project £64; forwarding of," 23 March 1950. p. 2. The memo listed a potential armament of *nine* torpedo tubes with stowage for about 30 torpedoes or an equivalent load of mines.
19. The term "keel" is an anachronism for modern nuclear submarines. President Truman welded his initials in a steel plate of the outer hull of the *Nautilus.* Partial and full double-hull submarines have a center vertical keel, connecting their outer and inner (pressure) hulls; this steel member also divides fuel tanks or ballast tanks. Single-hull sub-

marines have no such center member along their keel. Still, the bottom center of a submarine is referred to as the "keel," even though there is no distinguishing piece of structure.

20. The SQS-4 was a surface ship sonar fitted in early SSNs; earlier diesel-electric submarines and the *Triton* (SSRN 586) had the BQS-4 submarine variant of the same basic sonar. The early nuclear submarines also had BQR-4 passive ranging sonar.
21. Rickover was selected for promotion to rear admiral on 1 July 1953, at the direction of the Secretary of the Navy. He had previously been "passed over" for promotion by Navy selection boards because of his acerbic personality, coupled with the small number of engineering officers selected for flag rank. Although he had renounced Judaism in 1931, his Jewish origins were often raised (incorrectly) as the reason for his nonpromotion. He subsequently was promoted to vice admiral and full admiral while remaining in charge of the nuclear propulsion program until relieved in January 1982. See Polmar and Allen, *Rickover,* pp. 153, 186-205.
22. Ibid., pp. 222–49.
23. Hewlett and Duncan, *Nuclear Navy,* p. 192; the authors cite an article by Michael Amrine for the North American Newspaper Alliance on 20 February 1953. Massive reactors for producing plutonium had been built in Hanford, Wash., during World War II, a key factor in Jackson's interest in nuclear matters.
24. E. F. Noonan, BuShips, Memorandum for File, 11 March 1955. These and the following problems are discussed in detail in Gary E. Weir, *Forged in War* (Washington, D.C.: Naval Historical Center, 1993), pp. 178–86.
25. Memorandum from Comdr. Wilkinson to the Chief of Naval Operations, 6 August 1955.
26. First Endorsement by Rear Adm. Frank T. Watkins, USN, to Wilkinson memorandum of 6 August 1955; 9 August 1955.
27. Weir, *Forged in War,* p. 183.
28. Anderson recorded these adventures in *Nautilus 90 North* (Cleveland, Ohio: World Publishing, 1959).
29. During her 24 years in commission, the *Nautilus* nuclear plant was refueled three times; on her four nuclear cores she sailed:

| | |
|---|---|
| 1955–1957 | 62,559.6 n.miles (115,923 km) of which 36,498 n.miles (67,631 km) were submerged |
| 1957–1959 | 91,325 n.miles |
| 1959–1967 | 174,507 n.miles |
| 1967–1979 | 162,382 n.miles |

30. The *Seawolf* plant is described in Capt. Richard B. Laning, USN (Ret), "Thoughts from the Oral History of . . . ," *The Submarine Review* (January 1989), pp. 54–61, and Laning, "The *Seawolf*'s Sodium-Cooled Power Plant," *Naval History* (Spring 1992), pp. 45–48, and "The *Seawolf:* Going to Sea," *Naval History* (Summer 1992), pp. 55–58.
31. Capt. Robert Lee Moore Jr., USN (Ret), discussion with N. Polmar, Annapolis, Md., 1980.
32. Named for Soviet physicist Pavel Alekseevich Cherenkov.
33. The *Seawolf*'s 60-day submergence encountered no difficulties until about the 40th day, when the outer door to the trash disposal unit jammed. Laning conceived an ingenious and bold method of solving this and other problems; see Capt. Willis A. Matson, USN (Ret), "Dick Laning and *Seawolf,*" *The Submarine Review* (April 2001), pp. 101–107.
34. Capt. Richard B. Laning, USN (Ret), letter to N. Polmar, 12 September 1997.
35. The original, SIR/S2G reactor from the *Seawolf,* with the fuel core removed but still radioactive, was taken to sea on a barge to a point some 220 n.miles (222 km) due east of the Delaware-Maryland state line and scuttled in some 9,500 feet (2,896 m) of water. This was the only intentional dropping of a U.S. nuclear reactor into the sea.
36. See John Piña Craven, *The Silent War: The Cold War Battle Beneath the Sea* (New York: Simon & Schuster, 2001), pp. 137–38.
37. Deborah Shapley, "Nuclear Navy: Rickover Thwarted Research on Light Weight Reactors," *Science* (18 June 1978), pp. 1210–1213.
38. Hewlett and Duncan, *Nuclear Navy,* p. 278.
39. The estimated cost of the *Nautilus* was $64.9 million and the *Seawolf* $64.5 million (plus AEC development and fuel costs); the first two *Skate*-class submarines cost $50.6 million each, and the next two $47.9 million each.
40. See Comdr. James Calvert, USN, *Surface at the Pole* (New York: McGraw-Hill, 1960). Although unable to surface at the pole on 11 August 1958, later on that day the *Skate* was able to bring her sail structure up through an opening in the ice some 40 n.miles from the pole to transmit radio messages.
41. Arctic-sailing submarines were fitted with special fathometers and ice-detecting sonar, and the tops of their sails were strengthened to break through thin ice.
42. Based on "Naval Reactor Program, A Joint Navy-Atomic Energy Commission Program," 8 April 1959, pp. 14–15. Prepared by the Naval Reactors Branch, this annual statement was the basis for Rickover's extensive congressional testimony. These statements are cited in this volume as "Rickover [date]."
43. The fiscal 1956 program contained eight submarines, five of them with nuclear propulsion. It was the largest number of submarines authorized in

a single year since World War II:

| Fiscal Year | (Total) | Diesel-electric | Nuclear |
|---|---|---|---|
| 1948 | (5) | 4 SS | — |
| | | 1 SSK | — |
| 1949 | (4) | 2 SS | — |
| | | 2 SSK | — |
| 1950 | (1) | 1 AGSS | — |
| 1951 | (1) | 1 SST | — |
| 1952 | (4) | 2 SSR | 1 SSN |
| | | 1 SST | — |
| 1953 | (2) | 1 SSG | 1 SSN |
| 1954 | (1) | 1 SS | — |
| 1955 | (3) | 1 SSG | 2 SSN |
| 1956 | (8) | 3 SS | 3 SSN |
| | | — | 1 SSRN |
| | | — | 1 SSGN |

44. These were the largest non-nuclear submarines built by the United States after World War II—350½ feet (106.86 m) long, displacing 2,485 tons surfaced and 3,170 tons submerged; they had six bow torpedo tubes.

45. Rickover, 8 April 1959, p. 15.

46. Adm. I. J. Galantin, USN (Ret), *Submarine Admiral: From Battlewagons to Ballistic Missiles* (Urbana: University of Illinois Press, 1995), p. 234. Galantin served as head of Submarine Warfare Branch in the Office of the Chief of Naval Operations (1955–1957) and Director of Special Projects, i.e., Polaris-Poseidon (1962–1965).

47. Capt. Beach letter to N. Polmar, 31 March 1980. Capt. Beach already was well known for his book *Submarine!* (1952) and his best-selling novel *Run Silent, Run Deep* (1955).

48. Capt. Edward L. Beach, USN (Ret), *Salt and Steel: Reflections of a Submariner* (Annapolis, Md.: Naval Institute Press, 1999), p. 262. Also Beach discussion with N. Polmar, Annapolis, Md., 21 April 1999, and Beach letter to N. Polmar, 21 May 2000.

49. Capt. Beach letter to N. Polmar, 21 May 2000.

50. SPS = Surface [ship], P = radar, S = Search. The only other ship to carry the SPS-26 was the large ASW ship *Norfolk* (DL1, ex-CLK 2); it was installed in 1957.

51. Capt. Beach letter to N. Polmar, 8 February 1980.

52. By comparison, the authorized cost of the *Nautilus* was $64.9 million and the *Skate* just over $50 million; these costs do not include reactors, nuclear fuel, and related costs paid by the AEC. *Source:* Navy Department (OP-03B1), working papers, 19 March 1957. This source lists the *Triton* cost as $95 million (13 October 1960).

53. Speech by Vice Adm. Austin at *Triton* commissioning ceremonies, Groton, Conn., 10 November 1959.

54. The *Triton*'s conning tower (within her fairwater) permitted the submarine's pressure hull to be kept isolated from the atmosphere during the transfer. The cruiser had been in Argentine waters in conjunction with President Eisenhower's visit to that country.

55. Details of the *Triton*'s voyage are provided in Beach's *Around the World Submerged* (New York: Holt, Rinehart, Winston, 1962).

56. Enclosure to Capt. Beach letter to N. Polmar, 21 May 2000. Beach used this idea in his later novel *Cold Is the Sea* (1978).

## Chapter 5: Soviet Nuclear-Propelled Submarines

1. Molotov's speech was published in *Pravda,* 7 November 1945, p. 1; cited in Davis Holloway, *Stalin and the Bomb* (New Haven, Conn.: Yale University Press, 1994), p. 144.

2. Beginning in 1946 nuclear weapons design work was carried out at this new research facility near Arzamas, now called the Khariton Institute. It is located some 60 miles (96 km) south of Gor'kiy (now Nizhny Novgorod). Nuclear torpedoes are described in Kuzin and Nikol'skiy, *The Navy of the USSR,* and Shirokorad, *Soviet Submarines of Postwar Construction.*

3. Quoted in Capt. Lt. Yu. Stvolinsky, Soviet Navy (Ret), "The Beginning," *Leningradskaya Pravda* (15 March 1979), p. 3. The original Malinin manuscript is in the Central Naval Museum in St. Petersburg. Malinin died in 1949.

4. The development of the first Soviet nuclear submarine is described in I. D. Spassky (ed.), *The History of Indigenous Shipbuilding,* vol. V, pp. 133–44; A. M. Antonov (The First Generation of Nuclear Submarines: Special Design Bureau SDB-143) (St. Petersburg: Malachite, 1996); Antonov, "Atomic Submarine Project 627," *Sudostroenie* (no. 7, 1995), pp. 76–82; and unpublished Antonov manuscript "Birth of the Red November Submarine" (1994), an excerpt of which was published as "The Birth of the Red November," U.S. Naval Institute *Proceedings* (December 1995), pp. 79–81.

5. NII = *Naucho-Issledovatel'skii Institut* (scientific research institute).

6. At the time of his appointment, Peregudov was deputy director of the Shipbuilding Ministry's Central Scientific Research Institute No. 45.

7. Like many Soviet engineers, in the late 1930s Peregudov was imprisoned by the NKVD, beaten, and tortured. In 1952 he was assistant director of research institute NII-45, now the A. N. Krylov Research Institute (for shipbuilding). Peregudov was already seriously ill when named chief designer of Project 627; he died in 1967. See Capt. Lt. Yu. Stvolinsky, Soviet Navy (Ret), "In the Annals of the Fatherland: Designer of the Nuclear Submarine,"

*Krasnaya Zvezda* (15 October 1988), p. 4.

8. V. I. Barantsev discussions with N. Polmar, St. Petersburg, 3 October 1994 and 6 December 1996; direct quotation from Roy Davies, *Nautilus: The Story of Man Under the Sea* (London: BBC Books, 1995), p. 19. The book accompanied the BBC television series *Nautilus.*
9. The Obninsk research center was the location of the world's first atomic power station, put into operation in 1954. It was capable of producing a modest five magawatts of power.
10. Davies, *Nautilus,* p. 21. As an engineer-admiral, Kotov would serve as Deputy Commander-in-Chief of the Navy for Shipbuilding and Armaments from 1965 to 1986.
11. Davies, *Nautilus,* p. 22. Confirmed by Barantsev in discussion with N. Polmar, St. Petersburg, 6 December 1996.
12. V. P. Semyonov and CDB Rubin engineers discussions with N. Polmar, St. Petersburg, 5 December 1996.
13. Not all appraisals of the stern configuration were positive: "the schematic chosen, especially the stern extremity, turned out to be far from an improvement and in practice turned out to be worse than the U.S. nuclear submarine of the first generation which had . . . a usual diesel hull architecture" (Kuzin and Nikol'skiy, *The Navy of the USSR,* p. 74).
14. On trials the *K-3* reached a depth of 1,015 feet (310 m).
15. A. M. Antonov, et al., "40 Years of Nuclear Submarine Development: A View of a Designer," paper presented at Warship '96: International Symposium on Naval Submarines, London, 12 June 1996.
16. I. D. Spassky (ed.), *The History of Indigenous Shipbuilding,* vol. V, p. 135.
17. The *Narodovolets,* commissioned in 1931, was the second submarine designed and built under the Soviet regime. The submarine is preserved as a museum at Vasilyevskiy Island in St. Petersburg.
18. In 1957, after V. M. Molotov, former foreign affairs commissar, fell from favor with the Soviet political leadership, the town was renamed Severodvinsk.
19. Construction hall No. 42 was erected in the 1930s to permit the simultaneous, side-by-side construction of two battleships.
20. S. Bystrov, "A Reactor for Submarines," *Krasnaya Zvezda* (21 October 1989). Kuznetsov was head of the Soviet Navy from 28 April 1939 to 17 January 1947, and again from 20 July 1951 to 5 January 1956.
21. At the time Adm. Orël was head of the Board of Underwater Forces at Navy Headquarters.
22. Previous Soviet submarines could launch torpedoes at depths to 150 feet (45 m).
23. Quoted in Capt. 1/R George I. Sviatov, Soviet Navy (Ret), "First Soviet Nuclear Submarine," *The Submarine Review* (January 2000), p. 123.
24. Antonov, "The Birth of the Red November Submarine," p. 14.
25. Stovolinskiy, "In the Annals of the Fatherland," p. 4.
26. Adm. Fleet of the Soviet Union S. G. Gorshkov, *Vo Flotskom Stroyu* (Military Memoirs) (St. Petersburg: Logos, 1996), p. 175.
27. Gremikha is also known as the "Yokanga base." The base has no road or rail access and can be reached only by sea and air. Until August 1961 the *K-3* had been based at Severodvinsk, where she was built.
28. In this same period U.S. nuclear-propelled submarines avoided the Arctic ice pack, with only one submarine operation under the Arctic ice from 1962 to 1969. This limitation was due mainly to the loss of the USS *Thresher* (SSN 593) in April 1963.
29. Vice Adm. H. G. Rickover, USN, "Report on Russia" (Washington, D.C.: Committee on Appropriations, House of Representatives, 1959), p. 42. Rickover visited the *Lenin* while he was accompanying then-Vice President Richard M. Nixon on his 1959 trip to the Soviet Union. The *Lenin* was the world's first nuclear-propelled surface ship. Most of his report to Congress addressed Soviet education, with emphasis on the Soviet leadership over the United States in technical education.
30. Antonov, "Birth of the Red November Submarine," p. 15, and Kuzin and Nikol'skiy, *The Navy of the USSR,* p. 76. U.S. intelligence learned of these problems as early as 1959, based on discussions with Soviet engineers visiting the United States; see Henry S. Lowenhaupt, "How We Identified the Technical Problems of Early Russian Nuclear Submarines," *Studies in Intelligence* (Central Intelligence Agency) (vol. 18, Fall 1974), pp. 1–9.
31. Soviet submarine casualties are most recently described in N. G. Mormul, *Katastrofe pod Vodoe* [Catastrophe under Water] (Murmansk: Elteco, 1999).
32. Thirty men were lost with the submarine and 22 died in the water of exposure. Of those lost in the submarine, 12 were trapped in the after portion of the hull (aft of the burning No. 7 compartment), where the aftermost bulkhead was penetrated by gases and the aftermost escape hatch could not be opened.
33. "Accidents reported aboard Soviet subs," *The Baltimore Sun* (17 March 1982). The nuclear icebreaker *Lenin* suffered a major radiation accident in 1966 or 1967, with some reports citing a "reactor meltdown." Up to 30 crewmen are believed to have died of radiation poisoning, with others suffering injuries. The ship lay abandoned in an Arctic port for about five years, after which she was towed to a shipyard and rebuilt. She was returned to service.
34. John J. Engelhardt, "Soviet Sub Design Philosophy," U.S. Naval Institute *Proceedings* (October 1987), p. 197.

35. I. D. Spassky (ed.), *The History of Indigenous Shipbuilding,* vol. V, pp. 136–37; V. Bil'din, "Experimental Nuclear-Powered Torpedo Submarine of Project 645," *Morskoy Sbornik* (no. 8, 1993), pp. 59–62; and A. M. Antonov, "Atomic Submarine Project 645," *Sudostroenie* (no. 10, 1995), pp. 57–61. Project 645 was the only Soviet submarine to have a woman serve as the Navy supervisor, Engr. Capt. 1/R Alexandra Donchenko. She also was the only woman to be graduated from a senior naval school, the A. N. Krylov Naval Academy of Shipbuilding and Armament.

36. The Soviet calculation was 22-44 lbs/0.155 $in^2$ (10-20 $kg/cm^2$) for liquid metal versus 441-551 lbs/0.155 $in^2$ (200-250 $kg/cm^2$) for pressurized water.

37. Antonov, "Atomic Submarine Project 645," p. 60.

38. In 1981 the *K-27* was brought into a dock at the Zvezdochka shipyard at Severodvinsk and about 270 metric tons of bitumen were poured into the submarine's machinery spaces. When hardened, it would prevent any radioactive contamination of surrounding water. She was then towed into the Kara Sea off Novaya Zemlya in the fall of 1981 and sunk in water 246 feet (75 m) deep.

39. The Soviet term for World War II in Europe.

40. Adm. N. G. Kuznetsov, "Naval Preparedness" in Seweryn Bailer (ed.), *Stalin and His Generals* (London: Souvenir Press, 1969), p. 173. This essay first appeared as "Pered voinoy" [On the Eve], *Oktiabr'* (no. 11, 1965), pp. 141–44. However, in his posthumously published memoirs, Kuznetsov says that "somebody" made the comment about fighting close to American shores (*Memoirs of Wartime Minister of the Navy* [Moscow: Progress Publishers, 1990; in English], p. 126).

41. Kuznetsov, "Naval Preparedness," p. 173, and Kuznetsov, *Memoirs . . .*, p. 126.

42. The lone survivor was Lev Mikhailovich Galler, who served as vice commissar of the Navy during World War II; he was arrested on Stalin's order in February 1948 and died in prison in July 1950. Admiral ranks were introduced into the Soviet Navy on 7 May 1940; prior to that the most senior naval officers were "flagmen."

43. Arthur Koestler, *Darkness at Noon* (New York: Macmillan, 1941), pp. 150–151.

44. Nikita Khrushchev, *Khrushchev Remembers: The Last Testament* (Boston: Little, Brown, 1974), and Sergei N. Khrushchev, *Nikita Khrushchev and the Creation of a Superpower* (University Park: Pennsylvania State University, 2000).

45. Nikita Khrushchev, *Khrushchev Remembers: The Last Testament,* p. 30.

46. Ibid., p. 31.

47. Kuznetsov had suffered a heart attack in May 1955 and was confined to bed with Gorshkov becoming de facto CinC.

48. Nikita Khrushchev; *Khrushchev Remembers: The Last Testament,* p. 31.

49. The senior naval officer was both a deputy minister of defense and Commander-in-Chief of the Navy and thus the equivalent of both the U.S. Secretary of the Navy and Chief of Naval Operations, respectively.

50. The United States delivered two nuclear bombs in combat against Japan in early August 1945; a third weapon was available before the end of the month, with series production under way. Not until the early 1950s would the USSR have a similar nuclear-capable bomber force.

## Chapter 6: Cruise Missile Submarines

1. See Polmar, *Atomic Submarines,* pp. 183–84; based on Ernest A. Steinhoff letter to N. Polmar, 11 April 1962, and discussions with Steinhoff in 1962. The commanding officer of the *U-511* at the time was Steinhoff's brother, *Kapitanleutnant* Friedrich (Fritz) Steinhoff.

2. Rössler, *The U-Boat,* p. 145.

3. On this patrol the *Barb* also sent a team ashore to plant demolition charges on railroad tracks; these were the first U.S. "combat troops" to land in Japan.

4. The German Air Force designation was Fi 103 (for the Fiesler firm that designed the missile); the designation *V* for *Vergeltungswaffe* (vengeance force) was given to the missile by the German propaganda minister, Joseph Goebbels.

5. J = Jet; B = Bomb. The Navy's Loon designation was LTV-2; L= Launching; T = Test; V = Vehicle.

6. Max Rosenberg, *The Air Force and the National Guided Missile Program* (Washington, D.C.: U.S. Air Force Historical Division Liaison Office, June 1964), p. 9.

7. In contemporary Navy documents the *Cusk* was referred to as a "bombardment" submarine (SSB).

8. Capt. John H. Sides, USN, Office of the Deputy Chief of Naval Operations (Air), before the General Board of the Navy, 4 November 1948. The Loon program was terminated in 1953.

9. Triton missile was designated SSM-N-2; the Rigel was SSM-N-6; the Regulus I was originally designated SSM-N-8; and the Regulus II was SSM-N-9. In the multi-service designation system of 1963, the Regulus I was changed to RGM-6A and the Regulus II to RGM-15A; R = ship (launched); G = surface attack; M = missile.

   The history of the Regulus program is well told in David K. Stumpf, *Regulus: The Forgotten Weapon* (Paducah, Ky.: Turner Publishing, 1996).

10. TROUNCE = Tactical Radar Omnidirectional Underwater Navigation Control Equipment.

11. In a 1957 Navy evaluation, the Circular Error Probable (CEP)—the radius of a circle in which one-half of the warheads would fall—for the Regulus I was reported to be 300 yards (274 m) at a range of 500 n.miles (926.5 km). In addition to SSG/SSGNs, TROUNCE was fitted in at least five SSNs and six diesel-electric submarines.
12. The *Tunny* was a converted *Gato* (SS 212) class and the *Barbero* was a converted *Balao* (SS 285) class; they had similar dimensions, with test depths of 300 feet (91 m) and 400 feet (120 m), respectively.
13. This was the test ship *King County* (AG 157, ex-LST 857), converted under Project Lehi at the Mare Island Naval Shipyard in 1956–1958.
14. U.S. Navy, Office of the Chief of Naval Operations, *The Navy of the 1970 Era* (Washington, D.C., January 1958), p. 7. The two new construction Regulus diesel-electric submarines (SSG) were not listed in the document.
15. The Regulus II had flown 48 test flights—30 successful and 14 partially so; there were no apparent problems with the missile.
16. These were the SSGN 594–596 and 607; they were reordered as SSNs on 15 October 1959.
17. A rare, firsthand account of arming nuclear warheads on a Regulus patrol is found in Chief Warrant Officer Jerry E. Beckley, USN (Ret), "A Few Days in October," *The Submarine Review* (April 1999), pp. 54–57. The 41 patrols—each up to about 2½ months at sea—were made by the SSG *Tunny* (9), SSG *Barbero* (8), SSG *Grayback* (9), SSG *Growler* (8), and SSGN *Halibut* (7).
18. The *Long Beach* was completed in 1961, with three surface-to-air missile launchers and anti-submarine weapons. The Regulus was never provided to the ship, nor was the Polaris ballistic missile, which subsequently was proposed for installation.
19. See, for example, George Fielding Eliot, *Victory Without War, 1958-1961* (Annapolis, Md.: U.S. Naval Institute, 1958), pp. 66–67.
20. Adm. Elmo R. Zumwalt Jr., USN (Ret), *On Watch* (New York: Quadrangle/The New York Times Book Co., 1976), p. 81.
21. The first ballistic missile submarine patrol in the Pacific begin on 25 December 1964 when the USS *Daniel Boone* (SSBN 629) departed Guam armed with 16 Polaris A-3 missiles.
22. U.S. submarine special operations during the Cold War are described in Sherry Sontag and Christopher Drew, *Blind Man's Bluff: The Untold Story of American Submarine Espionage* (New York: PublicAffairs Press, 1998). Also see Roger C. Dunham, *Spy Sub* (Annapolis, Md.: Naval Institute Press, 1996); this is a thinly veiled account of the *Halibut*'s search for the hulk of the *K-129*. (See Chapter 9.)
23. There were reports of a German suicide force being planned to attack Soviet cities using *piloted* V-1 missiles, carried toward their targets by long-range bombers. See James Harford, *Korolev: How One Man Masterminded the Soviet Drive to Beat America to the Moon* (New York: John Wiley & Sons, 1997), p. 75.
24. Polikarpov had died of cancer in July 1944. His I-15 and I-16 were outstanding fighter aircraft of the late 1930s, and his subsequent Po-2 trainer was in Soviet service for 25 years.
25. Mikoyan worked with Mikhail I. Guryevich to form the MiG bureau.
26. NATO code names for Soviet surface-to-surface missiles began with the letter *S;* the designation indicated SS = Surface (or underwater) to Surface; N = Naval; 3 = third missile to be identified; c = third version. Western intelligence did not always identify and label Soviet missiles in the same order in which they were developed.
27. Sergei N. Khrushchev, son of the Soviet leader, worked as a guidance engineer at the Chelomei OKB from March 1958 to July 1968. Chelomei and his work are described in Sergei Khrushchev, *Nikita Khrushchev and the Creation of a Superpower.*
28. The development of the first cruise missile submarines is described in I. D. Spassky (ed.), *The History of Indigenous Shipbuilding,* vol. V, pp. 146–153.
29. A. M. Antonov, "Ships of Postwar Design Projects," *Morskoy Sbornik* (no. 6, 1995), pp. 72–74.
30. In this period Chelomei also was engaged in the development of land-based ballistic missiles.
31. I. D. Spassky (ed.), *The History of Indigenous Shipbuilding,* vol. V, p. 150.
32. F. G. Dergachev, "World's First Titanium High Speed Submarine, Design 661," *Gangut* (no. 14, 1998), p. 58.
33 Project 651 and 675 submarines fitted with the satellite receiver system were given the designation suffix K.
34. The smaller S-band Front Piece was the data-link guidance antenna; the larger UHF Front Door was the missile tracking and target acquisition antenna.
35. Rear Adm. V. Khramtsov, Soviet Navy (Ret), "Why Did Not the Nuclear Catastrophe in the Far East Prevent Chernobyl?" *Morskoy Sbornik* (no. 7, 1999), pp. 58–61.
36. Rear Adm. G. H. Miller, USN, Memorandum for Assistant Chief of Naval Operations (Intelligence), Subj: "Cruise Missile Threat," OP-97/aaa, Ser 000115P97, 28 October 1968, pp. 1–2.
37. K. J. Moore, Mark Flanigan, and Robert D. Helsel, "Developments in Submarine Systems, 1956–76," in Michael MccGwire and John McDonnell (ed.), *Soviet Naval Influence: Domestic and Foreign*

*Dimensions* (New York: Praeger Publishers, 1977), p. 155.

38. A. A. Sarkisov, *The Contribution of Russian Science to the Soviet Navy* (in Russian) (Moscow: Nauka Press [Russian Academy of Sciences], 1997), p. 232.
39. This feature was provided in U.S. torpedoes to avoid their homing on the propeller noises of ASW ships.
40. Adm. Fleet of the Soviet Union S. G. Gorshkov, *Morskaya Moshch Gosudarstva* (The Sea Power of the State), 2d ed. (Moscow: Military Publishing House, 1979); translation of U.S. Naval Intelligence Command, p. 255.

## Chapter 7: Ballistic Missile Submarines

1. The German Army designation was A-4 (A = *Aggregat* or Assembly); V = *Vergeltungswaffe* (vengeance force). The latter designation was given by propaganda minister Joseph Goebbels when the first missiles were launched.

   A ballistic missile relies on the initial thrust of its engine to launch it into a ballistic trajectory; propulsion is not sustained in flight. A ballistic missile does not have wings for lift and relies on tail fins and/or engine thrust for directional control.
2. "More people died producing [the V-2] than died from being hit by it," wrote Michael J. Neufeld in *The Rocket and the Reich: Peenemünde and the Coming of the Ballistic Missile Era* (New York: Free Press, 1995), p. 264. Neufeld estimated that some 10,000 concentration camp prisoners died while working on the V-2.
3. This warhead weight included 1,620 pounds (750 kg) of high explosives.
4. Bart J. Slattery Jr., Chief Information Officer, George C. Marshall Space Flight Center, Huntsville, Ala., letter to N. Polmar, 12 March 1962.
5. The preliminary drawings of the V-2 canisters are reproduced in Ernst Klee and Otto Merk, *The Birth of the Missile: The Secrets of Peenemünde* (New York: E. P. Dutton, 1965), pp. 106–107.
6. See, for example, Neufeld, *The Rocket and the Reich,* p. 255, and Frederick I. Ordway III and Mitchell R. Sharpe, *The Rocket Team* (New York: Thomas Y. Crowell, 1979), p. 55; these books are based on German documents and discussions with principals.
7. Walter Dornberger, *V-2* (New York: Ballantine Books, 1954), pp. 215–16.
8. *Rabe* was an abbreviation of *Raketenbau und Entwicklung* (missile production and development); it consisted of the main V-2 production plant at Nordhausen and several subsidiary facilities.
9. Upon relocation to the USSR, their living standards—by Soviet norms—generally were good, with large ruble bonuses being paid to some senior scientists for their successes. All were subsequently released. There is a graphic description of their "movement" to the USSR in Harford, *Korolev,* pp. 75–77.
10. At the time the *Midway* and her two sister carriers were the world's largest warships except for the four *Iowa*-class battleships. The carriers had a standard displacement of 45,000 tons and a full load displacement of 55,000 tons; the full load displacement of the battleships was a few hundred tons greater.
11. The Navy did consider providing a V-2 launch capability in the test ship *Norton Sound* (AVM 1), but Operation Sandy was the only effort to launch a V-2 from a ship. There was brief interest in the submerged-towed V-2 canisters; see Rear Adm. C. L. Brand, USN, and Capt. A. L. Dunning, USN, Bureau of Ships, testimony before the General Board of the Navy, 20 December 1948.
12. The U.S.-NATO designations were SS-1b Scud-A.
13. The Kapustan Yar test facility is located on the Volga River, near Volgograd (Stalingrad).
14. The development of the first ballistic missile submarines is described in I. D. Spassky (ed.), *The History of Indigenous Shipbuilding,* vol. V, pp. 140–146; Malachite SPMBM, *Istoriya Sankt-Peterburgskogo Morskogo Buro Mashinostroeniya "Malakhit"* (History of the Saint Petersburg Maritime Machinebuilding Bureau Malachite), vol. 2 (St. Petersburg: Malachite, 1995), pp. 150–166; A. A. Zapol'ski, *Raketi Startuyut s Morya* (Missiles Launch From the Sea) (St. Petersburg: Malachite, 1994); and V. I. Zharkov, "Creation of the First Submarines With Ballistic Missiles," *Gangut* (no. 14, 1998), pp. 103–17.
15. Isanin previously was chief designer of the *Stalingrad*-class battle cruisers (Project 69); that project was cancelled shortly after Stalin's death in March 1953.
16. V. Zharov, "Large Missile Diesel-Electric Submarines of the Base of Project 611," *Morskoy Sbornik* (no. 9, 1995), pp. 64–71.
17. Nikita Khrushchev observed an R-11FM launch by the *B-62* on 6 October 1959 from on board a destroyer.
18. Soviet submarine missile *systems* had D-series designations; see Appendix E.
19. Some accounts incorrectly cite Project 629 as being a derivative of Project 641; e.g., "In 1956 on the basis of Project 641 torpedo attack submarine plans for Project 629 were begun" (I. D. Spassky [ed.], *History of Indigenous Shipbuilding,* vol. V, p. 142).
20. "Ballistic Missile Submarines on a Conveyor Belt," *Sudostroenie* (no. 6, 1977), p. 68. Berezovskiy later became the first commanding officer of the *Leninets* (K-137), the first Project 667A/Yankee SSBN.

21. The Project 629 missile test configurations were:

| Project | Name | NATO | System | Missiles |
|---|---|---|---|---|
| 601 | *K-118* | Golf III | D-9 | 6 R-29/SS-N-8 |
| 605 | *K-102* | Golf IV | D-5 | 4 R-27K/ SS-NX-13 |
| 619 | *K-153* | Golf V | D-19 | 1 R-39/ SS-N-20 |

In addition, under Project 629E the *K-113* was to have been converted to a specialized minelayer (cancelled); and under Project 629R three submarines were converted to communication relay ships (SSQ): *B-42, K-61,* and *K-107* (renamed *BS-83, BS-167,* and *BS-107*), respectively.

22. The *Glomar Explorer*'s purpose and special features were highly secret; she was constructed using the cover story of a seafloor mining operation under the aegis of eccentric millionaire Howard Hughes.
23. Robert Gates, the U.S. Director of Central Intelligence, delivered a videotape of the burial to the Russian government when he visited Moscow in October 1992.
24. See, for example, Vice Adm. R. Golosov, Soviet Navy (Ret), "Was the Death of the *K-129* 'Quiet'?" *Morskoy Sbornik* (no. 3, 2000), pp. 94–96.
25. The Golf SSB was at Bahia de Nipe from 29 April to 6 May 1972. She was accompanied part of the time by a destroyer and submarine tender. The original Soviet planning in 1962 for the deployment of ballistic missiles and other military forces to Cuba included two tenders that were to establish a submarine base for both torpedo-attack and four Project 658/Hotel SSBNs.
26. The submarine visited Havana from 29 April to 7 May 1974.
27. Kuzin and Nikol'skiy, *The Navy of the USSR,* p. 50.
28. Quoted in "Ballistic Missile Submarines on a Conveyor Belt," p. 68.
29. Antonov, "Ships of Postwar Design Projects," pp. 75–76.
30. I.D. Spassky (ed.), *The History of Indigenous Shipbuilding,* vol. 5, p. 144.
31. Under Project 701 the *K-145* was modified to carry six RSM-40/SS-N-8 missiles; she was given the NATO designation Hotel III.
32. I. D. Spassky (ed.), *The History of Indigenous Shipbuilding,* vol. V, p. 146.

## Chapter 8: "Polaris—From Out of the Deep . . ."

1. Minutes of Submarine Officers Conference, 18 December 1946 (report of 2 January 1947), p. 3.
2. Naval versions of the IRBM initially were designated Fleet Ballistic Missile (FBM), a designation used for the first two decades of the Polaris program; FBM has since given way to the designation SLBM, which is used for consistency throughout this volume.
3. The *United States* (CVA 58) was laid down at Newport News Shipbuilding on 18 April 1949 and was cancelled by the Secretary of Defense on 23 April 1949. She was to have had a full-load displacement of approximately 80,000 tons.
4. Initially additional funds were provided to the Navy for SLBM development, but by 1959 the Navy was forced to cancel development of the Regulus II land-attack cruise missile and the P6M Seamaster flying-boat bomber, and to delay construction of an aircraft carrier to help pay for the Polaris project. At the time all three of these programs were viewed by the Navy as strategic strike weapons.
5. Dr. David A. Rosenberg, "Arleigh Albert Burke," in Robert William Love Jr. (ed.), *The Chiefs of Naval Operations* (Annapolis, Md.: Naval Institute Press, 1980), p. 277. Admiral Burke served an unprecedented six years as CNO, from 1955 to 1961.
6. The organizational/management history of SPO is found in Dr. Harvey M. Sapolsky, *The Polaris System Development* (Cambridge, Mass.: Harvard University Press, 1972). Also see Graham Spinardi, *From Polaris to Trident: The Development of U.S. Fleet Ballistic Missile Technology* (New York: Cambridge University Press, 1994). Also see the excellent article by Capt. D. A. Paolucci, USN (Ret), "The Development of Navy Strategic Offensive and Defensive Systems," in U.S. Naval Institute *Proceedings/Naval Review* (May 1970), pp. 204–23.
7. Polmar and Allen, *Rickover,* p. 539; based on interviews by the authors with Adm. Burke.
8. Hewlett and Duncan, *Nuclear Navy;* page 309 of this official history cites a Polaris warhead weight of 600 pounds compared to a Jupiter warhead of 1,600 pounds with a similar explosive yield-one megaton (i.e., the equivalent of 1,000,000 tons of TNT).
9. The Polaris submarine program is best described in two official histories: Hewlett and Duncan, *Nuclear Navy, 1946–1962,* pp. 307–17; and Gary E. Weir, *Forged in War,* pp. 243–66; also see Strategic Systems Program, *Facts/Chronology: Polaris-Poseidon-Trident* (Washington, D.C.: Navy Department, 2000), and previous editions.
10. The Polaris submarines were initially designated SSGN(FBM), but changed to SSB(N) on 26 June 1958.
11. Hewlett and Duncan, *Nuclear Navy,* p. 308.
12. Ibid., p. 314.
13. Ibid., p. 309.
14. There have been two surface launches from U.S. Polaris submarines, one A-2 missile from the USS *Henry Clay* (SSBN 625) on 20 April 1964, and one A-3 missile from the *Nathaniel Green* (SSBN 636) on 15 March 1965.

15. The British developed trailing-wire antenna to enable submarines to operate submerged near the surface and receive low-frequency communications without raising an antenna above the water. A trailing-wire antenna was shipped to the United States on the liner *Queen Elizabeth* in the spring of 1960 to serve as a prototype for U.S. SSBNs.
16. In addition to ports in the continental United States, during the 1960s and 1970s, U.S. strategic missile submarines were based at Holy Loch, Scotland; Rota, Spain; and Guam in the Marianas.
17. At the time Eisenhower was vacationing at Newport, R.I.
18. Polaris was given the missile system designation UGM-27; U = Underwater [to]; G = Ground; M = Missile.
19. This was the first Polaris SSBN visit to a foreign port other than Holy Loch, Scotland, which had served as a forward base for Polaris submarines since March 1961.
20. U.S. Navy, Office of the Chief of Naval Operations, *The Navy of the 1970 Era,* p. 7.
21. Edward C. Keefer and David W. Mabon (eds.), *Foreign Relations of the United States, 1958–1960,* vol. III, *National Security Policy; Arms Control and Disarmament* (Washington, D.C.: Government Printing Office, 1996), p. 5. Quarles statement in Exhibit 2, "Memorandum of Discussion at the 350th Meeting of the National Security Council," 6 January 1958.
22. David A. Rosenberg, Memorandum for the Record, telephone conversation with Adm. Burke, Subj: "How the number of 41 FBM submarines was arrived at," 1 December 1974.
23. Keefer and Mabon (eds.), *Foreign Relations of the United States, 1958–1960,* vol. III, p. 427. Burke statement in Exhibit 109, "Memorandum of Discussion at the 453d Meeting of the National Security Council," 25 July 1960.
24. pK = Probability of Kill.
25. Keefer and Mabon (eds.), *Foreign Relations of the United States, 1958–1960,* vol. VIII, p. 140. McNamara statement in Exhibit 46, "Draft Memorandum From Secretary of Defense McNamara to President Kennedy," 23 September 1961.
26. Robert S. McNamara, "Statement by Secretary of Defense Robert S. McNamara before the Senate Committee on Armed Services: The Fiscal Year 1963–67 Defense Program and 1963 Defense Budget" (1962), p. 20.
27. The British missiles carried warheads with two Multiple Re-entry Vehicles (MRVs).
28. The Polaris Circular Error of Probable (CEP) is classified; however, a declassified Secretary of the Navy memorandum of 30 January 1958 credited the Polaris A-1 (1,200 n.miles) with a CEP of three to four miles and the A-2 (1,500 n.miles) with a CEP of two miles.
29. At the time the U.S. nuclear-propelled, radar-picket submarine *Triton* was larger in dimensions with a greater submerged displacement because of her hull configuration:

| | Length | Beam | Surface | Submerged |
|---|---|---|---|---|
| *Triton* | 447½ ft | 37 ft | 5,950 tons | 7,780 tons |
| *G.W.* | 381⅔ ft | 33 ft | 5,900 tons | 6,700 tons |

30. The battleship conversions—initially for Jupiter liquid–fuel missiles and, subsequently, Polaris missiles—are described in William H. Garzke Jr., and Robert O. Dulin Jr., *Battleships: United States Battleships, 1935–1992* (Annapolis, Md.: Naval Institute Press, 1995), pp. 208–12.
31. The U.S. personnel would retain control of nuclear warheads and launch control.
32. President Kennedy observed the launch of an A-3 missile from the USS *Andrew Jackson* (SSBN 619) from a research ship on 16 November 1963; the launch occurred seven days before his assassination.
33. Dr. Harold Brown, Director of Defense Research and Engineering, Memorandum for the Secretary of Defense, Subj: "SecDef project List for CY [Calendar Year] 1964, Item III b." (15 September 1964); other sources list the A-3 warheads at 225 kilotons. By comparison, the atomic bombs exploded over Hiroshima and Nagasaki were 15 and 21 kilotons, respectively.
34. Poseidon was given the missile system designation UGM-73; U = Underwater [to]; G = Ground; M = Missile.
35. Dr. Brown subsequently served as Secretary of Defense from January 1977 to January 1981.
36. Polaris had another role in U.S. ABM development, when, from 15 April to 9 December 1971, five target vehicles adapted from the A-2 missiles were launched from the Polaris test ship *Observation Island* operating some 700 n.miles (1,300 km) northwest of Midway with the Kwajalein atoll as the target area. Intercepts were attempted with the Nike-X intercept missile on Kwajalein. At one point up to 65 Polaris-based targets were planned.
37. The ex-Polaris/Poseidon submarines converted to the transport role (SSN) are listed below; previously diesel-electric submarines were used in that role.

| Number | Name | Converted | Decommissioned |
|---|---|---|---|
| SSN 609 | *Sam Houston* | 1985 | Sep 1991 |
| SSN 611 | *John Marshall* | 1985 | July 1992 |
| SSN 642 | *Kamehameha* | 1994 | Apr 2002 |
| SSN 645 | *James K. Polk* | 1993 | July 1999 |

38. The term "Triad" is believed to have been coined about 1970 by Maj. Gen. Glenn Kent, USAF, a lead-

ing Air Force strategic planner. The first public use of the term on record was the Air Force Chief of Staff, Gen. John D. Ryan, using it on 30 October 1970 in a speech before the Los Angeles Chamber of Commerce. The use of "Triad" was quickly encouraged "throughout the Air Force to help explain the continuing need for a manned strategic bomber." *Source:* Col. Charles D. Cooper, USAF, Office of Public Affairs, Department of the Air Force, letter to N. Polmar, 5 February 1981.

39. Keefer and Mabon (eds.), *Foreign Relations of the United States, 1958–1960,* vol. VIII, p. 151.
40. Rear Adm. Vasiliy I. Platonov, Soviet Navy, "The Missions of the Navy and Methods of Carrying Them Out," *Voyennaya Mysl'* (*Military Thought*), 1961; this article appeared in the second 1961 edition of the top secret journal issued by the Ministry of Defense. In the Penkovsky collection, Central Intelligence Agency, dated 4 September 1961, case no. EO-1991-00231, document 12287.
41. Harold Brown, *Department of Defense Annual Report Fiscal Year 1979* (Washington, D.C.: 2 February 1978), p. 110.
42. Ibid.

## Chapter 9: The Quest for Speed

1. As completed from 1932 to 1945, the 62 S-class submarines had an underwater speed of ten knots; their conversion included providing higher-capacity batteries and upgraded motors, and streamlining the hull and conning tower.
2. This was one of several remarkable documents on advanced submarines produced in this period by Hendrix, in the Office of Research and Inventions, Navy Department; the paper is entitled "Submarine Future Developments: Ship Characteristics" (1945); located in Submarine Officers Conference files, NHC. Hendrix also was an early proponent of nuclear propulsion; see Chapter 4. (The Office of Research and Inventions was predecessor to the Office of Naval Research.)
3. Series 58 was the sequential ship project number assigned by the David Taylor Model Basin; the *Albacore* design was later assigned Ships Characteristics Board (SCB) No. 56.
4. For a chronology of these events see Richard P. Largess, "The Origins of *Albacore,*" *The Submarine Review* (January 1999), pp. 63–73; also, Robert P. Largess and James L. Mandleblatt, *U.S.S. Albacore: Forerunner of the Future* (Portsmouth, N.H.: Portsmouth Marine Society, 1999); Gary Weir, *Forged in War;* and Capt. Harry A. Jackson, USN (Ret), discussion with N. Polmar, Groton, Conn., 30 June 1997, and correspondence between Jackson and Polmar.
5. Rear Adm. George H. Fort, USN, hearings before the General Board of the Navy, 6 April 1950. At the time Fort was a member of the General Board; he assumed chairmanship of the board on 18 April 1950.
6. Rear Adm. Charles B. Momsen, USN, Hearings before the General Board of the Navy, 28 March 1950. At the time Adm. Momsen was the Assistant Chief of Naval Operations (Undersea Warfare).
7. NACA was the predecessor of the National Aeronautics and Space Administration (NASA), which was established in 1958.
8. The motor was rated at 7,500 horsepower; however, the power was limited to the capacity of the lead acid batteries to about 4,000 horsepower for short intervals; Harry A. Jackson letter to N. Polmar, 15 December 1999.
9. Galantin, *Submarine Admiral,* p. 202.
10. SST = target-training submarine; AG = miscellaneous auxiliary; AGSS = miscellaneous auxiliary submarine. (The Navy did construct two SSTs, small conventional submarines, completed in 1953.)
11. Vice Adm. Jon L. Boyes, USN (Ret), " 'Flying' the *Albacore,*" *The Submarine Review* (April 1987), pp. 16–17. Boyes was the second commanding officer of the *Albacore* (1955–1957),
12. Ibid., p. 17.
13. Vice Adm. Jon L. Boyes, USN (Ret), quoted in Largess and Mandelblatt, *U.S.S. Albacore,* p. 64.
14. Several German coastal submarine designs of World War II did not have bow-mounted diving planes.
15. Jack Hunter, "FAB, The First Submarine Towed Array?" *The Submarine Review* (January 2000), 129–32; James W. Fitzgerald, commentary on "FAB . . . ," *The Submarine Review* (April 2000), pp. 135–36.
16. Alfred J. Giddings, "*Albacore* X-Tail: Dead End?" U.S. Naval Institute *Proceedings* (February 1995), pp. 74–76; Lt. Comdr. Jack Hunter, USN (Ret), "Does Your Submarine Have Dive Brakes? Mine Did!" *The Submarine Review* (April 1999), p. 105.
17. With the lengthening of the *Albacore,* her displacement was increased to 1,631 tons surfaced and 1,837 tons submerged.
18. Largess and Mandelblatt, *U.S.S. Albacore,* pp. 132–34; Jackson discussions with N. Polmar, Groton, Conn., 30 June 1997.
19. Vice Adm. E. W. Grenfell, USN (Ret), Commander Submarine Force, Atlantic Fleet, cited in Largess and Mandelblatt, *U.S.S. Albacore,* p. 137.
20. Harry A. Jackson letter to N. Polmar, 23 August 1996.
21. Engelhardt, "Soviet Sub Design Philosophy," pp. 194–95.
22. The *Skipjack*'s reactor plant arrangement differed from the arrangement in the five subsequent submarines of the class.

23. For a description of the development of the S5W, see Hewlett and Duncan, *Nuclear Navy,* pp. 281–82.

24. Capt. Kearn interview with Dr. Gary E. Weir, Naval Historical Center, Washington, D.C., 12 September 1990.

25. Harry A. Jackson letter to N. Polmar, 23 August 1996.

26. The X-tail configuration subsequently was used in the diesel-electric submarines of the Australian, Dutch, German, and Swedish navies.

27. During the uncontrolled dive the turbines—driving the submarine forward—did not respond to the controls in the maneuvering room, because the mechanical linkage between the controls and the turbines had failed because of a pin dropping out; ultimately, a machinist's mate was able to manually reverse the turbines from a position in the turbine spaces. *Sources:* Robin B. Pirie, commanding officer of the *Skipjack,* 1969–1972, discussion with N. Polmar, Arlington, Va., 26 November 2001, and Commo. Robin W. Garson, RN, who was on board the *Skipjack* at the time of the uncontrolled dive, several telephone conversations with N. Polmar during 2001–2002.

28. Named for the crossing point between East and West Berlin from 1961 to 1989.

29. Adm. Arleigh A. Burke, USN, Chief of Naval Operations, speech at the launching of the *Skipjack,* Groton, Conn., 26 May 1958.

30. During the submerged circumnavigation by the *Triton* in 1960, that submarine twice transferred personnel with her sail awash and periodically used her snorkel, hence she did not maintain a sealed atmosphere.

31. See V. Bil'din, "Design 661 Nuclear Attack Submarine," *Morskoy Sbornik* (no. 4, 1993), pp. 64–66; L. Cherkashin, "Our Submarine Wins the Blue Ribbon," *Morskoy Sbornik* (no. 12, 1994), pp. 42–44; Capt. 1/R Igor Bogachenko, Soviet Navy (Ret), "Russian Submarines with Titanium Hulls," *The Submarine Review* (July 1997), pp. 93–97; and F. G. Dergachev, "World's First Titanium High Speed Submarine, Design 661," *Gangut* (no. 14, 1998), pp. 56–61.

32. Dr. George Sviatov discussions with N. Polmar, Bethesda, Md., 25 May 2001, and 12 June 2001.

33. Dergachev, "World's First Titanium High Speed Submarine, Design 661," p. 61.

34. Details of titanium processes provided by Malachite engineers at a meeting with N. Polmar, St. Petersburg, 7 May 1997. Building hall No. 42 had been erected in the 1950s and Project 627/November SSNs and Project 629/Golf SSBs had been constructed in that facility; subsequently, the titanium-hull Project 661/Papa SSGN, three Project 705/Alfa SSNs, and Project 685/Mike SSN were built in that hall.

35. "Rubin" was the name of the sonar; it was not related to the TsKB-18/Rubin submarine design bureau.

36. The Soviet designation for this missile was P-70; the U.S.-NATO designation for this weapon was SS-N-7 Starbright.

37. The alloy 48-OT3V had a specific weight of 4,500 kg/m$^3$ (280 lb/ft$^2$) and a yield of 6,000 kg/cm$^2$.

38. Cherkashin, "Our Submarine Wins the Blue Ribbon," p. 43.

39. Details of the reactor casualty are in Cherkashin, "Our Submarine Wins the Blue Ribbon," p. 44.

40. See R. A. Shmakov, "Ahead of Their Time . . ." *Morskoy Sbornik* (no. 7, 1996), pp. 57–61; R. A. Shmakov, "The Small, Fast, Automated Submarine-Interceptor of Projects 705 and 705K," *Tayfun* (no. 3, 1997), pp. 2–13; and Dr. Igor Sutyagin, "Alfa Class—The 1960's Dream Machine," *The Submarine Review* (January 1998), pp. 72–79.

41. Lothar-Günther Buchheim, *The Boat* (New York: Bantam Books, 1976), p. 125.

42. V. I. Barantsev discussion with N. Polmar, St. Petersburg, 6 December 1996; also A. M. Antonov, et al., "40 Years of Nuclear Submarine Development: A View of a Designer," paper presented at Warship '96: International Symposium on Naval Submarines, London, 12 June 1996.

43. Based on a speed of 40 knots; A. M. Antonov, et al., "40 Years of Nuclear Submarine Development: A View of a Designer," paper given at conference, London, 12 June 1996.

44. Ibid.

45. The escape chamber was tested with men on board from a special test platform in the Gulf of Finland in 1965.

46. Sudomekh (shipyard No. 196) became a major submarine construction yard in the 1930s. After World War II equipment removed from German submarine yards was installed in Sudomekh, and advanced submarine designs and propulsion systems became a hallmark of the yard. In 1972 the Sudomekh yard became part of the Admiralty shipyard complex.

47. The coveted Lenin Prize subsequently was conferred on Romin, and Rusanov was awarded the Order of Lenin for their work on Project 705.

48. The keel for the *K-123,* the first Severodvinsk unit, was laid down on 29 December 1967, i.e., before the *K-64,* but she was not completed until 1977. The cancelled unit was one of those assigned to Sudomekh; she was laid down as shipyard hull No. 930, but never launched.

49. Comdrs. William (Bill) Manthorpe and Steve Kime, also assistant U.S. naval attachés, while visiting Leningrad in the summer of 1973 retrieved additional titanium debris outside of the Sudomekh yard. *Source:* U.S. naval intelligence officers' discussions

with N. Polmar, Washington, D.C., 18 April 2001, Annapolis, Md., 25 April 2001, and Arlington, Va., 26 April 2001; also Capt. Steve Kime, USN (Ret), telephone conversation with N. Polmar, 29 May 2001.

50. Capt. Richard Brooks, USCG (Ret), telephone discussion with N. Polmar, 1 March 2002.
51. Vice Adm. William H. Rowden, USN, Deputy Chief of Naval Operations (Surface Warfare), hearings before the Appropriations Committee, Senate, 8 April 1981.
52. Capt. D. A. Paolucci, USN (Ret), discussions with N. Polmar, Alexandria, Va., 15 January 1980, and 26 June 1980; and Capt. S. D. Landersman, USN (Ret), discussions with N. Polmar, San Diego, Calif., 31 January 1980.
53. "The A-class SSN is the world's fastest submarine and probably the deepest diving (with an estimated operating depth of 640 meters)"; *source:* Director of Central Intelligence, *Soviet Naval Strategy and Programs Through the 1990s* (NIE 11-15-82/D, March 1983), p. 25.
54. A. M. Antonov, et al., "40 Years of Nuclear Submarine Development—A View of a Designer," paper given at conference Warship '96, London, 12 June 1996.
55. R. A. Shmakov, "Ahead of Their Time . . ." p. 58.
56. The *K-123* was renamed *B-123* on 3 June 1992.
57. The boundary layer is the fluid layer immediately adjacent to a boundary, such as a submarine's hull and, with a constant exchange of fluid, moves with it.

## Chapter 10: Second-Generation Nuclear Submarines

1. Rear Adm. C. L. Brand, USN, Bureau of Ships, testimony before the General Board of the Navy, 20 December 1948.
2. Capt. Richard B. Laning, USN (Ret), discussion with N. Polmar, Orlando, Fla., 23 December 1979.
3. Joint Committee on Atomic Energy, Congress of the United States, *Loss of the U.S.S. Thresher* (Washington, D.C.: Government Printing Office, 1965), pp. 78–79. This is the transcript of hearings held in June–July 1964. Also see N. Polmar, *Death of the USS Thresher* (New York: Lyons Press, 2001); this book, originally published in 1964, was written with the assistance of the *Thresher*'s first commanding officer, then-Comdr. Dean L. Axene, USN.
4. During barge tests angled torpedo tubes successfully launched torpedoes at speeds up to some 15 knots; subsequently, a 21-inch torpedo tube was installed in the underwater hull of the fleet oiler *Neosho* (AO 143), and she launched torpedoes at speeds of about 18 knots. The *Tullibee* was the first U.S. submarine to be completed with amidships torpedo tubes.
5. The Mk 37 had a 19-inch (482-mm) diameter but was launched from 21-inch torpedo tubes with guide rails; it carried a 330-pound (150-kg) warhead to a range of about 4.3 n.miles (8 km). The Mod 0 and 3 had acoustic guidance; the Mod 1 and 2 had combination wire-acoustic guidance. All previous U.S. nuclear submarines had six bow torpedo tubes except for the *Halibut* (SSGN 587), which had four plus two stern tubes.
6. The SUBROC missile designation was UUM-44A; U = Underwater [to], U = Underwater; M = missile.
7. HY-80 indicated High Yield steel capable of withstanding a working tensile stress of 80,000 pounds per square inch (approximately 60 kg/mm$^2$).
8. Capt. Donald Kern, USN (Ret), interview with Dr. Gary E. Weir, Naval Historical Center, Washington, D.C., 12 September 1990.
9. Joint Committee on Atomic Energy, *Loss of the U.S.S. Thresher,* p. 122.
10. The rafting concept was used in the same period in HMS *Valiant,* Britain's second nuclear-propelled submarine.
11. Hervey, *Submarines,* p. 195.
12. On her initial trials with her original propeller, the *Thresher* reached a speed of almost 33 knots; *source:* Harry A. Jackson interview with N. Polmar, Groton, Conn., 30 June 1997.
13. Harry A. Jackson interview with N. Polmar, Groton, Conn., 30 June 1997.
14. Vladimir Barantsev discussion with N. Polmar, St. Petersburg, 6 December 1996.
15. In addition to 12 officers and 96 enlisted men assigned to the *Thresher,* the ship went down with four naval officers and 14 civilian employees of the Navy, and three civilian factory representatives on board for sea trials.
16. Scram = Super Critical Reactor Ax Man. The word refers to the earliest days of reactor development when controls were crude and engineers joked that the only solution to a runaway rector was to "scram." Later, in the acronym era, nuclear-power trainees were told that in the first reactor (the Fermi pile outside of Chicago), control rods were suspended by rope and a man with an ax stood ready to cut the ropes if the reactor went critical.
17. See Capt. Frank Andrews, USN (Ret), "*Thresher*'s Debris Field," *The Submarine Review* (April 1987), p. 10.
18. At depth the full (emergency) blow of the ballast tanks produced pressure and temperature drops that caused moisture in the compressed air to freeze and block the valves and strainers in the ballast tank blow system. This situation had not been discovered because of the Navy's reluctance to execute a full blow at depth to avoid wear and tear on compressors and related factors.

19. L. A. Samarkin, "The Seventy-First," *Tayfun* (no. 4, 1997), p. 17.

20. During operations in the Mediterranean in 1978, the *Tullibee*'s propeller fell off, and the shaft, continuing to rotate, began admitting water into the submarine. The *Tullibee,* at shallow depth, was able to blow ballast and reach the surface, but without an auxiliary propulsion system, she was immobilized. The flooding was halted, and she was towed to Rota, Spain, for repairs.

21. Rear Adm. Dick Jortberg, USN (Ret), "The Legacy of the *Tullibee,*" *The Submarine Review* (January 1989), p. 82.

22. Ibid., p. 84.

23. One of the seized seafloor recording devices is on display in the KGB Museum in Moscow. These special operations to tap into the Sea of Okhotsk cable are described in Sontag and Drew, *Blind Man's Bluff,* pp. 158–83.

24. Hewlett and Duncan, *Nuclear Navy,* p. 282.

25. See William M. Leary, *Under Ice: Waldo Lyon and the Development of the Arctic Submarine* (College Station: Texas A&M University Press, 1999); and N. Polmar, "Sailing Under the Ice," U.S. Naval Institute *Proceedings* (June 1984), pp. 121–23.

26. Platonov, "The Missions of the Navy and Methods of Carrying Them Out."

27. L. A. Samarkin, "Multipurpose Nuclear Submarines of Project 671," *Morskoy Sbornik* (no. 2, 1995), pp. 72–76, republished as "The Seventy-First," *Tayfun* (no. 4, 1997), pp. 10–12; and R. A. Shmakov, "Building Atomic Submarine Projects 671, 671RT and 671RTM," *Sudostroenie* (no. 1, 2000), pp. 24–32.

28. Samarkin, "The Seventy-First," p. 11.

29. Samarkin, "The Seventy-First," p. 11.

30. The previous HEN-type nuclear submarines had direct-current, 320V electrical systems.

31. RT = *Raketnyy Torpeda* (rocket torpedo).

32. The designation 65-76 indicated a 65-cm diameter torpedo that entered service in 1976.

33. Nikolai Spassky (ed.), *Russia's Arms Catalog,* vol. III, *Navy* (Moscow: Military Parade, 1996), p. 71.

34. The Victor III was initially given the NATO code name "Uniform." RTM = *Raketnyy Torpeda Modifikatsirovanny.*

35. See, for example, David Brady and John Edyvane, "Propulsion in the Pod-Fact or Fiction?" *The Submarine Review* (April 1986), pp. 19–24.

36. Later named *60 Let Shefstva Vlksm.*

37. Rear Adm. Sumner Shapiro, USN, Director of Naval Intelligence, statement before Senate Armed Services Committee, 14 April 1978. U.S. nuclear attack submarines are similarly used to "delouse" possible Soviet SSN trailers of U.S. strategic missile submarines.

38. Rear Adm. John L. Butts, USN, Director of Naval Intelligence, statement before Senate Armed Services Committee, 26 February 1985.

39. Lt. Col. I. Altunin, Russian Navy, "An Ordinary Deployment," *Morskoy Sbornik* (no. 1, 1994), pp. 36–39. The article consists of excerpts from the diary of Capt. 2/R. A. Stakheyev, commanding officer of the submarine; reprinted in part in the U.S. Naval Institute *Proceedings* (July 1994), pp. 79–82.

40. I. D. Spassky (ed.), *The History of Indigenous Shipbuilding,* vol. V, pp. 294–96; and A. N. Gusev, *Submarines With Cruise Missiles* (in Russian), (St. Petersburg: Galeya Print, 2000), pp. 70–71.

41. "Malachite" was the name of the missile and was not related to the SKB-143/Malachite submarine design bureau.

42. Thus India became the sixth nation to operate nuclear-propelled submarines, after the United States, Soviet Union, Great Britain, France, and China.

43. Her commanding officer, Capt. 1/R N. M. Suvorov, was tried, found guilty, and sentenced to ten years in prison. He was amnestied after 2½ years of incarceration.

44. The two ships of this class were the *Moskva,* completed in 1967, and the *Leningrad,* completed in 1968.

45. The total includes the *Tullibee* (S2C reactor plant) and *Narwhal* (S5G reactor plant).

## Chapter 11: The Ultimate Weapon I

1.

| Class/Project | | Submarines | Missiles | Total |
|---|---|---|---|---|
| Zulu | V611, AV611 | 6 | 2 | 12 |
| Golf | 629, 629A | 22 | 3 | 66 |
| Golf | 629B | 1 | 2 | 2 |
| Hotel | 658 | 8 | 3 | 24 |

2. SRF = *Raketniye Voiska Strategicheskogo Naznacheniya.*

3. PVO = *Protivo-vozdushnoi Oborony Strany.*

4. Unlike the U.S. Strategic Air Command, which controlled U.S. land-based ICBM and strategic bomber forces, the Soviet Strategic Rocket Forces did not control Soviet long-range bombers. The bombers were under the operational control of the Soviet High Command and under the administrative control of the Air Forces.

5. S. N. Kovalev discussion with N. Polmar, St. Petersburg, 6 May 1997. Premier Khrushchev's son, Sergei, an engineer specializing in missile guidance, worked for Chelomei for ten years (March 1958 to July 1968). Chelomei, already noted for innovative cruise missile designs, further benefited from having the younger Khrushchev in his bureau. The

U.S.-NATO names "Sark" and "Serb" were applied to several SLBM variants of this period.

6. Michael MccGwire, "The Strength of the Soviet Navy," in MccGwire (ed.), *Soviet Naval Developments: Context and Capability* (Halifax, Nova Scotia: Dalhousie University [1973]), pp. 133–34. MccGwire was previously an analyst in British naval intelligence.
7. I. D. Spassky (ed.), *The History of Indigenous Shipbuilding,* vol. V, pp. 287–91; Kuzin and Nikol'skiy, *The Navy of the USSR,* pp. 51–61. The term "modern" denotes U.S. Polaris and Soviet Project 667A/Yankee and later SSBNs.
8. V. P. Semyonov discussion with N. Polmar, St. Petersburg, 6 May 1997. Semyonov was longtime principal deputy to Kovalev.
9. The ICBM had the Soviet designation RT-15 (NATO SS-X-14).
10. This device differed from later escape chambers in Soviet submarines by being a tethered, up-and-down, two-man device.
11. Capt. 1/R S. A. Novoselov, Soviet Navy (Ret), discussion with N. Polmar, St. Petersburg, 5 December 1996.
12. This is depth of water *above* the submarine (through which the missile travels); submarine depths normally are the distance from the submarine's keel to the surface.
13. This was also true of later U.S. submarines; see Rear Adm. Albert L. Kelln, USN, Director, Strategic Submarine Division and Trident Program Coordinator, testimony before the Committee on Armed Services, Subcommittee on Research and Development, U.S. Senate, 14 April 1978; also Douglas C. Waller, *Big Red: Three Months on Board a Trident Nuclear Submarine* (New York: HarperCollins, 2001), p. 205; and Jim Ring, *We Come Unseen: The Untold Story of Britain's Cold War Submariners* (London: John Murray, 2001), p. 89.
14. "Ballistic Missile Submarines on a Conveyer Belt," p. 70; also Kovalev discussion with N. Polmar, St. Petersburg, 6 May 1997; and I. D. Spassky (ed.), *The History of Indigenous Shipbuilding,* vol. V, p. 288.
15. See A. M. Ovcharenko, "Analysis of the Effectiveness of Groupings of Design 667A (AU) RPKSN [SSBN] in the System of the Soviet Union's Strategic Nuclear Forces," *Proceedings of the Scientific-Technical Conference "The Second Makeev Readings,"* series XIV, edition 1 (4), 1996, pp. 53–64.
16. Central Intelligence Agency, "Depressed Trajectories: Unlikely Role for Soviet SLBMs" (An Intelligence Memorandum) (SW 82-10075 (August 1982), p. iii.
17. Ibid., p. 6.
18. Kuzin and Nikol'skiy, *The Navy of the USSR,* p. 52.
19. A highly dramatic account of the *K-219*'s loss, containing major errors, appears in Capt. Peter Huchthausen, USN (Ret), Capt. 1/R Igor Kurdin, Russian Navy, and R. Alan White, *Hostile Waters* (New York: St. Martin's Press, 1997). Kurdin was executive officer of the *K-219* when she was lost. A much shorter version, that "leaves out the *Hollywood,*" was written by Kurdin and Lt. Wayne Grasdock, USN, "Loss of a Yankee SSBN," *The Submarine Review* (October 2000), p. 56–67.
20. The first major studies of this subject to appear in public were Donald C. Daniel, *Anti-Submarine Warfare and Superpower Strategic Stability* (Urbana: University of Illinois Press, 1986), and Tom Stefanick, *Strategic Anti-Submarine Warfare and Naval Strategy* (Lexington, Mass.: Lexington Books, 1987). Daniel was at the Center for Naval Warfare Studies at the U.S. Naval War College; Stefanick was with the Federation of American Scientists, a private organization.
21. See Chapter 2 for a description of SOSUS.
22. HF/DF-known as "huff-duff"—sought to detect Soviet submarine-to-shore communications to determine the location of submarines. The U.S. name for these facilities was *Wullenweber,* the name as well as the equipment being copied from the Germans; the Soviets had similar facilities to detect U.S. naval forces. HF/DF of submarine communications was a major factor in the Anglo-American victory over German U-boats in World War II.
23. Riste, *The Norwegian Intelligence Service,* p. 147.
24. Confidential source A discussion with N. Polmar, Washington, D.C., 22 August 1997.
25. Ibid. Ames was a Central Intelligence Agency counterintelligence officer who spied for the Soviets and, after the fall of the USSR, for Russia. When Ames was arrested in 1994, federal officials said that he had perpetrated the most costly breach of security in CIA history. During at least nine years as a Soviet agent, he had revealed more than 100 covert operations and betrayed more than 30 operatives spying for the CIA and other Western intelligence services.
26. Quoted in Melissa Healy, "Lehman: We'll Sink Their Subs," *Defense Week* (13 May 1985), p. 18. One of the first meaningful public discussions of this subject was Capt. John L. Byron, USN, "No Quarter for Their Boomers," U.S. Naval Institute *Proceedings* (April 1989), pp. 49–52.
27. Adm. James D. Watkins, "The Maritime Strategy," supplement to the U.S. Naval Institute *Proceedings* (January 1986), p. 11.
28. George C. Wilson, "Navy Is Preparing for Submarine Warfare beneath Coastal Ice," *The Washington Post* (19 May 1983), p. A5.
29. From Commanding Officer USS *Batfish* (SSN 681), to Chief of Naval Operations (Op-095), Subj:

Report of Mission LS-26, 2 March 1978-17 May 1978; 17 May 1978, ser LS-26-D-0006-T-78. Also see Thomas B. Allen, "Run Silent, Run Deep," *Smithsonian Magazine* (March 2001), pp. 51–61.

30. From Commanding Officer USS *Guardfish* (SSN 612), to Commander in Chief U.S. Pacific Fleet, Subj: Trail of Echo II nuclear submarine (Case Papa 07) during the period 12 May 1972-6 June 1972; 10 June 1972, ser 00015-72.
31. USS *Batfish* report, Enclosure (1) "Abstract," p. 1.
32. The active sonar was used every one to three hours throughout the Yankee's transit to patrol area and while in the alert area; there was one three-day period when the sonar was not intercepted after the Yankee began the home transit; USS *Batfish,* Enclosure (1) "Abstract," p. 7. The NATO term is derived from the "ping" of the active sonar, said to sound like the sharp clapping together of two blocks of wood.
33. Such baffle-clearing maneuvers at high speeds, sometimes involving a rapid descent to a deeper depth, are referred to as "crazy Ivan turns" by U.S. submariners.
34. Rear Adm. Thomas Evans, USN (Ret), discussion with N. Polmar, Washington, D.C., 26 January 2001.
35. See, for example, Milan Vego, *Soviet Naval Tactics* (Annapolis, Md.: Naval Institute Press, 1992), pp. 163–64, and B. N. Makeyev, *Voyenno-morskiye aspekty natsionalnoi bezopasnosti* (Naval Aspects of National Security) (Moscow: Nonproliferation and Critical Technologies Committee, 1997), pp. 63–67.
36. As part of the U.S. Navy's SSBN security program, the Anti-Launch phase Ballistic missile Intercept System (ALBIS) project culminated with the live firing of a Terrier surface-to-air missile against a submerged-launched Polaris A-2, reflecting a belief that the Soviets could employ a similar tactic. The attempted Polaris intercept failed.
37. One of the few discussions of these activities is N. Polmar and Raymond Robinson, "The Soviet Non-naval Force Multiplier," U.S. Naval Institute *Proceedings* (December 1987), pp. 66–69. The large nuclear icebreakers *Arktika* and *Rossiya* were armed on their trials; the KGB Border Guard icebreakers were armed and other naval icebreakers had provisions for weapons and naval electronics.
38. The first 39 submarines of the subsequent *Los Angeles* (SSN 688) class were not configured for under-ice operations.
39. There were eight missile tubes in compartment No. 4 and four tubes in No. 5.
40. SALT = Strategic Arms Limitation Talks. The SALT I negotiations extended from November 1969 to May 1972.
41. In addition to the Yankee/Delta SSBNs, the following older ballistic missile submarines were in service in 1974:

    7 SSBNProj. 658M/Hotel II

    1 SSBNProj. 701/Hotel III

    9 SSB Proj. 629/Golf I

    13 SSB Proj. 629A/Golf II

    *Source:* Defense Intelligence Agency, "Soviet Submarine Order of Battle: 1950-1974," DDB-1220-14-78 (16 August 1979), p. 300.
42. By 1972 the Soviet Union had exceeded the United States in total intercontinental strategic offensive delivery vehicles—ICBMs, SLBMs, and long-range bombers. At the time the United States had 1,054 operational ICBMs and none being produced; the Soviet Union had an estimated 1,618 ICBMs operational and being produced.
43. Pavel L. Podvig (ed.), *Strategicheskoye yadernoye vooruzhenie Rossi* (Russia's Strategic Nuclear Arms) (Moscow: IzdAT, 1998), p. 209.
44. Central Intelligence Agency, "Soviet Naval Strategy and Programs Through the 1990s," NIE 11-15-82/D (March 1983), p. 15.
45. See for example, Capt. 1/R Rank V. N. Parkhomenko, "Guarding the Decibels," *Morskoy Sbornik* (no. 6, 1996), pp. 64–67. Parkhomenko estimates that the cost of noise reduction measures as a percent of total *SSN* cost reached 20 percent in the 1990s.
46. Adm. Frank B. Kelso, II, USN, Commander-in-Chief, U.S. Atlantic Command, testimony before the House Appropriations Committee, Subcommittee on Defense, U.S. Congress, 7 February 1989.
47. Rear Adm. Thomas A. Brooks, USN (Ret), former Director of Naval Intelligence, discussion with K. J. Moore and N. Polmar, Arlington, Va., 29 October 2001.
48. See K. J. Moore and N. Polmar, "The Yankee and the SS-N-13," paper presented at Dalhousie University, Halifax, Nova Scotia, 8 September 1974.
49. The R-27K missile designation was 4K18. The NATO designation was SS-NX-13; the U.S. intelligence community initially used the designation KY-9, the prefix KY indicating the Kapustin Yar test facility.
50. I. D. Spassky discussion with N. Polmar, Arlington, Va., 25 October 1993.
51. See, for example, K. J. Moore, "Antisubmarine Warfare" in *Soviet Naval Influence,* pp. 196–98; also mentioned in "Industry Observer," *Aviation Week and Space Technology* (21 July 1975), p. 11.
52. N. S. Kovalev discussion with N. Polmar, St. Petersburg, 6 May 1997.
53. Confidential source MA letter to N. Polmar, 14 October 2001.
54. Gen. George S. Brown, USAF, *United States Military Posture for FY [Fiscal Year] 1978* (Washington, D.C.: Department of Defense, 20 January 1977), p. 16.

55. National Intelligence Daily, 6 June 1975, as quoted in Jack Anderson, "U.S. Kept Mum on Soviet Sub Patrol," *The Washington Post* (26 November 1980), p. B14.
56. Podvig, *Russia's Strategic Nuclear Arms,* p. 231.
57. These SSBNs were based at:

| Northern Fleet | Pacific Fleet |
|---|---|
| Yageinaya | Pavlovsk (near Vladivostok) |
| Olenya | Petropavlovsk |
| Ostrovnoy | |

58. Rear Adm. Malcolm I. Fages, USN, "Aboard a Delta I in the Russian Pacific Fleet," *The Submarine Review* (July 1997), p. 69.
59. "But our attack submarines also held the Soviet submarine force—and her other very capable forces—continually at risk, playing a key role in the Soviets shifting *their* Maritime Strategy from one of open ocean threat in the Atlantic and Pacific, to one of defensive hunkering in their own bastions, in the Barents and Sea of Okhotsk"; Adm. F. L. Bowman, USN, Director, Naval Nuclear Propulsion, Department of Energy, remarks to U.S. Navy League, Long Island Council, Melville, N.Y., 18 February 1999. *Bastions* was a Western term, and was not used by the Soviets.
60. Central Intelligence Agency, "Depressed Trajectories: Unlikely Role for Soviet SLBMs," p. iii.

## Chapter 12: The Ultimate Weapon II

1. WS = Weapon System.
2. Institute for Defense Analyses, *The Strat-X Report* (August 1967), p. 1.
3. See Spinardi, *From Polaris to Trident* pp. 113–63, and N. Polmar and Capt. D. A. Paolucci, USN (Ret), "Sea-Based 'Strategic' Weapons for the 1980s and Beyond," U.S. Naval Institute *Proceedings/Naval Review* (May 1978), pp. 99–113. Also Rear Adm. George H. Miller, USN, one of the two Navy participants in Strat-X, discussions with N. Polmar, Pentagon, Washington, D.C., during 1968–1969. In this period Adm. Miller was Director, Strategic Offensive and Defensive Systems (OP-97); Capt. Paolucci was his deputy.
4. A Minuteman ICBM was test launched from a C-5A transport on 24 October 1974; the missile was extracted by parachute and while in descent the solid-propellant rocket was ignited.
5. The original Polaris was proposed for launching from surface ships, both cruisers and specialized merchant-type ships, the latter to be operated by multi-national NATO crews.
6. An integrated Soviet satellite ocean surveillance system became operational by 1974. The Soviets had previously employed photo and weather satellites for ocean surveillance.
7. Adm. Thomas H. Moorer, USN, Chairman, Joint Chiefs of Staff, *United States Military Posture for FY [Fiscal Year] 1973* (8 February 1972), pp. 10–11.
8. Capt. 1/R Ye. Semenov, Soviet Navy, "On the Survivability of Submarines Under the Threat From the Air," *Morskoy Sbornik* (no. 1, 1988), p 23; and Rear Adm. Yu. Kviatkovskii, Soviet Navy, "Current Status and Development Prospects of Forces and Means for Combating Submarines," *Voennaya mysl'* (no. 1, 1988), p. 39.
9. Gen. Maj. M. A. Borchev, Soviet Army (Ret), "On the Military Organization of the Commonwealth of Independent States," *Voennaya mysl'* (no. 3, 1993), p. 7.
10. Hung P. Nguyen, *Submarine Detection From Space: Inferences From Soviet Writings* (Alexandria, Va.: Center for Naval Analyses, February 1992), CRM 92-29; and Nguyen, *Submarine Detection from Space: A Study of Russian Capabilities* (Annapolis, Md.: Naval Institute Press, 1993).
11. Vladislav Repin, "One More Time About Strategic Dilemmas," *Nezavisimaia gazeta* (24 September 1992), p. 2.
12. "Shuttle Flight Yields Data on Hiding Subs," *Washington Post* (22 March 1985), p. A10. The astronaut was Paul Scully-Power on board the shuttle STS-41G *Challenger* on 5–13 October 1984; he was an oceanographer at the Naval Underwater Systems Command in New London, Conn.
13. See, for example, Bob Woodward and Charles R. Babcock, "CIA Studies Sub Vulnerability," *The Washington Post* (6 June 1985), pp. 1, 16.
14. Anthony R. Battista statement before the Senate Armed Services Committee, 26 April 1994. Mr. Battista, a former Navy engineer, was a senior staff member of the Committee from 1974 until 1988; he has remained a consultant to the Senate on non-acoustic ASW.
15. Battista statement, 26 April 1994.
16. "Special access" programs are "black," i.e., very highly classified with strictly limited access.
17. Dr. Edward Whitman memo to N. Polmar, 23 January 2001.
18. Dr. Richard E. Twogood, Lawrence Livermore National Laboratory, statement before the Senate Armed Services Committee, 26 April 1994.
19. Craven, *The Silent War,* p. 265.
20. See Chapter 18 for a discussion of wake-homing torpedoes.
21. U.S. Naval Intelligence Support Center, "Soviet Surveillance Capabilities Against U.S. Naval Forces," DST-1280S-607-79 (1 August 1979), p. III-6.
22. The target satellites launched by the USSR during the anti-satellite development cycle had orbital characteristics similar to the U.S Transit satellites.
23. U.S. Polaris-Poseidon submarines were based at

Holy Loch, Scotland; Rota, Spain; and Guam as well as at continental U.S. locations.

24. Central Intelligence Agency, "Soviet Naval Strategy and Programs Through the 1990s," NIE 11-15-82/D (March 1983), pp. 16, 18.
25. Ibid., p. III-7.
26. Confidential source A memo to N. Polmar, 28 August 2001.
27. S. Eugene Poteat, "Counterintelligence Spy vs. Spy vs. Traitor," *American Intelligence Journal* (Winter 2000–2001), p. 57. Poteat, a retired CIA scientific officer, at the time was president of the Association of Former Intelligence Officers.
28. See P. Wittenbeck and K.J. Moore, "The Rising Vortex: A Source for NA [Non-Acoustic] ASW," presentation at UDT `93 conference, Cannes, France, 16 June 1993. The first Soviet ASW ship classes to mount the Top Sail radar were the helicopter carrier-missile cruiser *Moskva*, completed in 1967, and the missile cruiser *Kronshtadt* (NATO Kresta II), completed in 1969.
29. Central Intelligence Agency, "Soviet Naval Strategy and Programs Through the 1990s," p. 57.
30. See Raymond A. Robinson, "Incoming Ballistic Missiles at Sea," U.S. Naval Institute *Proceedings* (June 1987), pp. 66–71.
31. See Jerry Razmus, "SSBN Security," *The Submarine Review* (April 1996), pp. 25–35, and (July 1996), pp. 34–45.
32. The USSR deployed a limited ballistic missile defense system around Moscow in the mid-1960s; the United States briefly deployed such a system in the Mid-west to protect an ICBM base.
33. Lt. Comdr. Harlan Ullman, USN, "The Counter-Polaris Task," in Michael McGwire (ed.), *Soviet Naval Policy: Objectives and Constraints* (New York: Praeger, 1975), p. 597.
34. Ibid. Ullman retired from the U.S. Navy as a captain.
35. The Poseidon was 34 feet (10.4 m) long with a diameter of 74 inches (1.9 m); the new missile would have almost twice the volume of Poseidon.
36. Institute for Defense Analyses, *The Strat-X Report,* p. 84.
37. Smith officially retired from the Navy in 1972 when he reached the statutory retirement age, but was recalled to active service to remain the director of SSPO until 1977.
38. Spinardi, *From Polaris to Trident,* p. 115. FBM = Fleet Ballistic Missile.
39. Zumwalt, at age 50, was the youngest CNO and the youngest full admiral in U.S. Navy history.
40. Project Trident in the early 1960s addressed research and development for ocean surveillance, active and passive acoustic as well as non-acoustic efforts.
41. Harold Brown, *Department of Defense Annual Report Fiscal Year 1981* (Washington, D.C.: Department of Defense, 29 January 1980), p. 131.
42. Trident C-4 was given the missile system designation UGM-96A and Trident D-5 was UGM-96B; U = Underwater [to]; G = Ground; M = Missile.
43. The president approved Brickbat priority for the Trident program on 14 December 1972.
44. The best public discussion of the chaotic state of affairs in nuclear submarine construction and at Electric Boat is Patrick Tyler, *Running Critical: The Silent War, Rickover, and General Dynamics* (New York: Harper and Row, 1986).
45. Newport News Shipbuilding constructed and overhauled/refueled nuclear-propelled aircraft carriers and missile cruisers as well as submarines.
46. See John F. Lehman Jr., *Command of the Seas: Building the 600 Ship Navy* (New York: Charles Scribner's Sons, 1988), pp. 1–8; Norman Polmar and Thomas B. Allen, "The Plot to Get Rickover," *Washingtonian* (April 1982), pp. 140–45. Under law, Rickover's retention on active duty was required to be extended every two years.
47. Subsequently, Secretary Lehman named an attack submarine the *Hyman G. Rickover* (SSN 709), preempting a possible move by some members of Congress to name an aircraft carrier for the admiral.
48. The 41 Polaris strategic missile submarines completed from 1960 to 1967 were named for "famous Americans," although several were in fact named for persons never in the American colonies or United States. The subsequent Trident submarines, completed from 1981 to 1997, were named for States of the Union. Previously, state names were assigned to battleships and, later, to guided missile cruisers (CGN 36–41). The exception to the state name source for Trident submarines was made on 27 September 1983, when the SSBN 730 was named for the late Senator Henry M. (Scoop) Jackson, longtime supporter of nuclear and defense programs.
49. Rear Admiral G. P. Nanos, USN, "Strategic Systems Update," *The Submarine Review* (April 1977), p. 12.
50. About 400 W88 warheads were produced for the D-5 missiles, each having an explosive force of 300 to 475 kilotons; the remaining D-5 missiles are fitted with the W76 warhead, having a yield of about 100 kilotons.
51. In addition, four British SSBNs of the *Vanguard* class, completed 1994–2000, have been fitted with the Trident D-5 missile.
52. I. D. Spassky (ed.), *History of Indigenous Shipbuilding,* vol. V, pp. 292–93; Podvig (ed.), *Russia's Strategic Nuclear Arms,* pp. 264–65; S. N. Kovalev and V. P. Semyonov discussions with N. Polmar, St. Petersburg, 1992–1997. Semyonov was principal deputy to Kovalev in the design of Project 941/Typhoon.

53. The Russian name for Project 941 was Akula (shark), the code name subsequently assigned by NATO to submarine Project 971 (Soviet name Bars); this has been highly confusing to Western observers. *Tayfun* was the term used for the overall missile submarine program.
54. The single-pressure-hull design would have required a a hull diameter of approximately 46 feet (14 m).
55. The diameter of the parallel pressure hulls vary by compartments.
56. TK = *Tyazhely Kreyser* (heavy [submarine] cruiser). These were the only Soviet submarines given the "heavy" designation.
57. Ray Robinson telephone conversations with N. Polmar, 16 May 2001 and 4 June 2001. Mr. Robinson served as a naval analyst in the U.S. intelligence community.
58. The RSM-52/R-39 unofficially is credited with a CEP of 1,640 feet (500 m).
59. The Trident SSBNs can launch all 24 missiles in six minutes; see Waller, *Big Red*, p. 228.
60. See Viktor Litovkin, "Three Days on the Typhoon," part 2, *Izvestiya* (2 March 1992), p. 3.
61. V. P. Semyonov discussion with N. Polmar, St. Petersburg, 11 May 1994.
62. One of these Typhoon SSBNs, on 3–4 December 1997, fired a full load of 20 SS-N-20 ballistic missiles—with nuclear warheads removed—from the Barents Sea. The missiles were blown up at an altitude of about one mile (1.5-2 km), being destroyed in compliance with the START treaty limiting nuclear weapons. A U.S. "on-site" inspection team observed the detonations from an accompanying Russian hydrographic survey ship. In addition, a U.S. nuclear submarine of the *Los Angeles* class was in the area.
63. Central Intelligence Agency, "The Soviet Typhoon Submarine—A Radical Innovation in Submarine Design" (An Intelligence Assessment), SW86-10002X (January 1986), pp. 7–9.
64. V. P. Semyonov discussion with N. Polmar, St. Petersburg, 5 December 1996.
65. Confidential source MA memorandum to N. Polmar (n.d.).
66. Eugene V. Miasnikov, *The Future of Russia's Strategic Nuclear Forces: Discussions and Arguments* (Moscow: Center for Arms Control, Energy and Environmental Studies, 1996), p. 35. This work contains an excellent discussion of Soviet SSB/SSBN noise levels.
67. The SALT agreements provided for a maximum Soviet level of 62 SSBNs; U.S. intelligence sources often list 63 in this period (see Secretary of Defense Dick Cheney, *Soviet Military Power 1990* (Washington, D.C.: Department of Defense, September 1990), p. 53.
68. All Project 667BDRM/Delta IV submarines are assigned to the Northern Fleet, based at Olen'ya on the Kola Peninsula.
69. Both navies had discarded some older ballistic missile subnames.
70. U.S. intelligence sources placed the number at 33 percent of Soviet offensive warheads (i.e., ICBM, SLBM, strategic aviation).
71. Cheney, *Soviet Military Power 1990*, p. 53.
72. Other nations with SSBNs were Britain, China, and France.
73. I. D. Spassky discussion with N. Polmar, Arlington, Va., 23 October 1993. Also see Kuzin and Nikol'skiy, *The Navy of the USSR*, p. 56.

## Chapter 13: "Diesel Boats Forever"

1. Office of the Chief of Naval Operations, *The Navy of the 1970 Era* (January 1958), enclosure to memo "The Navy of the 1970 Era," Op93G/ac ser 04P93 (13 January 1958).
2. The four *Upholder*-class submarines, the last diesel-electric undersea craft to be built in Britain, were completed in 1990–1993. All four boats were taken out of Royal Navy service in 1994 and later leased to Canada.
3. Capt. Thomas A. Brooks, USN, "(Soviet) Diesel Boats Forever," U.S. Naval Institute *Proceedings* (December 1980), p. 107. Brooks served as Director of U.S. Naval Intelligence from July 1988 to August 1991.
4. On 1 January 1972 the Admiralty and Sudomekh yards were merged into the United Admiralty Association. This was part of an industry-wide effort to improve planning while cutting administrative personnel at the yards.
5. The first four Indian submarines were Project I641; the later four I641K.
6. Sergei Khrushchev, *Nikita Khrushchev*, p. 566.
7. Ibid., pp. 538–39.
8. The *Pamir* subsequently was converted to an intelligence collection ship and renamed *Peleng.*
9. The submarines had departed their home port of Polyarny, near the mouth of the Tuloma River, on 27 September for the smaller base at Sayda Bay. There each submarine loaded a nuclear torpedo before departing for the Caribbean on 1 October.
10. Ernest R. May and Philip D. Zelikow, *The Kennedy Tapes: Inside the White House During the Cuban Missile Crisis* (Cambridge, Mass.: Harvard University Press, 1997), p. 353.
11. Robert F. Kennedy, *Thirteen Days: A Memoir of the Cuban Missile Crisis* (New York: W. W. Norton, 1969), p. 69.
12. Ibid., pp. 69–70.
13. Rear Adm. Georgi Kostev interview with William Howard, Moscow, 18 June 2002.

14. Capt. 1/R Aleksey Dubivko interview with William Howard, Moscow, 18 June 2002.

15. The commanding officer of the *Charles P. Cecil,* Commander Charles P. Rozier, received the Navy Commendation Medal for his "outstanding leadership and professional skill" in the operation against the submarine.

16. CUBEX = Cuban Exercise.

17. Commander, Antisubmarine Warfare Force, U.S. Atlantic Fleet, to Chief of Naval Operations, Subj.: "Summary, Analysis and Evaluation of CUBEX," ser 008187/43, 5 November 1963, p. I-3-1.

18. SKB-112 was renamed Sudoproyekt in 1968 and changed to Lazurit in 1974; see Appendix F.

19. Sr. Capt. Xinhua Li, Chinese Navy, discussion with N. Polmar, Arlington, Va., 31 August 2001. Li had researched the subject at Naval Headquarters in Beijing.

20. See A. D. Baker III, *Combat Fleets of the World 2000–2001* (Annapolis, Md.: Naval Institute Press, 2000), p. 108. Chinese sources also state that submarines were transferred to Albania, but this is likely to be an error.

21. The U.S. Navy constructed two small target-training submarines in the early 1950s, the *T1* (later named *Mackerel*) and *T2* (*Marlin*), given the hull numbers SST 1 and 2. These were primarily sonar targets and were not intended to sustain weapons impact.

22. A midget submarine to serve as a "mobile rescue chamber" had been proposed in the late 1940s; see Memorandum from Chairman of the ad hoc Working Group, Submarine Conference, established to study Small Submarines & Explosive Motor Boats, to Chairman, Submarine [Officers] Conference, "Small Submarine, Explosive Motor Boats and other Sneak Craft-Report of Study concerning," 27 July 1949, Op-414G/mt, ser: 00331P414, p. 7.

23. Eventually several British and French submarines also were fitted to carry and support the U.S. Navy's DSRVs.

24. The U.S. Navy deep submergence plan developed in 1964 in the aftermath of the *Thresher* (SSN 593) called for 12 rescue vehicles, each to carry 12 rescuees. The number of DSRVs was reduced when the carrying capacity was increased to 24 men (plus three crewmen) and, subsequently, because of funding shortfalls.

25. See Boris F. Dronov and Boris A. Barbanel', "Research Submarine-Laboratory, Design 1710," *Gangut* (no. 14, 1998), pp. 128–36; and Dronov and Barbanel', "Beluga: Soviet Project 1710 Submarine Laboratory," U.S. Naval Institute *Proceedings* (June 1999), pp. 72–76.

26. The *BS-411* (ex-*K-411*) completed conversion in 1990 to support deep-sea/submersible operations. She replaced a modified Echo II in that role. After conversion the later submarine was given the NATO code name Yankee Stretch.

27. The *BS-129* (ex-*K-129*) was relaunched on 29 December 2000 after conversion to support deep-sea/submersible operations; she replaced the modified Yankee Stretch in that role. The Yankee Stretch mid-section was fitted in the Delta conversion. After conversion the submarine was given the NATO code name Delta Stretch.

28. See, for example, Comdr. Art Van Saun, USN, "Tactical ASW: A Case for a Non-Nuclear Submarine," U.S. Naval Institute *Proceedings* (November 1978), pp. 147–51; and Capt. K. G. Schacht, USN (Ret), Commentary on "Diesel Boats Forever?" *Proceedings* (February 1983), pp. 25–26.

29. A total of five British SSNs were deployed to the South Atlantic during the Falklands War.

30. A second Argentine diesel-electric submarine, the 40-year-old, ex-U.S. GUPPY *Santa Fe,* was caught on the surface while being used as a supply ship. She was damaged by anti-ship missiles and sank at South Georgia island.

31. Secretary of the Navy, *Lessons of the Falklands: Summary Report* (Washington, D.C.: Department of the Navy, February 1983), p. 8.

32. Adm. Woodward talk at U.S. Navy Memorial Museum, Navy Yard, Washington, D.C., 18 June 1997.

33. Bath Iron Works and Lockheed Shipbuilding expressed interest in the program in the event that Todd Pacific Shipyards did not reach an agreement with all parties. At the time Todd owned two major shipbuilding yards, at San Pedro, Calif., and Seattle, Wash.

34. Joseph M. Murphy, PALMCO Corp., memorandum to James Blankhorn, Bath Iron Works, et al., 11 September 1984.

35. Representative Whitehurst letter to Secretary of Defense Caspar W. Weinberger, 20 September 1984.

36. See for example, Adm. Watkins, Memorandum for the Secretary of the Navy, Subj: "SSN-21 Assessment Paper Written by [deleted]," 11 February 1985, ser 00/5U300058; Vice Adm. DeMars, letter to Adm. Robert L. J. Long, USN (Ret), Chairman of the Board, Naval Submarine League, 29 January 1988; and Department of Defense, Office of the Inspector General-Special Inquiries, "Report of Investigation Allegations of Repression by Navy Officials of Work-Related Private Writing Activities by a Department of Defense Civilian Employee," case no. S88L00000161, 11 October 1990. These documents address specific efforts by Adm. Watkins and Adm. DeMars to "squash dissent" to views of the submarine leadership (Adm. DeMars statement to OP-03 staff meeting, Pentagon, 8 August 1986).

37. Former Secretary of the Navy Nitze, Adm. Zumwalt, and N. Polmar, proposal to Department of Defense "Developing a Lower-Cost Submarine Program," 24 September 1990.
38. Capt. Schacht, "Diesel Boats Forever?" p. 26.
39. Lt. Jack Shriver, USN, "Developing Real Anti-Diesel Tactics," *The Submarine Review* (April 1998), p. 91.
40. Lt. Comdr. Carey Matthews, USNR, "Anti-sub warfare calls for 2 Russian diesels," *Navy Times* (26 February 1996), p. 33. A specific acquisition proposal for six non-nuclear submarines for this role is found in Congressional Budget Office, *Budget Options* (Washington, D.C.: Government Printing Office, February 2001), pp. 137–38.
41. Vice Adm. Grossenbacher speech at Naval Submarine League seminar, Alexandria, Va., 13 June 2001.
42. These submarines will carry Tomahawk land-attack missiles; see Chapter 18.
43. Newt Gingrich, Speaker of the House, discussion with N. Polmar, Washington, D.C., 6 September 1995.
44. For the foreseeable future the United States will maintain a submarine tender in the Mediterranean, at La Maddalena in Sardinia, while U.S. surface ships are based at two Japanese ports. Nuclear submarines are being based at Guam. Previously U.S. diesel-electric submarines were based in the Philippines and Japan. The non-provocative character of non-nuclear submarines tends to make them acceptable for forward basing in other countries.
45. I. D. Spassky (ed.), *The History of Indigenous Shipbuilding,* vol. V, pp. 302–3.
46. This submarine modification was given the designation 877V.
47. Two mines could be carried in each of the six torpedo tubes plus 12 on reload racks.
48. Chief of Naval Operations Adm. J. M. Boorda, USN, told a Senate committee in 1995 that China had agreed to purchase ten Kilos with additional options for a total of 22; see Barbara Starr, "China's SSK aspirations detailed by USN chief," *Jane's Defence Weekly* (18 March 1995), p. 3.
49. Comdr. Powess, "U.K.'s *Upholder*-Class Subs Go to Canada," U.S. Naval Institute *Proceedings* (October 2002), p. 101. Also Comdr. Powess discussion with N. Polmar, Washington, D.C., 30 June 1999.
50. Richard Scott, "Fuel cell AIP planned for new Russian submarine," *Jane's Navy International* (July/August 1996), p. 4.
51. Yevgenia Borisova and Lloyd Donaldson, "Russian yard lays keels for 'super subs,' " *Lloyds List* (5 January 1998), p. 5.
52. The *Sankt Petersburg* was being funded in part by a $100 million payment from the Inkombank of Moscow. Total cost per submarine (without AIP) at the time was estimated at approximately $300 million.

## Chapter 14: Unbuilt Giants

1. See Don Everitt, *The K Boats* (New York: Holt, Rinehart and Winston, 1963). The world's first practical steam-propelled submarine was the British *Swordfish,* also a relatively large submarine, completed in 1916.
2. Everitt, *The K Boats,* pp. 160–61.
3. Capt. Richard Compton-Hall, RN, *Submarine Warfare: Monsters & Midgets* (Poole, Dorset: Blandford Press, 1985), pp. 58–69.
4. As a submarine carrier the *M2* embarked personnel from the Fleet Air Arm and Royal Air Force; it is possible that Aircraftsman T. E. Shaw—Lawrence of Arabia—was a member of the RAF contingent on board for a short time.
5. The *X1* had a surface range of 12,500 n.miles (23,160 km) at 12 knots. In comparison, the first British post–World War II fleet submarines—the O class—had a range of 8,500 n.miles (15,750 km) at ten knots.
6. The *Surcouf*'s unusual torpedo armament provided four 21.7-inch tubes in the bow; four 21.7-inch tubes fixed in the after casing, external to the pressure hull; and four 15.75-inch tubes in a traversing mount amidships, external to the pressure hull (but reloadable). The small-diameter torpedoes were for use against merchant ships.
7. The best English-language descriptions of the *Deutschland* are Francis Duncan, "*Deutschland*-Merchant Submarine," U.S. Naval Institute *Proceedings* (April 1965), pp. 68–75, and Dwight R. Messimer, *The Merchant U-Boat* (Annapolis, Md.: Naval Institute Press, 1988). Also see Paul König, *Fahrten der U-Deuschland im Weltkrieg* (Berlin: Ullstein, 1916), and *The Voyage of the Deutschland* (New York: Hearst International Library, 1916); there are significant differences in these two volumes, which are largely propaganda tracts.
8. Simon Lake, *Submarine: The Autobiography of Simon Lake* (New York: D. Appleton-Century, 1938), pp. 256–57.
9. After surrendering to British forces on 24 November 1918, the *Deutschland* was put on display in London. Later, after being stripped of weapons, equipment, and machinery, her hulk was towed around Britain on a pay-to-visit basis.
10. Lake, *Submarine,* pp. 264–66.
11. See Rössler, *The U-boat,* pp. 71–75.
12. Ibid., p. 75; data from Erich Gröner, *Die deutschen Kriegschiffe 1815–1945,* vol. 1 (Munich: J. F. Lehmanns, 1966), p. 376.

13. Alden, *The Fleet Submarine in the U.S. Navy,* pp. 26–27.
14. For an English-language description of these submarines see Dorr Carpenter and Norman Polmar, *Submarines of the Imperial Japanese Navy* (Annapolis, Md.: Naval Institute Press, 1986).
15. F. H. Hinsely, et al., *British Intelligence in the Second World War,* vol. 2 (London: Her Majesty's Stationery Office, 1981), p. 612.
16. A recent work on this subject is John F. White, *U Boat Tankers 1941–45: Submarine Suppliers to Atlantic Wolf Packs* (Annapolis, Md.: Naval Institute Press, 1998).
17. Most of these sinkings were brought about by Allied codebreaking and high-frequency radio direction finding, which revealed the planned rendezvous of the tankers and U-boats. See Hinsely, et al., *British Intelligence,* vol. 3, part 1, pp. 212–14, et. seq.
18. Data primarily based on Rössler, *The U-boat.*
19. The largest U-boat actually produced by Germany was the Type XB minelayer, which carried 66 mines in addition to having two stern torpedo tubes and deck guns. Eight of these submarines were built.
20. P. Z. Golosovskii, *Proektirovaniye i stroitelstvo podvodnikh lodok, Tom II, Povodni Lodki: perioda 1926–1945 godov* (Design and Construction of Submarines), vol. II, *1926–1945* (Leningrad: Rubin [design bureau], 1979), p. 217.
21. Adm. William H. Standley, USN (Ret), *Admiral Ambassador to Russia* (Chicago, Ill.: Henry Regnery, 1955), p. 156.
22. The K class could carry 20 mines in special chutes, and the L class could carry 18 mines plus their torpedo armament. The *Argonaut* had a capacity for 60 mines in addition to her torpedo armament. The *Argonaut* also carried the designations SS 166 and APS 1 (transport).
23. I. D. Spassky (ed.), *The History of Indigenous Shipbuilding,* vol. V, p. 93.
24. V. I. Zharkov and Capt. 1/R B. Tyurin, Russian Navy, "Project 632 and 648 Mine-Laying Submarines," *Morskoy Sbornik* (no. 3, 1996), pp. 65–69.
25. The suffix M indicated *Modifitsirovannay* (modification).
26. Malachite SPMBM. *Istoriya Sankt-Peterburgskogo Morskogo Buro Mashinostroeniya "Malakhit"* (History of Saint-Petersburg's Marine Machine-building Bureau Malachite), *vol. 2, 1949–1974* (St. Petersburg: Malachite, 1995), pp. 174–79.
27. Estimated to feed 100 troops for 90 days.
28. See P. Wittenbeck and K.J. Moore, "The Rising Vortex: A Source for NA [Non-Acoustic] ASW," presentation at UDT '93 conference, Cannes, France, 16 June 1993. The first Soviet ASW ship classes to mount the Top Sail radar were the helicopter carrier-missile cruiser Moskva, completed in 1967, and the missile cruiser Kronshtadt (NATO Kresta II), completed in 1969.
29. In addition to cargo, Project 648 would have four 533-mm torpedo tubes and four 400-mm tubes and carry four and 22 torpedoes, respectively.
30. Malachite SPMBM, *History of Saint-Petersburg's Marine Machine-building Bureau Malachite* vol. 2, pp. 180–86.
31. Adm. Yu. Panteleyev, "The Submarine Operation of the Navy—the Naval Operation of the Future," *Voyennaya Mysl'* (10 July 1961); U.S. Central Intelligence Agency translation, 20 February 1962, p. 22.
32. Ibid., pp. 21–22.
33. V. I. Zharkov and Capt. 1/R B. Tyurin, "The Amphibious Troop Transport and Minelaying Nuclear Powered Submarines of Projects 748 and 717," *Morskoy Sbornik* (no. 1, 1998), pp. 63–67.
34. The OK-300 reactor plants were used in the Project 671/ Victor-class SSNs.
35. Malachite SPMBM. *Istoriya Sankt-Peterburgskogo Morskogo Buro Mashinostroeniya "Malakhit"* (History of Saint-Petersburg's Marine Machine-building Bureau Malachite) pp. 189–93.
36. B. F. Dronov, et al., "Nuclear-Powered Transport Ships for the Arctic," *Sudostroenie* (no. 1, 1992), p. 7. Also Malachite fact sheets "Underwater Tanker" and "Underwater Transport Container Carrier" [n.d.]. The Russian *Norilsk*-class ships are combination container/ vehicle/heavy lift cargo ships of 19,943 deadweight tons, 574 feet (174.0 m) in length, capable of carrying 576 standard containers. They are configured for Arctic operation.
37. The executive officer of the raiders was Maj. James Roosevelt, USMC, son of the president.
38. Thirty Marines were left behind—probably 25 dead and 5 alive; the latter were captured and executed by the Japanese.
39. Details of fleet boat conversions are found in Alden, *The Fleet Submarine in the U.S. Navy.*
40. Her designation initially was changed to SSA 317 and later to ASSA 317.
41. Initially redesignated SSO 362, subsequently changed to AGSS 362 (miscellaneous auxiliary) and later AOSS 362 (submarine oiler).
42. At various times these ex-fleet boats and the USS *Grayback,* after conversion to transports, had the designations SSP, ASSP, APSS, and LPSS.
43. During this period the U.S. Navy also considered "assault-transport" submarines in addition to these troop-carrying submarines. See minutes of Submarine Officers Conference, 18 December 1946 (report of 2 January 1947, p. 4).
44. It was planned to similarly convert her sister ship *Growler* (SSG 577) to the transport role, but that

effort was cancelled because of funding shortfalls during the Vietnam War.

45. The designations used for U.S. transport submarines were SSP, APSS, and LPSS; the later, nuclear-propelled submarines employed in this role were given attack submarine designations (SSN).

46. The *Kamehameha* "stood down" in October 2001, in preparation for decommissioning and strike; at the time she had been in commission for 36 years, longer than any other U.S. nuclear submarine.

47. Comdr. William R. Anderson, USN, "The Arctic as a Sea Route of the Future," *National Geographic Magazine* (January 1959), p. 21.

48. Electric Boat Division, General Dynamics Corp., *The Feasibility and Design of Nuclear-Powered Submarine Tankers* (1 December 1958).

49. Vito L. Russo, Harlan Turner Jr., and Frank W. Wood, "Submarine Tankers," The Society of Naval Architects and Marine Engineers *Transactions* (vol. 68, 1960), pp. 693–742. Mr. Russo was with the Maritime Administration and Messrs. Turner and Wood were with Electric Boat.

50. Ibid., p. 696.

51. Christopher Pala, "Turning Russian Subs into . . . Cargo Carriers?" *Business Week Online* (31 August 2000); also Pala telephone discussion with N. Polmar, 26 August 2000. The Russian name for Project 941 is *Akula.*

52. E. L. Baranov, S. L. Karlinsky, and S. O. Suchanov, " 'Tayfun' Changes Profession," *Sudostroenie* (no. 2, 2001), pp. 113–16.

## Chapter 15: Aircraft-Carrying Submarines

1. The designation E9W1 indicated the aircraft's mission (E = reconnaissance seaplane) and producer (W = Watanabe); Type 96 indicated the last two digits of the year since the founding of the Japanese state (2596-equivalent to 1936).

2. These submarines are described in Carpenter and Polmar, *Submarines of the Imperial Japanese Navy.*

3. See U.S. Naval Technical Mission to Japan, "Ships and Related Targets: Characteristics of Japanese Naval Vessels: Submarines," no. S-01-1 (January 1946), and Supplements I and II (January 1946); "*I-400* Class Submarines Just After War," *Ships of the World* (Tokyo) (no. 5, 1980), pp. 58–63; Christopher C. Wright (ed.), "The U.S. Navy's Operation of the Former Imperial Japanese Navy Submarines *I-14, I-400,* and *I-401,* 1945–1946," *Warship International* (no. 4, 2000), pp. 348–56, 362–401; Compton-Hall, *Submarine Warfare,* pp. 66, 71–79; and Thomas S. Momiyama, "All and Nothing," *Air & Space* (October/November 2001), pp. 22–31.

4. Only one other unit of this class was launched, the *I-404;* work on her stopped in March 1945 when she was 90 percent complete. The *I-404* was sunk at Kure on 28 July 1945 by U.S. carrier aircraft.

5. M = special aircraft. The M6A1 was developed in secrecy, without Allied knowledge, hence no Allied code name was assigned; *Seiran* meant Mountain Haze. The M6A1-K *Nanzan* variant, built for test and training, had a retractable undercarriage; *Nanzan* meant Southern Mountain.

6. Seven of the AM-type submarines were planned; four were launched, but only two were completed. The *I-13* was sunk by U.S. ships and aircraft in July 1945; the *I-14* was scrapped.

7. U.S. Naval Technical Mission to Japan, *Ships and Related Targets,* p. 29.

8. Rear Adm. A. R. McCann USN, Commander, Submarine Force, U.S. Pacific Fleet, Memorandum for the Chief of Naval Operations, subj: "Ex-Japanese Submarines *I-14, I-400,* and *I-400*-Status Of," FF12-10/L9-3, Ser 287, 1 February 1946. The document is reprinted in Wright, "The U.S. Navy's Operation of . . . ," pp. 391–93.

9. Project Rand, *Vulnerability of U.S. Strategic Air Power to a Surprise Enemy Attack in 1956,* Special Memorandum SM-15 (Santa Monica, Calif.: Rand Corp., 15 April 1953), p. ii. The Tupolev Tu-4 (NATO code name Bull) was a direct copy of the U.S. B-29 Superfortress, a long-range strategic bomber powered by four piston engines. The Tu-4 became operational in the Soviet Air Forces in 1949 and had a nuclear bombing capability from 1953.

10. The speed of sound (i.e., Mach 1.0) is 762 m.p.h. at sea level, 735 m.p.h. at 10,000 feet, 707 m.p.h. at 20,000 feet, 678 m.p.h. at 30,000 feet, and 662 m.p.h. at 40,000 feet. The term Mach is derived from Ernst Mach, an Austrian physicist (1838–1916).

11. Project Rand, *Vulnerability of U.S. Strategic Air Power,* p. 10.

12. The Rand estimate of damage to the U.S. bomber force was:

| **Type** | **No Warning** | | **With One-hour Warning** | |
|---|---|---|---|---|
| | Heavy Bombers | Medium Bombers | Heavy Bombers | Medium Bombers |
| Sub-launched | 100% | 76% | 100% | 73% |
| Tu-4 low-altitude | 100% | 82% | 100% | 72% |
| Tu-4 high-altitude | 90% | 64% | 43% | 42% |

13. Minutes of Submarine Officers Conference, 18 December 1946 (report of 2 January 1947, p. 3).

14. Among Heinemann's more notable aircraft were the Douglas SBD Dauntless, AD Skyraider, A3D Skywarrior, A4D Skyhawk, D-558 research aircraft; see Edward H. Heinemann and Capt. Rosario Rausa, USNR (Ret), *Ed Heinemann: Combat Aircraft Designer* (Annapolis, Md.: Naval Institute Press, 1980).

15. R. A. Mayer and W. A. Murray, *Flying Carpet Feasibility Study: Submarine Carrier,* Document No. D3-1870-8 (Wichita, Kansas: Boeing Airplane Company, 18 December 1958).
16. The F11F weighed 25,858 pounds (11,729 kg) and the flying carpet booster weighed 14,370 pounds (6,518 kg).
17. The *Grayback* had two hangars faired into her forward superstructure, each capable of accommodating one Regulus II missile; the *Halibut* had a single hangar for up to four Regulus II missiles. (See Chapter 6.)
18. Mayer and Murray, *Flying Carpet Feasibility Study,* p. 52.
19. Ibid., p. 56.
20. E. N. Greuniesen, Patent No. 1,828,655, filed 13 May 1930.
21. J. Seitzman, Patent No. 2,711,707, filed 8 March 1948.

## Chapter 16: Midget, Small, and Flying Submarines

1. See C. E. T. Warren and James Benson, *Above Us the Waves: The Story of Midget Submarines and Human Torpedoes* (London: George G. Harrap, 1953); Paul Kemp, *Underwater Warriors* (Annapolis, Md.: Naval Institute Press, 1996), and the overlapping Kemp, *Midget Submarines of the Second World War* (London: Chatham Publishing, 1999); and Carpenter and Polmar, *Submarines of the Imperial Japanese Navy,* pp. 130–36.
2. These A-type submarines each had a two-man crew. Nine crewmen were lost and one became the first prisoner of war to be held by the Americans.
3. Vice Adm. Matome Ugaki, IJN, *Fading Victory: The Diary of Admiral Matome Ugaki, 1941–1945* (Pittsburgh, Pa.: University of Pittsburgh Press, 1991), p. 56. Recent interpretative efforts of a photograph taken during the Pearl Harbor attack has led some analysts to conclude that a midget submarine may have launched two torpedoes at the battleships. However, someone who has seen a first-generation photo is certain that the object believed to be a midget submarine sail is in fact a motor launch (Di Virgilio, below). See Comdr. John Rodgaard, USNR (Ret), et al., "Pearl Harbor: Attack from Below," *Naval History* (December 1999), pp. 16–23, and commentary by John Di Virgilio, *Naval History* (February 2000), p. 6; Comdr. John A. Rodgaard, USNR (Ret), "Japanese Midget Submarines at Pearl Harbor," Naval Intelligence Professionals *Quarterly* (Spring 1995), pp. 1–3; and Burl Burlingame, *Advance Force Pearl Harbor* (Kailua, Hawaii: Pacific Monograph, 1992).
4. While being towed, the X-craft were submerged and manned by three-man "passage" crews; they were relieved by the combat crew (on the surface) when the submarines neared their objective.
5. Losses included the three-man passage crew on board the *X9* when the towline broke and the craft sank, probably because of the weight of the water-logged manila line.
6. Arabic numbers were used because of the Holerith punch-card system used for logistic support of these craft.
7. Unlike previous X-type midgets, these craft were given names: *Stickleback, Shrimp, Sprat,* and *Minnow,* respectively.
8. The Soviet Union was the only other nation known to develop nuclear mines; up to four of those mines were planned for shipment to Cuba during the missile crisis of October 1962.
9. The *Stickleback* was sold to Sweden in 1958; she was returned in 1977 and is on display at the Imperial War Museum's Duxford facility. The older *XE24* is preserved at HMS Dolphin, the submarine museum at Gosport in Portsmouth.
10. This was an NB-36H, which made 47 flights from 1955 to 1957 to test the effects of radiation on an aircraft as the Air Force sought to develop a nuclear-propelled bomber.
11. See Carolyn C. James, "The Politics of Extravagance: The Aircraft Nuclear Propulsion Project," Naval War College *Review* (Spring 2000), pp. 158–90; and Lt. Col. Kenneth F. Gantz, USAF (ed.), *Nuclear Flight: The United States Air Force Programs for Atomic Jets, Missiles, and Rockets* (New York: Duell, Sloan and Pearce, 1960).
12. Commo. Robin W. Garson, RN (Ret), telephone conversation with N. Polmar, 13 April 2002. Garson attended the briefing.
13. The *X-1* was officially classified as a "submersible service craft" and was assigned the hull number SSX 1; see, for example, Chief of Naval Operations to Secretary of the Navy, "Strike of Submersible Service Craft X-1 (SSX-1)," 13 December 1972.
14. Memorandum from Chairman of the ad hoc Working Group, Submarine Conference, established to study Small Submarines & Explosive Motor Boats, to Chairman, Submarine [Officers] Conference, "Small Submarine, Explosive Motor Boats and other Sneak Craft-Report of Study concerning," 27 July 1949, Op-414G/mt, ser 00331P414, p. 8. The study report was written by Capt. G. E. Peterson, USN.
15. These midget submarines were transported to the United States as deck cargo aboard freighters.
16. Capt. L. R. Daspit, USN, statement to the General Board of the Navy, 18 September 1950.
17. Ibid.
18. Comdr. R. H. Bass, USN, statement to the General Board of the Navy, 18 September 1950.
19. The *X-1* was placed "in service" vice in commission

and had an "officer-in-charge" vice commanding officer.

20. The expectation of eight knots was based on the expected 2,050 diesel r.p.m. at full power; in practice the engine was restricted to 1,200 r.p.m. A detailed description of her unique propulsion plant appears in Richard Boyle, "USS *X-1* Power Plant, 1956–57," *Naval Engineers Journal* (April 1972), pp. 42–50. Also, Boyle E-mail to N. Polmar, 16 May 2001.
21. Portsmouth Naval Shipyard, "Historical Account of *X-1* Power Plant Development, 24 May 1956–20 May 1957" [n.d.], p. 11.
22. Richard Boyle E-mail to N. Polmar, 19 August 1997. Boyle was engineer of the *X-1* from May 1956 to June 1957, and officer-in-charge from June 1957 until she was put out of service in December 1957.
23. The *X-1* subsequently was placed on exhibit at the U.S. Naval Academy in Annapolis, Md.; in 2001 she was moved to the Submarine Museum at Groton (New London), Conn.
24. Electric Boat Division, General Dynamics Corp., *Preliminary Design Study of a Small, High-Density, Hydro-Ski Equipped Seaplane,* Report No. U419-66-013 (September 1966); Eugene H. Handler, "Submersible Seaplane" (unpublished paper) (n.d.); and C. R. Tuttle, "The Submarine and Airplane as an Integrated Vehicle" (unpublished paper) (October 1966). Handler was a Navy engineer and Tuttle was a Convair project engineer; both were involved in the submersible seaplane program.
25. "Submersible Airplanes," *The New York Herald-Tribune* (18 September 1946), p. 11.
26. All American Engineering was considered one of the world's most experienced firms involved in water-ski research.
27. Eugene Handler, "The Flying Submarine," U.S. Naval Institute *Proceedings* (September 1964), p. 144.
28. Ibid.
29. Convair San Diego (Calif.) was a division of the General Dynamics Corporation, which also owned the Electric Boat submarine yard.
30. Convair Division, General Dynamics Corp., "Flying Submersible Vehicle" (Technical Proposal), (April 1964), p. 1.
31. The M507 was a twin M504 diesel-engine plant.
32. A. M. Antonov E-mail to N. Polmar, 21 December 2001.
33. Ministry of Defence, *Countering the Submarine Threat: Submarine violations and Swedish security policy* (Stockholm: Report of the Submarine Defence Commission, 1983), p. 32.
34. Thomas Ginsberg, " 'Subs' were frisky minks, Sweden admits," *Philadelphia Inquirer* (9 February 1995), p. 13.
35. I. D. Spassky (ed.), *The History of Indigenous Shipbuilding,* vol. V, pp. 306–8; Yu. K. Mineev and K. A. Nikitin, "Unique Piranya," *Gangut* (no. 14, 1991), pp. 77–84.
36. U.S. District Court, Southern District of Florida, *Indictment* (Messrs. Ludwig Fainberg, Juan Almeida, and Nelson Yeser), filed 28 January 1997, p. 5. Also see David Kidwell and Manny Garcia, "Feds Allege Wild Scheme: Drugs Hauled by *Submarine,*" *The Miami Herald* (6 February 1997), p. 1A.
37. SEAL = Sea-Air-Land (team), i.e., U.S. Navy special operations forces.

## Chapter 17: Third-Generation Nuclear Submarines

1. The Bureau of Ships was changed to the Naval Ship Systems Command (NavShips) on 1 May 1966, when the Navy's bureau organization—which dated from 1842—was abolished. On 1 July 1974 NavShips was merged with the Naval Ordnance Systems Command to form the Naval Sea Systems Command (NavSea), with that designation being used in the remainder of this volume.
2. Capt. Myron Eckhart Jr., USN (Ret), comment on "Running Critical," U.S. Naval Institute *Proceedings* (August 1987), p. 23.
3. Henry A. Jackson, discussion with N. Polmar, Groton, Conn., 30 June 1997. A direct-drive system alleviated the need for reduction gears.
4. Folding masts and periscopes are used in the U.S. Navy's Advanced SEAL Delivery System (ASDS); see Chapter 16.
5. Rickover (May 1974), p. 24.
6. Postwar U.S. submarines were assigned fish or other marine-life names until 1971, when Vice Adm. Rickover instituted the practice of naming attack submarines for deceased members of Congress who had supported nuclear programs; four SSNs were so named—*Glenard P. Lipscomb, L. Mendel Rivers, Richard B. Russell,* and *William H. Bates.* The naming source for attack submarines was changed to city names in 1974, when the SSN 688 was named *Los Angeles.* A popular expression at the time the congressional names were assigned was, "fish don't vote."
7. Rickover (1974), p. 24.
8. The agreement, dated 20 September 1967, is reprinted in Tyler, *Running Critical,* p. 53n.
9. Capt. Donald H. Kern, USN (Ret), discussion with N. Polmar, Saunderstown, R.I., 1 July 1997.
10. Mark Henry, head, submarine preliminary submarine design branch, NavSea, discussion with N. Polmar, Arlington, Va., 24 June 1999.
11. Capt. Eckhart telephone conversation with N. Polmar, 10 June 1999.

12. Capt. Kern interview with Dr. Gary E. Weir, Naval Historical Center, Washington, D.C., 12 September 1990.
13. Tyler, *Running Critical,* p. 63. Also Foster discussions with N. Polmar, Washington, D.C., 18 March 1997.
14. Ibid., p. 69.
15. Clark Clifford, *The Fiscal 1970 Defense Budget and Defense Program for Fiscal Years 1970–74* (Washington, D.C.: Department of Defense, 15 January 1969), p. 96.
16. Ibid.
17. The *Lipscomb* was authorized in fiscal 1968 and commissioned in 1974; the *Los Angeles* was authorized in fiscal 1970 and commissioned in 1976.
18. Eckhart comment on "Running Critical," p. 24.
19. The machinations of killing CONFORM are discussed in Tyler, *Running Critical,* pp. 55–72.
20. Rickover supporters in Congress undertook a massive letter-writing campaign to force the *Lipscomb* on DoD; see "Naval Nuclear Propulsion Program [Fiscal Year] 1969," Hearings before the Joint Committee on Atomic Energy, Congress, 23 April 1969, pp. 163–74.
21. See Tyler, *Running Critical,* p. 66. Tyler credits the reactor plant with the need to reduce pressure hull thickness to save weight; he lists the S6G reactor compartment of the *Los Angeles* as 1,050 tons compared to 650 tons for the S5W reactor.
22. Newport News constructed five submarines for Russia based on Simon Lake's *Protector* design; they were shipped to Libau for final assembly.
23. The crisis in submarine construction is described in Tyler, *Running Critical,* and in Lehman, *Command of the Seas,* pp. 196–227.
24. At the same time, Newport News Shipbuilding had six SSN 688s under contract, with EB having underbid that yard for the other submarines.
25. Vice Adm. C. R. Bryan, USN, testimony before the Joint Economic Committee, Congress, 19 May 1978.
26. Lehman, *Command of the Seas,* p. 207.
27. The *Ohio* was placed in commission at Groton, Conn., on Friday, 11 November 1981; the next day Secretary of Defense Caspar Weinberger announced that Adm. Rickover would leave active duty the following January.
28. Adm. Rickover, "Department of Defense Appropriations for [Fiscal Year] 1979," Hearings before a Subcommittee on Appropriations, House of Representatives, part 6, 16 March 1978, p. 490. The fiscal 1979 request for one SSN 688 was $433 million.
29. The "Submarine Alternatives Study" was initiated as a result of a Senate request in the Fiscal Year 1979 Defense Authorization Bill.
30. Robert C. Chapman, "Attack Submarine Research and Development: Recent Trends and Projected Needs" (Washington, D.C.: Office of the Director Defense Research and Engineering, July 1976), p. 29.
31. This concept is discussed in detail in Reuven Leopold, Otto P. Jöns, and John T. Drewry, "Design-to-Cost of Naval Ships," The Society of Naval Architects and Marine Engineers *Transactions* (vol. 82, 1974), pp. 211–43.
32. Chapman, "Attack Submarine Research and Development," p. 31.
33. Mann quoted in George C. Wilson, "New Submarine Proposals Challenge Rickover's Rule," *The Washington Post* (21 May 1979), p. A11.
34. "Fat Albert" was a character developed by comedian Bill Cosby. That Fat Albert was a short, fat character.
35. Vice Adm. H. G. Rickover, USN, "Analysis of Nuclear Attack Submarine Development" (Washington, D.C.: Naval Reactors Branch, Department of the Navy, 31 August 1979), pp. 13–14.
36. Rickover, "Analysis of Nuclear Attack Submarine Development," p. 49.
37. Vice Adm. Rickover, "Nuclear Submarines of Advanced Design," report of Hearings before the Joint Committee on Atomic Energy, Congress, 21 June 1968, p. 76.
38. Mark Henry discussion with N. Polmar, Potomac, Md., 18 March 2002. Henry was head of submarine preliminary design in NavSea from 1985 to 1999.
39. *Report of the President's Commission on Strategic Forces* (April 1983), p. 11. The commission was chaired by Lt. Gen. Brent Scowcroft, U.S. Air Force (Ret).
40. The term STAWS (Submarine Tactical Anti-ship Weapons System) also was used.
41. Vice Adm. Rickover testimony to Committee on Appropriations, Senate, 10 May 1972.
42. Vice Adm. Rickover testimony to Joint Committee on Atomic Energy, Congress, 10 March 1971.
43. See Capt. Robert F. Fox, USN (Ret), "Build It & They Will Come," U.S. Naval Institute *Proceedings* (April 2001), pp. 44–47.
44. Vice Adm. Rickover in "Naval Nuclear Propulsion Program [Fiscal Year] 1967–68," Hearings before the Joint Committee on Atomic Energy, Congress (1968), p. 30. Also see Lee Vyborny and Don Davis, *Dark Waters: An Insider's Account of the NR-1, the Cold War's Undercover Nuclear Sub* (New York: New American Library, 2003).
45. Untitled White House press release (Austin, Texas), 18 April 1965.
46. Vice Adm. Rickover, "Naval Nuclear Propulsion Program—1972–73," hearings before the Joint Committee on Atomic Energy, Congress, 8 February 1972 (1974), pp. 80–81.

47. I. D. Spassky (ed.), *The History of Indigenous Shipbuilding,* vol. V, pp. 296–97; and Kuzin and Nikol'skiy, *The Navy of the USSR,* pp. 67, 69.

48. From 1977 the chief designer of Project 949 was I. L. Baranov.

49. The lead Oscar SSGN, *K-525,* had only two 650-mm torpedo tubes in addition to 533-mm tubes.

50. Renamed *Arkhangelsk* in 1991 when Minsk became the capital of Beloruss after the breakup of the USSR.

51. The carrier *Kuznetsov* displaces some 55,000 tons standard and 59,000 tons full load.

52. Kuzin and Nikol'skiy, *The Navy of the USSR,* p. 69.

53. The 118 men consisted of 5 personnel from the staff of the 7th Submarine Division, 2 torpedo specialists, and a crew of 43 officers and 68 enlisted men. It was the worst submarine disaster in history after the USS *Thresher* (SSN 593), which was lost with 129 men on 10 April 1963. In a remarkable salvage operation performed by two Dutch firms, the *Kursk* was lifted from the ocean floor, a depth of some 350 feet (105 m), and towed into a floating dry dock in October 2001.

54. The yield strength of 48-OT3V is 6,000 kg/cm$^2$.

55. V. P. Semyonov discussion with N. Polmar, St. Petersburg, 5 March 1997.

56. Capt. 1/R Igor Bogachenko, Russian Navy (Ret), "Russian Nuclear Submarines with Titanium Hulls," *The Submarine Review* (July 1997), p. 95.

57. The USS *Baton Rouge* (SSN 689) collided with the *K-239* in the Barents Sea on 11 February 1992. U.S. officials claimed that the incident occurred in international waters, beyond the 12-n.mile (22.2-km) territorial zone recognized by the United States; Russian officials declared that the collision, off Murmansk, was within their territorial waters. Both submarines were damaged, although no injuries were reported. The *Baton Rouge* was taken out of service in January 1993; she was the first *Los Angeles*-class SSN to be decommissioned, having served 15½ years, little more than one-half the predicted service life. The *K-239* was laid up in 1997 because of funding limitations. The incident is described in Rear Adm. V. Aleksin, Russian Navy, "Incidents in the Barents Sea," *Morsky Sbornik* (no. 5-6, 1992), pp. 21–22.

58. Capt. 1/R George Sviatov, Soviet Navy (Ret), "Akula Class Russian Nuclear Attack Submarines," *The Submarine Review* (October 1997), pp. 60–65, and A. Alekseyev and L. Samarkin, "Bars Class Submarines Pose Problems to Other Navies," *Morskoy Sbornik* (no. 4, 1997), pp. 51–56.

59. AK-32 is reported to have a yield strength of 100 kg/mm$^2$.

60. Assignment of the NATO name Akula was confusing; the prolific Soviet submarine program had exhausted the alphabet of NATO code names, the Project 705/NATO Alfa having been the first *A* submarine. The Soviet name for the Project 941/NATO Typhoon is *Akula* (shark).

61. This torpedo-like MG-74 can be programmed to produce the noise of a moving submarine and to transmit selective jamming signals. Its diameter is 533 mm, and it weighs 1,757 pounds (797 kg). A total of 12 MG-74s are carried by each submarine.

62. Russian president Vladimir Putin participated in the *Gepard* ceremonies at Sevmash on 4 December 2001.

63. Adm. Boorda, statement at meeting of the Naval & Maritime Correspondents Circle, Washington, D.C., 27 February 1995; also Boorda discussion with N. Polmar, Pentagon, 14 September 1995.

64. Walker was convicted and sentenced to life imprisonment. "The Soviets gained access to weapons and sensor data and naval tactics, terrorist threats and surface, submarine and airborne training, readiness, and tactics," said Secretary of Defense Caspar Weinberger in assessing the Walker spy ring damage. "We now have clear signals of dramatic Soviet gains in all areas of naval warfare, which must now be interpreted in the light of the Walker conspiracy."

65. "Quieter Soviet subs will cost U.S. at least $30 billion," *Navy News & Undersea Technology* (14 March 1988), p. 8.

66. GKP = *Glavnyi Komandnyi Punkt.*

67. The unfinished Akula IIs at Sevmash were the *Kuguar* (K-337) and *Rys'* (K-333); the Akula I at Komsomol'sk was the *Kaban.*

68. Dmitriy A. Romanov discussion with N. Polmar, St. Petersburg, 9 December 1996; and D. A. Romanov, *Tragediya Podvodnoy Lodki Komsomolets* (Tragedy of the Submarine *Komsomolets*) (St. Petersburg: Publishing Association of St. Petersburg, 1993). Romanov was the deputy chief designer of the *Komsomolets.* Also Capt. 1/R V. Krapivin, "Tragedy of a Ship and Honor of a Crew," *Morskoy Sbornik* (no. 4, 1994), pp. 46–51 (this is a review of the Romanov book), and Yevgeniy Solomenko, "The Battle of the Admirals Over the *Komsomolets,*" *Krasnaya Zvezda* (22 March 1994), p. 7.

69. Klimov died of cancer on 20 February 1977.

70. The deepest diving U.S. submarine is the small, diesel-electric research ship *Dolphin;* she is rated at 3,000 feet (915 m).

71. The Russian nickname for the submarine was *plavnik* (fin).

72. Central Intelligence Agency, "The Soviet M-Class Submarine: A Preliminary Assessment" (An Intelligence Assessment), SOV 84-10134/SW 84-10049 (August 1984), p. iii.

73. Ibid.

74. Romanov discussion with N. Polmar, St. Petersburg, 9 December 1996.

75. Romanov, *Tragedy of the Submarine Komsomolets,* p. 17.

76. The chief designer of the X-ray was Sergei M. Bavilin, who was one of the chief designers of the Project 865/Piranya small submarine.

77. Adm. James Watkins, USN, speech at change of command for Commander, Submarine Force, Atlantic Fleet, Norfolk, Va., 27 June 1983. Watkins was Chief of Naval Operations from 1982 to 1986.

78. Capt. Harry A. Jackson, USN (Ret), Comdr. William D. Needham, USN, Lt. Dale E. Sigman, USN, "ASW: Revolution or Evolution," U.S. Naval Institute *Proceedings* (September 1986), p. 68.

79. Ronald O'Rourke, "Maintaining the Edge in US ASW," *Navy International* (July/August 1988), p. 354. This article is an excellent appraisal of U.S. ASW thinking at the time.

## Chapter 18: Submarine Weapons

1. Sources for this chapter, in addition to those noted, include Baker, *Combat Fleets of the World* (various editions); A. B. Karpyenko, *Rosseskoye Raketnoyi Oreze 1943–1993* (Russian Missile Weapons, 1943–1993) (St. Petersburg: PEKA, 1993); Kuzin and Nikol'skiy, *The Navy of the USSR;* Podvig, *Russia's Strategic Arms;* and Polmar, *Ships and Aircraft of the U.S. Fleet* (various editions).

2. Also replaced by the Mk 48 were the last Mk 4 torpedoes in U.S. Navy service. That 21-inch torpedo had been in service since 1938(!) and was the principal U.S. anti-ship weapon of World War II. The 3,210-pound torpedo carried 643 pounds of high explosive; its performance was 9,000 yards at 31 knots and 4,500 yards at 46 knots. See Chapter 2 for a description of the Mk 45 ASTOR.

3. The Mk 37 torpedo is described in Chapter 2.

4. The German Navy developed the first wire-guided torpedo, the *Lerche,* late in World War II.

5. See, for example, General Accounting Office, "Navy Torpedo Programs: MK-48 ADCAP Upgrades Not Adequately Justified," GAO/NSIAD-95-104 (June 1995).

6. These views were discussed on a classified basis in Dr. Donald M. Kerr and N. Polmar, "Torpedo Warheads: Technical and Tactical Issues in ASW," *Technology Review* (Los Alamos National Laboratory) (January–February 1984), pp. 26–38; also, Kerr and Polmar, "Nuclear Torpedoes," U.S. Naval Institute *Proceedings* (August 1986), pp. 62–68.

7. See, for example, Capt. Charles Arnest, USN, Commanding Officer, Naval Technical Intelligence Center, Memorandum to Naval Investigative Service, Subj.: "Possible Public Media Compromise," 27 April 1990.

8. The Sea Lance was given the designation UUM-125B; U = Underwater [to], U = Underwater, M = Missile.

9. Previously U.S. submarines had 18-inch (457-mm) torpedo tubes.

10. Fleet Adm. E. J. King at the end of World War II proposed a 24-inch (610-mm) torpedo for U.S. submarines; however, no action was taken on the proposal. See Fleet Adm. King, Commander in Chief U.S. Fleet, Memorandum for the Chairman, General Board, Subj: "Submarine Characteristics; Combat Vessel Program to Supplement Existing Program," FF1/A-3, Ser 01531, 28 May 1945, p. 1. Also, the CONFORM studies addressed large-diameter weapon tubes.

11. The Harpoon was given the designation UGM-84A; U = Underwater [to], G = surface, M = Missile.

12. The changing missions of U.S. submarines and the high support costs of Harpoons led to the missiles being removed from U.S. submarines in 1997.

13. The Tomahawk was the designation BGM-109; B = multiple launch platforms [to], G = surface, M = Missile.

14. The Navy's inventory goal for TLAM/N missiles before the president's statement was reported as 637, with 399 funded through fiscal 1991.

15. R. A. Shmakov, "On the History of the Development of Anti-Submarine Weapon Systems 'V'yuga,' 'Vodopad,' and 'Shkval,' " *Gangut* (no. 14, 1998), pp. 119–28; and Nikolai Spassky (ed.), *Russia's Arms Catalog* vol. III *Navy,* pp. 428–37.

16. Richard Scott, "Russia's 'Shipwreck' missile enigma solved, *Jane's Defence Weekly* (5 September 2001), p. 28.

17. The SS-N-19 was also fitted in the four *Kirov*-class battle cruisers (20 missiles) and the single aircraft carrier *Admiral Kuznetsov* (12 missiles).

18. The air-launched version of the SS-N-21 is the AS-15 Kent; a ground-launched version designated SSC-X-4 was developed but not deployed. (See above.)

19. Grom is the Soviet name for this weapon, Grom is the NATO code name for the RSM-52/R-39M strategic missile (NATO SS-N-28).

20. A planned air-launched version was given the NATO designation AS-X-19 Koala and the land-launched version SSC-X-15.

21. See N. Polmar, *Ships and Aircraft,* 14th edition (1987), p. 476.

22. The United States and Germany both had passive acoustic homing torpedoes in service during World War II. The German *U-250,* salvaged by the Soviets in 1944, provided them with acoustic homing torpedoes. (See Chapter 2.)

23. I. D. Spassky (ed.), *The History of Indigenous Shipbuilding,* vol. V, p. 130.

24. Kuzin and Nikol'skiy, *The Navy of the USSR,* p. 376.

25. Vice Adm. Joseph Metcalf, USN, Deputy Chief of Naval Operations (Surface Warfare), at conference sponsored by the U. S. Naval Institute, Charleston, S.C., 26 February 1987.

26. Chief of Naval Operations (Staff Op-03K), "Report of the Ship Operational Characteristics Study on the Operational Characteristics of the Surface Combatant of the Year 2010" (26 April 1988), p. 11.

27. Ibid.

28. Two nuclear torpedoes were on board at least two of the four Soviet nuclear submarines lost before 1991, the *K-219* (1986) and *Komsomolets* (1989). Two Mk 45 ASTOR nuclear torpedoes were in the USS *Scorpion* (SSN 589) when she sank in 1968.

29. NII-24 developed the famed *Katusha* rocket launchers of World War II. Merkulov was succeeded as chief designer by V. R. Serov, and the work was completed under Ye. D. Rakov. See Shmakov, op. cit., and A. B. Shirokorad, *Oruzhiye Otechestvennogo Flota* (Weapons of Our Fleet), *1945–2000* (Moscow: AST, 2001), pp. 322–24.

30. The conventional-warhead variant has been offered to Third World countries. Several former Soviet republics, including Ukraine, have the missile; 50 have been reported sold to China.

31. Quoted in Patrick E. Tyler, "Behind Spy Trial in Moscow: A Superfast Torpedo," *The New York Times* (1 December 2000), p. A3. Former U.S. naval intelligence officer Edmond Pope was arrested by the Russian government in 1999 on charges of attempting to procure classified information on the Shkval; sentenced to life imprisonment, he was released immediately after sentencing because of medical problems. See Edmond D. Pope and Tom Shachtman, *Torpedoed* (Boston: Little, Brown, 2001).

32. Baker, *Combat Fleets of the World 2000–2001,* p. 581.

## Chapter 19: Fourth-Generation Submarines

1. "SSN 21" was originally a *program* rather than a submarine designation, i.e., a submarine for the 21st Century; subsequently the U.S. submarine community adopted SSN 21 as the hull number for the lead submarine (followed by the SSN 22 and SSN 23). Although the Centurion/NSSN/*Virginia* (SSN 774) class was initiated during the Cold War as a low-cost alternative to the *Seawolf.* Construction of the *Virginia* class was not begun until 1997, several years after the end of the Cold War.

2. Vice Adm. Earl B. Fowler Jr., USN, Commander Naval Sea Systems Command, testimony before the Committee on Defense, House of Representatives, 30 July 1981.

3. Three previous CNOs had been submarine qualified: Fleet Adm. Ernest J. King (CNO 1942–1945), Fleet Adm. Chester W. Nimitz (1945–1947), and Adm. Louis E. Denfeld (1947–1949). From 1982 three successive nuclear submarine officers served as Chief of Naval Operations: Adms. Watkins (1982–1986), Carlisle A. H. Trost (1986–1990), and Frank B. Kelso, II (1990–1994).

4. Lehman discussion with N. Polmar, Washington, D.C. (Pentagon), 12 March 1985.

5. Adm. Rickover was succeeded as head of Naval Reactors by Adm. Kinnard McKee (1982–1988) and then Adm. Bruce DeMars (1988–1996). Previously, as a vice admiral, DeMars was Assistant CNO for Undersea Warfare, i.e., for submarines (1985–1988).

6. Adm. Alfred J. Whittle, USN (Ret), statement at Georgetown Center for Advanced International Studies, Washington, D.C., 26 September 1983.

7. Vice Adm. Thunman discussion with N. Polmar, Washington, D.C. (Pentagon), 20 September 1985. Also see Merrill and Wyld, *Meeting the Submarine Challenge,* pp. 321–2.

8. The *Seawolf*'s design speed of 35 knots was revealed by Adm. Watkins in testimony before the Committee on Appropriations, House of Representatives, Congress, 5 March 1985.

9. Ibid. Some unofficial sources placed the *Seawolf's* tactical (acoustic) speed at 25 knots; see Stan Zimmerman, *Submarine Technology for the 21st Century* (Arlington, Va.: Pasha Publications, 1997), p. 164.

10. Gerald Cann, Principal Deputy Assistant Secretary of the Navy for Research, Engineering and Systems, testimony before Committee on Appropriations, House of Representatives, Congress, 2 April 1985.

11. HMS *Churchill,* commissioned in 1970, was the world's first operational submarine to have a pump-jet propulsion system.

12. Beyond British and U.S. submarines, ducted propulsors are in use in French nuclear-propelled submarines and has been fitted in the Russian Project 877/Kilo submarine *B-871.*

13. Capt. 1/R Lev I. Khudiakov, Russian Navy, Chief Deputy, First Central Research Institute (Naval Shipbuilding), discussion with K. J. Moore, St. Petersburg, 14 November 1994.

14. Carderock Division, Naval Sea Systems Command, "Advanced Polymer Drag Reduction and Quieting," Navy web site: www.dt.navy.mil/div/capabilities/accomplishments/C-3.html, accessed.

15. B = submarine system, S = Special, Y = multi-purpose.

16. GAO, "Navy Acquisition: SUBACS Problems May Adversely Affect Navy Attack Submarine Programs" (November 1985), p. 10.

17. John Landicho, GAO staff, statement to Committee on Armed Services, Senate, 24 March 1987.

18. GAO, "Submarine Combat System: Status of Selected Technical Risks in the BSY-2 Development" (May 1991), p. 2.
19. GAO, "Submarine Combat System: BSY-2 Development Risks Must Be Addressed and Production Schedule Reassessed" (August 1991), pp. 1–2.
20. GAO, "Navy Acquisition: AN/BSY-1 Combat System Operational Evaluation" (November 1992), p. 2.
21. "SSN-21 Assessment," report prepared for the Deputy Under Secretary of Defense (Plans and Policy), 10 January 1985.
22. Constructing three submarines per annum with an expected service life of just over 30 years (3 x 30+ = 90+). Under the Lehman 600-ship plan in 1982 the SSN force goal was increased from 90 to 100 submarines.
23. Adm. Watkins letter to Secretary Lehman, Subj: "SSN-21 Assessment Paper Written by [deleted]," ser. 00/5U300058, 11 February 1985.
24. Stan Zimmerman, "Navy junior varsity losing ground in *Seawolf* drive," *Navy News* (30 July 1990), p. 3.
25. Adm. DeMars letter to Adm. Robert L. J. Long, USN (Ret), Chairman of the Board, Naval Submarine League, 29 January 1988.
26. Susan J. Crawford, Inspector General, Department of Defense, letter to John J. Engelhardt, 11 October 1990.
27. Referring to the *Seawolf*, Adm. DeMars stated, "Today the country has the fastest, the quietest, the most heavily armed submarine in the world." *Source:* "Seawolf completes initial sea trials," NAVNEWS (Navy News) by E-mail, 11 July 1996 (navnews @opnav-emh.navy.mil).
28. The first edition of the Naval Sea Systems Command's "Quarterly Progress Report for Shipbuilding and Conversion" issued after the construction contract was awarded listed the SSN 21 projected completion date as 26 May 1995.
29. From June to December 2001 the *Seawolf* operated in the North Atlantic and Mediterranean.
30. The SSN 22 was the first U.S. attack submarine to be named for a state; previously states of the Union were honored by battleships, guided missile cruisers (6 ships), and Trident missile submarines (17 ships).
31. The SSN 23 was the first U.S. attack submarine to be named for a former president; other former presidents have been honored by aircraft carriers (8 ships) and Polaris missile submarines (12 ships). President Franklin D. Roosevelt was honored with the aircraft carrier CVB 42 and, subsequently, with his wife by the missile destroyer *Roosevelt* (DDG 80).
32. SEAL = Sea-Air-Land (team), i.e., special forces.
33. Confidential source J to N. Polmar, Washington, D.C., 8 September 1999.
34. There has never been a U.S. Navy ship with the name Centurion. The Royal Navy has had nine ships named *Centurion* since 1650, including two battleships.
35. Capt. David Burgess, USN, Program Manager, New Attack Submarine, briefing to Messrs. Anthony R. Battista and N. Polmar, "Submarine Research, Development and Technology Insertion," 6 March 1997.
36. "Report of the Advisory Panel on Submarine and Antisubmarine Warfare to the House Armed Services Subcommittee on Research and Development and Seapower and Strategic Critical Materials," Congress (21 March 1989) (unclassified edition), p. 1.
37. Aspen subsequently served as Secretary of Defense from 1994 to 1997, during the Clinton administration.
38. Vice Adm. Burkhalter was the only submarine officer on the panel; before retiring from the Navy, he had been Director of the Intelligence Community Staff (1982–1986).
39. DARPA is a semi-independent defense agency established in 1958 in response to the Soviet Union orbiting Sputnik; the agency is charged with developing advanced technologies for military and civil use.
40. "Report of the Advisory Panel," (21 March 1989), p. 5.
41. Bruce Wooden, discussion with N. Polmar, Arlington, Va., 5 March 2002.
42. The four were Anthony Battista, former staff director of the House Armed Services Committee (changed to National Security Committee from January 1995 to December 1998); Dr. John Foster, former Director Defense Research and Engineering; Dr. Lowell Wood, technologist; and Norman Polmar.
43. The four met individually and collectively with Messrs. Gingrich and Duncan, and testified at National Security Committee hearings on submarine issues and the Navy submarine program conducted on 7 September 1995, 27 March 1996, and 18 March 1997.
44. Vice Adm. A. J. Baciocco, Jr., USN (Ret), "Statement of the Submarine Technology Assessment Panel," presentation to Armed Services Committee, Senate (27 March 1996), pp. 3–4.
45. Vice Adm. Baciocco, "Statement of the Submarine Technology Assessment Panel," p. 5.
46. The argument has been made that this division of labor was a means of keeping both shipyards in the submarine business. There were, however, several other solutions, such as alternating the submarines between the two yards; this would have kept both in business while developing a cost comparison between them for future competition.
47. Office of the Under Secretary of Defense for Acquisition & Technology, "Report of the Defense Science Board Task Force on Submarine of the Future"

(Washington, D.C.: Department of Defense, July 1998).

48. The exception is that the three *Seawolf*-class submarines have eight 26½-inch torpedo tubes.

49. Navy-DARPA contractors conference, Arlington, Va., 10 December 1998.

50. Confidential source B, discussion with N. Polmar, Falls Church, Va., 8 February 2001.

51. COTS = Commercial Off-The-Shelf (components).

52. Aleksandr Nadzharov, "Russian Army. New Times," *Nezavisimaya Gazeta* (Moscow) (8 June 1993), p. 5.

53. Yuri Golotyuk, "The 'Vavorok': Russia's Nuclear Might Is Maintained Free at the Northern Machine Building Enterprise: Nuclear Powered Submarine Builders Have Not yet Coped With Conversion," *Segodnya* (Moscow) (7 July 1994), p. 3.

54. By comparison, in 2000 the Electric Boat yard had some 12,000 employees, while Newport News Shipbuilding, with a total employment of some 20,500, has about 3,200 men and women engaged in submarine construction.

55. See Chapter 18.

56. Ernest B. Blazar, "The quiet controversy," *Navy Times* (5 September 1994), p. 30.

57. Ibid.

58. Malachite designers discussions with N. Polmar, St. Petersburg, 7 May 1997.

59. Yuri Vladimirovich Dolgorukiy was the founder of Moscow in 1147; the family name translates "Long Arm."

60. Podvig, *Russia's Strategic Nuclear Arms,* p. 213.

61. *Topol* is Russian for poplar tree.

62. A. V. Kuteinikov discussions with K. J. Moore, Washington, D.C., 21–27 September 1998.

63. Capt. 2/R Nikolay Litkovets, Russian Navy, "Fleet in Financial Squeeze," *Krasnaya Zvezda* (Moscow) (1 April 1994), p. 1. Also, the commander-in-chief of the Pacific Fleet was fired in 1993 because of recruit food and housing conditions in the Far East, which caused the deaths of several from emaciation, pneumonia, and related diseases.

## Chapter 20: Soviet Versus U.S. Submarines

1. The Royal Navy pursued closed-cycle (Walter) propulsion, with full access to U.S. naval officers, and the U.S. Navy briefly experimented with shore-based close-cycle plants. (See Chapter 3.)

2. Although HY-80 steel was used in the *Skipjack,* the submarine was not designed nor fitted with systems for deep operation.

3. Of 55 attack submarines in commission in the U.S. Navy in 2002, all but three were of the *Los Angeles* series.

4. See, for example, Sontag and Drew, *Blind Man's Bluff.* The Forty Committee was assigned the task of approving covert actions conducted by the Central Intelligence Agency, but became a de facto intelligence planning group.

5. Cited in Polmar and Allen, *Rickover,* pp. 665–6.

6. Gerhardt B. Thamm comments on "The Quest for the Quiet Submarine" and "Newest Akula II Goes on Trials," U.S. Naval Institute *Proceedings* (May 1996), p. 17. Mr. Thamm was a former submarine analyst at the Naval Intelligence Support Center.

7. Ibid.

8. Adm. Gorshkov's tenure had spanned that of 13 U.S. Secretaries of the Navy and nine Chiefs of Naval Operations, the equivalent U.S. officials. The longest-serving U.S. Chief of Naval Operations since that position was established in 1915 has been Admiral Arleigh Burke, who served as CNO from August 1955 to August 1961.

9. Spassky joined Rubin (then TsKB-18) in 1953.

10. Prof. Rear Adm. I. G. Zakharov, the head of the First Central Research Institute, presentation "Today's Trends in the Development of Fighting Ships" during the conference celebrating the 300th anniversary of the Russian Navy, St. Petersburg, 27 February 1996.

11. See, for example, Academician A. V. Kuteinikov, paper "Emerging Technology and Submarines of the 21st Century," presented at the Royal Institution of Naval Architects, London, 14 June 1996. At the time Kuteinikov was head of the Malachite design bureau.

12. Few Americans have knowledgeably written on the subject of Soviet submarine design; highly recommended is John J. Engelhardt, "Soviet Sub Design Philosophy," U.S. Naval Institute *Proceedings* (October 1987), pp. 193–200. Also see V. M. Bukalov and A. A. Narusbayev, (Design of Nuclear Submarines) (Leningrad: Sudostroenie Publishing House, 1968).

13. G. M. Novak and B. A. Lapshin, *Survivability of Warships* (in Russian) (Moscow: Military Publishing House, 1959), pp. 20, 182.

14. Little has been written addressing these factors; see Rear Adm. Thomas A. Brooks, USN, "Soviet Weaknesses Are U.S. Strengths," U.S. Naval Institute *Proceedings* (August 1992), p. 68.

15. Soviet shipyards and related industries along the Black Sea coast were captured and then destroyed by German forces; those in the Leningrad area were under German siege and largely inactive for almost three years and suffered heavy damage.

16. Sergei Khrushchev, *Nikita Khrushchev,* p. 272.

17. Ibid.

18. Sergei Khrushchev discussion with N. Polmar, Providence, R.I., 3 August 2001.

19. On occasion, however, major U.S. firms refused to provide components for the nuclear submarine program; see, for example, "Naval Nuclear Propulsion Program-[Fiscal Year] 1967–68," Hearings before the Joint Committee on Atomic Energy, Congress, 16 March 1967, pp. 19–29.
20. Officials of the First Central Research Institute, Aurora Automation Group, and Malachite design bureau, discussions with K. J. Moore, St. Petersburg, 11 August 1993.
21. During World War II the submarine force comprised less than two percent of the U.S. Navy's personnel.
22. Adm. Frank L. Bowman, USN, speech at Naval Submarine League Symposium, Alexandria, Va., 13 June 2001; printed as "Opening Remarks at Annual NSL Symposium," *The Submarine Review* (July 2001), p. 12.
23. The Soviet universal military service required conscripts to serve for two years on naval shore duty or three years if assigned to ships and coastal logistic support units; before 1967 conscription was for three years on shore duty or four years of service on shipboard duty. Conscripts normally could not be retained beyond the stipulated period. An effort to develop a cadre of long-term, professional sailors was begun at the end of the Soviet era.
24. Rear Adm. Thomas A. Brooks, USN (Ret), former Director of Naval Intelligence, telephone conversation with N. Polmar, 5 October 2001.
25. The Sviatov meeting with Ustinov was on 15 January 1965; see Sviatov, "Recollections of a Maverick," *The Submarine Review* (January 2002), pp. 91–96. Ustinov was Soviet Minister of Defense from April 1976 until his death in December 1984.
26. These eight classes were:

| | Project | |
|---|---|---|
| SSBN | 667BDRM | Delta IV |
| SSN | Project 671RTM | Victor III |
| SS | Project 877 | Kilo |
| SSBN | Project 941 | Typhoon |
| SSN | Project 945M | Sierra II |
| SSGN | Project 949A | Oscar II |
| SSN | Project 971 | Akula |
| SSAN | Project 1083.1 | Paltus |

27. Capt. William H. J. Manthorpe Jr., USN (Ret), Deputy Director of Naval Intelligence, "The Submarine Threat of the Future," presentation to the Naval Submarine League, Alexandria, Va., 14 June 1990. Soviet military spending had been secretly capped at zero growth in 1988 (revealed in 1990).
28. Rear Adm. Nikolay P. V'yunenko, Capt. 1/R Boris N. Makeyev, and Capt. 1/R Valentin D. Skugarev. *Voenno-Morskoi Flot: Rol, Perspektivi Razvitiya, Izpolzovaniye* (The Navy: Its Role, Prospects for Deployment and Employment), (Moscow: Voennoe Izdatelstvo 1988). Gorshkov was retired as head of the Soviet Navy in December 1985.
29. A concept similar to that used, on a much smaller scale, in the U.S. Navy's Deep Submergence Rescue Vehicles (DSRV), which have an operating depth of 5,000 feet (1,525 m). This configuration is considerably more expensive to construct than the "stiffened cylinders" used in conventional submarine hulls, which, using the same materiel, could not reach such great depths.
30. Apparently using Project 1710/Beluga as a test bed; see Chapter 13.
31. Manthorpe, "The Submarine Threat of the Future."

## Appendix A: U.S. Submarine Construction, 1945–1991

1. The U.S. Navy's ship designation system was established in 1920, assigning letters for ship types and sequential hull numbers. The USS *Holland*, commissioned in 1900, was assigned the hull number SS 1 in 1920, initiating the U.S. submarine designation series.
2. SCB = Ship Characteristics Board.
3. One reactor in each submarine except for the USS *Triton*.
4. Hull number SSX 1.
5. The *Nautilus* did not get underway until January 1955.
6. Replaced in 1958–1960 by an S2Wa reactor plant.
7. S3W in *Skate* (SSN 578) and *Sargo* (SSN 583); S4W in *Swordfish* (SSN 579) and *Seadragon* (SSN 584).
8. SCB series was changed in 1964 from sequential to block numbers; 300-series indicated submarines; changed from 188A to SCB-300.65 with the SSN 665.
9. Fifteen *Los Angeles*-class submarines were completed after 1991.
10. Six *Ohio*-class submarines completed after 1991.
11. Three *Seawolf*-class submarines were to be completed from 1997 to 2005.

## Appendix B: Soviet Submarine Construction, 1945–1991

1. "Whale" was not an official NATO name in the alphabetical series; "w" or "Whiskey" was the first code name in the series.
2. One additional Project 667BDRM submarine completed after 1991.
3. The ship name was *Komsomolets.*
4. One Project 945A submarine was completed after 1991.

5. Five project 949A submarines were completed after 1991.
6. Six Project 971 submarines were completed after 1991.
7. One Project 1851 submarine was completed after 1991.
8 Two Project 1910 submarines were completed after 1991.

## Appendix C: U.S. Submarine Reactor Plants

1. Prefix: S = Submarine; manufacturer letters: C = Combustion Engineering, G = General Electric, and W = Westinghouse. The *NR-1* reactor was developed by the Knolls Atomic Power Laboratory.
2. Bettis = Bettis Atomic Laboratory, West Mifflin, Pa.; Knolls = Knolls Atomic Power Laboratory, Schenectady, N.Y.
3. Developed by Combustion Engineering, Inc.
4. Developed as a "rodless" reactor to use water tubes to control level of radioactivity; fitted with control rods in the late 1980s. Not intended for submarine use.
5. Also used in first British nuclear submarine, HMS *Dreadnought.*

## Appendix D: Soviet Submarine Design Bureaus

1. See Glossary (page xix) for definition of abbreviations.

## Appendix E: Soviet Ballistic Missile Systems

1. System cancelled.
2. The NATO name Serb, initially assigned on the basis of observing Soviet missiles being displayed at national holidays, was also assigned by some documents to the SS-N-6 missile.

# Selected Bibliography

## Books in English


Alden, Comdr. John D., USN (Ret). *The Fleet Submarine in the U.S. Navy.* Annapolis, Md.: Naval Institute Press, 1979.

Anderson, Capt. William R., USN. *Nautilus 90 North.* Cleveland, Ohio: World Publishing, 1959.

Baker, A. D., III. *Combat Fleets of the World 2000–2001.* Annapolis, Md.: Naval Institute Press, 2000.

Beach, Capt. Edward L. USN (Ret). *Around the World Submerged.* New York: Holt, Rinehart, Winston, 1962.

———. *Salt and Steel: Reflections of a Submariner.* Annapolis, Md.: Naval Institute Press, 1999.

Buchheim, Lothar-Günther. *The Boat.* New York: Bantam Books, 1976.

Burleson, Clyde W. *The Jennifer Project: Howard Hughes . . . the CIA . . . A Russian submarine . . . the intelligence coup of the decade.* Englewood Cliffs, N.J.: Prentice-Hall, 1977.

Burlingame, Burl. *Advance Force Pearl Harbor.* Kailua, Hawaii: Pacific Monograph, 1992.

Calvert, Comdr. James, USN. *Surface at the Pole.* New York: McGraw-Hill, 1960.

Carpenter, Dorr, and Norman Polmar. *Submarines of the Imperial Japanese Navy.* Annapolis, Md.: Naval Institute Press, 1986.

Churchill, Winston S. *The Hinge of Fate.* Boston, Mass.: Houghton Mifflin, 1950.

———. *Their Finest Hour.* Boston, Mass.: Houghton Mifflin, 1949.

Compton-Hall, Capt. Richard, RN. *Submarine Warfare: Monsters & Midgets.* Poole, Dorset: Blandford Press, 1985.

Craven, John Piña. *The Silent War: The Cold War Battle Beneath the Sea.* New York: Simon & Schuster, 2001.

Daniel, Donald C. *Anti-Submarine Warfare and Superpower Strategic Stability.* Urbana: University of Illinois Press, 1986.

Davies, Roy. *Nautilus: The Story of Man under the Sea,* London: BBC Books, 1995.

Doenitz, Adm. Karl. *Memoirs: Ten Years and Twenty Days.* Cleveland, Ohio: World Publishing, 1959; reprinted in 1990 by Naval Institute Press.

Dunham, Roger C. *Spy Sub.* Annapolis, Md.: Naval Institute Press, 1996.

Eliot, George Fielding. *Victory Without War, 1958–1961.* Annapolis, Md.: U.S. Naval Institute, 1958.

Everitt, Don. *The K Boats.* New York: Holt, Rinehart and Winston, 1963.

Fahey, James C. *The Ships and Aircraft of the U.S. Fleet*/6th edition (1954) and 7th edition (1958); reprinted by Naval Institute Press.

Friedman, Norman. *U.S. Submarines since 1945: An Illustrated Design History.* Annapolis, Md.: Naval Institute Press, 1994.

Galantin, Adm. I. J., USN (Ret). *Submarine Admiral: From Battlewagons to Ballistic Missiles.* Urbana: University of Illinois Press, 1995.

Gantz, Lt. Col. Kenneth F., USAF (ed.). *Nuclear Flight: The United States Air Force Programs for Atomic Jets, Missiles, and Rockets.* New York: Duell, Sloan and Pearce, 1960.

Gardiner, Robert (ed.). *Conway's All the World's Fighting Ships, 1947–1982,* part I: *The Western Powers.* London: Conway Maritime Press, 1983.

Garzke, William H., Jr., and Robert O. Dulin, Jr. *Battleships: United States Battleships, 1935–1992.* Annapolis, Md.: Naval Institute Press, 1995.

Hackmann, Willem. *Seek & Strike: Sonar, anti-submarine warfare and the Royal Navy, 1914–54.* London: Her Majesty's Stationery Office, 1984.

Harford, James. *Korolev: How One Man Masterminded the Soviet Drive to Beat America to the Moon.* New York: John Wiley & Sons, 1997.

Hervey, Rear Adm. John, RN (Ret). *Submarines.* London: Brassey's, 1994.

Hewlett, Richard G., and Francis Duncan, *Nuclear Navy, 1946–1962.* Chicago: University of Chicago Press, 1974.

Heinemann, Edward H., and Capt. Rosario Rausa, USNR (Ret). *Ed Heinemann: Combat Aircraft Designer.* Annapolis, Md.: Naval Institute Press, 1980.

Hinsley, F. H., et al., *British Intelligence in the Second World War.* London: Her Majesty's Stationery Office, vol. 2 (1981); vol. 3, part 1 (1984); and vol. 3, part 2 (1988).

Huchthausen, Capt. Peter, USN (Ret). *October Fury.* Hoboken, N.J: John Wiley & Sons, 2002.

Huchthausen, Capt. Peter, USN (Ret), Capt. 1/R Igor Kurdin, Russian Navy, and R. Alan White. *Hostile Waters.* New York: St. Martin's Press, 1997.

Kemp, Paul. *Midget Submarines of the Second World War.* London: Chatham Publishing, 1999.

———. *Underwater Warriors.* Annapolis, Md.: Naval Institute Press, 1996.

Kennedy, Robert F. *Thirteen Days: A Memoir of the Cuban Missile Crisis.* New York: W. W. Norton, 1969.

Khrushchev, Nikita S. *Khrushchev Remembers: The Last Testament.* Boston: Little, Brown, 1974.

Khrushchev, Sergei N. *Nikita Khrushchev and the*

*Creation of a Superpower.* University Park, Pa.: Pennsylvania State University, 2000.

Klee, Ernst, and Otto Merk. *The Birth of the Missile: The Secrets of Peenemünde.* New York: E. P. Dutton, 1965.

Koestler, Arthur. *Darkness at Noon.* New York: Macmillan, 1941.

König, Paul. *The Voyage of the Deutschland.* New York: Hearst International Library, 1916.

Kuznetsov, Adm. N. G. *Memoirs of Wartime Minister of the Navy.* Moscow: Progress Publishers, 1990.

Lake, Simon. *Submarine: The Autobiography of Simon Lake.* New York: D. Appleton-Century, 1938.

Largess, Robert P., and James L. Mandleblatt. *U.S.S. Albacore: Forerunner of the Future.* Portsmouth, N.H.: Portsmouth Marine Society, 1999.

Leary, William M. *Under Ice: Waldo Lyon and the Development of the Arctic Submarine.* College Station, Texas A&M University Press, 1999.

Lehman, John F., Jr. *Command of the Seas: Building the 600 Ship Navy.* New York: Charles Scribner's Sons, 1988.

Lockwood, Vice Adm. Charles A., USN (Ret). *Down to the Sea in Subs: My Life in the U.S. Navy.* New York: W. W. Norton, 1967.

May, Ernest R., and Philip D. Zelikow. *The Kennedy Tapes: Inside the White House During the Cuban Missile Crisis.* Cambridge, Mass.: Harvard University Press, 1997.

Merrill, John, and Lionel D. Wyld. *Meeting the Submarine Challenge: A Short History of the Naval Underwater Systems Center.* Washington, D.C.: Government Printing Office, 1997.

Messimer, Dwight R. *The Merchant U-Boat.* Annapolis, Md.: Naval Institute Press, 1988.

Miasnikov, Eugene V. *The Future of Russia's Strategic Nuclear Forces: Discussions and Arguments.* Moscow: Center for Arms Control, Energy and Environmental Studies, 1996.

Moore, Robert. *A Time to Die: The Untold Story of Kursk Tragedy.* New York: Crown, 2002.

Neufeld, Michael J. *The Rocket and the Reich: Peenemünde and the Coming of the Ballistic Missile Era.* New York: Free Press, 1995.

Nguyen, Hung P. *Submarine Detection from Space: A Study of Russian Capabilities.* Annapolis, Md.: Naval Institute Press, 1993.

Ordway, Frederick I., III, and Mitchell R. Sharpe. *The Rocket Team.* New York: Thomas Y. Crowell, 1979.

Polmar, Norman. *Atomic Submarines.* Princeton, N.J.: D. Van Nostrand, 1963.

———. *Death of the Thresher.* Philadelphia, Pa.: Chilton Books, 1964; revised edition *The Death of the USS Thresher.* New York: Lyons Press, 2001.

———. *Ships and Aircraft of the U.S. Fleet.* Annapolis, Md.: Naval Institute Press, 14th edition (1987), 17th edition (2001).

Polmar, Norman, and Thomas B. Allen, *Rickover: Controversy and Genius.* New York: Random House, 1982.

Pope, Edmond D., and Tom Shachtman. *Torpedoed.* Boston: Little, Brown, 2001.

Ries, Thomas, and Johnny Skorve. *Investigating Kola: A Study of Military Bases using Satellite Photography.* London: Brassey's Defence Publishers, 1987.

Ring, Jim. *We Come Unseen: The Untold Story of Britain's Cold War Submariners.* London: John Murray, 2001.

Riste, Olav. *The Norwegian Intelligence Service 1945–1970.* London: Frank Cass, 1999.

Roskill, Capt. S. W. Roskill, RN (Ret). *The War at Sea,* vol. II. London: Her Majesty's Stationery Office, 1957.

Rössler, Eberhard. *The U-boat: The Evolution and Technical History of German Submarines.* Annapolis, Md.: Naval Institute Press, 1981.

Sapolsky, Harvey M. *The Polaris System Development.* Cambridge, Mass.: Harvard University Press, 1972.

Sharpe, Mitchell R. *The Rocket Team.* New York: Thomas Y. Crowell, 1979.

Sontag, Sherry, and Christopher Drew. *Blind Man's Bluff: The Untold Story of American Submarine Espionage.* New York: PublicAffairs Press, 1998.

Spassky, Nikolai (ed.). *Russia's Arms Catalog,* vol. III, *Navy.* Moscow: Military Parade, 1996.

Spinardi, Graham. *From Polaris to Trident: The Development of US Fleet Ballistic Missile Technology.* New York: Cambridge University Press, 1994.

Standley, Adm. William H., USN (Ret). *Admiral Ambassador to Russia.* Chicago, Ill.: Henry Regnery, 1955.

Steele, Comdr. George P. *Seadragon: Northwest Under the Ice.* New York: E. P. Dutton, 1962.

Stefanick, Tom. *Strategic Anti-Submarine Warfare and Naval Strategy.* Lexington, Mass.: Lexington Books, 1987.

Stumpf, David K. *Regulus: The Forgotten Weapon.* Paducah, Ky.: Turner Publishing, 1996.

Smyth, Henry DeWolf. *Atomic Energy for Military Purposes.* Princeton, N.J.: Princeton University Press, 1945.

Treadwell, Terry C. *Submarines With Wings: The Past, Present and Future of Aircraft-Carrying Submarines.* London: Conway Maritime Press, 1985.

Tyler, Patrick. *Running Critical: The Silent War, Rickover, and General Dynamics.* New York: Harper and Row, 1986.

Ugaki, Vice Adm. Matome, IJN. *Fading Victory: The Diary of Admiral Matome Ugaki, 1941–1945.* Pittsburgh, Pa.: University of Pittsburgh Press, 1991.

Varner, Roy, and Wayne Collier. *A Matter of Risk: The Incredible Inside Story of the CIA's Hughes Glomar Explorer Mission to Raise a Russian Submarine.* New York: Random House, 1978.

Vego, Milan. *Soviet Naval Tactics.* Annapolis, Md.: Naval Institute Press, 1992.

Vyborny, Lee, and Don Davis. *Dark Waters: An Insider's Account of the NR-1, the Cold War's Undercover Nuclear Sub.* New York: New American Library, 2003.

Waller, Douglas C. *Big Red: Three Months on Board a Trident Nuclear Submarine.* New York: HarperCollins, 2001.

Warren, C. E. T., and James Benson. *Above Us the Waves: The Story of Midget Submarines and Human Torpedoes.* London: George G. Harrap, 1953.
Weir, Gary E. *Forged in War.* Washington, D.C.: Naval Historical Center, 1993.
White, John F. *U Boat Tankers 1941–45: Submarine Suppliers to Atlantic Wolf Packs.* Annapolis, Md.: Naval Institute Press, 1998.
Zaloga, Steven J. *The Kremlin's Nuclear Sword: The Rise and Fall of Russia's Strategic Nuclear Forces, 1945–2000.* Washington, D.C. Smithsonian Institution Press, 2002.
Zimmerman, Stan. *Submarine Design for the 21st Century.* Arlington, Va.: Pasha Publications, 1993.
———. *Submarine Technology for the 21st Century.* Arlington, Va.: Pasha Publications, 1st edition 1990; 2nd edition, 1997.
Zumwalt, Adm. Elmo R., Jr. *On Watch.* New York: Quadrangle, The New York Times Book Co., 1976.

## Books in German

Gröner, Erich. *Die deutschen Kriegschiffe 1815–1945,* vol. 1. Munich: J. F. Lehmanns, 1966.
König, Paul. *Fahrten der U-Deuschland im Weltkrieg.* Berlin: Ullstein, 1916.
Rössler, Eberhard. *Geschichte des deutschen Ubotbaus.* Munich: J.F. Lehmanns, 1975.
———. *U-Boottyp XXI.* Berlin [privately published], 1 April 1966; revised and expanded as *U-Boottyp XXI.* Munich: Bernard & Graefe Verlag, 1980.

## Books in Russian

Antonov, A. M. *Pervoe Pokoleniye Atomokholov SKV-143* (The First Generation of Nuclear Submarines: Special Design Bureau SDB-143), St. Petersburg: Malachite, 1996.
Bukalov, V. M., and A. A. Narusbayev. *Proyektirovaniye Atomnykh Podvodnykh Lodok* (Design of Nuclear Submarines). Leningrad: Sudostroenie Publishing House, 1968.
Gorshkov, Adm. Fleet of the Soviet Union S. G. *Morskaya Moshch Gosudarstva* (The Sea Power of the State), 2d ed. Moscow: Military Publishing House, 1979; translation of U.S. Naval Intelligence Command.
———. *Vo Flotskom Stroyu* (Military Memoirs). St. Petersburg: Logos, 1996.
Gusev, A. N. *Submarines with Cruise Missiles (in Russian).* St. Petersburg: Galeya Print, 2000.
Karpyenko, A.B. *Rossiskoye Raketnoye Oruzhiye 1943–1993* (Russian Missile Weapons, 1943–1993). St. Petersburg: PEKA, 1993.
Kuzin, V. P., and V. I. Nikol'skiy. *Voyenno-morskoy Flot SSSR, 1945–1991* (The Navy of the USSR, 1945–1991). St. Petersburg: Historical Oceanic Society, 1996.
Kuznetsov, Adm. N. G., Soviet Navy (Ret), "Naval Preparedness" in Seweryn Bailer (ed.), *Stalin and His Generals.* London: Souvenir Press, 1969.
Makeyev, B. N. *Voyenno-morskoy aspekty natsionalnoi bezopasnosti* (Naval Aspects of National Security). Moscow: Nonproliferation and Critical Technologies Committee, 1997.
Malachite SPMBM. *Istoriya Sankt-Peterburgskogo Morskogo Buro Mashinostroeniya "Malakhit"* (History of the Saint Petersburg Maritime Machinebuilding Bureau Malachite), vol. 2. St. Petersburg: Malachite, 1995.
Mormul, N. G. *Katastrofe pod Vodoe* (Catastrophe under Water). Murmansk: Elteco, 1999.
Novak, G. M., and B. A. Lapshin. *Survivability of Warships (in Russian)* (Moscow: Military Publishing House, 1959).
Podvig, Pavel L. (ed.). *Strategicheskoye yadernoye vooruzhenie Rossi* (Russia's Strategic Nuclear Arms). Moscow: IzdAT, 1998. Revised English-language edition is Podvig, *Russian Strategic Nuclear Forces,* Cambridge, Mass.: MIT Press, 2001.
Romanov, D. A. *Tragediya Podvodnoy Lodki Komsomolets* (Tragedy of the Submarine *Komsomolets*). St. Petersburg: Publishing Association of St. Petersburg, 1993.
Sarkisov, A. A. *The Contribution of Russian Science to the Soviet Navy* (in Russian). Moscow: Nauka Press (Russian Academy of Sciences), 1997.
Shirokorad, A. B. *Oruzhiye Otechestvennogo Flota 1945–2000* (Weapons of Our Fleet). Moscow: AST, 2001.
———. *Sovetskiye Podvodnyye Lodki Poslevoyennov Postryki* (Soviet Submarines of Postwar Construction). Moscow: Arsenal Press, 1997.
Spassky, I. D. (ed.). *Istoriya Otechestvennogo Sudostroeniya* (The History of Indigenous Shipbuilding), vol. V *1946–1991.* St. Petersburg: Sudostroenie, 1996.
Sviatov, G.I. *Atomnyye Podvodnyye Lodki* (Atomic Submarines). Moscow: Military Publishing House, 1969.
V'yunenko, Rear Adm. Nikolay P., Capt. 1/R Boris N. Makeyev, and Capt. 1/R Valentin D. Skugarev. *Voenno-Morskoi Flot: Rol, Perspektivi Razvitiya, Izpolzovaniye* (The Navy: Its Role, Prospects for Deployment and Employment). Moscow: Voennoe Izdatelstvo, 1988.
Zapol'ski, A. A. *Raketi Startuyut s Morya* (Missiles Launch from the Sea). St. Petersburg: Malachite, 1994.

## Articles in English*

"Accidents reported aboard Soviet subs," *The Baltimore Sun* (17 March 1982).
Allen, Thomas B. "Run Silent, Run Deep," *Smithsonian Magazine* (March 2001).
Anderson, Jack. "U.S. Kept Mum on Soviet Sub Patrol," *The Washington Post* (26 November 1980).
Anderson, Comdr. William R., USN. "The Arctic as a Sea Route of the Future," *National Geographic Magazine* (January 1959).
Andrews, Capt. Frank A., USN (Ret). "Submarine Against Submarine" in *Naval Review, 1966.* Annapolis, Md.: U.S. Naval Institute, 1965.

———. "*Thresher*'s Debris Field," *The Submarine Review* (April 1987).
Antonov, A. M. "The Birth of the Red November," USNI *Proceedings* (December 1995).
Beckley, Chief Warrant Officer Jerry E., USN (Ret). "A Few Days in October," *The Submarine Review* (April 1999).
Blazar, Ernest B. "The quiet controversy," *Navy Times* (5 September 1994).
Bogachenko, Capt. 1/R Igor, Soviet Navy (Ret). "Russian Submarines with Titanium Hulls," *The Submarine Review* (July 1997).
Boyle, Richard. "USS *X-1* Power Plant, 1956–57," *Naval Engineers Journal* (April 1972).
Boyes, Vice Adm. Jon L., USN (Ret). " 'Flying' the *Albacore*," *The Submarine Review* (April 1987).
Brady, David, and John Edyvane. "Propulsion in the Pod—Fact or Fiction?" *The Submarine Review* (April 1986).
Brooks, Capt. Thomas A., USN. "(Soviet) Diesel Boats Forever," USNI *Proceedings* (December 1980).
———. "Soviet Weaknesses Are U.S. Strengths," USNI *Proceedings* (August 1992).
Bowman, Adm. Frank L. "Opening Remarks at the Annual NSL [Naval Submarine League] Symposium," *The Submarine Review* (July 2001).
Byron, Capt. John L., USN. "No Quarter for Their Boomers," USNI *Proceedings* (April 1989).
Chatham, Lt. Comdr. Ralph E., USN. "A Quiet Revolution," USNI *Proceedings* (January 1984).
Duncan, Francis. "*Deutschland*–Merchant Submarine," USNI *Proceedings* (April 1965).
Durbo, Alec. "Sounding Out Soviet Subs," *Science Digest* (October 1983).
Eckhart, Capt. Myron, Jr., USN (Ret). Commentary on "Running Critical," USNI *Proceedings* (August 1987).
Engelhardt, John J. "Soviet Sub Design Philosophy," USNI *Proceedings* (October 1987).
Fages, Rear Adm. Malcolm I., USN. "Aboard a Delta I in the Russian Pacific Fleet," *The Submarine Review* (July 1997).
Fitzgerald, James W. Commentary on "FAB . . . ," *The Submarine Review* (April 2000).
Fox, Capt. Robert F., USN (Ret). "Build It & They Will Come," USNI *Proceedings* (April 2001).
Friedman, Norman. "US ASW SSK Submarines," *Warship* (vol. 8, 1984).
Giddings, Alfred J. "*Albacore* X-Tail: Dead End?" USNI *Proceedings* (February 1995).
Ginsberg, Thomas. " 'Subs' were frisky minks, Sweden admits," *Philadelphia Inquirer* (9 February 1995).
Graham, Lt. Comdr. M. T. "*U-2513* Remembered," *The Submarine Review* (July 2001).
Hadeler, Wilhelm. "The Ships of the Soviet Navy," in M. G. Saunders (ed.), *The Soviet Navy.* New York: Frederick A. Praeger, 1958.
Handler, Eugene. "The Flying Submarine," USNI *Proceedings* (September 1964).
Healy, Melissa. "Lehman: We'll Sink Their Subs," *Defense Week* (13 May 1985).
Holzer, Robert. "DoD Orders Navy to Conduct Study of Non-Nuclear Sub," *Defense News* (10 February 1997).
Hunter, Lt. Comdr. Jack, USN (Ret). "Does Your Submarine Have Dive Brakes? Mine Did!" *The Submarine Review* (April 1999).
———. "FAB, The First Submarine Towed Array?" *The Submarine Review* (January 2000).
James, Carolyn C. "The Politics of Extravagance: The Aircraft Nuclear Propulsion Project," Naval War College *Review* (Spring 2000).
Jortberg, Rear Adm. Dick, USN (Ret). "The Legacy of the *Tullibee*," *The Submarine Review* (January 1989).
Kellogg, Capt. Ned, USN (Ret). "Operation Hardtack as a Submerged Target." *The Submarine Review* (October 2002).
Kerr, Donald M., and N. Polmar. "Nuclear Torpedoes," USNI *Proceedings* (August 1986).
———. "Torpedo Warheads: Technical and Tactical Issues in ASW," *Technology Review* [Los Alamos National Laboratory] (January–February 1984).
Kidwell, David, and Manny Garcia. "Feds Allege Wild Scheme: Drugs Hauled by *Submarine*," *The Miami Herald* (6 February 1997).
Kozhevnikov, Vladimir. "The Best of . . . Delta-Class Submarines," *Military Parade* (September-October 2002).
Kurdin, Capt. 1/R Igor, Russian Navy, and Lt. Wayne Grasdock, USN. "Loss of a Yankee SSBN," *The Submarine Review* (October 2000).
Kuznetsov, Adm. N. G. "Naval Preparedness," in Seweryn Bailer (ed.), *Stalin and His Generals,* London: Souvenir Press, 1969.
Laning, Capt. Richard B., USN (Ret). "The *Seawolf*: Going to Sea," *Naval History* (U.S. Naval Institute) (Summer 1992).
———. "The *Seawolf*'s Sodium-Cooled Power Plant," *Naval History* (U.S. Naval Institute)(Spring 1992).
———. "Thoughts from the Oral History of . . . ," *The Submarine Review* (January 1989).
Largess, Richard P. "The Origins of the *Albacore*," *The Submarine Review* (January 1999).
Litovkin, Viktor. "Three Days on the Typhoon," part 2, *Izvestiya* (2 March 1992).
Lowenhaupt, Henry S. "How We Identified the Technical Problems of Early Russian Nuclear Submarines," *Studies in Intelligence* (CIA) (vol. 18, Fall 1974).
Matson, Capt. Willis A., USN (Ret). "Dick Laning and the *Seawolf*," *The Submarine Review* (April 2001).
Matthews, Lt. Comdr. Carey, USNR. "Anti-Sub Warfare Calls for 2 Russian Diesels," *Navy Times* (26 February 1996).
McCGwire, Michael. "Soviet Naval Procurement" in *The Soviet Union in Europe and the Near East: Her Capabilities and Intentions.* London: Royal United Service Institution, 1970.
———. "The Strength of the Soviet Navy" in McCGwire (ed.), in *Soviet Naval Developments: Context and Capa-*

*bility.* Halifax, Nova Scotia: Dalhousie University, 1973.
Momiyama, Thomas S. "All and Nothing," *Air & Space* (October/November 2001).
Moore, K. J. "Antisubmarine Warfare" in Michael McGwire and John McDonnell (ed.), *Soviet Naval Influence: Domestic and Foreign Dimensions.* New York: Praeger Publishers, 1977.
Moore, K. J., Mark Flanigan, and Robert D. Helsel, "Developments in Submarine Systems, 1956–76," in Michael McCGwire and John McDonnell (ed.), *Soviet Naval Influence: Domestic and Foreign Dimensions.* New York: Praeger Publishers, 1977.
"Mystery of Missing Submarines," *U.S. News & World Report* (11 December 1953).
Mozgovoi, Alexander. "Finest Whiskey," *Military Parade* (November 2001).
Nanos, Rear Adm. G. P., USN. "Strategic Systems Update," *The Submarine Review* (April 1977).
O'Rourke, Ronald. "Maintaining the Edge in US ASW," *Navy International* (July/August 1988).
Pala, Christopher. "Turning Russian Subs into .... Cargo Carriers?" *Business Week Online* (31 August 2000).
Paolucci, Capt. D. A., USN (Ret). "The Development of Navy Strategic Offensive and Defensive Systems," in USNI *Proceedings/Naval Review* (May 1970).
Polmar, Norman. "Diesel Submarines: New Option for the Fleet?" *Navy Times* (18 April 1983).
———. "Sailing Under the Ice," USNI *Proceedings* (June 1984).
———. "Submarines: All Ahead—Very, Very Slowly," USNI *Proceedings* (December 1998).
Polmar, Norman, and Thomas B. Allen. "The Plot to Get Rickover," *Washingtonian* (April 1982).
Polmar, Norman, and Capt. D. A. Paolucci, USN (Ret). "Sea-Based 'Strategic' Weapons for the 1980s and Beyond," in USNI *Proceedings/Naval Review* (May 1978).
Polmar, Norman, and Raymond Robinson. "The Soviet Non-Naval Force Multiplier," USNI *Proceedings* (December 1987).
Poteat, S. Eugene. "Counterintelligence Spy vs. Spy vs. Traitor," *American Intelligence Journal* (Winter 2000–2001).
Powess, Comdr. Jonathan, RN. "U.K.'s *Upholder*-Class Subs Go to Canada," USNI *Proceedings* (October 2002).
"Quieter Soviet subs will cost U.S. at least $30 billion," *Navy News & Undersea Technology* (14 March 1988).
Razmus, Jerry. "SSBN Security," *The Submarine Review* (April 1996).
Robinson, Raymond A. "Incoming Ballistic Missiles at Sea," USNI *Proceedings* (July 1987).
Rodgaard, Comdr. John, USNR (Ret). "Japanese Midget Submarines at Pearl Harbor," Naval Intelligence Professionals *Quarterly* (Spring 1995).
Rodgaard, Comdr. John, USNR (Ret), Peter K. Hsu, Carroll L. Lucas, and Capt. Andrew Biache, Jr., USNR (Ret). "Pearl Harbor: Attack from Below," *Naval History* (U.S. Naval Institute) (December 1999). Also commentary by John Di Virgilio, *Naval History* (February 2000).
Rosenberg, Dr. David A. "Arleigh Albert Burke" in Robert William Love, Jr. (ed.), *The Chiefs of Naval Operations.* Annapolis, Md.: Naval Institute Press, 1980.
Schhacht, Capt. K. G., USN (Ret), Commentary on "Diesel Boats Forever?" USNI *Proceedings* (February 1983).
Scott, Richard. "Russia's 'Shipwreck' missile enigma solved," *Jane's Defence Weekly* (5 September 2001).
Shapley, Deborah. "Nuclear Navy: Rickover Thwarted Research on Light Weight Reactors," *Science* (18 June 1978).
Shriver, Lt. Jack, USN. "Developing Real Anti-Diesel Tactics," *The Submarine Review* (April 1998).
Sieche, Erwin. "The Walter Submarine–1," *Warship* (no. 20, 1981).
Spassky, I. D., and V. P. Semyonon. "Project 617: The Soviet 'Whale'," *Naval History* (U.S. Naval Institute) (November/December 1994).
Starr, Barbara. "China's SSK aspirations detailed by USN chief," *Jane's Defence Weekly* (18 March 1995).
Steele, Comdr. George P. USN. "Killing Nuclear Submarines," USNI *Proceedings* (November 1960).
"Submersible Airplanes," *The New York Herald-Tribune* (18 September 1946).
Sutyagin, Igor. "Alfa Class—The 1960's Dream Machine," *The Submarine Review* (January 1998).
Sviatov, Capt. 1/R George I., Soviet Navy (Ret). "Akula Class Russian Nuclear Attack Submarines," *The Submarine Review* (October 1997).
———. "First Soviet Nuclear Submarine," *The Submarine Review* (January 2000).
———. "Recollections of a Maverick," *The Submarine Review* (January 2002).
Thamm, Gerhard. "It was all about ALFA," Naval Intelligence Professionals *Quarterly* (Fall 2002).
———. Comments on "The Quest for the Quiet Submarine" and "Newest Akula II Goes on Trials," USNI *Proceedings* (May 1996).
Tyler, Patrick E. "Behind Spy Trial in Moscow: A Superfast Torpedo," *The New York Times* (1 December 2000).
Ullman, Lt. Comdr. Harlan, USN. "The Counter-Polaris Task," in Michael McCGwire (ed.), *Soviet Naval Policy: Objectives and Constraints.* New York: Praeger, 1975.
Van Saun, Comdr. Art, USN. "Tactical ASW: A Case for a Non-Nuclear Submarine," USNI *Proceedings* (November 1978).
Warren, Chris L., and Mark W. Thomas. "Submarine Hull Optimization Case Study," *Naval Engineers Journal* (November 2000).
Weatherup, Comdr. R. A., USN. "Defense Against Nuclear-Powered Submarines," USNI *Proceedings* (December 1959).
Weir, Gary E. Review of *Hitler's U-Boat War: The Hunters, 1939–1942*" in *The Journal of Military History* (July 1997).

Wilson, George C. "Navy Is Preparing for Submarine Warfare beneath Coastal Ice," *The Washington Post* (19 May 1983).

———. "New Submarine Proposals Challenge Rickover's Rule," *The Washington Post* (21 May 1979).

Wilson, Michael. "The Walter Submarine–2," *Warship* (no. 20, 1981).

Woodward, Bob, and Charles R. Babcock. "CIA Studies Sub Vulnerability," *The Washington Post* (6 June 1985).

Wright, Christopher (ed.). "The U.S. Navy's Operation of the Former Imperial Japanese Navy Submarines *I-14, I-400,* and *I-401,* 1945–1946," *Warship International* (no. 4, 2000).

Zimmerman, Stan. "Navy junior varsity losing ground in *Seawolf* drive," *Navy News* (30 July 1990).


## Articles in German


Breyer, Siegfried. "The Soviet Submarines of the W-Class as a Family of Classes," *Soldat und Technik* (no. 1, 1971).


## Articles in Japanese


"*I-400* Class Submarines Just After the War," *Ships of the World* (Tokyo)(no. 5, 1980).


## Articles in Russian


Aframeev, Eduard. "Diving Missile-Boats," *Voennaya Parade* (May-June 1998).

Alekseyev, A., and L. Samarkin. "The Present and Future of Our Best Multipurpose Nuclear Submarines," *Morskoy Sbornik* (no. 4, 1997).

Aleksin, Rear Adm. V., Russian Navy. "Incidents in the Barents Sea," *Morsky Sbornik* (no. 5–6, 1992).

Altunin, Lt. Col. I., Russian Navy. "An Ordinary Deployment," *Morskoy Sbornik* (no. 1, 1994); reprinted in part in the USNI *Proceedings* (July 1994).

Antonov, A. M. "Atomic Submarine Project 627," *Sudostroenie* (no. 7, 1995).

———. "Multipurpose Submarines on the Threshold of the 21st Century," *Gangut* (no. 14, 1991).

———. "Ships of Postwar Design Projects," *Morskoy Sbornik* (no. 6, 1995).

Badin, V.A., and L. Yu. Khudyakov. "Submarines with the Single Engine—Water Exhaust Diesel Power Plant," *Sudostroenie* (no. 4, 1996).

Badin, V.A., V.P. Semyonov, and L. Yu. Khudyakov. "Design 615 and 615A Submarines," *Sudostroenie* (no. 4, 1995).

"Ballistic Missile Submarines on a Conveyor Belt," *Sudostroenie* (no. 6, 1997).

Baranov, E. L., S. L. Karlinsky, and S. O. Suchanov. "'Tayfun' Changes Profession," *Sudostroenie* (no. 2, 2001).

Bil'din, V. "Design 661 Nuclear Attack Submarine," *Morskoy Sbornik* (no. 4, 1993).

———. "Experimental Nuclear-Propelled Torpedo Submarine of Project 645," *Morskoy Sbornik* (no. 8, 1993).

Borchev, Gen. Maj. M. A., Soviet Army (Ret). "On the Military Organization of the Commonwealth of Independent States," *Voennaya mysl'* (no. 3, 1993).

Bystrov, S. "A Reactor for Submarines," *Krasnaya Zvezda* (21 October 1989).

Cherkashin, L. "Our Submarine Wins the Blue Ribbon," *Morskoy Sbornik* (no. 12, 1994).

Dergachev, F. G. "World's First Titanium High Speed Submarine, Design 661," *Gangut* (no. 14, 1998).

Dronov, Boris F. "Nuclear-Powered Transport Ships for the Arctic," *Sudostroenie* (no. 1, 1992).

Dronov, Boris F., and Boris A. Barbanel'. "Research Submarine-Laboratory, Design 1710," *Gangut* (no. 14, 1988). Reprinted as "Beluga: Soviet Project 1710 Submarine-Laboratory," USNI *Proceedings* (June 1999).

Golosov, Vice Adm. R., Soviet Navy (Ret). "Was the Death of the *K-129* 'Quiet'?" *Morskoy Sbornik* (no. 3, 2000).

Golotyuk, Yuri. "The 'Vavorok': Russia's Nuclear Might Is Maintained Free at the Northern Machine Building Enterprise: Nuclear Powered Submarine Builders Have Not yet Coped With Conversion," *Segodnya* (Moscow) (7 July 1994).

Khramtsov, Rear Adm. V., Soviet Navy (Ret). "Why Did Not the Nuclear Catastrophe in the Far East Prevent Chernobyl?" *Morskoy Sbornik* (no. 7, 1999).

Krapivin, Capt. 1/R V. "Tragedy of a Ship and Honor of a Crew," *Morskoy Sbornik* (no. 4, 1994).

Kucher, V. A., Yu. V. Manuylov, and V. P. Semyonov. "Project 611 Submarines," *Sudostroenie* (no. 5–6, 1995).

Kuznetsov, Adm. N. G., Soviet Navy (Ret). "On the Eve," *Oktiabr'* (no. 11, 1965).

Kviatkovskii, Rear Adm. Yu., Soviet Navy. "Current Status and Development Prospects of Forces and Means for Combating Submarines," *Voennaya mysl'* (no. 1, 1988).

Litkovets, Capt. 2/R Nikolay, Russian Navy. "Fleet in Financial Squeeze," *Krasnaya Zvezda* (Moscow) (1 April 1994).

Mineev, Yu. K., and K. A. Nikitin. "Unique Piranya," *Gangut* (no. 14, 1991).

Nadzharov, Aleksandr. "Russian Army. New Times," *Nezavisimaya Gazeta* (Moscow) (8 June 1993).

Panteleyev, Adm. Yu. "The Submarine Operation of the Navy—the Naval Operation of the Future," *Voyennaya Mysl'* (10 July 1961).

Parkhomenko, Capt. 1/R V. N. "Guarding the Decibels," *Morskoy Sbornik* (no. 6, 1996).

Platonov, Rear Adm. Vasiliy I., Soviet Navy. "The Missions of the Navy and Methods of Carrying Them Out," *Voyennaya Mysl'* (no. 2, 1961).

Postnov, A. A. "Experimental Submarine Design 613Eh with Electrochemical Generators," *Sudostroenie* (no. 2, 1998).

Repin, Vladislav. "One More Time About Strategic Dilemmas," *Nezavisimaia gazeta* (24 September 1992).

Samarkin, L. A. "Multipurpose Nuclear Submarines of Project 671," *Morskoy Sbornik* (no. 2, 1995); republished as "The Seventy-First," *Tayfun* (no. 4, 1997).

Semenov, Capt. 1/R Ye., Soviet Navy. "On the Survivability of Submarines Under the Threat From the Air," *Morskoy Sbornik* (no. 1, 1998).
Shmakov, R. A. "Ahead of Their Time . . ." *Morskoy Sbornik* (no. 7, 1996).
———. "Building Atomic Submarine Projects 671, 671RT and 671RTM," *Sudostroenie* (no. 1, 2000).
———. "On the History of the Development of Anti-Submarine Weapon Systems 'V'yuga,' 'Vodopad,' and 'Shkval,' " *Gangut* (no. 14, 1998).
———. "The Small, Fast, Automated Submarine-Interceptor of Projects 705 and 705K," *Tayfun* (no. 3, 1997).
———. "Our Submarines' Anti-Submarine Missiles," *Morskoy Sbornik* (no. 5, 1997).
Solomenko, Yevgeniy. "The Battle of the Admirals Over the *Komsomolets,*" *Krasnaya Zvezda* (22 March 1994).
Stvolinsky, Capt. Lt. Yu., Soviet Navy (Ret). "The Beginning," *Leningradskaya Pravda* (15 March 1979).
———. "In the Annals of the Fatherland: Designer of the Nuclear Submarine," *Krasnaya Zvezda* (15 October 1988).
Tyurin, Capt. 1/R B., Russian Navy. "Medium Diesel-Electric Submarines of Project 613," *Morskoy Sbornik* (no. 8, 1995).
Zharkov, V. I. "Creation of the First Submarines With Ballistic Missiles," *Gangut* (no. 14, 1998).
———. "Large Missile Diesel-Electric Submarines of the Base of Project 611," *Morskoy Sbornik* (no. 9, 1995).
Zharkov, V. I., and Capt. 1/R B. Tyurin, Russian Navy. "The Amphibious Troop Transport and Minelaying Nuclear Powered Submarines of Projects 748 and 717," *Morskoy Sbornik* (no. 1, 1998).
———. "Project 632 and 648 Mine-Laying Submarines," *Morskoy Sbornik* (no. 3, 1996).

## Studies and Presentations

Antonov, A. M., V. I. Barantsev, B. F. Dronov, and A. V. Kuteinikov. "40 Years of Nuclear Submarine Development: A View of a Designer," paper presented at Warship '96: International Symposium on Naval Submarines, London, 12 June 1996.
———. "Soviet and Russian Attack Submarines: Evolution of Concept," paper presented at Warship '99: International Symposium on Naval Submarines, London, 14 June 1999.
Brooks, Rear Adm. Thomas A., USN. "Commentary on Three Views of Roles and Missions of the Soviet Navy," paper given at Center for Naval Analyses conference, Arlington, Va., 13 November 1985.
Bruins, Berend Derk. "U.S. Naval Bombardment Missiles, 1940–1958: A Study of the Weapons Innovation Process," PhD dissertation, Columbia University, 1981.
Dronov, Boris F., and Boris A. Barbanel. "Early Experience of BLC [Boundry Layer Control] Techniques Usage in Underwater Shipbuilding," paper presented at Warship '99: International Symposium on Naval Submarines, London, 16 June 1999.
Convair Division, General Dynamics Corp. "Flying Submersible Vehicle" (Technical Proposal), April 1964.
Electric Boat Division, General Dynamics Corp. *The Feasibility and Design of Nuclear-Powered Submarine Tankers,* 1 December 1958.
———. *Preliminary Design Study of a Small, High-Density, Hydro-Ski Equipped Seaplane,* Report No. U419-66-013, September 1966.
Hellqvist, K. "Submarines with Air Independent Propulsion," paper presented at Warship '93, International Symposium on Naval Submarines, London, 13 May 1993.
Hoppe, Herbert, Norman Polmar, and A.C. Trapold. *Measures and Trends US and USSR Strategic Force Effectiveness.* Alexandria, Va.: Santa Fe Corp., March 1978. Prepared for Director, Defense Nuclear Agency.
Jackson, Capt. H. A., USN (Ret). "The Influence of the USS Albacore on Submarine Design," paper presented at Warship '93: International Symposium on Naval Submarines, London, 11 May 1993.
Kerros, P., P. Leroy, and D. Grouset. "MESMA: AIP System for Submarines," paper presented at Warship '93: International Symposium on Naval Submarines, London, 13 May 1993.
Kuteinikov, A. V., "Emerging Technology and Submarines of the 21st Century," paper presented at the Royal Institution of Naval Architects, London, 14 June 1996. At the time Kuteinikov was head of the Malachite design bureau.
Leopold, Reuven, Otto P. Jöns, and John T. Drewry. "Design-to-Cost of Naval Ships" in The Society of Naval Architects and Marine Engineers *Transactions,* vol. 82, 1974.
Manthorpe, Capt. William H. J., Jr., USN (Ret). "The Submarine Threat of the Future," presentation to the Naval Submarine League, Alexandria, Va., 14 June 1990.
Mayer, R. A., and W. A. Murray. *Flying Carpet Feasibility Study: Submarine Carrier,* Document No. D3-1870-8. Wichita, Kan.: Boeing Airplane Company, 18 December 1958.
Moore, K. J. "AIP Versus Efficiency—An Investment Strategy," paper presented at Warship '93: International Symposium on Naval Submarines, London, 11 May 1993.
———. "Designing For Efficiency 21st Century Exploitation of 20th Century Miscues," paper presented at Warship '99: International Symposium on Naval Submarines, London, 16 June 1999.
Moore, K. J., Gary Jones, and Ejike Ndefo. "Vortex Control in Submarine Design," paper presented at Warship '91: International Symposium on Naval Submarines, London, 14 May 1991.
Ndefo, Dr. Ejike, and K. J. Moore. "Traveling Wave Research," paper presented at Warship '91: Interna-

tional Symposium on Naval Submarines, London, 14 May 1991.

Nguyen, Hung P. *Submarine Detection from Space: Inferences from Soviet Writings*, CRM 92-29. Alexandria, Va.: Center for Naval Analyses, February 1992.

Nitze, Paul H., Adm. Elmo R. Zumwalt, USN (Ret), and Norman Polmar. "Developing a Lower-Cost Submarine Program," proposal to the Department of Defense, 24 September 1990.

Polmar, Norman. "Cultural Aspects of U.S. and Soviet Submarine Development" presentation at conference on the Undersea Dimension of Maritime Strategy, Dalhousie University, Halifax, Nova Scotia, 21–24 June 1989.

———. "The Polaris—A Revolutionary Missile System and Concept" presentation at the American Military Institute, Washington Navy Yard, 9 April 1988.

Portsmouth Naval Shipyard. "Historical Account of *X-1* Power Plant Development, 24 May 1956–20 May 1957."

RAND Corp. Project Rand. *Vulnerability of U.S. Strategic Air Power to a Surprise Enemy Attack in 1956*, Special Memorandum SM-15. Santa Monica, Calif.: Rand Corp., 15 April 1953.

Russo, Vito L., Harlan Turner, Jr., and Frank W. Wood. "Submarine Tankers" in The Society of Naval Architects and Marine Engineers *Transactions*, vol. 68, 1960.

Smith, Rear Adm. Levering, USN. "Polaris as an Element of Strategic Deterrence" presentation to the New York Academy of Sciences, 12 January 1965.

Spassky, I. D. "The Role and Missions of Soviet Navy Submarines in the Cold War," paper given at the Naval History Symposium, U.S. Naval Academy, Annapolis, Md., 23 October 1993.

Watkins, Adm. James D., USN. "The Maritime Strategy," supplement to the USNI *Proceedings* (January 1986).

Wittenbeck, P., and K.J. Moore. "The Rising Vortex: A Source for NA [Non-Acoustic] ASW," paper presented at UDT '93 conference: Cannes, France, 16 June 1993.

## Official Documents and Reports in English

Beyond materials made available by the Rubin and Malachite design bureaus and to a lesser degree by the Lazurit bureau, their several books and monographs describing their respective submarine designs have been very useful. Various British, German, Soviet/Russian, and U.S. records were made available by personal contacts in the countries concerned, as well as in Britain's Public Record Office in Kew, the U.S. Navy's Operational Archives (Naval Historical Center), and the National Archives in Washington, D.C., and College Park, Maryland.

Of particular value were the minutes of the U.S. Navy's Submarine Officers Conferences and Undersea Warfare Conferences held from 1945 to 1950, and the Hearings of the General Board of the Navy for the same period. The conference minutes as well as numerous studies and special reports are contained in Boxes 1 through 29 of the Submarines/Undersea Warfare Division files in the Operational Archives, Naval Historical Center, Washington, D.C.; the General Board hearings are in Record Group 19 at the National Archives, Washington, D.C. Also useful were the reports of the U.S. Naval Technical Mission in Europe and Japan, 1945 (Navy Operational Archives).

Also valuable to this volume have been hearings in which U.S. Department of Defense and Navy officials testified before various committees of the Senate and House of Representatives and, especially, the Joint Committee on Atomic Energy.

Brown, Gen. George S., USAF. *United States Military Posture for FY [Fiscal Year] 1978.* Washington, D.C.: Department of Defense, 20 January 1977.

Brown, Harold. *Department of Defense Annual Report Fiscal Year 1979.* Washington, D.C.: Department of Defense, 2 February 1978.

———. *Department of Defense Annual Report Fiscal Year 1981.* Washington, D.C.: Department of Defense, 29 January 1980.

Cheney, Dick. *Soviet Military Power 1990.* Washington, D.C.: Department of Defense, September 1990.

Clifford, Clark. *The Fiscal 1970 Defense Budget and Defense Program for Fiscal Years 1970–74.* Washington, D.C.: Department of Defense, 15 January 1969.

Keefer, Edward C., and David W. Mabon (eds.). *Foreign Relations of the United States, 1958–1960,* vol. III, *National Security Policy; Arms Control and Disarmament.* Washington, D.C.: Government Printing Office, 1996.

Moorer, Adm. Thomas H., USN. *United States Military Posture for FY [Fiscal Year] 1973.* Washington, D.C.: Department of Defense, 8 February 1972.

Rickover, Vice Adm. H. G., USN. "Analysis of Nuclear Attack Submarine Development." Washington, D.C.: Naval Reactors Branch, Department of the Navy, 31 August 1979.

———. "Naval Nuclear Propulsion Program–1972–73," hearings before the Joint Committee on Atomic Energy, Congress, 8 February 1972 (1974).

———. "Nuclear Submarines of Advanced Design," report of Hearings before the Joint Committee on Atomic Energy, Congress, 21 June 1968.

———. "Report on Russia." Washington, D.C.: Committee on Appropriations, House of Representatives, 1959.

Rosenberg, Max. *The Air Force and the National Guided Missile Program.* Washington, D.C.: U.S. Air Force Historical Division Liaison Office, June 1964.

Royal Navy. *German Naval History: The U-Boat War in the Atlantic, 1939–1945,* vol. III. London: Her Majesty's Stationery Office, 1989.

Sweden, Ministry of Defence. *Countering the Submarine Threat: Submarine violations and Swedish security policy.* Stockholm: Report of the Submarine Defence Commission, 1983.

U.S., Central Intelligence Agency. "Depressed Trajectories: Unlikely Role for Soviet SLBMs" (An Intelligence Memorandum). SW 82-10075, August 1982.

———. "Molotivsk Shipyard No. 402 in Molotivsk Archangel Oblast," CIA/SC/RR 79, 27 October 1954.

———. "National Intelligence Estimate: Soviet Naval Policy and Programs," NIE 11-15-74, 23 December 1974.

———. "The Soviet M-Class Submarine: A Preliminary Assessment" (An Intelligence Assessment), SOV 84-10134/SW 84-10049, August 1984.

———. "Soviet Naval Strategy and Programs Through the 1990s." NIE 11-15-82/D, March 1983.

———"The Soviet Typhoon Submarine—A Radical Innovation in Submarine Design" (An Intelligence Assessment), SW86-10002X, January 1986.

U.S., Congress, Joint Committee on Atomic Energy. *Loss of the U.S.S. Thresher.* Washington, D.C.: Government Printing Office, 1965.

U.S., Congress, "Report of the Advisory Panel on Submarine and Antisubmarine Warfare to the House Armed Services Subcommittee on Research and Development and Seapower and Strategic Critical Materials," 21 March 1989 (unclassified edition).

U.S., Congressional Budget Office. *Budget Options.* Washington, D.C.: Government Printing Office, February 2001.

U.S., Defense Intelligence Agency. "Papa Class Cruise Missile-Launching Nuclear Submarine (Weapon System)–U.S.S.R.," DST-1222S-193-77, 12 December 1977.

———. "Soviet Submarine Order of Battle: 1950–1974," DDB-1220-14-78, 16 August 1979.

———. "Soviet Surveillance Capabilities Against U.S. Naval Forces," DST-1280S-607-79, 1 August 1979.

U.S., Deputy Under Secretary of Defense (Plans and Policy). "SSN-21 Assessment," 10 January 1985.

U.S. District Court, Southern District of Florida, *Indictment* (Messrs. Ludwig Fainberg, Juan Almeida, and Nelson Yeser), filed 28 January 1997.

U.S., General Accounting Office. "Attack Submarines: Alternatives for a More Affordable SSN Force Structure," GAO/NSIAD-95-16, October 1994.

———. "Navy Acquisition: AN/BSY-1 Combat System Operational Evaluation," GAO/NSIAD-93-81, November 1992.

———. "Navy Acquisition: Cost, Schedule, and Performance of New Submarine Combat Systems," GAO NSIAD-90-72, January 1990.

———. "Navy Acquisition: SUBACS Problems May Adversely Affect Navy Attack Submarine Programs," GAO/NSIAD-86-12, November 1985.

———. "Navy Ships: Concurrency Within the SSN-21 Program," GAO/NSIAD-90-297, September 1990.

———. "Navy Ships: Problems Continue to Plague *Seawolf* Submarine Program," GAO/NSIAD-93-171, August 1993.

———. "Navy Ships: *Seawolf* Cost Increases and Schedule Delays Continue," GAO/NSIAD-94-201BR, June 1994.

———. "Navy Torpedo Programs: MK-48 ADCAP Upgrades Not Adequately Justified," GAO/NSIAD-95-104, June 1995.

———. "Submarine Combat System: BSY-2 Development Risks Must Be Addressed and Production Schedule Reassessed," GAO/NSIAD-91-30, August 1991.

———. "Submarine Combat System: Status of Selected Technical Risks in the BSY-2 Development," GAO/IMTEC-91-46BR, May 1991.

U.S., Joint Intelligence Committee. "Review of Certain Significant Assumptions, Summaries, and Conclusions as Presented in Joint Intelligence Committee (Intelligence Section, JIG), Estimates of the Capabilities of the USSR," January 1948.

U.S., Naval Intelligence Support Center. "Soviet Surveillance Capabilities Against U.S. Naval Forces," DST-1280S-607-79, 1 August 1979.

U.S., Naval Sea Systems Command. "Quarterly Progress Report for Shipbuilding and Conversion" (various editions).

U.S., Naval Technical Mission in Europe. "Technical Report No. 312-45, German Submarine Design, 1935–1945," July 1945.

U.S., Naval Technical Mission to Japan, "Ships and Related Targets: Characteristics of Japanese Naval Vessels: Submarines," no. S-01-1, January 1946.

U.S. Navy, Office of the Chief of Naval Operations. *The Navy of the 1970 Era,* January 1958.

———. "Report of the Ship Operational Characteristics Study on the Operational Characteristics of the Surface Combatant of the Year 2010," 26 April 1988.

U.S., Navy, Portsmouth Naval Shipyard, "Surrendered German Submarine Report, Type XXI," July 1946.

U.S. Navy, Strategic Systems Program. *Facts/Chronology: Polaris-Poseidon-Trident.* Washington, D.C.: Navy Department, 2000. Also previous editions of U.S. Polaris, Poseidon, Trident fact sheets and chronologies.

U.S., Office of the Under Secretary of Defense for Acquisition & Technology, "Report of the Defense Science Board Task Force on Submarine of the Future." Washington, D.C.: Department of Defense, July 1998.

U.S. *Report of the President's Commission on Strategic Forces,* April 1983.

U.S., Secretary of the Navy. *Lessons of the Falklands: Summary Report.* Washington, D.C.: Department of the Navy, February 1983.

## Official Documents and Reports in German

Germany, Supreme Command of the Navy, 2d Division Naval Staff B.d.U., *Ueberlegungen zum Einsatz des Typ XXI* (Considerations on the Use of Type XXI), 10 July 1944.

## Official Documents and Reports in Russian

Golosovskii, P. Z. *Proektirovaniye i stroitelstvo podvodnikh lodok, Tom II, Povodni Lodki: perioda 1926–1945 godov* (Design and Construction of Submarines), vol. II, *1926–1945.* Leningrad: Rubin [design bureau], 1979.

Malachite SPMBM, "300th Anniversary of Russian Navy–SPMBM 'Malachite' For Submarine Forces of Russia, 1948–1996," 1996.

———. *Istoriya Sankt-Peterburgskogo Morskogo Buro Mashinostroeniya "Malakhit"* (History of the Saint-Petersburg Maritime Machinebuilding Bureau Malachite), vol. 2. St. Petersburg: Malachite, 1995.

Ovcharenko, A. M. "Analysis of the Effectiveness of Groupings of Design 667A (AU) RPKSN [SSBN] in the System of the Soviet Union's Strategic Nuclear Forces," *Proceedings of the Scientific-Technical Conference "The Second Makeev Readings,"* series XIV, edition 1 (4), 1996.

Zakharov, Prof. Rear Adm. I. G. "Today's Trends in the Development of Fighting Ships" during the conference celebrating the 300th anniversary of the Russian Navy, St. Petersburg, 27 February 1996.

Note: * USNI = U.S. Naval Institute.

# Index

Note: Ranks indicated are the highest mention in text. All ships and submarines are Soviet or U.S. unless otherwise indicated.

# Author Biographies

**NORMAN POLMAR** is an internationally known author and analyst in the naval intelligence, and aviation fields. He has consulted to three Secretaries of the Navy, three U.S. Senators, and two members of the House of Representatives, as well as to officials in the Department of Defense and the aerospace and shipbuilding industries. Mr. Polmar is the author of more than 30 books, several about submarines. He is a graduate of The American University in Washington, D.C.

He has visited Russia nine times in addition to having interviewed Soviet-Russian submarine officers and designers in the United States.

**KENNETH J. MOORE,** a former U.S. submarine officer, is founder and president of the Cortana Corporation, a submarine technology firm. A graduate of the University of Texas, as a naval officer he served in diesel-electric submarines and then as a tactical and technology analyst in the Department of Defense addressing Soviet submarine development.

He has held extensive technical discussions in Russia and Ukraine related to submarine technology.